VOLUME ONE HUNDRED AND FORTY

CURRENT TOPICS IN DEVELOPMENTAL BIOLOGY

Maternal Effect Genes in Development

CURRENT TOPICS IN DEVELOPMENTAL BIOLOGY

"A meeting-ground for critical review and discussion of developmental processes"
A.A. Moscona and Alberto Monroy (Volume 1, 1966)

VOLUME ONE HUNDRED AND FORTY

CURRENT TOPICS IN DEVELOPMENTAL BIOLOGY

Maternal Effect Genes in Development

Edited by

FLORENCE L. MARLOW

Associate Professor, Department of Cell, Developmental and Regenerative Biology. Icahn School of Medicine at Mount Sinai, NY, United States

ELSEVIER

ACADEMIC PRESS

An imprint of Elsevier

Academic Press is an imprint of Elsevier
50 Hampshire Street, 5th Floor, Cambridge, MA 02139, United States
525 B Street, Suite 1650, San Diego, CA 92101, United States
The Boulevard, Langford Lane, Kidlington, Oxford OX5 1GB, United Kingdom
125 London Wall, London, EC2Y 5AS, United Kingdom

First edition 2020

ISBN: 978-0-12-815220-1
ISSN: 0070-2153

For information on all Academic Press publications
visit our website at https://www.elsevier.com/books-and-journals

Publisher: Zoe Kruze
Editorial Project Manager: Shellie Bryant
Production Project Manager: Denny Mansingh
Cover Designer: Greg Harris

Typeset by SPi Global, India

Contents

Contributors *xi*

Preface *xv*

Part I
Meiosis and production of the oocyte and egg

1. Post-transcriptional gene regulation regulates germline stem cell to oocyte transition during *Drosophila* oogenesis **3**

Patrick Blatt, Elliot T. Martin, Shane M. Breznak, and Prashanth Rangan

1. Introduction 4
2. Alternative splicing ensures accurate production of critical germline mRNAs that regulate sex determination and differentiation 6
3. RNA modifications direct splicing of sex determinants and translation of differentiation promoting genes in the germline 10
4. Production of ribosomes is finely tuned to facilitate differentiation 12
5. Hand-off mechanisms facilitated by combinatorial RNA binding proteins dynamically shape the translational landscape during oogenesis 18
6. Summary 23

References 23

Further reading 34

2. Maternal inheritance of centromeres through the germline **35**

Arunika Das, Ben E. Black, and Michael A. Lampson

1. Introduction 36
2. How is CENP-A chromatin maintained? 38
3. Biased inheritance of centromeres through female meiosis: Centromere drive 42
4. Regulating centromere size in early development 45
5. Conclusions and future perspectives 48

References 49

3. Signaling between somatic follicle cells and the germline patterns the egg and embryo of *Drosophila* **55**

Julie A. Merkle, Julia Wittes, and Trudi Schüpbach

1. Introduction 56
2. Formation of the egg chamber 56
3. Selection of the oocyte 59
4. Germline-soma and soma-soma signaling pattern the polar and stalk cells and pre-polarize the newly formed oocyte 60
5. Anterior–posterior axis formation involves signaling between germline and follicle cells 62
6. Dorsal-ventral polarity establishment requires oocyte nuclear migration and signaling between the oocyte and follicle cells 67
7. Terminal patterning of the embryo requires follicle cell activity 72
8. Open questions in germline-follicle cell communication during *Drosophila* oogenesis and early development 74

Acknowledgments 75

References 76

4. Organizing the oocyte: RNA localization meets phase separation **87**

Sarah E. Cabral and Kimberly L. Mowry

1. Introduction 88
2. Functional advantages of RNA localization 89
3. Mechanisms of RNA localization 91
4. RNA localization in *Xenopus* oocytes 94
5. RNPs and localization 100
6. Biomolecular condensates and phase separation 102
7. Biomolecular condensates in oocytes and embryos 106
8. Conclusions 109

Acknowledgments 110

References 110

Part II
Fertilization and germline specification

5. Species-specific mechanisms during fertilization **121**

Krista R. Gert and Andrea Pauli

1. Introduction 122
2. The sperm's journey to the egg 123
3. Sperm-egg binding 126

4. Sperm-egg fusion: A lingering question mark 136
5. Conclusion 136
References 137
Further reading 143

6. The role of the cytoskeleton in germ plasm aggregation and compaction in the zebrafish embryo 145

Cara E. Moravec and Francisco Pelegri

1. Introduction 146
2. Localization routes of germ plasm ribonucleoparticles during oogenesis and early embryogenesis 147
3. Composition of germ plasm masses 149
4. Stages of animal germ plasm aggregation in the early zebrafish blastodisc 151
5. Interaction of astral microtubules and dynamic F-actin in RNP pre-aggregation 156
6. Recruitment of germ plasm during furrow induction 162
7. Medial-to-distal dynamics mediating germ plasm furrow compaction 164
8. Conclusion 172
References 173

7. Dead end and Detour: The function of the RNA-binding protein Dnd in posttranscriptional regulation in the germline 181

Theresa Gross-Thebing and Erez Raz

1. Introduction 182
2. Dead end: A conserved vertebrate-specific germ cell protein 182
3. Dead end is important for development and function of the germline 187
4. The role of Dead end in germ cell fate 189
5. Tumor formation associated with loss of Dead end 192
6. Regulation of RNA expression, stability and translation by Dead end 193
7. Dead end acts as a molecular hub for diverse posttranscriptional regulation events 199
Acknowledgments 201
References 201

8. From mother to embryo: A molecular perspective on zygotic genome activation 209

Edlyn Wu and Nadine L. Vastenhouw

1. Introduction 210
2. The many-layers of transcriptional regulation in ZGA 211
3. The molecular landscape of ZGA 217

4. Compartmentalizing transcription 238
5. Concluding remarks and future perspectives 240
Acknowledgments 242
References 242

Part III
Maternal regulation of embryonic patterning and morphogenesis

9. Maternal regulation of seed growth and patterning in flowering plants 257

Allison R. Phillips and Matthew M.S. Evans

1. Introduction 258
2. Female gametophyte development 259
3. Seed development 260
4. Gametophytic maternal effects 262
5. Sporophytic maternal effects on seed development 269
6. Conclusion 274

Acknowledgments 274
References 274

10. Maternal factors regulating symmetry breaking and dorsal–ventral axis formation in the sea urchin embryo 283

Maria Dolores Molina and Thierry Lepage

1. Introduction 284
2. Maternal determination of the D/V axis in sea urchin embryos: The heritage from experimental embryology 285
3. Nodal signaling initiates D/V axis specification of the sea urchin embryo 288
4. Studies on the Nodal promoter identify maternal transcription factors and signaling molecules required for the initiation of Nodal expression 291
5. Yan, a maternal ETS domain transcription factor represses the *nodal* promoter 294
6. The maternal function of the transcriptional repressor Yan/Tel is essential for spatial regulation of *nodal* 295
7. Stability of sea urchin Yan/Tel is regulated by MAPK and maternal GSK3 and β-TRCP 297
8. Activity of the maternal type I BMP receptors Alk1/2 and Alk3/6 is required for the early spatial restriction of *nodal* 300

9. Identification of Panda as a factor required to restrict *nodal* expression during dorsal–ventral axis formation in the sea urchin embryo 302
10. *panda* mRNA is distributed asymmetrically in the oocyte and unfertilized egg 303
11. Increasing or decreasing the levels of Panda locally orients the D/V axis 304
12. Spatially restricted Panda signaling specifies the dorsal–ventral axis 305
13. Panda and Yan/Tel act upstream of the Lefty-dependent reaction–diffusion mechanism to initiate the spatial restriction of *nodal* 306
14. Panda's mechanism of action 306
15. Conclusion: Maternal determinants of D/V axis formation and developmental plasticity of the early blastomeres 310
Acknowledgments 312
References 312

11. Maternal factors regulating preimplantation development in mice 317

Di Wu and Jurrien Dean

1. Introduction 318
2. Initiation of embryonic development 318
3. Transition to independence: Zygotic genome activation 322
4. Other preimplantation events regulated by maternal factors 327
5. Maternal factors require timely clearance 330
6. Concluding remarks 333
Acknowledgments 333
References 333

12. The maternal coordinate system: Molecular-genetics of embryonic axis formation and patterning in the zebrafish 341

Ricardo Fuentes, Benjamin Tajer, Manami Kobayashi, Jose L. Pelliccia, Yvette Langdon, Elliott W. Abrams, and Mary C. Mullins

1. Polarity and cytoplasmic reorganization prime embryonic axis formation 342
2. Maternal control of DV axial specification 351
3. Signaling pathways and the patterning of the embryonic axes 357
4. Nodal signaling in germ layer formation and left-right patterning 368
5. Concluding remarks 373
Acknowledgments 373
References 374

13. Maternal contributions to gastrulation in zebrafish **391**

Lilianna Solnica-Krezel

1. Introduction 392
2. Maternal-to-zygotic transition and its timing relative to gastrulation: A genomic perspective 397
3. Requirement for maternal and zygotic gene products for gastrulation: A genetic perspective 401
4. Instructive and permissive contributors to gastrulation from the maternal and zygotic genomes: An embryo perspective 411
5. Perspective 418

Acknowledgments 419

References 419

Contributors

Elliott W. Abrams
Department of Biology, Purchase College, State University of New York, Harrison, NY, United States

Ben E. Black
Department of Biochemistry and Biophysics, Perelman School of Medicine, University of Pennsylvania, Philadelphia, PA, United States

Patrick Blatt
Department of Biological Sciences/RNA Institute; University at Albany SUNY, Albany, NY, United States

Shane M. Breznak
Department of Biological Sciences/RNA Institute; University at Albany SUNY, Albany, NY, United States

Sarah E. Cabral
Department of Molecular Biology, Cell Biology, and Biochemistry, Brown University, Providence, RI, United States

Arunika Das
Department of Biology; Department of Biochemistry and Biophysics, Perelman School of Medicine, University of Pennsylvania, Philadelphia, PA, United States

Jurrien Dean
Laboratory of Cellular and Developmental Biology, NIDDK, National Institutes of Health, Bethesda, MD, United States

Matthew M.S. Evans
Department of Plant Biology, Carnegie Institution for Science, Stanford, CA, United States

Ricardo Fuentes
Departamento de Biología Celular, Facultad de Ciencias Biológicas, Universidad de Concepción, Concepción, Chile

Krista R. Gert
Research Institute of Molecular Pathology (IMP), Vienna BioCenter (VBC), Vienna, Austria

Theresa Gross-Thebing
Institute of Anatomy and Vascular Biology, University of Münster, Münster, Germany

Manami Kobayashi
Department of Cell and Developmental Biology, University of Pennsylvania, Perelman School of Medicine, Philadelphia, PA, United States

Michael A. Lampson
Department of Biology, University of Pennsylvania, Philadelphia, PA, United States

Yvette Langdon
Millsaps College, Jackson, MS, United States

Thierry Lepage
Institut de Biologie Valrose, Université Côte d'Azur, Nice, France

Elliot T. Martin
Department of Biological Sciences/RNA Institute; University at Albany SUNY, Albany, NY, United States

Julie A. Merkle
Department of Biology, University of Evansville, Evansville, IN, United States

Maria Dolores Molina
Institut de Biologie Valrose, Université Côte d'Azur, Nice, France

Cara E. Moravec
Laboratory of Genetics, University of Wisconsin—Madison, Madison, WI, United States

Kimberly L. Mowry
Department of Molecular Biology, Cell Biology, and Biochemistry, Brown University, Providence, RI, United States

Mary C. Mullins
Department of Cell and Developmental Biology, University of Pennsylvania, Perelman School of Medicine, Philadelphia, PA, United States

Andrea Pauli
Research Institute of Molecular Pathology (IMP), Vienna BioCenter (VBC), Vienna, Austria

Francisco Pelegri
Laboratory of Genetics, University of Wisconsin—Madison, Madison, WI, United States

Jose L. Pelliccia
Department of Cell and Developmental Biology, University of Pennsylvania, Perelman School of Medicine, Philadelphia, PA, United States

Allison R. Phillips
Biology Department, Wisconsin Lutheran College, Milwaukee, WI, United States

Prashanth Rangan
Department of Biological Sciences/RNA Institute; University at Albany SUNY, Albany, NY, United States

Erez Raz
Institute of Cell Biology, University of Münster, Münster, Germany

Trudi Schüpbach
Department of Molecular Biology, Princeton University, Princeton, NJ, United States

Lilianna Solnica-Krezel
Department of Developmental Biology and Center of Regenerative Medicine, Washington University School of Medicine, St. Louis, MO, United States

Benjamin Tajer
Department of Cell and Developmental Biology, University of Pennsylvania, Perelman School of Medicine, Philadelphia, PA, United States

Nadine L. Vastenhouw
Max Planck Institute of Molecular Cell Biology and Genetics, Dresden, Germany

Julia Wittes
Department of Biological Sciences, Columbia University, New York, NY, United States

Di Wu
Laboratory of Cellular and Developmental Biology, NIDDK, National Institutes of Health, Bethesda, MD, United States

Edlyn Wu
Max Planck Institute of Molecular Cell Biology and Genetics, Dresden, Germany

Preface

"When you start with a portrait and search for a pure form, a clear volume, through successive eliminations, you arrive inevitably at the egg. Likewise, starting with the egg and following the same process in reverse, one finishes with the portrait" (Pablo Picasso). Like Picasso's portrait, if one starts with an embryo and tries to understand its form you arrive at production of the maternal gamete, the oocyte or egg. This meiotic cell forms from the differentiating divisions of mitotically dividing germline stem cells in the ovary of females. When I began assembling this volume on maternal control of development, I envisioned that the chapters in the first section would begin with that "pure form" reviewing what is known of the cellular and molecular mechanisms underlying formation of the egg, including the mitotic to meiotic switch and oogenesis, RNA regulation, and inheritance of maternal factors in model organisms, including plants. From there, the second section would focus on fertilization of the egg, specification of the germline, clearance of maternal factors, and activation of zygotic genome. Finally, the third part of the volume would feature chapters on the cellular and molecular mechanisms underlying embryonic axis determination, including maternal factors that induce and pattern the germ layers across model systems. Although the details are somewhat varied the molecular players are remarkably conserved. The volume concludes with processes that are coordinated by both maternal and zygotic programs. With contributions from many experts, 13 chapters of this volume of *Current Topics in Developmental Biology* collectively paint a clear portrait of the current state of the field.

Our understanding of how the immortal germline transits through distinct cell cycle and differentiation programs has flourished considerably over the years from genetic and biochemical studies in model systems. Based on the work of many labs utilizing various model systems, it is apparent that much of this regulation is accomplished in large part at the posttranscriptional level, via regulation of translation, localization, and stability of RNA. It is well appreciated that vast stores of maternal products, including RNAs and proteins are amassed in the oocyte and that these maternal stores are transmitted to the progeny to support the earliest developmental events. Yet, we still do not fully understand how this maternal dowry is regulated, which machinery and factors ensure that the reserves are preserved until they are needed and are only discarded when their function has been fulfilled. In Chapter 1, Patrick Blatt, Elliot T. Martin, Shane M. Breznak, and Prashanth Rangan review recent advances in our understanding of the contributions of evolutionarily conserved factors and pathways to posttranscriptional control of RNA production, turnover, and expression afforded by high-throughput sequencing and genetic studies with a focus on the stem cell to oocyte transition in *Drosophila melanogaster*.

The contribution of factors that are transmitted independent of the genome one inherits, or epigenetic regulation has become an increasingly recognized and important feature of the mechanisms underlying developmental and homeostatic processes. In the context of maternal-effects, extragenomic or DNA sequence independent influencers of regulation have been considered to include all factors with transgenerational influences on the immediate progeny or grandchildren, including maternal diet, metabolism, or health, as well as the centromere-based mechanisms that ensure faithful separation of the chromosomes during meiotic divisions. In Chapter 2, Arunika Das, Ben E. Black, and Michael A. Lampson explore new developments in our understanding of the mechanisms governing inheritance of the centromeres in the maternal germline and pose unsolved and emerging questions about meiotic drive, centromere architecture, and propagation of centromeres from diverse species.

Of course, no successful endeavor is realized without tremendous support, and production of a developmentally competent oocyte or egg is no exception. Throughout the course of its development, the oocyte is fully supported by the somatic cells that surround, signal to, and nourish the developing gamete. These support cells that comprise the somatic gonad are commonly known to reproductive biologists as follicle cells because they, together with the germline cell, form the developing follicle. Follicle cells come in several types ranging from a couple of distinct cell types found in invertebrate systems to the increasingly complex multilayered and multicell types found in maturing vertebrate and mammalian ovarian follicles. Regardless of the degree of relative complexity, oocyte development and egg production absolutely depend on the follicle cells of the somatic gonad, as well as the extragonadal signals that these cells receive and transmit from other organ systems to maintain reproductive health. In Chapter 3, Julie A. Merkle, Julia Wittes, and Trudi Schüpbach review the genetic mechanisms that establish distinct follicle cell types, and the contributions of these unique somatic fates to signaling and patterning of the *Drosophila* egg.

In many species, the cytoplasm of the oocyte is not homogeneous; proteins, RNAs, and organelles have been observed to be asymmetrically distributed at specific hallmark stages in nearly every species examined thus far. Notably, although these organelles and molecules are sequestered within cells, they are not bound by membrane enclosed compartments. Instead, they reside in biochemically unique environments that are thought to form through a process of phase separation. Importantly, these subcellular environments are not unique quirks of oocytes, but rather represent a broader mechanism to control expression of the cellular transcriptome, potentially allowing for more precise and rapid responses to changing cellular environments. Chapter 4 of this volume by Sarah E. Cabral and Kimberly L. Mowry provides insight into this regulatory mechanism gleaned from studies of RNA localization and phase separation in vertebrate oocytes.

With the exception of parthenogenesis that occurs in some invertebrates and plants, even a perfectly formed egg will not develop further unless it is fertilized. Gametes of sexually reproducing species are haploid, each with only half the number of chromosomes necessary to make an individual. In female gametes of some species, the final meiotic division occurs in tandem with egg activation or fertilization. In all sexually reproducing animals, mechanisms are in place to ensure that an egg is fertilized by only one sperm, thus restoring ploidy and generating an individual with roughly half the chromosomes of maternal origin and the other of paternal origin. These mechanisms involve processes that are intrinsic to the egg as well as processes governed by the zona or eggshell structure that is produced in partnership between the germline and somatic follicle cells, as well as interactions between the egg and sperm. In Chapter 5, Krista R. Gert and Andrea Pauli examine the mechanisms regulating species-specific interactions between gametes for animals with external fertilization as well as the molecules and factors mediating activation of the egg to promote successful fertilization and prevent polyspermy.

The newly fertilized egg has the potential to make every cell type that comprises the individual, including all the germ layers and cell types that make up the organ systems of the species. Given its vital role in perpetuation of the species, it is no surprise that specification of the germline occurs during early development in many organisms. In some animals, but not all, the germline is specified by and fully dependent on a maternally inherited program. In others a maternal program is present but can be replaced by an inductive zygotic mechanism if the maternal program fails. Finally, some species, including mammals lack a maternal program for germline specification and instead rely on an inductive zygotic pathway. Chapter 6 by Cara E. Moravec and Francisco Pelegri focuses on the cellular and molecular mechanisms and maternal factors, including the cytoskeleton and molecular motors that protect and enrich germ cell determinants subcellularly and ultimately limit germline fate to just a few cells of the early zebrafish embryo. Once specified, by whatever mechanism, the newly specified germ cells must retain their unique fate as they migrate through a myriad of signaling environments in the gastrulating embryo to their eventual home in the prospective gonad. As with the other transitions in the life of the germline, maintenance of germline identity relies heavily on RNA regulation. In Chapter 7, Theresa Gross-Thebing and Erez Raz review the role of an evolutionarily conserved RNA binding protein in navigating the voyage of the nascent germline to the gonad while preventing it from losing its unique character and identity along the way.

Although the mechanism of germline specification can be maternally or zygotically regulated, the initial events following fertilization, including the early cleavages that transform the giant oocyte into the smaller cells that make up the blastula, and nongermline fates are completely under maternal control. However, after just a few cleavage cycles, the embryo must wake up its sleeping genome and become an active player in its own development by producing its own genetic resources and weaning itself from maternal control to achieve normal development. In Chapter 8, conserved and species-specific factors and mechanisms underlying the transition from maternal to zygotic control are reviewed by Edlyn Wu and Nadine L. Vastenhouw.

The mechanisms that form the scaffold for setting up pattern or breaking symmetry in the early embryo are governed by both maternal and zygotic mechanisms. In some species, the foundation is set up by coordinates or spatial asymmetries established in the ovary well before fertilization occurs. In others, the first asymmetries are manifest through a combination of cellular arrangements and gene expression after zygotic expression has begun. In the third section of this volume the mechanisms that pattern the axes of diverse embryos are explored beginning with maternal control of growth and seed control in flowering plants by Allison R. Phillips and Matthew M.S. Evans in Chapter 9. In Chapter 10, specification of the dorsal-ventral axis of sea urchins is reviewed by Maria Dolores Molina and Thierry Lepage, and in Chapter 11, Di Wu and Jurrien Dean present maternal control of the earliest events and fate decisions of the preimplantation mouse embryo. In Chapter 12, Ricardo Fuentes, Mary C. Mullins, and colleagues provide analysis of patterning in zebrafish with comparisons to patterning in other vertebrate embryos. Finally, with the necessary cells, activation of the genome, and initial pattern in place, maternal factors collaborate with zygotic programs in some animals to ensure that the cells and tissues are aligned to give the embryo its species-specific form during the process of gastrulation. In Chapter 13, Lilianna Solnica-Krezel examines maternal contributions to gastrulation with an emphasis on genetic studies in zebrafish and comparisons to other model systems.

Altogether this collection of reviews written by experts in each area highlights the profound impact of combining genetic, cell biology, biochemistry, and powerful sequencing approaches to unraveling the maternal brushstrokes, the factors and processes, that sketch the embryo.

Acknowledgments

I am indebted and grateful to all colleagues for sharing their expertise and time and for their thoughtful contributions to this volume. I am grateful to Paul Wasserman for being an incredible colleague and advisor, for his visits to check in on me even after evicting me from my office ☺, and for sending me on this editorial journey. I appreciate the editors at Elsevier, Shellie Bryant and Zoe Kruze, for their patience, professionalism, and efforts to make sure that the chapters and volume were completed in a timely manner. I am thankful for the members of my research group who share, appreciate, and/or indulge my passion for puzzles and trying to understand maternally regulated processes. Work in my laboratory and my scholarly endeavors, such as this volume, were supported in part by R01 support from the National Institute of General Medical Sciences/National Institutes of Health, and Icahn School of Medicine at Mount Sinai.

Florence L. Marlow
Associate Professor,
Department of Cell, Developmental and Regenerative Biology.
Icahn School of Medicine at Mount Sinai,
NY, United States

PART I

Meiosis and production of the oocyte and egg

CHAPTER ONE

Post-transcriptional gene regulation regulates germline stem cell to oocyte transition during *Drosophila* oogenesis

Patrick Blatt[a,b], Elliot T. Martin[a,b], Shane M. Breznak[a,b], Prashanth Rangan[a,b,*]
[a]Department of Biological Sciences/RNA Institute, University at Albany SUNY, Albany, NY, United States
[b]University at Albany SUNY, Albany, NY, United States
*Corresponding author: e-mail address: prangan@albany.edu

Contents

1. Introduction 4
2. Alternative splicing ensures accurate production of critical germline mRNAs that regulate sex determination and differentiation 6
3. RNA modifications direct splicing of sex determinants and translation of differentiation promoting genes in the germline 10
4. Production of ribosomes is finely tuned to facilitate differentiation 12
5. Hand-off mechanisms facilitated by combinatorial RNA binding proteins dynamically shape the translational landscape during oogenesis 18
6. Summary 23
References 23
Further reading 34

Abstract

During oogenesis, several developmental processes must be traversed to ensure effective completion of gametogenesis including, stem cell maintenance and asymmetric division, differentiation, mitosis and meiosis, and production of maternally contributed mRNAs, making the germline a salient model for understanding how cell fate transitions are mediated. Due to silencing of the genome during meiotic divisions, there is little instructive transcription, barring a few examples, to mediate these critical transitions. In *Drosophila*, several layers of post-transcriptional regulation ensure that the mRNAs required for these processes are expressed in a timely manner and as needed during germline differentiation. These layers of regulation include alternative splicing, RNA modification, ribosome production, and translational repression. Many of the molecules and pathways involved in these regulatory activities are conserved from *Drosophila* to humans making the *Drosophila* germline an elegant model for studying the role of post-transcriptional regulation during stem cell differentiation and meiosis.

Current Topics in Developmental Biology, Volume 140
ISSN 0070-2153
https://doi.org/10.1016/bs.ctdb.2019.10.003

1. Introduction

Gametogenesis gives rise to eggs or sperm in all sexually-reproducing organisms (Cinalli, Rangan, & Lehmann, 2008; Ellis & Kimble, 1994; Lesch & Page, 2012; Seydoux & Braun, 2006). Thus, understanding how gametogenesis is regulated is critical to comprehending this essential phenomenon that dictates fertility. Post-fertilization, the zygote gives rise to an entire organism, thus understanding how gametogenesis is regulated also has implications for the field of regeneration (Lasko, 2012; Lee & Lee, 2014; Magnúsdóttir & Surani, 2014; Soldner & Jaenisch, 2018; Tadros & Lipshitz, 2009; Theunissen & Jaenisch, 2017). *Drosophila melanogaster* has been one of the central organisms used to study heritability and gametogenesis for nearly a century due to its rapid generation time and genetic tractability (Mattox, Palmer, & Baker, 1990; Spradling, 1993; Spradling et al., 1997; Spradling, Fuller, Braun, & Yoshida, 2011; Spradling & Rubin, 1981; Xie & Li, 2007). These traits have facilitated the establishment of an extensive collection of informative and useful mutant and transgenic flies (Hales, Korey, Larracuente, & Roberts, 2015). In addition, many of the gametogenic regulatory factors described in the *Drosophila* germline are conserved to mammals and also play critical roles in other tissues, such as neurons (Goldstrohm, Hall, & McKenney, 2018; Lin & Spradling, 1997; Reichardt et al., 2018; Vessey et al., 2010; Zamore, Bartel, Lehmann, & Williamson, 1999; Zhang & Smith, 2015). While both male and female *Drosophila* undergo meiosis to give rise to gametes, here we focus on the female germline as regulation of gametogenesis in males has been reviewed elsewhere (Barreau, Benson, Gudmannsdottir, Newton, & White-Cooper, 2008; Fuller, 1998; Spradling et al., 2011; Yamashita & Fuller, 2005; Zhao & Garbers, 2002).

The spatiotemporal stages of *Drosophila* oogenesis are discrete and can be easily identified by their morphology and molecular markers (Gáspár & Ephrussi, 2017; Jia, Xu, Xie, Mio, & Deng, 2016; Spradling et al., 2011). At the anterior end of the ovary, germline stem cells (GSCs) reside in a structure known as the germarium and initiate differentiation to give rise to gametes (Kai, Williams, & Spradling, 2005; Twombly et al., 1996; Xie, 2000; Xie & Li, 2007; Xie & Spradling, 1998) (Fig. 1A). GSCs are maintained by signaling from the surrounding somatic niche. GSCs undergo asymmetric mitotic division, producing a stem cell daughter, or cystoblast (CB) which will begin the process of differentiation by expressing the essential

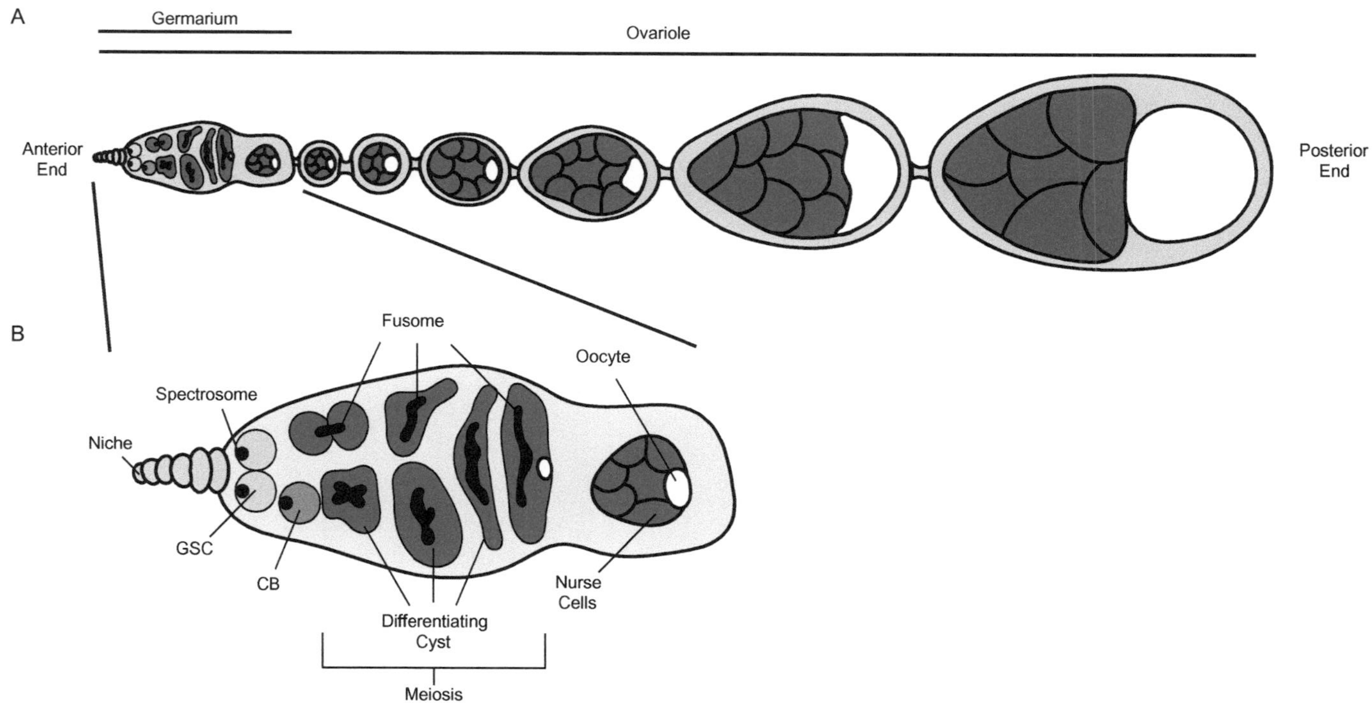

Fig. 1 (A) Schematic of *Drosophila* an ovariole. *Drosophila* females have two ovaries consisting of 16–20 ovarioles, which are assembly lines for producing mature eggs. The germarium, the structure that houses the germline stem cell (GSC), is present at anterior tip of the ovariole. The germline stem cell asymmetrically divides, giving rise to another GSC and a GSC daughter. The daughter cell then will undergo four incomplete rounds of mitosis, giving rise to a 16-cell cyst. Of the 16 cells one will be specified as the egg while the others serve as polyploid nurse cells that support oocyte and egg development. The surrounding somatic cells encapsulate the 16-cell cyst creating egg chambers. As development proceeds, the nurse cells provide mRNAs and proteins allowing the oocyte to grow in size and to eventually become a mature egg. (B) Inset of a germarium showing the developing germline, with the GSC located at the most anterior tip. Upon differentiation, the CB will undergo 4 incomplete mitotic divisions giving rise to a 16-cell cyst. Only one cell of the 16 cells completes meiosis and is destined to become the oocyte.

differentiation factor *bag of marbles* (*bam*) (Chen & McKearin, 2003b; McKearin & Ohlstein, 1995). The differentiating CB then undergoes four incomplete mitotic divisions, giving rise to an interconnected 16-cell cyst (McKearin & Ohlstein, 1995; McKearin & Spradling, 1990). In this cyst, one cell is designated to become the oocyte and the other 15 cells take on the role of nurse cells, which generate proteins and mRNAs that are provided to the developing oocyte (Navarro, Puthalakath, Adams, Strasser, & Lehmann, 2004; Spradling et al., 1997). The specified oocyte and its associated nurse cells are then encapsulated by somatic cells to form an egg chamber that buds off from the germarium (Fig. 1B) (Gilboa & Lehmann, 2004; Margolis & Spradling, 1995). The nurse cells will enter into a unique state in which they undergo a modified version of the cell cycle without undergoing mitosis, creating polyploid nuclei capable of fulfilling the high transcriptional demand required to transcribe all of the mRNAs necessary for the egg (Lilly & Duronio, 2005; Royzman & Orr-Weaver, 1998). As this process ensues, the egg chambers and oocyte increase in size as the supply of mRNAs and proteins is created and deposited into the mature egg (Fig. 1A) (Lasko, 2012; Richter & Lasko, 2011).

Oocyte development entails multiple processes that ensure effective completion of gametogenesis and fertility. Among these are stem cell maintenance and asymmetric division, differentiation, mitosis and meiosis, and production of the maternal mRNA contribution, thus the germline is a salient model for understanding how cells navigate fate transitions (Chen & McKearin, 2003b; Fu et al., 2015; Harris, Pargett, Sutcliffe, Umulis, & Ashe, 2011; Lasko, 2012). During oogenesis, there is little instructive transcription, barring a few examples, to mediate these critical transitions (Cinalli et al., 2008; Rangan, DeGennaro, & Lehmann, 2008). Instead, the germline relies highly on post-transcriptional regulatory mechanisms to coordinate gametogenesis (Slaidina & Lehmann, 2014). These include: alternative splicing, RNA modifications to modulate splicing, protein-RNA interactions, small RNA biology, and organization of the translation machinery to control the output of gene expression to mediate cell fate transitions. Here we focus on post-transcriptional processing of germline mRNAs and translational regulation both of which are required for successful oogenesis.

2. Alternative splicing ensures accurate production of critical germline mRNAs that regulate sex determination and differentiation

Splicing decisions are crucial during the generation of mature mRNAs post-transcriptionally and significantly contribute to germline development.

Splicing is mediated by a large ribonucleoprotein catalytic complex called the spliceosome, the core of which is made up of five small nuclear RNAs (snRNA), U1, U2, U4, U5 and U6, that work with spliceosomal proteins to form a small nuclear ribonucleoprotein complex (snRNP) (Madhani, Bordonne, & Guthrie, 1990; Wahl, Will, & Lührmann, 2009; Will & Lührmann, 2001, 2011). This complex removes introns from newly synthesized pre-mRNAs and links exonic sequences together (Wahl et al., 2009). Initially, U1 snRNP recognizes the donor site, which is located at the 5′ end of the intron, and U2 snRNP binds the branch site located at the 3′ end, leading to structural rearrangements of the complex and its associated substrate pre-mRNA (Matera & Wang, 2014). Catalytic actions of pre-mRNA splicing occur in two main steps. Cleavage at the 5′ splice site forms a lariat-like structure such that a 2′,5′-phosphodiester bond is created between the first nucleotide of the donor site and a conserved adenosine residue at the branch site (Rymond & Rosbash, 1985). Next, a second cleavage event occurs at the 3′ splice site and is followed by ligation of flanking exons to complete splicing (Umen & Guthrie, 1995; Wahl et al., 2009).

Alternative splicing is a process by which a single locus can give rise to many unique mRNA isoforms and their resulting protein variants (Black, 2000). The selection of the splice sites is exquisitely regulated to determine which exons will be included in the resulting alternatively spliced transcripts (Wang et al., 2015). Alternative splicing is highly regulated and is critical to germline development (Hager & Cline, 1997; Kalsotra & Cooper, 2011). There are a myriad of RNA targets that must be differentially spliced, and a complex web of interacting proteins orchestrate production of their splice variants (Lee & Rio, 2015). One of the first described instances of alternative splicing in *Drosophila* females is the splicing of the sex determination gene *sex-lethal* (*sxl*) (Bell, Maine, Schedl, & Cline, 1988). *sxl* is alternatively spliced to generate isoforms that control sex determination in somatic tissues (Chang, Dunham, Nuzhdin, & Arbeitman, 2011). In females, an autoregulatory loop forms between Sxl protein, U2AF splicing factor and U1 snRNP (Nagengast, Stitzinger, Tseng, Mount, & Salz, 2003). In *Drosophila*, the protein component of the U1 and U2 snRNPs are encoded by a gene called *sans fille* (*snf*) (Cline, Rudner, Barbash, Bell, & Vutien, 1999). Loss of *snf* results in a sterility phenotype in females that specifically affects germline *sxl* splicing and leads to a tumor comprised of undifferentiated cells (Johnson, Nagengast, & Salz, 2010). When correctly spliced, the resulting Sxl protein recognizes its own pre-mRNAs by binding both upstream and downstream of Exon 3 (Penalva & Sánchez, 2003). In addition, Sxl protein interacts with the U2AF and U1 snRNP to block the recognition of splice sites at Exon 3 (Nagengast et al., 2003). As a result, exon 3 is spliced out of the pre-mRNA in the final transcript that is capable of being translated into a fully functional protein (Penalva & Sánchez, 2003). In contrast, males include exon three in the final *sxl* transcript. Exon 3 contains a premature stop codon within the *sxl* transcript that results in a truncated protein that lacks the activity of the female-specific variant (Inoue, Hoshijima, Sakamoto, & Shimura, 1990). Thus, *sxl* is differentially expressed in the male and the female gonad due to alternative splicing events.

In addition to control of *sxl* via alternative splicing, *sxl* expression is controlled at the level of transcription by several transcription factors, such as Ovo (Salles, Mével-Ninio, Vincent, & Payre, 2002). Ovo is a zinc finger DNA binding protein that is required in the germline for proper gametogenesis (Andrews et al., 2000). *ovo* is also alternatively spliced and each of its isoforms have different implications for *sxl* expression. Ovo-A and Ovo-B were the first splice variants of *ovo* shown to be expressed in the female germline during oogenesis (Salles et al., 2002). In addition to differences due to alternative exon usage, Ovo-A, unlike Ovo-B, contains a 381 amino acid N-terminal extension which arises due to alternative transcription start sites (Andrews et al., 2000). Use of these promoters generates distinct Ova isoforms with unique temporal requirements during oocyte development; Ovo-B was found to be necessary and sufficient during early oogenesis and Ovo-A is critical in the later stages of egg development for a fully functional egg. The *ovo-B* gene has two characterized isoforms, Ovo +2B and Ovo-2B, which were discovered through a transposon insertion that disrupts exon splicing of *ovo-B*. This transposition event prevents inclusion of the exon 2b extension, producing a nonfunctional protein that accumulates during oogenesis. In the absence of retrotransposon insertion, the 178-amino acid extension encoded by exon 2b is included forming a fully functional Ovo protein, known as Ovo+2B (Salles et al., 2002). Interestingly, Ovo-B promotes transcription of *ovarian tumor* (*otu*), which enhances *sxl* expression (Fig. 2) (Lu & Oliver, 2001). The mechanism by which Otu regulates sxl expression is unknown but various mutations in *otu* lead to a myriad of phenotypes such as loss of germ cell proliferation, and inability to complete the differentiation process. The *otu* gene produces two cytoplasmic protein isoforms, a 104-kDa isoform (Otu-104) and a 98-kDa isoform (Otu-98) (Tirronen, Lahti, Heino, & Roos, 1995). Strikingly, only Otu-104 is capable of rescuing all the *otu* mutant phenotypes, indicating its requirement during oogenesis, while Otu-98 is

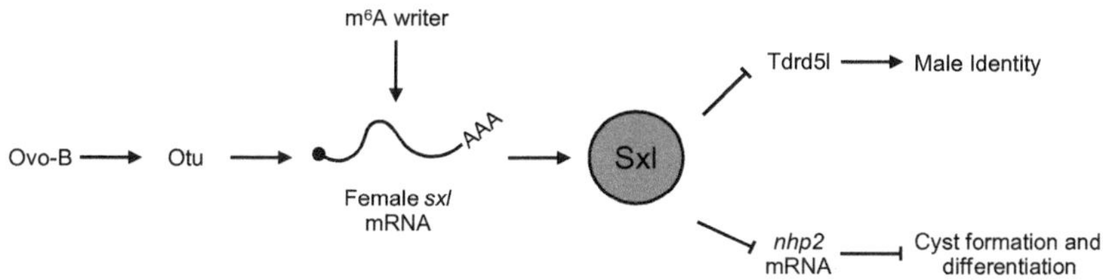

Fig. 2 Schematic of the pathway that promotes alternative splicing of *sxl* to generate the female sex determining variant in the germline. Ovo-B promotes the transcription of *otu,* which enhances splicing of *sxl.* The female-specific splice form of *sxl* is further enhanced by RNA modification by the m6A writer. Formation of the female-specific form generates a functional Sxl protein. Sxl represses Tdrd5l, a protein that promotes male identify. Additionally, Sxl post-transcriptionally represses *nhp2* to promote cyst formation during differentiation.

dispensable during this process (Tirronen et al., 1995). Despite the lack of insight into how the *otu* splice forms regulate GSC development, its alternative splicing is critical for oogenesis (Sass, Comer, & Searles, 1995). Thus, a cascade of alternative splicing events regulate production of Sxl in the female germline to promote oogenesis (Fig. 2).

Sxl expression in the female gonad regulates both sex determination as well as differentiation (Chau, Kulnane, & Salz, 2012). One critical task of Sxl is to represses Tudor domain containing protein 5-like (*tdrd5l*) (Primus, Pozmanter, Baxter, & Van Doren, 2019). Tdrd5l is present in the cytoplasm of the male germline, localizing to granules associated with RNA regulation, to promote male identity and differentiation. Sxl expression the female gonad represses translation of Tdrd5l to promote female identity (Primus et al., 2019). In addition, female Sxl has been found to regulate transcription of *PHD finger protein 7* (*phf7*), a key regulator of male identity (Yang, Baxter, & Van Doren, 2012). Sxl was found to indirectly promote recruitment of SETDB1, a chromatin writer, to deposit trimethylated H3K9 (H3K9me3) repressing transcription of *phf7* (Smolko, Shapiro-Kulnane, & Salz, 2018). Thus, alternative splicing of *sxl* results in different sexes helps promote proper sex determination in the germline (Fig. 2). Sxl also fulfills additional functions outside of sex determination. Sxl is required in the female germline for germline stem cell GSC differentiation. Loss of Sxl protein causes an accumulation of single cells and two cell cysts (Chau, Kulnane, & Salz, 2009). It is thought that Sxl binds *nanos* (*nos*) mRNA, an RNA binding protein that is necessary for GSC self-renewal, using a canonical Sxl binding sequence in the 3′ UTR (Chau et al., 2012). Loss of Sxl leads to an accumulation of excess of Nanos protein, which is thought to limit GSC differentiation (Chau et al., 2012). While regulation by Sxl is beginning to be deciphered, several aspects remain to be discovered. For example, Sxl, a splicing factor, is predominantly cytoplasmic in undifferentiated cells but becomes nuclear as differentiation proceeds (Chau et al., 2009), yet, how it works as translational regulator while in the cytoplasm and how it is transported to the nucleus to function as splicing factor during differentiation are not known.

Polypyrimidine tract binding proteins (PTBs) promote splicing by binding polypyrimidine tracts that are ~10 nt long and bring splice sites together by means of protein dimerization to promote alternative splicing (Polydorides, Okano, Yang, Stefani, & Darnell, 2000; Romanelli, Diani, & Lievens, 2013). A PTB, *half pint* (*hfp*), a homolog of human PUF60, is important for oogenesis (Maniatis & Tasic, 2002). Loss of *hfp* results in missplicing of the *otu* transcripts described above (Van Buskirk & Schüpbach, 2002). In addition, *hfp* also regulates alternative splicing of *eukaryotic initiation factor 4E* (*eIF4E*) during development through 3′ splice site selection (Reyes & Izquierdo, 2008). Hfp is required to increase the relative abundance of the longer *eIF4E* transcript (Van Buskirk & Schüpbach, 2002). Lastly, *hfp* also regulates splicing of *gurken*, a critical regulator of dorsal-ventral patterning (Kalifa, Armenti, & Gavis, 2009). Thus, sex determination, differentiation and production of the determinants of embryonic patterning for the next generation are all regulated by mechanisms involving alternative splicing in the female germline.

3. RNA modifications direct splicing of sex determinants and translation of differentiation promoting genes in the germline

Post-transcriptional RNA modifications are abundant and conserved in all branches of life (Yi & Pan, 2011). There have been over 100 described RNA modifications that can alter stability, function and splicing of RNAs (Licht & Jantsch, 2016; Roundtree, Evans, Pan, & He, 2017). A well-known example of an mRNA modification is the 5′ methylguanosine cap that is added to all mRNAs to promote their stability and aid in translation initiation (Mitchell et al., 2010; Mukherjee et al., 2012). A variety of RNA modifications have been linked to developmental transitions, such as those affecting GSC fate (Batista et al., 2014; Roundtree et al., 2017). Specifically during oogenesis, N6A-methyladenosine (m^6A) has been shown to be important for differentiation of germline stem cell daughter cells in females by ensuring proper female-specific splicing of *sxl* (Haussmann et al., 2016). Additionally, the H/ACA box complex, an RNP complex responsible for depositing pseudouridine on rRNA, has been suggested to be regulated by Sxl during the germline stem cell to daughter cell transition and is required for proper cyst differentiation (Kiss, Fayet-Lebaron, & Jády, 2010; Morita, Ota, & Kobayashi, 2018).

m^6A is prevalent on mRNA and is mediated by a methyltransferase complex that deposits a methyl-group at the sixth nitrogen on adenosine (Yang, Hsu, Chen, & Yang, 2018). In *Drosophila,* m^6A is placed by a m^6A writer complex consisting of Xio, Virilizer (Vir), Spenito (Nito), female lethal d (fl(2)d), Methyltransferase like 3 (Mettl3) and Methyltransferase like 14 (Mettl14) (Yan & Perrimon, 2015). Some described roles of m^6A involve modulating RNA-structure, facilitating mRNA degradation, promoting translation initiation and mediating alternative splicing (Roundtree et al., 2017). Interestingly, the m^6A writer complex has been linked to *sxl* splicing during *Drosophila* oogenesis (Kan et al., 2017). miCLIP data revealed that m^6A must be placed at intergenic regions of the *sxl* mRNA in order to produce the female-specific isoform (Kan et al., 2017). Accordingly, loss of m^6A complex members such as *spenito* result in expression of the male specific isoform of *sxl*, and tumors of undifferentiated cells, similar to loss of *sxl* (Kan et al., 2017; Mattox et al., 1990). This suggests that m^6A enables proper splicing of female-specific *sxl*, which allows for proper differentiation of germline stem cells into cystoblast daughter cells (Fig. 2).

Pseudouridine is one of the most abundant RNA modifications (Zhao & He, 2015). Although most commonly found on tRNAs, pseudouridine is also found on mRNAs as well as rRNA (Penzo & Montanaro, 2018). Unlike the canonical nucleoside uridine which is attached to the sugar via a nitrogen–carbon bond, pseudouridine is a uridine isomer attached through a carbon–carbon bond (Cohn, 1960). Pseudouridine can be placed by two different classes of enzymes; either by a sequence specific pseudouridine synthase or a small RNA guided complex called the box H/ACA ribonucleoprotein (De Zoysa & Yu, 2017). Depletion of the H/ACA box complex member Nucleolar Protein Family A Member 2 (NHP2) in the germline leads to an accumulation of 4- and 8-cell cysts that do not transition to the 16-cell cyst stage (Morita et al., 2018). Interestingly, the accumulation of single cells due to loss of *sxl* is partially rescued by loss of *NHP2* indicating that this *sxl* phenotype is due to excess NHP2 (Morita et al., 2018). Consistent with this notion, Sxl interacts with *nhp2* mRNA suggesting that Sxl may impose a regulatory function, in this case likely repression of *nhp2* to allow initiation of the differentiation program (Fig. 2) (Morita et al., 2018). Thus, although it is clear that RNA modifications help to ensure proper splicing of sex determination factors, but the pathway, mechanism, and direct targets remain unresolved.

4. Production of ribosomes is finely tuned to facilitate differentiation

While splicing mediates proper mRNA production, access of the mature mRNAs to ribosomes controls their translation. Once mRNAs are gated for translation, proper ribosome levels control protein production. The levels of ribosomes during early oogenesis are strictly regulated and shockingly dynamic. Ribosome biogenesis is the process of transcribing and processing the ribosomal RNA (rRNA) components, as well as transcribing and translating the protein constituents of the ribosome (Granneman & Baserga, 2004; Nazar, 2004; Teng, Thomas, & Mercer, 2013; Yelick & Trainor, 2015). This process is exquisitely regulated as ribosome biogenesis is one of the most energy intensive tasks of maintaining cell homeostasis and is even more crucial in proliferative cells (Phipps, Charette, & Baserga, 2011). In addition to the high energy requirement of ribosome biogenesis, all of the components of the ribosome must be coordinated in their production. The process of ribosome biogenesis involves a series of coordinated steps of processing and assembly that involve dozens of non-coding RNAs and proteins and the molecular details of this process have been thoroughly covered in detail in several recent reviews (Granneman & Baserga, 2004; Yelick & Trainor, 2015). Briefly, ribosomal DNA (rDNA) is present in multicopy stretches within the genome; these areas of DNA are localized to a subnuclear organelle called the nucleolus (Karpen, Schaefer, & Laird, 1988; Ritossa & Spiegelman, 1965; Schwarzacher & Wachtler, 1993). rDNA is transcribed into rRNA in the nucleolus and processing steps begin co-transcriptionally (Koš & Tollervey, 2010) to remove internal and external spacers found in immature rRNA (Granneman & Baserga, 2004; Granneman, Petfalski, Tollervey, & Hurt, 2011; Schäfer, Strauß, Petfalski, Tollervey, & Hurt, 2003; Tafforeau et al., 2013). As these processing steps occur, the rRNA is covalently modified and ribosomal proteins begin to interact with the partially processed rRNA (Agalarov, Sridhar, Funke, Stout, & Williamson, 2000; Deshmukh, Tsay, Paulovich, & Woolford, 1993; Gumienny et al., 2017; Jády & Kiss, 2001; Kiss, Jady, Bertrand, & Kiss, 2004). When the rRNA is mostly mature it is exported from the nucleus to the cytoplasm where the small and large subunits of the ribosome fully mature and assemble (Lo et al., 2010; Schäfer et al., 2003; Sloan et al., 2017; Tschochner & Hurt, 2003; Zemp & Kutay, 2007). Errors at any of these steps can result in ribosome biogenesis defects which in humans result in disease states known as ribosomopathies (Armistead & Triggs-Raine, 2014; Barlow et al., 2010; Brooks et al., 2014; Higa-Nakamine et al., 2012; Mills & Green, 2017; Sloan et al., 2017).

Curiously, despite the presence of ribosomes across cell types and sharing similar molecular origins, ribosomopathies manifest as tissue-specific defects rather than pleiotropic phenotypes (Brooks et al., 2014; Higa-Nakamine et al., 2012; Mills & Green, 2017; Pereboom, van Weele, Bondt, & MacInnes, 2011; Yelick & Trainor, 2015). The reasons behind the unique, tissue-specific manifestations are still being investigated but in several cases it seems that stem cells may be particularly sensitive to perturbations in ribosome biogenesis (Brooks et al., 2014; Morgado-Palacin, Llanos, & Serrano, 2012; Pereboom et al., 2011; Watanabe-Susaki et al., 2014). Indeed, a growing body of evidence is beginning to suggest that *Drosophila* GSCs not only have a specific requirement for ribosome biogenesis, but also that ribosome biogenesis, as well as global translation, vary greatly over the course of GSC differentiation and are uncoupled during early oogenesis (Sanchez et al., 2016; Zhang, Shalaby, & Buszczak, 2014). These attributes make *Drosophila* oogenesis an excellent system to address how perturbations of ribosome levels affects stem cell differentiation.

In order to maintain stem cell fate, GSCs asymmetrically partition factors required for ribosome biogenesis by retaining more of this machinery than they pass on to daughter cells (Fichelson et al., 2009; Zhang et al., 2014). In particular, Underdeveloped (Udd), an rRNA transcription factor segregates asymmetrically to the GSC during mitosis and seems to promote a high rate of rRNA synthesis within the GSC (Zhang et al., 2014). Furthermore, Wicked (Wcd), a U3 snoRNP complex member required for rRNA maturation, is also asymmetrically partitioned to GSCs and associates with the original spectrosome, an ER rich organelle found in GSCs and CBs (Spradling et al., 1997), of the dividing GSC. How GSCs carry out this specialized cellular division requires further investigation, however, asymmetric stem cell division is crucial for proper differentiation (Chen & McKearin, 2003a, 2003b; Lin & Spradling, 1997). Consistent with this loss of *wcd* results in premature differentiation of GSCs (Fichelson et al., 2009). Nascent rRNA production, measured by BrUTP incorporation, and presumably

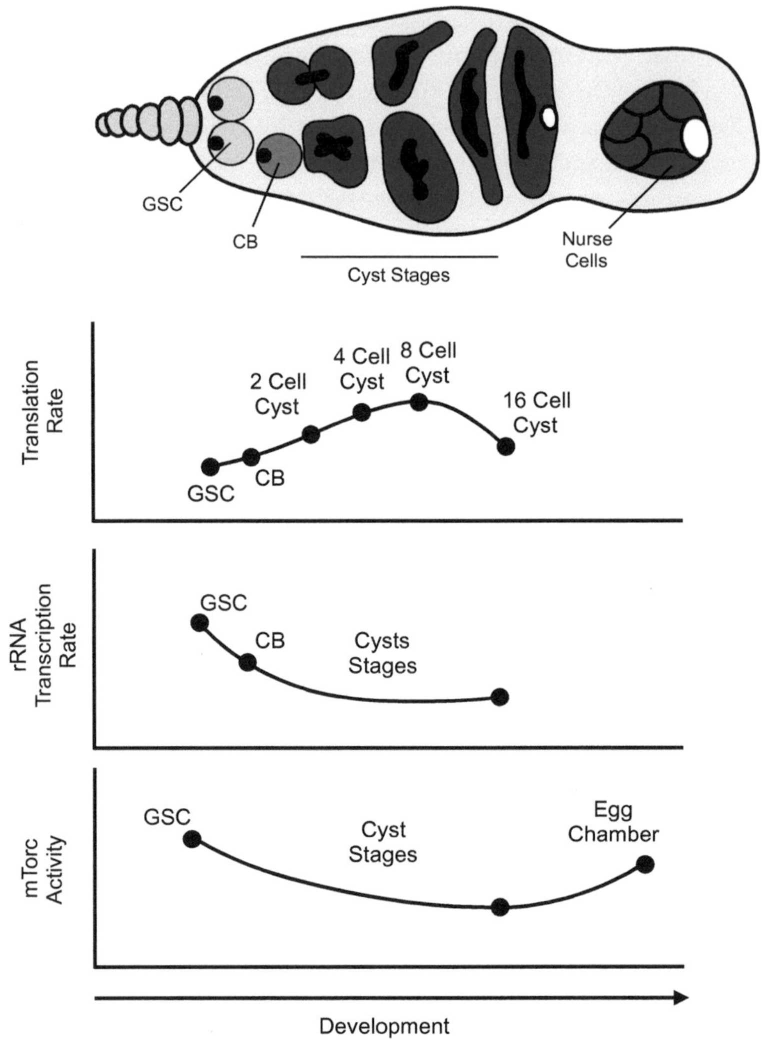

Fig. 3 Schematic representing the germarium and plots representing relative changes in global translation rate, rRNA transcription rate, and mTorc1 activity during development at the developmental stages indicated. As germline stem cell differentiation occurs rRNA production decreases, while global translation initially increases as differentiation occurs then falls off post-differentiation. A global regulator of both translation and rRNA production, mTorc1 activity decreases during differentiation and increases post-differentiation.

ribosomes, are produced at high levels in GSCs but this production drops in CBs and in subsequent stages (Fig. 3) (Zhang et al., 2014). Additionally, it has been observed that certain ribosome biogenesis components are expressed at high levels specifically in the germline (Kai et al., 2005).

In particular, RNA exonuclease 5 (Rexo5) is an RNA exonuclease that facilitates ribosome biogenesis by trimming snoRNAs as well as rRNAs (Gerstberger et al., 2017). Depletion of *rexo5* in the germline results in an accumulation of egg chambers that bud off from the germarium, but do not grow in size, and causes defects in GSC proliferation (Gerstberger et al., 2017). These observations suggest that the machinery for ribosome biogenesis is not only critical for germline development but is also dynamically regulated.

Sanchez et al. demonstrated that the dynamic nature of rRNA transcription during germline development is not simply a consequence of the differentiation process (Sanchez et al., 2016). Instead, lowering ribosome biogenesis is required for timely differentiation, but severe loss of ribosome biogenesis causes formation of stem-cysts, a product of perturbed cytokinesis of GSC daughters (Mathieu et al., 2013; Matias, Mathieu, & Huynh, 2015; Sanchez et al., 2016). Somewhat surprisingly, despite their increased retention of ribosome biogenesis components, GSCs exhibit a lower rate of translation compared to daughter cells and cyst stages (Fig. 3) (Sanchez et al., 2016). This finding invokes the hypothesis that despite the GSCs elevated capacity for ribosome biogenesis, GSCs do not intrinsically require higher ribosome levels for translation. Instead, the data is suggestive of the possibility that GSCs produce high levels of ribosomes in order to pass them on to and facilitate differentiation of their daughter cells. We thus hypothesize that a ribosome biogenesis checkpoint could couple ribosome production to cell cycle progression to ensure a sufficient ribosome concentration is passed from the GSC to the daughter CB. Conversely, increasing ribosome biogenesis via overexpression of TIF-IA, an RNA Pol I transcription initiation factor that is required for rRNA synthesis (Grewal, Evans, & Edgar, 2007), results in a failure of germ cells to differentiate, causing a marked overproliferation of undifferentiated GSC daughters (Zhang et al., 2014). This overproliferation may be caused by bypassing or rapid progression through the proposed ribosome biogenesis checkpoint such that the cell cycle is hastened in response to elevated ribosome biogenesis. The overproliferation of undifferentiated germ cells when ribosome levels are elevated is consistent with observations that high ribosome levels lead to rapidly growing cancers (Belin et al., 2009; Deisenroth & Zhang, 2010; Vlachos & Muir, 2010).

Although reducing ribosome biogenesis tends to result in the formation of a stem-cyst as previously described, some factors that play a role in ribosome biogenesis have a less severe phenotypes. For example, some mutants of the ribosomal protein S2 (rps2) gene have a repeating egg chamber mid-oogenesis defect, wherein ovarian development halts at stage 5 and successive egg chambers do not grow in size and eventually die, resulting in sterility (Cramton & Laski, 1994). This phenotype may be the consequence of incomplete loss of function as the allele that results in the repeating egg chamber phenotype reduces mRNA expression of *rps2*, incompletely, by 60–70%, while other allelic combinations result in embryonic lethality (Cramton & Laski, 1994). Incomplete loss of function alleles for another ribosomal protein, *ribosomal protein S3*, result in a similar repeating egg chamber phenotype (Sæbøe-Larssen, Lyamouri, Merriam, Oksvold, & Lambertsson, 1998). These observations suggest that partial loss of ribosome biogenesis during oogenesis may be tolerated during differentiation but results in phenotypes at a later phase of egg production, consistent with the model that high levels of biogenesis in early stages supply the ribosomes for subsequent differentiation and development.

Not only do ribosome levels vary but a class of ribosomal protein paralogs are enriched specifically in early germ cells (Xue & Barna, 2012). Several variant ribosomal proteins such as *ribosomal proteins S5b* (*rps5b*), *s10a*, *s19b*, and *l22*-like are enriched in the germline and others are enriched during early oogenesis (Kai et al., 2005). The role of these ribosomal proteins has not been thoroughly explored, but their presence indicates either a role for specialized ribosomes early during germline development or as a way to further increase the availability of ribosomal proteins to facilitate the high level of ribosome production in GSCs. One of these ribosomal protein paralogs, RpS5b, has recently been characterized (Kong et al., 2019). *rps5b* is most highly expressed in ovaries in contrast to its paralog, *ribosomal protein S5a* (*rps5a*), which is expressed at high levels ubiquitously (Kong et al., 2019). Loss of *rps5a* in the germline does not cause a germline phenotype, however, loss of *rps5b* results in a mid-oogenesis defect that is further exacerbated when *rps5a* is depleted in a *rps5b* mutant background (Kong et al., 2019). This could suggest that RpS5a and RpS5b are functionally similar and that the RpS5b phenotype results from lowering the overall amount of RpS5 available during oogenesis. However, RpS5b was also found to interact preferentially with mRNAs that encode proteins involved in mitochondrial electron transport, in contrast to RpS5a which binds mRNAs from a broad spectrum of gene categories (Kong et al., 2019). In accordance with the binding data, *rps5b* depleted ovaries expressed lower levels of proteins involved in oxidative phosphorylation and mitochondrial respiration (Kong et al., 2019). This evidence suggests that the expression of ribosomal protein paralogs may be a part of specialized ribosomes that translate specific groups of mRNAs; however, these ribosomal protein paralogs must be carefully analyzed to determine if they make up bonafide special ribosomes or instead have ribosome independent functions (Dinman, 2016).

What regulates ribosome biogenesis to allow for it to be dynamic during early *Drosophila* germline development? The best understood regulator of ribosome biogenesis is the Target of Rapamycin (TOR) pathway (Chymkowitch, Aanes, Robertson, Klungland, & Enserink, 2017; Magnuson, Ekim, & Fingar, 2012; Wei & Zheng, 2009; Yerlikaya et al., 2016) TOR is a kinase that is part of two distinct subcomplexes, TOR complex 1 (TORC1) and TOR complex 2 (TORC2) (Wullschleger, Loewith, & Hall, 2006). These complexes have distinct biological roles. TORC2 has been shown to function as an important regulator of the cytoskeleton (Wullschleger et al., 2006). Whereas, TORC1 receives and integrates several different signals including nutritional and growth factors and its activity promotes pro-proliferative activities such as global translation, ribosomal protein translation, and cell cycle progression (Kim, Goraksha-Hicks, Li, Neufeld, & Guan, 2008; Magnuson et al., 2012; Texada et al., 2019). TORC1 activity also helps to coordinate the transcription and translation of the components required for ribosome biogenesis (Grewal et al., 2007; Magnuson et al., 2012; Martin, Powers, & Hall, 2006). In *Drosophila,* TORC1 activity is high in GSCs through the 4-cell cyst, but TORC1 activity dips in 8- and 16-cell cysts and subsequently increases after the cyst stages (Wei, Bettedi, Kim, Ting, & Lilly, 2018). Interestingly, the landscape of TORC1 activity resembles the landscape of ribosome biogenesis, but not global translation (Fig. 3) (Sanchez et al., 2016; Zhang et al., 2014). However, loss of TORC1 components does not phenocopy perturbation of ribosome biogenesis (Sanchez et al., 2016). This is possibly because TORC1 plays a broader role in early oogenesis given the myriad of regulatory functions TORC1 is known to play in other systems (Kim et al., 2008; Li, Minor, Park, McKearin, & Maines, 2009; Li, Zhang, Takemori, Zhou, & Xiong, 2009; Moreno-Torres, Jaquenoud, & De Virgilio, 2015; Noda, 2017; Wei & Zheng, 2009). A downstream effector of mTORC1, La related protein 1 (Larp1) is known to silence ribosomal protein translation in mammals through binding to terminal oligopyrimidine tracts in the 5′ UTR of its targets (Fonseca et al., 2015; Hong et al., 2017; Lahr et al., 2017; Tcherkezian et al., 2014); however, the same has yet to be demonstrated for the *Drosophila* ortholog, La related protein (Larp). Tantalizingly, Larp is required for male and female fertility in *Drosophila*, but details of Larp's precise role in the female and oogenesis are lacking (Blagden et al., 2009; Ichihara, Shimizu, Taguchi, Yamaguchi, & Inoue, 2007). In contrast, in males Larp is required for proper spindle pole formation as well as proper cytokinesis (Blagden et al., 2009). Given the regulatory role Larp plays in ribosome biogenesis in mammals and the data from *Drosophila* spermatogenesis, Larp could facilitate the dynamic nature of ribosome biogenesis during GSC differentiation and meiosis. However, further study is required to understand the role of Larp during GSC differentiation and oogenesis to determine its function in this context.

The process of differentiation requires major cellular reprogramming. Surprisingly, despite being required for cell viability ribosome biogenesis and global translation are two key programs that are modulated to shape GSC differentiation (Sanchez et al., 2016; Zhang et al., 2014). When ribosome production is improperly modulated during GSC differentiation it results in characteristic phenotypes, accumulation of single cells if biogenesis components are overexpressed and formation of a stem-like cyst if ribosome biogenesis components are knocked down in the germline (Sanchez et al., 2016; Zhang et al., 2014). Additionally, several ribosomal protein variants are highly enriched in ovaries and they may perform special functions, however, these variants are just beginning to be studied. Additionally, based on what we know of the mechanisms and networks that control ribosome biogenesis in *Drosophila* oocytes, the dynamic nature of ribosome biogenesis seems likely to be conserved; however, further investigation is required to determine and compare the basis of ribosome biogenesis control.

5. Hand-off mechanisms facilitated by combinatorial RNA binding proteins dynamically shape the translational landscape during oogenesis

While some mRNAs are translated post-transcriptionally, other critical mRNAs are translationally regulated. For efficient translation of mRNAs, it is thought that the mRNAs must be circularized—bringing their 5′ cap and 3′ poly-A tail in close proximity to each other (Fukao et al., 2009; Martineau et al., 2008; Preiss & Hentze, 1998). This interaction is mediated by cap-binding proteins such as eukaryotic initiation factor 4E (eIF4E) and the poly-A binding protein (PABP) (Eichhorn et al., 2016; Kronja et al., 2014; Subtelny, Eichhorn, Chen, Sive, & Bartel, 2014; Tarun, Wells, Deardorff, & Sachs, 1997). A longer poly-A tail and uninhibited access to the 5′ cap for eIF4E is believed to promote efficient translation (Jalkanen, Coleman, & Wilusz, 2014). A major mode of translational regulation is that RNA binding proteins (RBPs) recognize cognate sequences in the 3′ UTRs of their target mRNAs (Harvey et al., 2018). The binding of the RBP prevents circularization of the mRNA and inhibits efficient translation initiation, leading to reduced translation (Mazumder, Seshadri, Imataka, Sonenberg, & Fox, 2001). RBP binding to the 3′ UTR can mediate translation inhibition by recruiting cofactors to inhibit circularization (Szostak & Gebauer, 2013).This inhibition of circularization can be achieved by RBP binding to the cap and competing with eIF4E, removal of the cap by the decapping machinery, or recruitment of factors such as the CCR4-Not complex to shorten poly-A tail length (Rissland, 2017). In some cases, RBPs can both block initiation as well as mediate shortening of the poly-A tail (Neve, Patel, Wang, Louey, & Furger, 2017).

As mentioned in the germline several developmental processes such as stem cell maintenance, differentiation, mitosis and meiosis are coordinated and successful transition through these diverse programs relies on precise translational control (Fig. 4) (Joshi, Riddle, Djabrayan, & Rothman, 2010; Slaidina & Lehmann, 2014). As factors that interfere with translation such as the decapping machinery and the poly-A tail shortening CCR4-Not complex are expressed continuously during oogenesis, and cannot support dynamic translational control on their own, a dynamic and diverse landscape of translational regulators has evolved to allow for fine-scale temporal control of mRNA translation (Eichhorn et al., 2016; Flora et al., 2018). To add an additional layer of complexity, the expression or abundance of several RBPs that regulate translational control oscillate as oogenesis progresses (Fig. 4) (Flora et al., 2018; Rangan et al., 2009; Richter & Lasko, 2011). As the levels of RBPs decrease, their bound mRNA targets are licensed for translation (Flora et al., 2018; Lasko, 2000; Linder & Lasko, 2006). There are three major themes that work to control mRNA translation: 1. RBPs collaborate in a combinatorial manner to regulate mRNAs, 2. Target mRNAs are handed off from one RBP complex to another as levels oscillate during oogenesis to consistently repress or promote target mRNA translation, and 3. Multiple feedback mechanisms operate to mediate each transition (Fig. 4) (Flora, Wong-Deyrup, et al., 2018). The feedback mechanism has been extensively reviewed elsewhere and is not the focus of this chapter (Slaidina & Lehmann, 2014). Here, we outline how RBPs both collaborate as well hand off mRNAs during the transition from GSC to mature oocyte.

GSCs rely on several translational control factors to maintain self-renewal, two of the main factors are Pumilio (Pum) and Nanos (Nos), which work in a combinatorial fashion to repress the translation of differentiation-promoting mRNAs (Fig. 4) (Forbes & Lehmann, 1998;

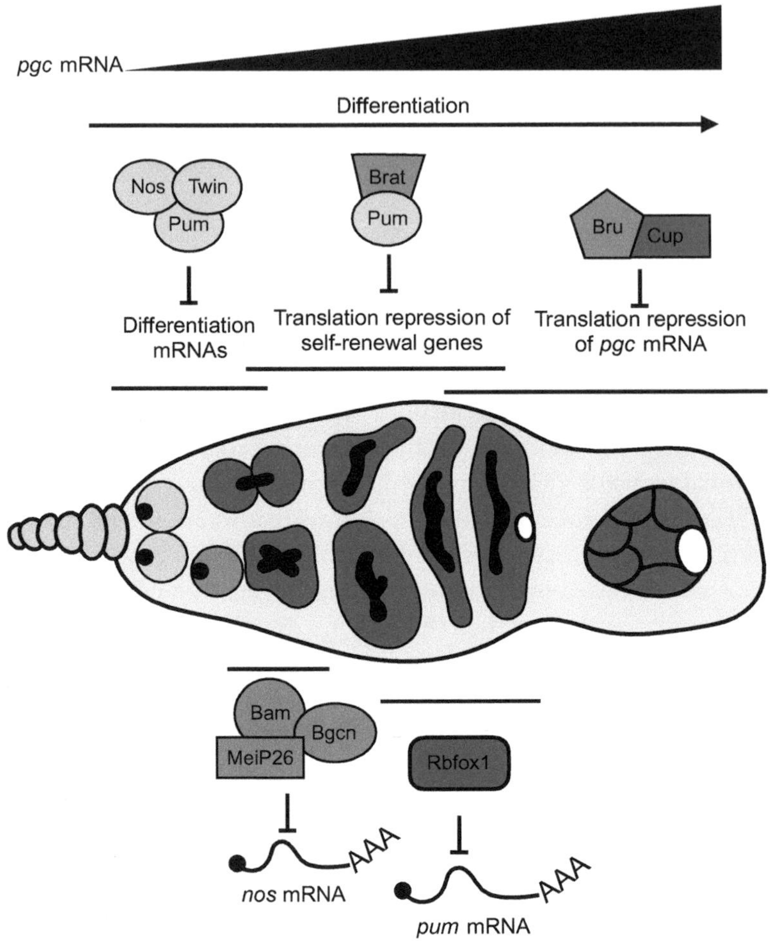

Fig. 4 Schematic of combinatorial and dynamic translation regulation in the *Drosophila* germarium. In the GSCs Nos, Pum and Twin form a complex to inhibit the translation of differentiation mRNAs such as *pgc*, which increases throughout oogenesis. Expression of Bam in the CB initiates differentiation by interacting with its partner Bgcn and Mei-P26 to repress the translation of GSC-expressed mRNAs, specifically *nos*. As Nos protein levels decrease in the CB, Pum is available to partner with Brat to repress the translation of self-renewal genes and *pgc*. In cyst stages, Rbfox1 binds the *pum* 3′ UTR to inhibit its translation. Throughout oogenesis Bru and Cup continuously block translation of *pgc*.

Gilboa & Lehmann, 2004; Joly, Chartier, Rojas-Rios, Busseau, & Simonelig, 2013; Lin & Spradling, 1997). Pum, a member of the conserved Pum- and Fem-3-binding factor (PUF) family of proteins, is present at high levels in the undifferentiated germline cells of the ovary, including GSCs, CBs, and early-differentiating cysts (Forbes & Lehmann, 1998; Kai et al., 2005). Independent of other factors, Pum can directly bind mRNA, but it requires the catalytic activity of other proteins to regulate translation of its targets in the *Drosophila* germline (Sonoda & Wharton, 1999; Tadauchi, Matsumoto, Herskowitz, & Irie, 2001). Pum is known to have dynamic interactions with two critical regulators, Nos in GSCs, and Brain tumor (Brat) in CBs (Fig. 4) (Arvola, Weidmann, Tanaka Hall, & Goldstrohm, 2017; Goldstrohm et al., 2018; Harris et al., 2011; Reichardt et al., 2018; Sonoda & Wharton, 2001, 1999). Nos, a conserved RNA binding protein, has the ability to bind mRNA, albeit at low affinity and requires the presence of Pum to recognize its targets (Arvola et al., 2017; Zamore et al., 1999). Nanos directly interacts with Not1, a member of the CCR4-Not complex, recruiting it to target mRNAs, such as *meiotic P26* (*mei-p26*) and *brat*, to regulate their translation (Bhandari, Raisch, Weichenrieder, Jonas, & Izaurralde, 2014; Raisch et al., 2016; Temme, Simonelig, & Wahle, 2014). While in some systems Pum can directly recruit the CCR4-Not complex, activity of *nos* is required for this interaction in the *Drosophila* germline (Joly et al., 2013; Temme et al., 2014). Upon loss of Pum, Nanos or Twin, GSCs fail to maintain stem cell fate and differentiate into stem cell daughters, resulting in the inability to sustain oogenesis as outlined below.

An example of distinct, stage-specific translational control by Pum/Nos/CCR4-Not complex in the germline is the mechanism by which *polar granule component* (*pgc*), a germline-specific transcriptional repressor, is controlled (Fig. 4) (Flora, Wong-Deyrup, et al., 2018). Pgc interacts with the Positive Transcription Elongation Factor (P-TEFb) complex and inhibits the phosphorylation of the Serine-2 residue that is critical for transcriptional elongation, resulting in global transcriptional silencing (Hanyu-Nakamura, Sonobe-Nojima, Tanigawa, Lasko, & Nakamura, 2008). A single pulse of expression of Pgc protein in the CB allows for epigenetic and transcriptomic reprogramming during differentiation (Flora et al., 2018). While *pgc* mRNA is expressed highly and ubiquitously throughout oogenesis, translation of *pgc* mRNA is tightly regulated to mitigate the effects of its potent transcriptional silencing activity. The *pgc* 3′ UTR contains a conserved consensus sequence that is transiently and sequentially bound by multiple distinct, developmentally regulated RBPs (Flora, Wong-Deyrup, et al., 2018). This 3′ UTR sequence is required for post-transcriptional control of *pgc* as Pgc protein expression is restricted to the CB. In the GSCs, Pum and Nos bind the *pgc* 3′ UTR and recruit Twin a component of the CCR4-Not complex to deadenylate *pgc* mRNA and inhibit its translation (Fig. 4)

(Flora, Wong-Deyrup, et al., 2018). In addition to *pgc*, Pum/Nos and Twin also regulate Brain tumor (Brat) (Joly et al., 2013). Brat is a TRIM-NHL domain protein expressed in the germline that represses translation by engaging with d4EHP and competing with the cap-binding protein eIF4E to prevent translation initiation (Fig. 4) (Arvola et al., 2017; Harris et al., 2011; Sonoda & Wharton, 2001). While *brat* mRNA is expressed in the GSC, it is specifically repressed by Nos and Pum. In addition to these targets, several differentiation-promoting mRNAs such as *meiP26* are also repressed (Joly et al., 2013). Thus, in the GSCs, a combination of Pum, Nos and CCR4-Not complex are required for repressing translation of several critical differentiation promoting mRNAs (Flora, Wong-Deyrup, et al., 2018; Lasko, 2012, 2000; Slaidina & Lehmann, 2014).

Subsequent differentiation of the GSC daughters relies on several factors to repress expression of *nos* mRNA (Lasko, 2000, 2012; Slaidina & Lehmann, 2014). Differentiation is initiated upon Bam expression in the CB, where Bam and its binding partner benign gonial cell neoplasm (Bgcn) act through a sequence in the *nos* 3′ UTR to its inhibit translation (Fig. 4) (Li, Minor, et al., 2009; Li, Zhang, et al., 2009; McCarthy, Deiulio, Martin, Upadhyay, & Rangan, 2018). This repression mechanism includes deadenylation activity by Twin, which works in conjunction with Bam and Bgcn (Fu et al., 2015). As Nos protein levels decrease in the CB, *pgc* and *brat* mRNAs are translated (Flora, Wong-Deyrup, et al., 2018). The expressed Brat protein now partners with Pum to repress translation of GSC self-renewal genes (Fig. 4) (Harris et al., 2011). In addition, expression of Mei-P26 increases initiating interactions with Bam, Bgcn and Sxl. Mei-P26 then promotes translational repression of GSC fate promoting genes such as *nos*, allowing for further differentiation by cooperating with Bam and Bgcn (Li et al., 2013; Reichardt et al., 2018). As the CB differentiates into 2-, 4-, 8- and 16-cell cysts, levels of Nanos protein rebound. However, in spite of the presence of Nos, Pum partners with Brat to suppress *pgc* translation in the 4- to 16-cell cyst stages (Fig. 4) (Flora, Wong-Deyrup, et al., 2018). Thus, in CBs, absence of Nos allows for Pum to complex with a different subset of proteins as well as license expression of new translational regulators to promote differentiation.

After cyst differentiation, Pum protein levels decrease and expression of another translational repressor, Bruno (Bru), increases (Kim-Ha, Kerr, & Macdonald, 1995; Schupbach & Wieschaus, 1989, 1991; Webster, Liang, Berg, Lasko, & Macdonald, 1997). Down-regulation of Pum expression is critical for the transition from GSC to an oocyte (Carreira-Rosario et al., 2016;

Forbes & Lehmann, 1998). Rbfox1, an RBP whose cytoplasmic isoform regulates the translation of specific mRNAs in the germline is responsible for repressing Pum translation through binding of a consensus sequence in the *pum* 3′ UTR (Fig. 4) (Carreira-Rosario et al., 2016). Loss of Rbfox1 leads to an expansion of Pum protein expression and a disruption of differentiation (Carreira-Rosario et al., 2016). Repression of Pum levels by Rbfox1 allows for Bru expression (Carreira-Rosario et al., 2016). Surprisingly, Bru can bind to a sequence in the 3′ UTR that is very similar to Pum binding sequence (Fig. 4) (Reveal, Garcia, Ellington, & Macdonald, 2011). Bruno blocks translation initiation by interacting with Cup, a conserved eIF4E binding protein (Kim et al., 2015; Nakamura, Sato, & Hanyu-Nakamura, 2004). In fact, Bru binds the same sequence in the *pgc* 3′ UTR as Nos/Pum to prevent *pgc* translation (Flora, Wong-Deyrup, et al., 2018). This mode of translation repression is not restricted to *pgc*, rather a cohort of maternal mRNAs are co-regulated by Pum and Bru representing a hand-off mechanism for repression of maternal mRNAs (Flora, Wong-Deyrup, et al., 2018).

6. Summary

Decades of work using elegant genetics has revealed several paradigms in which splicing machinery, RNA modifying enzymes, ribosome levels, and translational regulation mediates the transition from GSC to oocyte fate. However, several critical details such as the direct targets and mechanisms still need to be deciphered. Together the advent of cost-effective sequencing technologies combined with the increasing ability to easily create mutants in previously uncharacterized genes will allow us to further elucidate the regulatory logic underlying this critical transition.

References

Agalarov, S. C., Sridhar, G., Funke, P. M., Stout, C. D., & Williamson, J. R. (2000). Structure of the S15, S6, S18-rRNA complex: Assembly of the 30S ribosome central domain. *Science*, *288*, 107–112.

Andrews, J., Garcia-Estefania, D., Delon, I., Lu, J., Mével-Ninio, M., Spierer, A., et al. (2000). OVO transcription factors function antagonistically in the Drosophila female germline. *Development*, *127*, 881–892.

Armistead, J., & Triggs-Raine, B. (2014). Diverse diseases from a ubiquitous process: The ribosomopathy paradox. *FEBS Letters*, *588*, 1491–1500. https://doi.org/10.1016/j.febslet.2014.03.024.

Arvola, R. M., Weidmann, C. A., Tanaka Hall, T. M., & Goldstrohm, A. C. (2017). Combinatorial control of messenger RNAs by Pumilio, Nanos and Brain Tumor Proteins. *RNA Biology*, *14*, 1445–1456. https://doi.org/10.1080/15476286.2017.1306168.

Barlow, J. L., Drynan, L. F., Trim, N. L., Erber, W. N., Warren, A. J., & Mckenzie, A. N. J. (2010). New insights into 5q-syndrome as a ribosomopathy. *Cell Cycle, 9*, 4286–4293. https://doi.org/10.4161/cc.9.21.13742.

Barreau, C., Benson, E., Gudmannsdottir, E., Newton, F., & White-Cooper, H. (2008). Post-meiotic transcription in Drosophila testes. *Development, 135*, 1897–1902.

Batista, P. J., Molinie, B., Wang, J., Qu, K., Zhang, J., Li, L., et al. (2014). m6A RNA modification controls cell fate transition in mammalian embryonic stem cells. *Cell Stem Cell, 15*, 707–719.

Belin, S., Beghin, A., Solano-Gonzàlez, E., Bezin, L., Brunet-Manquat, S., Textoris, J., et al. (2009). Dysregulation of ribosome biogenesis and translational capacity is associated with tumor progression of human breast cancer cells. *PLoS One, 4*, e7147.

Bell, L. R., Maine, E. M., Schedl, P., & Cline, T. W. (1988). Sex-lethal, a Drosophila sex determination switch gene, exhibits sex-specific RNA splicing and sequence similarity to RNA binding proteins. *Cell, 55*, 1037–1046.

Bhandari, D., Raisch, T., Weichenrieder, O., Jonas, S., & Izaurralde, E. (2014). Structural basis for the nanos-mediated recruitment of the CCR4-NOT complex and translational repression. *Genes & Development, 28*, 888–901. https://doi.org/10.1101/gad.237289.113.

Black, D. L. (2000). Protein diversity from alternative splicing: A challenge for bioinformatics and post-genome biology. *Cell, 103*, 367–370.

Blagden, S. P., Gatt, M. K., Archambault, V., Lada, K., Ichihara, K., Lilley, K. S., et al. (2009). Drosophila Larp associates with poly (A)-binding protein and is required for male fertility and syncytial embryo development. *Developmental Biology, 334*, 186–197. https://doi.org/10.1016/J.YDBIO.2009.07.016.

Brooks, S. S., Wall, A. L., Golzio, C., Reid, D. W., Kondyles, A., Willer, J. R., et al. (2014). A novel ribosomopathy caused by dysfunction of RPL10 disrupts neurodevelopment and causes X-linked microcephaly in humans. *Genetics, 198*, 723–733. https://doi.org/10.1534/genetics.114.168211.

Carreira-Rosario, A., Bhargava, V., Hillebrand, J., Kollipara, R. K. K., Ramaswami, M., & Buszczak, M. (2016). Repression of Pumilio protein expression by Rbfox1 promotes germ cell differentiation. *Developmental Cell, 36*, 562–571. https://doi.org/10.1016/j.devcel.2016.02.010.

Chang, P. L., Dunham, J. P., Nuzhdin, S. V., & Arbeitman, M. N. (2011). Somatic sex-specific transcriptome differences in Drosophila revealed by whole transcriptome sequencing. *BMC Genomics, 12*, 364.

Chau, J., Kulnane, L. S., & Salz, H. K. (2009). Sex-lethal facilitates the transition from germline stem cell to committed daughter cell in the Drosophila ovary. *Genetics, 182*, 121–132.

Chau, J., Kulnane, L. S., & Salz, H. K. (2012). Sex-lethal enables germline stem cell differentiation by down-regulating Nanos protein levels during Drosophila oogenesis. *Proceedings of the National Academy of Sciences of the United States of America, 109*, 9465–9470.

Chen, D., & McKearin, D. (2003a). Dpp signaling silences bam transcription directly to establish asymmetric divisions of germline stem cells. *Current Biology, 13*, 1786–1791. https://doi.org/10.1016/J.CUB.2003.09.033.

Chen, D., & McKearin, D. M. (2003b). A discrete transcriptional silencer in the bam gene determines asymmetric division of the Drosophila germline stem cell. *Development, 130*, 1159–1170. https://doi.org/10.1242/dev.00325.

Chymkowitch, P., Aanes, H., Robertson, J., Klungland, A., & Enserink, J. M. (2017). TORC1-dependent sumoylation of Rpc82 promotes RNA polymerase III assembly and activity. *Proceedings of the National Academy of Sciences of the United States of America, 114*, 1039–1044.

Cinalli, R. M., Rangan, P., & Lehmann, R. (2008). Germ cells are forever. *Cell, 132*, 559–562. https://doi.org/10.1016/j.cell.2008.02.003.
Cline, T. W., Rudner, D. Z., Barbash, D. A., Bell, M., & Vutien, R. (1999). Functioning of the Drosophila integral U1/U2 protein Snf independent of U1 and U2 small nuclear ribonucleoprotein particles is revealed by snf+ gene dose effects. *Proceedings of the National Academy of Sciences of the United States of America, 96*, 14451–14458.
Cohn, W. E. (1960). Pseudouridine, a carbon-carbon linked ribonucleoside in ribonucleic acids: Isolation, structure, and chemical characteristics. *The Journal of Biological Chemistry, 235*, 1488–1498.
Cramton, S. E., & Laski, F. A. (1994). String of pearls encodes Drosophila ribosomal protein S2, has minute-like characteristics, and is required during oogenesis. *Genetics, 137*, 1039–1048.
De Zoysa, M. D., & Yu, Y.-T. (2017). Posttranscriptional RNA pseudouridylation. In *The enzymes* (pp. 151–167). Elsevier.
Deisenroth, C., & Zhang, Y. (2010). Ribosome biogenesis surveillance: Probing the ribosomal protein-Mdm2-p53 pathway. *Oncogene, 29*, 4253–4260. https://doi.org/10.1038/onc.2010.189.
Deshmukh, M., Tsay, Y. F., Paulovich, A. G., & Woolford, J. L. (1993). Splice variants of the SWR1-type nucleosome remodeling factor Domino have distinct functions during *Drosophila melanogaster* oogenesis. *Molecular and Cellular Biology, 13*, 2835–2845.
Dinman, J. D. (2016). Pathways to specialized ribosomes: The brussels lecture. *Journal of Molecular Biology, 428*, 2186–2194.
Eichhorn, S. W., Subtelny, A. O., Kronja, I., Kwasnieski, J. C., Orr-Weaver, T. L., & Bartel, D. P. (2016). mRNA poly(A)-tail changes specified by deadenylation broadly reshape translation in Drosophila oocytes and early embryos. *eLife, 5*. e16955. https://doi.org/10.7554/eLife.16955.
Ellis, R. E., & Kimble, J. (1994). Control of germ cell differentiation in *Caenorhabditis elegans*. *Ciba Foundation Symposium, 182*, 179–192.
Fichelson, P., Moch, C., Ivanovitch, K., Martin, C., Sidor, C. M., Lepesant, J.-A., et al. (2009). Live-imaging of single stem cells within their niche reveals that a U3snoRNP component segregates asymmetrically and is required for self-renewal in Drosophila. *Nature Cell Biology, 11*, 685.
Flora, P., Schowalter, S., Wong-Deyrup, S., DeGennaro, M., Nasrallah, M. A., & Rangan, P. (2018). Transient transcriptional silencing alters the cell cycle to promote germline stem cell differentiation in Drosophila. *Developmental Biology, 434*, 84–95. https://doi.org/10.1016/j.ydbio.2017.11.014.
Flora, P., Wong-Deyrup, S. W., Martin, E. T., Palumbo, R. J., Nasrallah, M., Oligney, A., et al. (2018). Sequential regulation of maternal mRNAs through a conserved cis-acting element in their 3' UTRs. *Cell Reports, 25*, 3828–3843.e9. https://doi.org/10.1016/j.celrep.2018.12.007.
Fonseca, B. D., Zakaria, C., Jia, J.-J., Graber, T. E., Svitkin, Y., Tahmasebi, S., et al. (2015). La-related protein 1 (LARP1) represses terminal oligopyrimidine (TOP) mRNA translation downstream of mTOR complex 1 (mTORC1). *The Journal of Biological Chemistry, 290*, 15996–16020.
Forbes, A., & Lehmann, R. (1998). Nanos and Pumilio have critical roles in the development and function of Drosophila germline stem cells. *Development, 125*, 679–690.
Fu, Z., Geng, C., Wang, H., Yang, Z., Weng, C., Li, H., et al. (2015). Twin promotes the maintenance and differentiation of germline stem cell lineage through modulation of multiple pathways. *Cell Reports, 13*, 1366–1379. https://doi.org/10.1016/j.celrep.2015.10.017.

Fukao, A., Sasano, Y., Imataka, H., Inoue, K., Sakamoto, H., Sonenberg, N., et al. (2009). The ELAV protein HuD stimulates cap-dependent translation in a Poly (A)- and eIF4A-dependent manner. *Molecular Cell, 36*, 1007–1017. https://doi.org/10.1016/j.molcel.2009.11.013.

Fuller, M. T. (1998). Genetic control of cell proliferation and differentiation in Drosophilaspermatogenesis. In *Seminars in cell & developmental biology* (pp. 433–444). Elsevier.

Gáspár, I., & Ephrussi, A. (2017). RNA localization feeds translation. *Science, 357*, 1235–1236. https://doi.org/10.1126/science.aao5796.

Gerstberger, S., Meyer, C., Benjamin-Hong, S., Rodriguez, J., Briskin, D., Bognanni, C., et al. (2017). The conserved RNA exonuclease Rexo5 is required for 3′ end maturation of 28S rRNA, 5S rRNA, and snoRNAs. *Cell Reports, 21*, 758–772.

Gilboa, L., & Lehmann, R. (2004). Repression of primordial germ cell differentiation parallels germ line stem cell maintenance. *Current Biology, 14*, 981–986. https://doi.org/10.1016/j.cub.2004.05.049.

Goldstrohm, A. C., Hall, T. M. T., & McKenney, K. M. (2018). Post-transcriptional regulatory functions of mammalian Pumilio proteins. *Trends in Genetics, 34*, 972–990. https://doi.org/10.1016/j.tig.2018.09.006.

Granneman, S., & Baserga, S. J. (2004). Ribosome biogenesis: Of knobs and RNA processing. *Experimental Cell Research, 296*, 43–50.

Granneman, S., Petfalski, E., Tollervey, D., & Hurt, E. C. (2011). A cluster of ribosome synthesis factors regulate pre-rRNA folding and 5.8S rRNA maturation by the Rat1 exonuclease. *The EMBO Journal, 30*, 4006–4019. https://doi.org/10.1038/emboj.2011.256.

Grewal, S. S., Evans, J. R., & Edgar, B. A. (2007). Drosophila TIF-IA is required for ribosome synthesis and cell growth and is regulated by the TOR pathway. *The Journal of Cell Biology, 179*, 1105–1113. https://doi.org/10.1083/jcb.200709044.

Gumienny, R., Jedlinski, D. J., Schmidt, A., Gypas, F., Martin, G., Vina-Vilaseca, A., et al. (2017). High-throughput identification of C/D box snoRNA targets with CLIP and RiboMeth-seq. *Nucleic Acids Research, 45*, 2341–2353. https://doi.org/10.1093/nar/gkw1321.

Hager, J. H., & Cline, T. W. (1997). Induction of female sex-lethal RNA splicing in male germ cells: Implications for Drosophila germline sex determination. *Development, 124*, 5033–5048.

Hales, K. G., Korey, C. A., Larracuente, A. M., & Roberts, D. M. (2015). Genetics on the fly: A primer on the Drosophila model system. *Genetics, 201*, 815–842. https://doi.org/10.1534/genetics.115.183392.

Hanyu-Nakamura, K., Sonobe-Nojima, H., Tanigawa, A., Lasko, P., & Nakamura, A. (2008). Drosophila Pgc protein inhibits P-TEFb recruitment to chromatin in primordial germ cells. *Nature, 451*, 730–733. https://doi.org/10.1038/nature06498.

Harris, R. E., Pargett, M., Sutcliffe, C., Umulis, D., & Ashe, H. L. (2011). Brat promotes stem cell differentiation via control of a bistable switch that restricts BMP signaling. *Developmental Cell, 20*, 72–83. https://doi.org/10.1016/j.devcel.2010.11.019.

Harvey, R. F., Smith, T. S., Mulroney, T., Queiroz, R. M. L., Pizzinga, M., Dezi, V., et al. (2018). Trans-acting translational regulatory RNA binding proteins. *Wiley Interdisciplinary Reviews RNA, 9*, e1465. https://doi.org/10.1002/wrna.1465.

Haussmann, I. U., Bodi, Z., Sanchez-Moran, E., Mongan, N. P., Archer, N., Fray, R. G., et al. (2016). m6A potentiates Sxl alternative pre-mRNA splicing for robust Drosophila sex determination. *Nature, 540*, 301.

Higa-Nakamine, S., Suzuki, T. T., Uechi, T., Chakraborty, A., Nakajima, Y., Nakamura, M., et al. (2012). Loss of ribosomal RNA modification causes developmental defects in zebrafish. *Nucleic Acids Research, 40*, 391–398. https://doi.org/10.1093/nar/gkr700.

Hong, S., Freeberg, M. A., Han, T., Kamath, A., Yao, Y., Fukuda, T., et al. (2017). LARP1 functions as a molecular switch for mTORC1-mediated translation of an essential class of mRNAs. *eLife*, *6*, e25237.

Ichihara, K., Shimizu, H., Taguchi, O., Yamaguchi, M., & Inoue, Y. H. (2007). A Drosophila orthologue of larp protein family is required for multiple processes in male meiosis. *Cell Structure and Function*, *32*(2), 89–100.

Inoue, K., Hoshijima, K., Sakamoto, H., & Shimura, Y. (1990). Binding of the Drosophila sex-lethal gene product to the alternative splice site of transformer primary transcript. *Nature*, *344*, 461.

Jády, B. E., & Kiss, T. (2001). A small nucleolar guide RNA functions both in 2′-O-ribose methylation and pseudouridylation of the U5 spliceosomal RNA. *The EMBO Journal*, *20*, 541–551. https://doi.org/10.1093/emboj/20.3.541.

Jalkanen, A. L., Coleman, S. J., & Wilusz, J. (2014). Determinants and implications of mRNA poly(A) tail size—does this protein make my tail look big? *Seminars in Cell & Developmental Biology*, *34*, 24–32. https://doi.org/10.1016/j.semcdb.2014.05.018.

Jia, D., Xu, Q., Xie, Q., Mio, W., & Deng, W.-M. (2016). Automatic stage identification of Drosophila egg chamber based on DAPI images. *Scientific Reports*, *6*, 18850. https://doi.org/10.1038/srep18850.

Johnson, M. L., Nagengast, A. A., & Salz, H. K. (2010). PPS, a large multidomain protein, functions with sex-lethal to regulate alternative splicing in Drosophila. *PLoS Genetics*, *6*, e1000872.

Joly, W., Chartier, A., Rojas-Rios, P., Busseau, I., & Simonelig, M. (2013). The CCR4 deadenylase acts with Nanos and Pumilio in the fine-tuning of Mei-P26 expression to promote germline stem cell self-renewal. *Stem Cell Reports*, *1*, 411–424. https://doi.org/10.1016/j.stemcr.2013.09.007.

Joshi, P. M., Riddle, M. R., Djabrayan, N. J. V., & Rothman, J. H. (2010). *Caenorhabditis elegans* as a model for stem cell biology. *Developmental Dynamics*, *239*, 1539–1554. https://doi.org/10.1002/dvdy.22296.

Kai, T., Williams, D., & Spradling, A. C. (2005). The expression profile of purified Drosophila germline stem cells. *Developmental Biology*, *283*, 486–502.

Kalifa, Y., Armenti, S. T., & Gavis, E. R. (2009). Glorund interactions in the regulation of gurken and oskar mRNAs. *Developmental Biology*, *326*, 68–74.

Kalsotra, A., & Cooper, T. A. (2011). Functional consequences of developmentally regulated alternative splicing. *Nature Reviews Genetics*, *12*, 715.

Kan, L., Grozhik, A. V., Vedanayagam, J., Patil, D. P., Pang, N., Lim, K.-S., et al. (2017). The m 6 A pathway facilitates sex determination in Drosophila. *Nature Communications*, *8*, 15737.

Karpen, G. H., Schaefer, J. E., & Laird, C. D. (1988). A Drosophila rRNA gene located in euchromatin is active in transcription and nucleolus formation. *Genes & Development*, *2*, 1745–1763.

Kim, E., Goraksha-Hicks, P., Li, L., Neufeld, T. P., & Guan, K.-L. (2008). Regulation of TORC1 by Rag GTPases in nutrient response. *Nature Cell Biology*, *10*, 935.

Kim, G., Pai, C.-I., Sato, K., Person, M. D., Nakamura, A., & Macdonald, P. M. (2015). Region-specific activation of oskar mRNA Translation by inhibition of Bruno-mediated repression. *PLoS Genetics*, *11*, e1004992.

Kim-Ha, J., Kerr, K., & Macdonald, P. M. (1995). Translational regulation of oskar mRNA by Bruno, an ovarian RNA-binding protein, is essential. *Cell*, *81*, 403–412. https://doi.org/10.1016/0092-8674(95)90393-3.

Kiss, T., Fayet-Lebaron, E., & Jády, B. E. (2010). Box H/ACA small ribonucleoproteins. *Molecular Cell*, *37*, 597–606.

Kiss, A. M., Jady, B. E., Bertrand, E., & Kiss, T. (2004). Human box H/ACA pseudouridylation guide RNA machinery. *Molecular and Cellular Biology, 24*, 5797–5807. https://doi.org/10.1128/MCB.24.13.5797-5807.2004.

Kong, J., Han, H., Bergalet, J., Bouvrette, L. P. B., Hernández, G., Moon, N.-S., et al. (2019). *Drosophila ribosomal protein S5b is essential for oogenesis and interacts with distinct RNAs,* (p. 600502). bioRxiv.

Koš, M., & Tollervey, D. (2010). Yeast pre-rRNA processing and modification occur cotranscriptionally. *Molecular Cell, 37*, 809–820.

Kronja, I., Yuan, B., Eichhorn, S. W. W., Dzeyk, K., Krijgsveld, J., Bartel, D. P. P., et al. (2014). Widespread changes in the posttranscriptional landscape at the Drosophila oocyte-to-embryo transition. *Cell Reports, 7*, 1495–1508. https://doi.org/10.1016/j.celrep.2014.05.002.

Lahr, R. M., Fonseca, B. D., Ciotti, G. E., Al-Ashtal, H. A., Jia, J.-J., Niklaus, M. R., et al. (2017). La-related protein 1 (LARP1) binds the mRNA cap, blocking eIF4F assembly on TOP mRNAs. *eLife, 6*, e24146.

Lasko, P. (2000). The *Drosophila melanogaster* genome: Translation factors and RNA binding proteins. *The Journal of Cell Biology, 150*, 51–56. https://doi.org/10.1083/jcb.150.2.F51.

Lasko, P. (2012). mRNA localization and translational control in Drosophila oogenesis. *Cold Spring Harbor Perspectives in Biology, 4*, a012294. https://doi.org/10.1101/cshperspect.a012294.

Lee, K.-A., & Lee, W.-J. (2014). Drosophila as a model for intestinal dysbiosis and chronic inflammatory diseases. *Developmental and Comparative Immunology, 42*, 102–110. https://doi.org/10.1016/j.dci.2013.05.005.

Lee, Y., & Rio, D. C. (2015). Mechanisms and regulation of alternative pre-mRNA splicing. *Annual Review of Biochemistry, 84*, 291–323.

Lesch, B. J., & Page, D. C. (2012). Genetics of germ cell development. *Nature Reviews. Genetics, 13*, 781.

Li, Y., Minor, N. T., Park, J. K., McKearin, D. M., & Maines, J. Z. (2009). Bam and Bgcn antagonize;Nanos-dependent germ-line stem cell maintenance. *Proceedings of the National Academy of Sciences of the United States of America, 106*, 9304–9309. https://doi.org/10.1073/pnas.0901452106.

Li, Y., Zhang, Q., Carreira-Rosario, A., Maines, J. Z., McKearin, D. M., & Buszczak, M. (2013). Mei-P26 cooperates with Bam, Bgcn and Sxl to promote early germline development in the Drosophila ovary. *PLoS One, 8*, e58301.

Li, S., Zhang, C., Takemori, H., Zhou, Y., & Xiong, Z.-Q. (2009). TORC1 regulates activity-dependent CREB-target gene transcription and dendritic growth of developing cortical neurons. *The Journal of Neuroscience, 29*, 2334–2343.

Licht, K., & Jantsch, M. F. (2016). Rapid and dynamic transcriptome regulation by RNA editing and RNA modifications. *The Journal of Cell Biology, 213*, 15–22.

Lilly, M. A., & Duronio, R. J. (2005). New insights into cell cycle control from the Drosophila endocycle. *Oncogene, 24*, 2765–2775. https://doi.org/10.1038/sj.onc.1208610.

Lin, H., & Spradling, A. C. (1997). A novel group of pumilio mutations affects the asymmetric division of germline stem cells in the Drosophila ovary. *Development, 124*, 2463–2476.

Linder, P., & Lasko, P. (2006). Bent out of shape: RNA unwinding by the DEAD-Box helicase vasa. *Cell, 125*, 219–221. https://doi.org/10.1016/j.cell.2006.03.030.

Lo, K.-Y., Li, Z., Bussiere, C., Bresson, S., Marcotte, E. M., & Johnson, A. W. (2010). Defining the pathway of cytoplasmic maturation of the 60S ribosomal subunit. *Molecular Cell, 39*, 196–208.

Lu, J., & Oliver, B. (2001). Drosophila OVO regulates ovarian tumor transcription by binding unusually near the transcription start site. *Development, 128*, 1671–1686.

Madhani, H. D., Bordonne, R., & Guthrie, C. (1990). Multiple roles for U6 snRNA in the splicing pathway. *Genes & Development*, *4*, 2264–2277.

Magnúsdóttir, E., & Surani, M. A. (2014). How to make a primordial germ cell. *Development*, *141*, 245–252. https://doi.org/10.1242/dev.098269.

Magnuson, B., Ekim, B., & Fingar, D. C. (2012). Regulation and function of ribosomal protein S6 kinase (S6K) within mTOR signalling networks. *The Biochemical Journal*, *441*, 1–21. https://doi.org/10.1042/BJ20110892.

Maniatis, T., & Tasic, B. (2002). Alternative pre-mRNA splicing and proteome expansion in metazoans. *Nature*, *418*, 236.

Margolis, J., & Spradling, A. (1995). Identification and behavior of epithelial stem cells in the Drosophila ovary. *Development*, *121*, 3797–3807.

Martin, D. E., Powers, T., & Hall, M. N. (2006). Regulation of ribosome biogenesis: Where is TOR? *Cell Metabolism*, *4*, 259–260.

Martineau, Y., Derry, M. C., Wang, X., Yanagiya, A., Berlanga, J. J., Shyu, A.-B., et al. (2008). Poly(A)-binding protein-interacting protein 1 binds to eukaryotic translation initiation factor 3 to stimulate translation. *Molecular and Cellular Biology*, *28*, 6658–6667. https://doi.org/10.1128/MCB.00738-08.

Matera, A. G., & Wang, Z. (2014). A day in the life of the spliceosome. *Nature Reviews. Molecular Cell Biology*, *15*, 108.

Mathieu, J., Cauvin, C., Moch, C., Radford, S. J., Sampaio, P., Perdigoto, C. N., et al. (2013). Aurora B and cyclin B have opposite effects on the timing of cytokinesis abscission in Drosophila germ cells and in vertebrate somatic cells. *Developmental Cell*, *26*, 250–265.

Matias, N. R., Mathieu, J., & Huynh, J.-R. (2015). Abscission is regulated by the ESCRT-III protein shrub in Drosophila germline stem cells. *PLoS Genetics*, *11*, e1004653.

Mattox, W., Palmer, M. J., & Baker, B. S. (1990). Alternative splicing of the sex determination gene transformer-2 is sex-specific in the germ line but not in the soma. *Genes & Development*, *4*, 789–805.

Mazumder, B., Seshadri, V., Imataka, H., Sonenberg, N., & Fox, P. L. (2001). Translational silencing of ceruloplasmin requires the essential elements of mRNA circularization: Poly(A) tail, poly(A)-binding protein, and eukaryotic translation initiation factor 4G. *Molecular and Cellular Biology*, *21*, 6440–6449. https://doi.org/10.1128/mcb.21.19.6440-6449.2001.

McCarthy, A., Deiulio, A., Martin, E. T., Upadhyay, M., & Rangan, P. (2018). Tip60 complex promotes expression of a differentiation factor to regulate germline differentiation in female Drosophila. *Molecular Biology of the Cell*, *29*, 2933–2945. https://doi.org/10.1091/mbc.E18-06-0385.

McKearin, D., & Ohlstein, B. (1995). A role for the Drosophila bag-of-marbles protein in the differentiation of cystoblasts from germline stem cells. *Development*, *121*, 2937–2947.

McKearin, D. M., & Spradling, A. C. (1990). Bag-of-marbles: A Drosophila gene required to initiate both male and female gametogenesis. *Genes & Development*, *4*, 2242–2251. https://doi.org/10.1101/gad.4.12b.2242.

Mills, E. W., & Green, R. (2017). Ribosomopathies: There's strength in numbers. *Science*, *358*, eaan2755. https://doi.org/10.1126/SCIENCE.AAN2755.

Mitchell, S. F., Walker, S. E., Algire, M. A., Park, E.-H., Hinnebusch, A. G., & Lorsch, J. R. (2010). The 5′-7-methylguanosine cap on eukaryotic mRNAs serves both to stimulate canonical translation initiation and to block an alternative pathway. *Molecular Cell*, *39*, 950–962.

Moreno-Torres, M., Jaquenoud, M., & De Virgilio, C. (2015). TORC1 controls G 1–S cell cycle transition in yeast via Mpk1 and the greatwall kinase pathway. *Nature Communications*, *6*, 8256.

Morgado-Palacin, L., Llanos, S., & Serrano, M. (2012). Ribosomal stress induces L11-and p53-dependent apoptosis in mouse pluripotent stem cells. *Cell Cycle*, *11*, 503–510.

Morita, S., Ota, R., & Kobayashi, S. (2018). Downregulation of NHP 2 promotes proper cyst formation in Drosophila ovary. *Development, Growth & Differentiation*, *60*, 248–259.

Mukherjee, C., Patil, D. P., Kennedy, B. A., Bakthavachalu, B., Bundschuh, R., & Schoenberg, D. R. (2012). Identification of cytoplasmic capping targets reveals a role for cap homeostasis in translation and mRNA stability. *Cell Reports*, *2*, 674–684.

Nagengast, A. A., Stitzinger, S. M., Tseng, C.-H., Mount, S. M., & Salz, H. K. (2003). Sex-lethal splicing autoregulation in vivo: Interactions between SEX-LETHAL, the U1 snRNP and U2AF underlie male exon skipping. *Development*, *130*, 463–471.

Nakamura, A., Sato, K., & Hanyu-Nakamura, K. (2004). Drosophila cup is an eIF4E binding protein that associates with Bruno and regulates oskar mRNA translation in oogenesis. *Developmental Cell*, *6*, 69–78. https://doi.org/10.1016/S1534-5807(03)00400-3.

Navarro, C., Puthalakath, H., Adams, J. M., Strasser, A., & Lehmann, R. (2004). Egalitarian binds dynein light chain to establish oocyte polarity and maintain oocyte fate. *Nature Cell Biology*, *6*, 427–435. https://doi.org/10.1038/ncb1122.

Nazar, R. (2004). Ribosomal RNA processing and ribosome biogenesis in eukaryotes. *IUBMB Life*, *56*, 457–465. https://doi.org/10.1080/15216540400010867.

Neve, J., Patel, R., Wang, Z., Louey, A., & Furger, A. M. (2017). Cleavage and polyadenylation: Ending the message expands gene regulation. *RNA Biology*, *14*, 865–890. https://doi.org/10.1080/15476286.2017.1306171.

Noda, T. (2017). Regulation of autophagy through TORC1 and mTORC1. *Biomolecules*, *7*, 52.

Penalva, L. O. F., & Sánchez, L. (2003). RNA binding protein sex-lethal (Sxl) and control of Drosophila sex determination and dosage compensation. *Microbiology and Molecular Biology Reviews*, *67*, 343–359.

Penzo, M., & Montanaro, L. (2018). Turning uridines around: Role of rRNA pseudouridylation in ribosome biogenesis and ribosomal function. *Biomolecules*, *8*, 38.

Pereboom, T. C., van Weele, L. J., Bondt, A., & MacInnes, A. W. (2011). A zebrafish model of dyskeratosis congenita reveals hematopoietic stem cell formation failure resulting from ribosomal protein-mediated p53 stabilization. *Blood*, *118*, 5458–5465.

Phipps, K. R., Charette, J. M., & Baserga, S. J. (2011). The small subunit processome in ribosome biogenesis—Progress and prospects. *Wiley Interdisciplinary Reviews. RNA*, *2*, 1–21.

Polydorides, A. D., Okano, H. J., Yang, Y. Y. L., Stefani, G., & Darnell, R. B. (2000). A brain-enriched polypyrimidine tract-binding protein antagonizes the ability of Nova to regulate neuron-specific alternative splicing. *Proceedings of the National Academy of Sciences of the United States of America*, *97*, 6350–6355. https://doi.org/10.1073/pnas.110128397.

Preiss, T., & Hentze, M. W. (1998). Dual function of the messenger RNA cap structure in poly(A)-tail-promoted translation in yeast. *Nature*, *392*, 516–520. https://doi.org/10.1038/33192.

Primus, S., Pozmanter, C., Baxter, K., & Van Doren, M. (2019). Tudor-domain containing protein 5-like promotes male sexual identity in the Drosophila germline and is repressed in females by sex lethal. *PLoS Genetics*, *15*, e1007617.

Raisch, T., Bhandari, D., Sabath, K., Helms, S., Valkov, E., Weichenrieder, O., et al. (2016). Distinct modes of recruitment of the CCR4-NOT complex by Drosophila and vertebrate nanos. *The EMBO Journal*, *35*, 974–990. https://doi.org/10.15252/embj.201593634.

Rangan, P., DeGennaro, M., Jaime-Bustamante, K., Coux, R.-X. X., Martinho, R. G., & Lehmann, R. (2009). Temporal and spatial control of germ-plasm RNAs. *Current Biology*, *19*, 72–77. https://doi.org/10.1016/j.cub.2008.11.066.

Rangan, P., DeGennaro, M., & Lehmann, R. (2008). Regulating gene expression in the drosophila germ line. *Cold Spring Harbor Symposia on Quantitative Biology*, *73*, 1–8. https://doi.org/10.1101/sqb.2008.73.057.

Reichardt, I., Bonnay, F., Steinmann, V., Loedige, I., Burkard, T. R., Meister, G., et al. (2018). The tumor suppressor brat controls neuronal stem cell lineages by inhibiting deadpan and delda. *EMBO Reports*, *19*, 102–117. https://doi.org/10.15252/embr.201744188.

Reveal, B., Garcia, C., Ellington, A., & Macdonald, P. M. (2011). Multiple RNA binding domains of Bruno confer recognition of diverse binding sites for translational repression. *RNA Biology*, *8*, 1047–1060. https://doi.org/10.4161/rna.8.6.17542.

Reyes, R., & Izquierdo, J. M. (2008). Half pint couples transcription and splicing of eIF4E-1, 2 gene during fly development. *Biochemical and Biophysical Research Communications*, *374*, 758–762.

Richter, J. D., & Lasko, P. (2011). Translational control in oocyte development. *Cold Spring Harbor Perspectives in Biology*, *3*, a002758. https://doi.org/10.1101/cshperspect.a002758.

Rissland, O. S. (2017). The organization and regulation of mRNA–protein complexes. *Wiley Interdisciplinary Reviews. RNA 8*, e1369. https://doi.org/10.1002/wrna.1369.

Ritossa, F. M., & Spiegelman, S. (1965). Localization of DNA complementary to ribosomal RNA in the nucleolus organizer region of *Drosophila melanogaster*. *Proceedings of the National Academy of Sciences of the United States of America*, *53*, 737.

Romanelli, M., Diani, E., & Lievens, P. (2013). New insights into functional roles of the polypyrimidine tract-binding protein. *International Journal of Molecular Sciences*, *14*, 22906–22932.

Roundtree, I. A., Evans, M. E., Pan, T., & He, C. (2017). Dynamic RNA modifications in gene expression regulation. *Cell*, *169*, 1187–1200.

Royzman, I., & Orr-Weaver, T. L. (1998). S phase and differential DNA replication during Drosophila oogenesis. *Genes to Cells*, *3*, 767–776. https://doi.org/10.1046/j.1365-2443.1998.00232.x.

Rymond, B. C., & Rosbash, M. (1985). Cleavage of 5′ splice site and lariat formation are independent of 3′ splice site in yeast mRNA splicing. *Nature*, *317*, 735.

Sæbøe-Larssen, S., Lyamouri, M., Merriam, J., Oksvold, M. P., & Lambertsson, A. (1998). Ribosomal protein insufficiency and the minute syndrome in Drosophila: A dose-response relationship. *Genetics*, *148*, 1215–1224.

Salles, C., Mével-Ninio, M., Vincent, A., & Payre, F. (2002). A germline-specific splicing generates an extended ovo protein isoform required for Drosophila oogenesis. *Developmental Biology*, *246*, 366–376.

Sanchez, C. G., Teixeira, F. K., Czech, B., Preall, J. B., Zamparini, A. L., Seifert, J. R. K., et al. (2016). Regulation of ribosome biogenesis and protein synthesis controls germline stem cell differentiation. *Cell Stem Cell*, *18*, 276–290. https://doi.org/10.1016/J.STEM.2015.11.004.

Sass, G. L., Comer, A. R., & Searles, L. L. (1995). The ovarian tumor protein isoforms of *Drosophila melanogaster* exhibit differences in function, expression, and localization. *Developmental Biology*, *167*, 201–212.

Schäfer, T., Strauß, D., Petfalski, E., Tollervey, D., & Hurt, E. (2003). The path from nucleolar 90S to cytoplasmic 40S pre-ribosomes. *The EMBO Journal*, *22*, 1370–1380.

Schupbach, T., & Wieschaus, E. (1989). Female sterile mutations on the second chromosome of *Drosophila melanogaster*. I. Maternal effect mutations. *Genetics*, *121*, 101–117.

Schupbach, T., & Wieschaus, E. (1991). Female sterile mutations on the second chromosome of *Drosophila melanogaster*. II. Mutations blocking oogenesis or altering egg morphology. *Genetics*, *129*, 1119–1136.

Schwarzacher, H. G., & Wachtler, F. (1993). The nucleolus. *Anatomy and Embryology (Berlin)*, *188*, 515–536.

Seydoux, G., & Braun, R. E. (2006). Pathway to totipotency: Lessons from germ cells. *Cell*, *127*, 891–904. https://doi.org/10.1016/j.cell.2006.11.016.

Slaidina, M., & Lehmann, R. (2014). Translational control in germline stem cell development. *The Journal of Cell Biology*, *207*, 13–21. https://doi.org/10.1083/jcb.201407102.

Sloan, K. E., Warda, A. S., Sharma, S., Entian, K. D., Lafontaine, D. L. J., & Bohnsack, M. T. (2017). Tuning the ribosome: The influence of rRNA modification on eukaryotic ribosome biogenesis and function. *RNA Biology*, *14*, 1138–1152. https://doi.org/10.1080/15476286.2016.1259781.

Smolko, A. E., Shapiro-Kulnane, L., & Salz, H. K. (2018). The H3K9 methyltransferase SETDB1 maintains female identity in Drosophila germ cells. *Nature Communications*, *9*, 4155.

Soldner, F., & Jaenisch, R. (2018). Stem cells, genome editing, and the path to translational medicine. *Cell*, *175*, 615–632. https://doi.org/10.1016/j.cell.2018.09.010.

Sonoda, J., & Wharton, R. P. (1999). Recruitment of Nanos to hunchback mRNA by Pumilio. *Genes & Development*, *13*, 2704–2712. https://doi.org/10.1101/gad.13.20.2704.

Sonoda, J., & Wharton, R. P. (2001). Drosophila brain tumor is a translational repressor. *Genes & Development*, *15*, 762–773. https://doi.org/10.1101/gad.870801.

Spradling, A. C. (1993). Germline cysts: Communes that work. *Cell*, *72*, 649–651. https://doi.org/10.1016/0092-8674(93)90393-5.

Spradling, A. C., De Cuevas, M., Drummond-Barbosa, D., Keyes, L., Lilly, M., Pepling, M., et al. (1997). The Drosophila germarium: Stem cells, germ line cysts, and oocytes. *Cold Spring Harbor Symposia on Quantitative Biology*, *62*, 25–34. https://doi.org/10.1101/SQB.1997.062.01.006.

Spradling, A., Fuller, M. T., Braun, R. E., & Yoshida, S. (2011). Germline stem cells. *Cold Spring Harbor Perspectives in Biology*, *3*, a002642.

Spradling, A. C., & Rubin, G. M. (1981). Drosophila genome organization: Conserved and dynamic aspects. *Annual Review of Genetics*, *15*, 219–264. https://doi.org/10.1146/annurev.ge.15.120181.001251.

Subtelny, A. O., Eichhorn, S. W., Chen, G. R., Sive, H., & Bartel, D. P. (2014). Poly(A)-tail profiling reveals an embryonic switch in translational control. *Nature*, *508*, 66.

Szostak, E., & Gebauer, F. (2013). Translational control by 3'-UTR-binding proteins. *Briefings in Functional Genomics*, *12*, 58–65. https://doi.org/10.1093/bfgp/els056.

Tadauchi, T., Matsumoto, K., Herskowitz, I., & Irie, K. (2001). Post-transcriptional regulation through the HO 3'-UTR by Mpt5, a yeast homolog of Pumilio and FBF. *The EMBO Journal*, *20*, 552–561. https://doi.org/10.1093/emboj/20.3.552.

Tadros, W., & Lipshitz, H. D. (2009). The maternal-to-zygotic transition: A play in two acts. *Development*, *136*, 3033–3042. https://doi.org/10.1242/dev.033183.

Tafforeau, L., Zorbas, C., Langhendries, J.-L., Mullineux, S.-T., Stamatopoulou, V., Mullier, R., et al. (2013). The complexity of human ribosome biogenesis revealed by systematic nucleolar screening of pre-rRNA processing factors. *Molecular Cell*, *51*, 539–551. https://doi.org/10.1016/J.MOLCEL.2013.08.011.

Tarun, S. Z., Jr., Wells, S. E., Deardorff, J. A., & Sachs, A. B. (1997). Translation initiation factor eIF4G mediates in vitro poly(A) tail-dependent translation. *Proceedings of the National Academy of Sciences of the United States of America*, *94*, 9046–9051. https://doi.org/10.1073/pnas.94.17.9046.

Tcherkezian, J., Cargnello, M., Romeo, Y., Huttlin, E. L., Lavoie, G., Gygi, S. P., et al. (2014). Proteomic analysis of cap-dependent translation identifies LARP1 as a key regulator of 5′TOP mRNA translation. *Genes & Development*, *28*, 357–371. https://doi.org/10.1101/gad.231407.113.

Temme, C., Simonelig, M., & Wahle, E. (2014). Deadenylation of mRNA by the CCR4-NOT complex in Drosophila: Molecular and developmental aspects. *Frontiers in Genetics*, *5*, 143. https://doi.org/10.3389/fgene.2014.00143.

Teng, T., Thomas, G., & Mercer, C. A. (2013). Growth control and ribosomopathies. *Current Opinion in Genetics & Development, 23*, 63–71. https://doi.org/10.1016/J.GDE.2013.02.001.

Texada, M. J., Jørgensen, A. F., Christensen, C. F., Koyama, T., Malita, A., Smith, D. K., et al. (2019). A fat-tissue sensor couples growth to oxygen availability by remotely controlling insulin secretion. *Nature Communications, 10*, 1955. https://doi.org/10.1038/s41467-019-09943-y.

Theunissen, T. W., & Jaenisch, R. (2017). Mechanisms of gene regulation in human embryos and pluripotent stem cells. *Development, 144*, 4496–4509. https://doi.org/10.1242/dev.157404.

Tirronen, M., Lahti, V.-P., Heino, T. I., & Roos, C. (1995). Two otu transcripts are selectively localised in Drosophila oogenesis by a mechanism that requires a function of the otu protein. *Mechanisms of Development, 52*, 65–75.

Tschochner, H., & Hurt, E. (2003). Pre-ribosomes on the road from the nucleolus to the cytoplasm. *Trends in Cell Biology, 13*, 255–263.

Twombly, V., Blackman, R. K., Jin, H., Graff, J. M., Padgett, R. W., & Gelbart, W. M. (1996). The TGF-beta signaling pathway is essential for Drosophila oogenesis. *Development, 122*, 1555–1565.

Umen, J. G., & Guthrie, C. (1995). The second catalytic step of pre-mRNA splicing. *RNA, 1*, 869.

Van Buskirk, C., & Schüpbach, T. (2002). Half pint regulates alternative splice site selection in Drosophila. *Developmental Cell, 2*, 343–353.

Vessey, J. P., Schoderboeck, L., Gingl, E., Luzi, E., Riefler, J., Di Leva, F., et al. (2010). Mammalian Pumilio 2 regulates dendrite morphogenesis and synaptic function. *Proceedings of the National Academy of Sciences of the United States of America, 107*, 3222–3227. https://doi.org/10.1073/pnas.0907128107.

Vlachos, A., & Muir, E. (2010). How I treat Diamond-Blackfan anemia. *Blood, 116*, 3715–3723. https://doi.org/10.1182/blood-2010-02-251090.

Wahl, M. C., Will, C. L., & Lührmann, R. (2009). The spliceosome: Design principles of a dynamic RNP machine. *Cell, 136*, 701–718.

Wang, Y., Liu, J., Huang, B. O., Xu, Y., Li, J., Huang, L., et al. (2015). Mechanism of alternative splicing and its regulation. *Biomedical Reports, 3*, 152–158.

Watanabe-Susaki, K., Takada, H., Enomoto, K., Miwata, K., Ishimine, H., Intoh, A., et al. (2014). Biosynthesis of ribosomal RNA in nucleoli regulates pluripotency and differentiation ability of pluripotent stem cells. *Stem Cells, 32*, 3099–3111.

Webster, P. J., Liang, L., Berg, C. A., Lasko, P., & Macdonald, P. M. (1997). Translational repressor bruno plays multiple roles in development and is widely conserved. *Genes & Development, 11*, 2510–2521. https://doi.org/10.1101/gad.11.19.2510.

Wei, Y., Bettedi, L., Kim, K., Ting, C.-Y., & Lilly, M. (2018). *The GATOR complex regulates an essential response to meiotic double-stranded breaks in Drosophila.* (p. 421206). bioRxiv.

Wei, Y., & Zheng, X. F. S. (2009). Sch 9 partially mediates TORC1 signaling to control ribosomal RNA synthesis. *Cell Cycle, 8*, 4085–4090.

Will, C. L., & Lührmann, R. (2001). Spliceosomal UsnRNP biogenesis, structure and function. *Current Opinion in Cell Biology, 13*, 290–301.

Will, C. L., & Lührmann, R. (2011). Spliceosome structure and function. *Cold Spring Harbor Perspectives in Biology, 3*, a003707.

Wullschleger, S., Loewith, R., & Hall, M. N. (2006). TOR signaling in growth and metabolism. *Cell, 124*, 471–484.

Xie, T. (2000). A niche maintaining germ line stem cells in the Drosophila ovary. *Science, 290*, 328–330. https://doi.org/10.1126/science.290.5490.328.

Xie, T., & Li, L. (2007). Stem cells and their niche: An inseparable relationship. *Development, 134*, 2001 LP–2006. https://doi.org/10.1242/dev.002022.

Xie, T., & Spradling, A. C. (1998). Decapentaplegic is essential for the maintenance and division of germline stem cells in the Drosophila ovary. *Cell, 94*, 251–260. https://doi.org/10.1016/S0092-8674(00)81424-5.

Xue, S., & Barna, M. (2012). Specialized ribosomes: A new frontier in gene regulation and organismal biology. *Nature Reviews Molecular Cell Biology, 13*, 355–369. https://doi.org/10.1038/nrm3359.

Yamashita, Y. M., & Fuller, M. T. (2005). Asymmetric stem cell division and function of the niche in the Drosophila male germ line. *International Journal of Hematology, 82*, 377–380. https://doi.org/10.1532/IJH97.05097.

Yan, D., & Perrimon, N. (2015). Spenito is required for sex determination in *Drosophila melanogaster. Proceedings of the National Academy of Sciences of the United States of America, 112*, 11606–11611.

Yang, S. Y., Baxter, E. M., & Van Doren, M. (2012). Phf7 controls male sex determination in the Drosophila germline. *Developmental Cell, 22*, 1041–1051.

Yang, Y., Hsu, P. J., Chen, Y.-S., & Yang, Y.-G. (2018). Dynamic transcriptomic m 6 A decoration: Writers, erasers, readers and functions in RNA metabolism. *Cell Research, 28*, 616.

Yelick, P. C., & Trainor, P. A. (2015). Ribosomopathies: Global process, tissue specific defects. *Rare Diseases, 3*, e1025185. https://doi.org/10.1080/21675511.2015.1025185.

Yerlikaya, S., Meusburger, M., Kumari, R., Huber, A., Anrather, D., Costanzo, M., et al. (2016). TORC1 and TORC2 work together to regulate ribosomal protein S6 phosphorylation in *Saccharomyces cerevisiae. Molecular Biology of the Cell, 27*, 397–409.

Yi, C., & Pan, T. (2011). Cellular dynamics of RNA modification. *Accounts of Chemical Research, 44*, 1380–1388.

Zamore, P. D., Bartel, D. P., Lehmann, R., & Williamson, J. R. (1999). The PUMILIO-RNA interaction: A single RNA-binding domain monomer recognizes a bipartite target sequence. *Biochemistry, 38*, 596–604. https://doi.org/10.1021/bi982264s.

Zemp, I., & Kutay, U. (2007). Nuclear export and cytoplasmic maturation of ribosomal subunits. *FEBS Letters, 581*, 2783–2793.

Zhang, Q., Shalaby, N. A., & Buszczak, M. (2014). Changes in rRNA transcription influence proliferation and cell fate within a stem cell lineage. *Science, 343*, 298–301.

Zhang, K., & Smith, G. W. (2015). Maternal control of early embryogenesis in mammals. *Reproduction, Fertility, and Development, 27*, 880–896. https://doi.org/10.1071/RD14441.

Zhao, G.-Q., & Garbers, D. L. (2002). Male germ cell specification and differentiation. *Developmental Cell, 2*, 537–547.

Zhao, B. S., & He, C. (2015). Pseudouridine in a new era of RNA modifications. *Cell Research, 25*, 153.

Further reading

Boerner, K., & Becker, P. B. (2016). Splice variants of the SWR1-type nucleosome remodeling factor domino have distinct functions during *Drosophila melanogaster* oogenesis. *Development, 143*, 3154–3167.

You, K. T., Park, J., & Kim, V. N. (2015). Role of the small subunit processome in the maintenance of pluripotent stem cells. *Genes & Development, 29*, 2004–2009. https://doi.org/10.1101/gad.267112.115.

CHAPTER TWO

Maternal inheritance of centromeres through the germline

Arunika Das[a,b], Ben E. Black[b,*], Michael A. Lampson[a,*]
[a]Department of Biology, University of Pennsylvania, Philadelphia, PA, United States
[b]Department of Biochemistry and Biophysics, Perelman School of Medicine, University of Pennsylvania, Philadelphia, PA, United States
*Corresponding authors: e-mail address: blackbe@pennmedicine.upenn.edu; lampson@sas.upenn.edu

Contents

1. Introduction 36
   1.1 Epigenetic specification of centromeres 36
   1.2 Self-propagating nature of CENP-A nucleosomes 37
2. How is CENP-A chromatin maintained? 38
   2.1 CENP-A stability 39
   2.2 Continual replenishment of CENP-A in germline 41
3. Biased inheritance of centromeres through female meiosis: Centromere drive 42
4. Regulating centromere size in early development 45
   4.1 Centromere-mediated genome elimination: Importance of equivalent centromeres 45
   4.2 Resolving centromere differences in early embryos 48
5. Conclusions and future perspectives 48
References 49

Abstract

The centromere directs chromosome segregation but is not itself genetically encoded. In most species, centromeres are epigenetically defined by the presence of a histone H3 variant CENP-A, independent of the underlying DNA sequence. Therefore, to maintain centromeres and ensure accurate chromosome segregation, CENP-A nucleosomes must be inherited across generations through the germline. In this chapter we discuss three aspects of maternal centromere inheritance. First, we propose mechanisms for maintaining CENP-A nucleosomes through the prolonged prophase arrest in mammalian oocytes. Second, we review mechanisms by which selfish centromeres bias their transmission through female meiosis. Third, we discuss regulation of centromere size through early embryonic development.

Current Topics in Developmental Biology, Volume 140
ISSN 0070-2153
https://doi.org/10.1016/bs.ctdb.2020.03.004

1. Introduction

1.1 Epigenetic specification of centromeres

Accurate segregation of eukaryotic chromosomes is directed by a locus called the centromere. In most eukaryotes, with budding yeast as a notable exception, centromeres are not encoded by a particular DNA sequence, but are instead epigenetically specified by the presence of a histone H3 variant centromere protein A (CENP-A), independent of the underlying DNA sequence (Allshire & Karpen, 2008; Black & Cleveland, 2011). However, centromeres are typically associated with characteristic DNA sequences ranging from a 125 bp sequence necessary and sufficient for "point" centromeres in budding yeast (Bloom & Carbon, 1982; Fitzgerald-Hayes, Clarke, & Carbon, 1982; Panzeri & Philippsen, 1982; Saunders, Fitzgerald-Hayes, & Bloom, 1988) to highly diverged repetitive or a mixture of repetitive and non-repetitive sequences (Locke, Segraves, Carbone, et al., 2003; Piras, Nergadze, Magnani, et al., 2010; Shang, Hori, Toyoda, et al., 2010) in "regional" centromeres. Regional centromeres in humans, for example, contain up to 5 Mb of 171 bp alpha-satellite repeats (Waye & Willard, 1987).

Initial evidence for epigenetic specification of centromeres was the existence of naturally occurring neocentromeres, found on complex DNA sequences (i.e., not repetitive) (Barry, Howman, Cancilla, et al., 1999; Choo, 1997; Depinet, Zackowski, Earnshaw, et al., 1997; Scott & Sullivan, 2014), that can be inherited through multiple generations, indicating their function in mitosis and meiosis (Amor, Bentley, Ryan, et al., 2004; Tyler-Smith, Gimelli, Giglio, et al., 1999). Neocentromeres recruit CENP-A and other centromere proteins (Bassett, Wood, Salimian, et al., 2010; Stellfox, Bailey, & Foltz, 2013; Voullaire, Slater, Petrovic, & Choo, 1993), suggesting that typical repetitive centromere DNA sequences are unnecessary for centromere specification and that centromere propagation is epigenetic and conferred by the presence of CENP-A nucleosomes. In fact, targeting CENP-A or its dedicated histone chaperone to non-centromeric chromatin is sufficient to form functional centromeres and recruit other kinetochore proteins in *Drosophila* and human cells (Barnhart, Kuich, Stellfox, et al., 2011; Logsdon, Gambogi, Liskovykh, et al., 2019; Mendiburo, Padeken, Fulop, et al., 2011). Taken together, these studies support the centrality of CENP-A nucleosomes in specifying and maintaining the centromere locus through multiple cell cycles and across generations.

1.2 Self-propagating nature of CENP-A nucleosomes

CENP-A chromatin is maintained at the same locus on each chromosome from one cell cycle to the next to preserve centromere identity, preventing loss from a single chromosome or duplication to make a functional "dicentric" chromosome that leads to chromosome breakage at mitosis (McClintock, 1941). In fact, centromeric chromatin is quite immobile on the timescale of organismal generations. For example, two different higher order α-satellite arrays on human chromosome 17 act as epialleles (Maloney, Sullivan, Matheny, et al., 2012). Each one inherits the parental centromere location, which is then maintained independently on the maternal and paternal chromosomes. To maintain centromeric chromatin through each cell cycle, CENP-A is partitioned equally between sister chromatids during S-phase DNA replication, reducing levels to half at each centromere, and subsequently restored to full levels by deposition of new CENP-A by a dedicated chaperone (HJURP in mammals, Scm3 in yeast and CAL1 in flies) (Jansen, Black, Foltz, & Cleveland, 2007; Lagana, Dorn, De Rop, et al., 2010; Schuh, Lehner, & Heidmann, 2007) (Fig. 1). Assembly of new CENP-A nucleosomes near the existing CENP-A nucleosomes preserves the location and is achieved by specific interactions of CENP-A assembly factors with constitutive centromere proteins (e.g., CENP-C or CENP-A itself). This process is restricted to G1 as phosphorylation of CENP-A assembly factors (HJURP and Mis18BP1) by CDK1/2 inhibits assembly in late S (following CENP-A protein expression; Shelby, Vafa, & Sullivan, 1997), G2 and M (Dunleavy, Roche, Tagami, et al., 2009; Foltz, Jansen, Bailey, et al., 2009; Hayashi, Fujita, Iwasaki, et al., 2004; Silva, Bodor, Stellfox, et al., 2012) (Fig. 1).

Because existing CENP-A nucleosomes direct the deposition of new ones, centromeres are self-propagating in nature. In current models of centromere inheritance, interactions between the CENP-A nucleosome and its assembly factors are expected to be stoichiometric, so that recruitment of assembly factors would be proportional to the number of CENP-A nucleosomes. Thus, after equal partitioning between sister chromatids in S-phase, each CENP-A nucleosome would direct deposition of a new one, maintaining constant level over many cell cycles. This view of cell cycle-coupled assembly of CENP-A chromatin has emerged largely from experiments in tissue culture cells (Chen & Mellone, 2016; De Rop, Padeganeh, & Maddox, 2012; Erhardt, Mellone, Betts, et al., 2008;

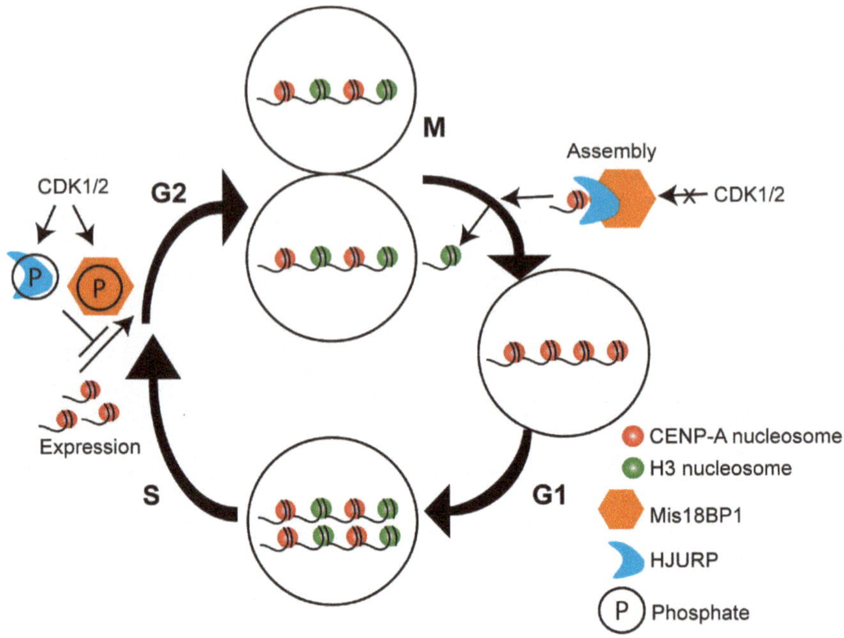

Fig. 1 Epigenetic inheritance cycle of CENP-A as established in cycling somatic cells. CENP-A levels are reduced to 50% in S and then restored by cell cycle-coupled chromatin assembly upon exit from mitosis, in G1. The CENP-A specific chaperone HJURP and its interacting partner Mis18BP1 are inhibited at S, G2 and M by CDK1/2 mediated phosphorylation.

Falk & Black, 2012). However, there are additional complexities to the transmission of CENP-A chromatin across generations via the germline. In this chapter, we discuss three key aspects of centromere inheritance from parent to progeny:

1. How is CENP-A chromatin maintained?
2. Biased inheritance of centromeres in female meiosis.
3. Regulation of centromere size in early development.

2. How is CENP-A chromatin maintained?

Centromere propagation through the germline requires the maintenance of CENP-A nucleosomes through gametogenesis, posing a challenge for the existing paradigm of centromere transmission in cycling somatic cells. Mammalian oocytes arrest in prophase I until meiotic resumption occurs, which could be up to 2 years in mice or decades in humans.

Therefore, the mammalian maternal germline faces the challenge of maintaining CENP-A chromatin through this prolonged prophase arrest. A parallel challenge exists in the male germline as CENP-A nucleosomes must be preferentially retained when other bulk histones are exchanged for protamines in a dramatic genome-wide reorganization (Bao & Bedford, 2016; Gaucher, Reynoird, Montellier, et al., 2010; Rathke, Baarends, Awe, & Renkawitz-Pohl, 2014). Here we propose two possible mechanisms to maintain CENP-A nucleosomes through the extended prophase arrest in oocytes: (1) stability of the chromatin-assembled CENP-A nucleosomes with little or no decay and/or (2) continual replenishment of CENP-A nucleosomes by new CENP-A assembly to compensate for decay during the prophase arrest.

2.1 CENP-A stability

Evidence from mammalian cycling somatic cells demonstrates that CENP-A nucleosomes are remarkably stable and persist at centromeres over several cell cycles with no detectable turnover (Bodor, Mata, Sergeev, et al., 2014; Bodor, Valente, Mata, et al., 2013). Furthermore, experiments with tagged CENP-A in plants and flies corroborate the idea that G1 assembled CENP-A chromatin can stably propagate centromere identity from one somatic cell cycle to the next (Lermontova, Koroleva, Rutten, et al., 2011; Schuh, Lehner, & Heidmann, 2007). In the mouse germline, an oocyte-specific conditional knockout of CENP-A early in the prophase I arrest does not affect fertility or CENP-A levels at oocyte centromeres. This result indicates that CENP-A is retained at centromeres during the prophase I arrest with no detectable assembly of newly expressed CENP-A during the reproductive lifespan of the animal, and that centromeric chromatin assembled prior to meiotic entry is sufficient for centromere function and inheritance (Smoak, Stein, Schultz, et al., 2016). Taken together, these studies highlight the remarkable stability of CENP-A nucleosomes at the centromere, but the factors that contribute to this stability are unclear.

One candidate for providing CENP-A stability is its intrinsic structural rigidity. Biophysical and structural studies have revealed that the internal dynamics of the $(CENP\text{-}A/H4)_2$ heterotetramer are different compared to its counterpart histone complex, the $(H3/H4)_2$ heterotetramer (Black, Foltz, Chakravarthy, et al., 2004; Sekulic, Bassett, Rogers, & Black, 2010). So-called "hydrophobic stitches" generated by substitutions of six amino acids in CENP-A relative to its counterpart, histone H3, provide conformational rigidity to the interface of CENP-A with its histone partner, H4 (Sekulic et al., 2010). Replacement of these six amino acids with the counterpart H3 residues reduces the steady-state level of accumulation at centromeres while leaving interactions with the CENP-A chaperone, HJURP, intact (Bassett, DeNizio, Barnhart-Dailey, et al., 2012). Could this conformational rigidity help retain CENP-A at centromeres during the prolonged prophase arrest in mammalian oocytes? Some clues are available from experiments in other systems. For instance, mutating one of the hydrophobic stitch residues in plants results in severely impaired localization of centromeric CENP-A (CENH3), consistent with a role of the hydrophobic interface in stabilizing CENP-A nucleosomes (Karimi-Ashtiyani, Ishii, Niessen, et al., 2015). Another H3 variant, the testis specific histone H3.5, forms an unstable nucleosome that has been attributed to the presence of a Leu residue at the interface with histone H4 (position 103 in human) instead of the bulkier hydrophobic phenylalanine residue present in canonical H3 or other H3 variants (Schenk, Jenke, Zilbauer, et al., 2011; Tachiwana, Kagawa, Osakabe, et al., 2010; Urahama, Harada, Maehara, et al., 2016).

Non-histone centromere proteins, such as the 16 constitutive centromere associated network (CCAN) proteins, could also contribute to CENP-A stability (Cheeseman & Desai, 2008; Hori, Shang, Takeuchi, & Fukagawa, 2012; Perpelescu & Fukagawa, 2011). Two of these proteins, CENP-C and CENP-N, are direct binding partners of CENP-A (Carroll, Silva, Godek, et al., 2009; Guse, Carroll, Moree, et al., 2011; Kato, Jiang, Zhou, et al., 2013) and contribute to CENP-A retention at centromeres. CENP-C can reshape CENP-A nucleosomes, generating stability at multiple locations within the folded histone core, and is important for retaining the pool of assembled CENP-A nucleosomes in mitotic cell cycles (Falk, Guo, Sekulic, et al., 2015; Falk, Lee, Sekulic, et al., 2016). In addition, flies with impaired CENP-C function have reduced CENP-A at centromeres in spermatids, suggesting a potential role in centromere maintenance during early meiosis (Kwenda, Collins, Dattoli, & Dunleavy, 2016). CENP-N contacts both CENP-A and nucleosomal DNA, cross-bridging the molecules in a manner that likely imparts stability to CENP-A nucleosomes in mitotically-cycling cells (Guo, Allu, Zandarashvili, et al., 2017; Pentakota, Zhou, Smith, et al., 2017). The role of either CENP-C or CENP-N is yet to be determined in oocytes.

2.2 Continual replenishment of CENP-A in germline

An alternative to a structural maintenance mechanism is the possibility that CENP-A nucleosomes are replenished by new assembly during the meiotic prophase arrest to compensate for potential CENP-A decay, thereby maintaining CENP-A chromatin. CENP-A can gradually assemble at a slow rate of 2% per day in prophase-arrested starfish oocytes cultured *in vitro*. Furthermore, when the essential assembly component Mis18 binding protein 1 (Mis18BP1) was depleted, these oocytes experienced segregation errors, suggesting that CENP-A nucleosomes are normally replenished gradually over time (Swartz, McKay, Su, et al., 2018). This finding is consistent with studies in *Drosophila* and plants that report gradual CENP-A nucleosome assembly in prophase I in the male germline (Dunleavy, Beier, Gorgescu, et al., 2012; Raychaudhuri, Dubruille, Orsi, et al., 2012; Schubert, Lermontova, & Schubert, 2014). However, these studies appear inconsistent with the apparently stable CENP-A nucleosomes observed in prophase-arrested mouse oocytes. CENP-A injected into oocytes cultured *in vitro* on short timescales (up to 2 days) does not localize to centromeres (Smoak et al., 2016), arguing against new CENP-A nucleosome assembly during the prophase arrest. In addition, aged mice have ~30% lower levels of CENP-A nucleosomes than young mice (Smoak et al., 2016), inconsistent with regulation through a homeostasis mechanism including nascent deposition. This reduction over time is not affected by deletion of the CENP-A gene early in the prophase arrest, ruling out deposition of newly expressed CENP-A protein during the extended arrest. One possibility to reconcile these conflicting results is that a stable pool of CENP-A protein slowly cycles on and off centromeres during the prophase arrest, which would not require new transcription. Thus, further studies are necessary to test a requirement for the CENP-A assembly machinery in prophase-arrested mammalian oocytes.

In contrast to the organisms discussed above, some do not require CENP-A maintenance in the female germline. For instance, worms remove CENP-A nucleosomes in the female germline and re-assemble them *de novo* during embryogenesis (Monen, Maddox, Hyndman, et al., 2005). Similarly, CENP-A is undetectable on *Arabidopsis* egg chromatin and is assembled *de novo* in the zygote (Ingouff, Rademacher, Holec, et al., 2010). The loss of CENP-A in the female germline in worms and plants is certainly in contrast to the robust stability of CENP-A chromatin in mammalian oocytes

(Smoak et al., 2016), suggesting that plants and worms have evolved an active CENP-A removal pathway in the germline that mammals—and presumably many other eukaryotes—lack.

3. Biased inheritance of centromeres through female meiosis: Centromere drive

Female meiosis is asymmetric, in that only one of the four possible division products become the egg while the others are degraded, providing a clear opportunity for selfish genetic elements to cheat, as any chromosome that can preferentially segregate to the egg has a transmission advantage (Chmátal, Schultz, Black, & Lampson, 2017; Pardo-Manuel de Villena & Sapienza, 2001). The first example of preferential transmission was observed in maize (Rhoades, 1942), where motor proteins recruited to repetitive selfish DNA elements (Dawe, Lowry, Gent, et al., 2018) drive preferential meiotic segregation. The term "meiotic drive" was later introduced to emphasize the key role of asymmetric female meiosis in transmission ratio distortion (Sandler & Novitski, 1957). The centromere, which mediates attachment to the spindle and directs segregation, has the potential to cheat in female meiosis and bias its own inheritance. Drive by diverged repetitive centromeric satellite DNA and the antagonistic coevolution of centromere proteins was proposed to explain the "centromere paradox": the unexpected rapid evolution of centromere DNA and essential centromere proteins with conserved functions (Henikoff, Ahmad, & Malik, 2001). According to the centromere drive hypothesis, selfish centromere DNA sequences that increase their transmission through female meiosis also incur fitness costs, possibly due to linked deleterious alleles or mis-segregation in male meiosis. Consistent with this hypothesis, expansion of repetitive centromere DNA is associated with a driving locus in monkeyflowers that exhibits a strong transmission bias in female meiosis and significantly reduced pollen viability in males (Fishman & Saunders, 2008). These costs would provide selective pressure promoting rapid evolution of centromere proteins to suppress drive.

Another example of centromere drive is preferential segregation of Robertsonian (Rb) fusions, common chromosomal rearrangements formed by two telocentric chromosomes (centromere at the end) joining at their centromeres to create one metacentric chromosome (internal centromere) (White, Bordewich, & Searle, 2010). Preferential transmission of Rb fusions correlates with the size of the fusion centromere relative to centromeres of the homologous unfused telocentrics, defined by levels of centromere proteins (Chmátal, Gabriel, Mitsainas, et al., 2014). In oocytes heterozygous for a single Rb fusion, both CENP-A and the major microtubule-binding protein NDC80 kinetochore complex component (NDC80)/highly expressed in cancer (HEC1) are enriched on the telocentric chromosomes that preferentially remain in the egg, relative to the homologous metacentric fusion that preferentially segregates to the polar body. Moreover, in a natural metacentric population that accumulated Rb fusions (CHPO strain), the fusion centromeres are enriched for these same proteins relative to the telocentric chromosomes. Further support for the drive hypothesis comes from holocentric plants (in which the centromere is present all over the chromosome) like *Luzula* where centromeres are not rapidly evolving since no single DNA element can bias its segregation, and species with symmetric meiosis that do not show signs of adaptive centromere evolution (Zedek & Bureš, 2016a, 2016b).

Conceptually, selfish segregation of centromeres in female meiosis depends on functional differences between centromeres on homologous chromosomes and asymmetric interactions with the metaphase I spindle that favors preferential orientation of the driving centromere toward the egg. These mechanisms have been studied in mouse systems with large differences in centromere DNA. For example, in a cross between two mouse strains (CF-1 and CHPO) with widely different amounts of centromeric minor satellite repeats, the larger centromeres have ~10-fold larger arrays of satellites, build larger kinetochores, and preferentially orient toward the egg side of the spindle. The small CHPO satellite arrays limit CENP-A nucleosome assembly to maintain small centromeres relative to the CF-1 centromeres even when they share the same nucleoplasm (Iwata-Otsubo, Dawicki-McKenna, Akera, et al., 2017). These larger centromeres exploit an asymmetry in α-tubulin tyrosination within the meiosis I spindle. Spindle asymmetry arises after migration to the cortex and depends on cortical polarization by the RAN GTPase and activation of the membrane-associated CDC42 GTPase (Akera, Chmátal, Trimm, et al., 2017).

To exploit spindle asymmetry, the larger centromeres preferentially detach from the cortical side and re-orient (or flip) toward the egg side. The larger centromeres initiate flipping at a higher frequency than the smaller centromeres and recruit more microtubule destabilizing activities, such as the depolymerizing kinesin-13, MCAK (mitotic centromere associated kinesin). MCAK favors tyrosinated microtubules, potentially explaining preferential flipping from the cortical side and more stable orientation toward the egg side (Akera et al., 2017; Akera, Trimm, & Lampson, 2019) (Fig. 2A). This strategy of enriching destabilizing activity to win in female meiosis was also found in an interspecific cross between *Mus musculus* and *Mus spretus*. Although *spretus* centromeres have substantially more centromere DNA repeats, the recruitment of centromere proteins to *musculus* and *spretus* centromeres was similar in the hybrid. The repeats are not limiting in either case, as CENP-A nucleosomes number in the hundreds out of several thousand centromeric nucleosomes in mouse (Bodor et al., 2014; Iwata-Otsubo et al., 2017) (except for CHPO). Instead, the *spretus* centromeres recruit more MCAK by a mechanism based on differences in Condensin localization to gain a transmission advantage over the *musculus* centromeres (Akera et al., 2019). Thus, although winning centromeres in both the intraspecific and interspecific hybrid mouse models

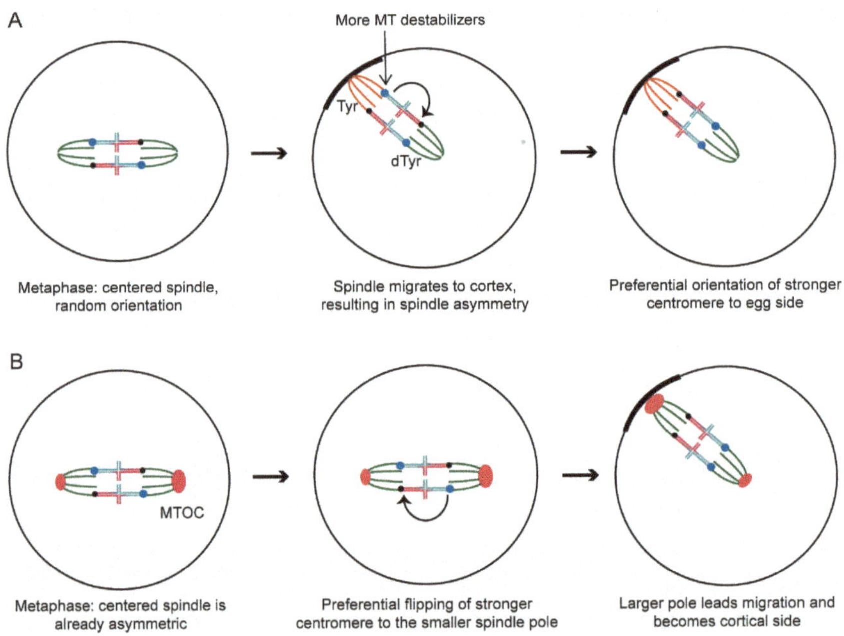

Fig. 2 Divergent mechanisms of centromere drive in female meiosis. (A) In the intraspecific CF1/CHPO hybrid oocyte, the spindle is asymmetric in tubulin tyrosination after cortical migration, and enrichment of microtubule destabilizers at stronger centromeres (blue) drives preferential flipping to the egg pole. (B) In the intraspecific BL6/SJL hybrid oocyte, the spindle poles (pink) are asymmetric prior to migration. In this case, stronger centromeres flip preferentially toward the smaller spindle pole and are retained in the egg as the larger pole leads migration toward the cortex.

exploit microtubule destabilizing activity to promote preferential orientation toward the egg pole in meiosis I, the mechanism for enriching destabilizers is distinct.

Another example of a diverged mechanism to achieve preferential inheritance was found in a different intraspecific cross (BL6 × SJL) with two chromosomes with driving centromeres (Fig. 2B). In this case the spindle was asymmetric with differences in tubulin and MTOC (microtubule organizing center) density between the two sides, prior to spindle migration to the cortex (Wu, Lane, Morgan, & Jones, 2018). The winning centromeres preferentially orient toward the pole with greater density of MTOCs, which is destined to become the egg side. These findings imply a different mechanism to establish spindle asymmetry, independent of cortical signals, as well as directional spindle migration to orient the larger spindle pole toward the cortex.

The centromere drive hypothesis predicts that systems with drive would also evolve suppression mechanisms (Henikoff et al., 2001). For example, any mechanism that reduces flipping events has the potential to negate drive if it prevents selfish centromeres from orienting toward the egg side of the spindle. This could be achieved by reducing the time available for flipping, which is borne out by the observation that the *spretus/musculus* hybrid does not show a bias in segregation unless anaphase is artificially delayed (Akera et al., 2019). As another potential suppression mechanism, centromere binding proteins could evolve to modulate binding affinity to expanded repetitive DNA and equalize centromeres. For example, if maternal and paternal centromeres differ in the zygote after fertilization, this asymmetry could be attenuated during early embryonic cell cycles to equalize centromeres. Therefore, evolution of centromere proteins to suppress centromere drive may be a key factor in determining centromere size, with a molecular "tug-of-war" between forces that generate expanded centromeres that can drive and those that limit centromere size while maintaining enough CENP-A chromatin to support centromere function.

4. Regulating centromere size in early development

4.1 Centromere-mediated genome elimination: Importance of equivalent centromeres

The importance of close equivalency in centromere size and function across chromosomes is exemplified in plant crosses with uniparental genome elimination, generating haploid plants (Comai & Tan, 2019; Ishii, Karimi-Ashtiyani, & Houben, 2016; Riera-Lizarazu, Rines, & Phillips, 1996; Sanei, Pickering, Kumke, et al., 2011). If parents are from different species, or one parent expresses a mutant CENP-A, centromere asymmetry in the zygote can result in elimination of the genome with the smaller/mutant centromere in early embryonic development (Ravi & Chan, 2010). This is not only an interesting phenomenon for understanding the role of centromeric chromatin in genome maintenance, but also useful as a rapid method to generate haploid plants to accelerate genetic manipulations (i.e., for agricultural biotech efforts) (Ishii et al., 2016; Ravi, Marimuthu, Tan, et al., 2014). The key factor driving centromere-mediated genome elimination appears to be an inability to resolve large differences in either size or function (i.e., wild type or mutant CENP-A) between the parental centromeres, whereby the larger/wild type centromere outcompetes the smaller/mutant counterpart (Fig. 3) (Wang & Dawe, 2018). During genome elimination, the presumed inability of the smaller/mutant centromere to recruit and assemble new CENP-A results in perturbed interactions with spindle microtubules and subsequent mis-segregation in the zygote (Sanei et al., 2011).

Several lines of evidence illustrate that large differences between parental centromeres can result in centromere-mediated genome elimination (Ishii et al., 2016; Wang & Dawe, 2018). For example, crossing an *Arabidopsis*

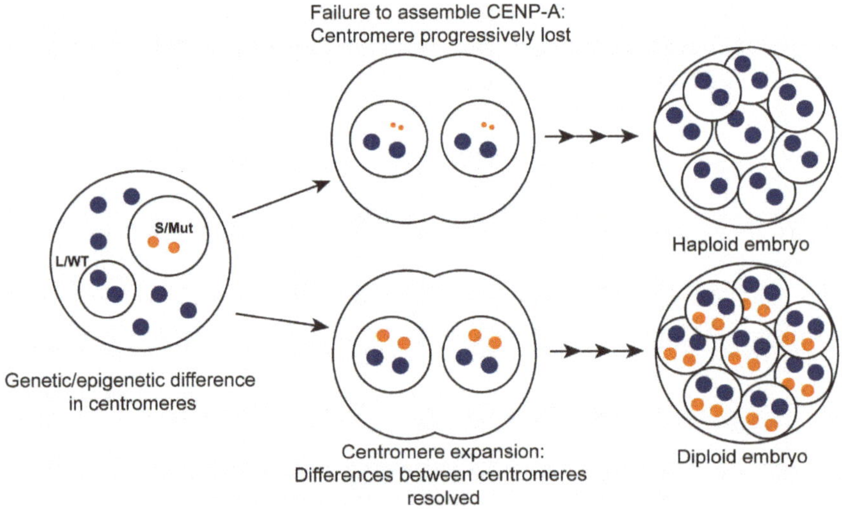

Fig. 3 Centromere-mediated genome elimination in embryos. Centromere dimorphism between parents can either be due to a size difference between two species (larger/L vs smaller/S) or due to one parent carrying a mutant CENP-A (WT/Mut). The smaller/mutant centromeres are often eliminated unless they expand during early development.

plant with mutant CENP-A to a wild type plant generates haploids containing only the genome with wild type CENP-A (Ravi & Chan, 2010). In this case the centromeres with mutant CENP-A nucleosomes lose to those with WT CENP-A in terms of nascent nucleosome assembly in each cell cycle or stability once assembled (Ravi, Shibata, Ramahi, et al., 2011). The embryos from this cross have high aneuploidy rates, along with haploidy, highlighting the fact that centromere size asymmetry in early development can lead to segregation errors. Point mutations in the histone fold domain of CENP-A (L130F in Arabidopsis or L106F in sugar beet, corresponding to the known hydrophobic stitch residue [see above] L91 in humans), which greatly reduce centromere localization of the mutant CENP-A nucleosomes in gametes, can also induce haploids but at a lower frequency (Karimi-Ashtiyani et al., 2015). Another classic case of genome elimination is a cross between *H. vulgare* and *H. bulbosum*, where despite having comparable centromere sizes in the two parents, haploids are still generated at a high rate with the *H. bulbosum* genome eliminated due to a parent-specific loss of CENP-A on the *H. bulbosum* chromosomes (Sanei et al., 2011). This is a unique example of genome elimination occurring due to differing properties but not size of CENP-A chromatin, such that one is incompatible for templating assembly of the other, leading to progressive loss of CENP-A on the *H. bulbosum* chromosomes.

What determines whether the genome with smaller/mutant centromeres will be eliminated? The balance between factors that influence either degradation of the smaller/mutant centromere or expansion through *de novo* CENP-A assembly may determine whether that genome will be lost or survive in the cross (Wang & Dawe, 2018). For example, a cross between oat (larger centromere species) and maize (smaller centromere species) usually results in haploid oat plants with complete elimination of the maize genome early in development (Kynast, Riera-Lizarazu, Vales, et al., 2001; Riera-Lizarazu et al., 1996), except in rare cases where the maize genome can survive at low frequencies in oat when the maize centromeres expand, thus reducing the size disparity (Wang, Wu, Zhang, et al., 2014). Taken together, these studies support the idea that differences between parental centromeres in the embryo arising from disparate centromere size, mutations in CENP-A nucleosomes or incompatibility between CENP-A from different species are detrimental to development of the organism. By extension, mechanisms that reduce centromere size dimorphism are likely crucial to avoid genome elimination or aneuploidy. Such mechanisms would also suppress centromere drive by equalizing maternal and paternal centromeres in the embryo.

4.2 Resolving centromere differences in early embryos

In theory, centromeres of disparate size might equalize by either a redistribution of the chromatin bound fraction to achieve comparable centromere sizes on all chromosomes or specific expansion of the smaller centromere by preferential assembly. Related to this point, following reduction in CENP-A levels by RNAi-mediated knockdown in *Drosophila* sperm, progeny maintain reduced levels of CENP-A chromatin even when CENP-A expression is restored (Raychaudhuri et al., 2012). This result is consistent with the idea that the abundance of CENP-A nucleosomes provides the epigenetic information to set the size of the centromere in the next generation. However, even though CENP-A is knocked down only in the sperm and not the egg, CENP-A levels in the progeny are reduced on both maternal and paternal centromeres of the autosomes, suggesting a mechanism to equalize CENP-A levels. Perhaps the fly example is a less extreme case of the instances in other species (*C. elegans* and *A. thaliana*, mentioned above) where CENP-A is lost during reproduction and subsequently "reset" during embryogenesis (Ingouff et al., 2010; Monen et al., 2005). Conversely, in centromere drive systems, asymmetry between centromeres of homologous chromosomes persists, ultimately leading to biased segregation in female meiosis. In such cases, for example the mouse hybrids discussed above, differences in repetitive centromere DNA may contribute to the persistent asymmetry (Akera et al., 2019; Iwata-Otsubo et al., 2017). Overall, these studies raise questions (see below) related to CENP-A chromatin inheritance and mechanisms to eliminate centromere asymmetry. In particular, the balance of genetic (centromere DNA sequence) and epigenetic (differences in CENP-A) contributions to centromere inheritance remain unclear.

5. Conclusions and future perspectives

Overall, although we know that centromere location is defined by an epigenetic mark that directs its own propagation through cell divisions (and through organismal generations in many eukaryotic species), the exact mechanisms by which centromeres persist in the germline are still unclear. Some outstanding issues for the field to address in the coming years are:

(1) What are the molecular factors that contribute to CENP-A nucleosome stability through the extended prophase arrest in the mammalian female germline, and how does nascent CENP-A assembly contribute to maintaining this epigenetic mark?

(2) How do centromeres bias their segregation in female meiosis? Mouse models reveal the importance of asymmetry both in microtubule destabilizing activity between centromeres of homologous chromosomes and between the two sides of the meiotic spindle. How winning centromeres preferentially destabilize interactions with the cortical side of the spindle is unclear. Furthermore, little is known about fitness costs associated with driving centromeres or mechanisms that may have evolved to minimize these costs. New cell biological and genetic model systems may provide insights into these questions.

(3) What are the mechanisms regulating centromere size in the embryo? Studies of uniparental genome elimination suggest that centromeres either reach comparable size or smaller centromeres are inactivated by loss of CENP-A chromatin (Wang & Dawe, 2018). Direct evidence for equalization in the embryo is limited, however, and exploring CENP-A assembly on maternal and paternal chromosomes in early embryos promises to provide insight into how CENP-A chromatin is regulated on the two parental genomes. A putative equalization mechanism must differ from the model of centromere propagation purely by stoichiometric interactions between CENP-A and its assembly factors, and may therefore reveal a new paradigm for centromere inheritance. Finally, an important model to test is whether the mechanisms that reduce centromere strength imbalances in embryos evolved to suppress centromere drive.

References

Akera, T., Chmátal, L., Trimm, E., et al. (2017). Spindle asymmetry drives non-Mendelian chromosome segregation. *Science*, *358*, 668–672. https://doi.org/10.1126/science.aan0092.

Akera, T., Trimm, E., & Lampson, M. A. (2019). Molecular strategies of meiotic cheating by selfish centromeres. *Cell*, *178*, 1132–1144, e10. https://doi.org/10.1016/j.cell.2019.07.001.

Allshire, R. C., & Karpen, G. H. (2008). Epigenetic regulation of centromeric chromatin: Old dogs, new tricks? *Nature Reviews. Genetics*, *9*, 923–937. https://doi.org/10.1038/nrg2466.

Amor, D. J., Bentley, K., Ryan, J., et al. (2004). Human centromere repositioning "in progress". *Proceedings of the National Academy of Sciences*, *101*, 6542–6547. https://doi.org/10.1073/pnas.0308637101.

Bao, J., & Bedford, M. T. (2016). Epigenetic regulation of the histone-to-protamine transition during spermiogenesis. *Reproduction*, *151*, R55–R70. https://doi.org/10.1530/REP-15-0562.

Barnhart, M. C., Kuich, P. H. J. L., Stellfox, M. E., et al. (2011). HJURP is a CENP-A chromatin assembly factor sufficient to form a functional de novo kinetochore. *The Journal of Cell Biology*, *194*, 229–243. https://doi.org/10.1083/jcb.201012017.

Barry, A. E., Howman, E. V., Cancilla, M. R., et al. (1999). Sequence analysis of an 80 kb human neocentromere. *Human Molecular Genetics*, *8*, 217–227. https://doi.org/10.1093/hmg/8.2.217.

Bassett, E. A., DeNizio, J., Barnhart-Dailey, M. C., et al. (2012). HJURP uses distinct CENP-A surfaces to recognize and to stabilize CENP-A/histone H4 for centromere assembly. *Developmental Cell*, *22*, 749–762. https://doi.org/10.1016/j.devcel.2012.02.001.

Bassett, E. A., Wood, S., Salimian, K. J., et al. (2010). Epigenetic centromere specification directs aurora B accumulation but is insufficient to efficiently correct mitotic errors. *The Journal of Cell Biology*, *190*, 177–185. https://doi.org/10.1083/jcb.201001035.

Black, B. E., & Cleveland, D. W. (2011). Epigenetic centromere propagation and the nature of CENP-A nucleosomes. *Cell*, *144*, 471–479. https://doi.org/10.1016/j.cell.2011.02.002.

Black, B. E., Foltz, D. R., Chakravarthy, S., et al. (2004). Structural determinants for generating centromeric chromatin. *Nature*, *430*, 578–582. https://doi.org/10.1038/nature02766.

Bloom, K. S., & Carbon, J. (1982). Yeast centromere DNA is in a unique and highly ordered structure in chromosomes and small circular minichromosomes. *Cell*, *29*, 305–317.

Bodor, D. L., Mata, J. F., Sergeev, M., et al. (2014). The quantitative architecture of centromeric chromatin. *eLife*, *3*, e02137. https://doi.org/10.7554/eLife.02137.

Bodor, D. L., Valente, L. P., Mata, J. F., et al. (2013). Assembly in G1 phase and long-term stability are unique intrinsic features of CENP-A nucleosomes. *Molecular Biology of the Cell*, *24*, 923–932. https://doi.org/10.1091/mbc.E13-01-0034.

Carroll, C. W., Silva, M. C. C., Godek, K. M., et al. (2009). Centromere assembly requires the direct recognition of CENP-A nucleosomes by CENP-N. *Nature Cell Biology*, *11*, 896–902. https://doi.org/10.1038/ncb1899.

Cheeseman, I. M., & Desai, A. (2008). Molecular architecture of the kinetochore–microtubule interface. *Nature Reviews. Molecular Cell Biology*, *9*, 33–46. https://doi.org/10.1038/nrm2310.

Chen, C.-C., & Mellone, B. G. (2016). Chromatin assembly: Journey to the CENter of the chromosome. *The Journal of Cell Biology*, *214*, 13–24. https://doi.org/10.1083/jcb.201605005.

Chmátal, L., Gabriel, S. I., Mitsainas, G. P., et al. (2014). Centromere strength provides the cell biological basis for meiotic drive and karyotype evolution in mice. *Current Biology*, *24*, 2295–2300. https://doi.org/10.1016/j.cub.2014.08.017.

Chmátal, L., Schultz, R. M., Black, B. E., & Lampson, M. A. (2017). Cell biology of cheating-transmission of centromeres and other selfish elements through asymmetric meiosis. *Progress in Molecular and Subcellular Biology*, *56*, 377–396.

Choo, K. H. (1997). Centromere DNA dynamics: Latent centromeres and neocentromere formation. *American Journal of Human Genetics*, *61*, 1225–1233. https://doi.org/10.1086/301657.

Comai, L., & Tan, E. H. (2019). Haploid induction and genome instability. *Trends in Genetics*, *35*, 791–803. https://doi.org/10.1016/j.tig.2019.07.005.

Dawe, R. K., Lowry, E. G., Gent, J. I., et al. (2018). A kinesin-14 motor activates neocentromeres to promote meiotic drive in maize. *Cell*, *173*, 839–850, e18. https://doi.org/10.1016/j.cell.2018.03.009.

De Rop, V., Padeganeh, A., & Maddox, P. S. (2012). CENP-A: The key player behind centromere identity, propagation, and kinetochore assembly. *Chromosoma*, *121*, 527–538. https://doi.org/10.1007/s00412-012-0386-5.

Depinet, T. W., Zackowski, J. L., Earnshaw, W. C., et al. (1997). Characterization of neocentromeres in marker chromosomes lacking detectable alpha-satellite DNA. *Human Molecular Genetics*, *6*, 1195–1204.

Dunleavy, E. M., Beier, N. L., Gorgescu, W., et al. (2012). The cell cycle timing of centromeric chromatin assembly in Drosophila meiosis is distinct from mitosis yet requires CAL1 and CENP-C. *PLoS Biology, 10,* 1–16. https://doi.org/10.1371/journal.pbio.1001460.

Dunleavy, E. M., Roche, D., Tagami, H., et al. (2009). HJURP is a cell-cycle-dependent maintenance and deposition factor of CENP-A at centromeres. *Cell, 137,* 485–497. https://doi.org/10.1016/j.cell.2009.02.040.

Erhardt, S., Mellone, B. G., Betts, C. M., et al. (2008). Genome-wide analysis reveals a cell cycle-dependent mechanism controlling centromere propagation. *The Journal of Cell Biology, 183,* 805–818. https://doi.org/10.1083/jcb.200806038.

Falk, S. J., & Black, B. E. (2012). Centromeric chromatin and the pathway that drives its propagation. *Biochimica et Biophysica Acta, 1819,* 313–321. https://doi.org/10.1016/j.bbagrm.2011.11.002.

Falk, S. J., Guo, L. Y., Sekulic, N., et al. (2015). Chromosomes. CENP-C reshapes and stabilizes CENP-A nucleosomes at the centromere. *Science, 348,* 699–703. https://doi.org/10.1126/science.1259308.

Falk, S. J., Lee, J., Sekulic, N., et al. (2016). CENP-C directs a structural transition of CENP-A nucleosomes mainly through sliding of DNA gyres. *Nature Structural & Molecular Biology, 23,* 204–208. https://doi.org/10.1038/nsmb.3175.

Fishman, L., & Saunders, A. (2008). Centromere-associated female meiotic drive entails male fitness costs in monkeyflowers. *Science, 322,* 1559–1562. https://doi.org/10.1126/science.1161406.

Fitzgerald-Hayes, M., Clarke, L., & Carbon, J. (1982). Nucleotide sequence comparisons and functional analysis of yeast centromere DNAs. *Cell, 29,* 235–244. https://doi.org/10.1016/0092-8674(82)90108-8.

Foltz, D. R., Jansen, L. E. T., Bailey, A. O., et al. (2009). Centromere-specific assembly of CENP-A nucleosomes is mediated by HJURP. *Cell, 137,* 472–484. https://doi.org/10.1016/j.cell.2009.02.039.

Gaucher, J., Reynoird, N., Montellier, E., et al. (2010). From meiosis to postmeiotic events: The secrets of histone disappearance. *The FEBS Journal, 277,* 599–604. https://doi.org/10.1111/j.1742-4658.2009.07504.x.

Guo, L. Y., Allu, P. K., Zandarashvili, L., et al. (2017). Centromeres are maintained by fastening CENP-A to DNA and directing an arginine anchor-dependent nucleosome transition. *Nature Communications, 8,* 15775. https://doi.org/10.1038/ncomms15775.

Guse, A., Carroll, C. W., Moree, B., et al. (2011). In vitro centromere and kinetochore assembly on defined chromatin templates. *Nature, 477,* 354–358. https://doi.org/10.1038/nature10379.

Hayashi, T., Fujita, Y., Iwasaki, O., et al. (2004). Mis16 and Mis18 are required for CENP-A loading and histone deacetylation at centromeres. *Cell, 118,* 715–729. https://doi.org/10.1016/j.cell.2004.09.002.

Henikoff, S., Ahmad, K., & Malik, H. S. (2001). The centromere paradox: Stable inheritance with rapidly evolving DNA. *Science, 293,* 1098–1102. https://doi.org/10.1126/science.1062939.

Hori, T., Shang, W.-H., Takeuchi, K., & Fukagawa, T. (2012). The CCAN recruits CENP-A to the centromere and forms the structural core for kinetochore assembly. *The Journal of Cell Biology, 200,* 45–60. https://doi.org/10.1083/jcb.201210106.

Ingouff, M., Rademacher, S., Holec, S., et al. (2010). Zygotic resetting of the HISTONE 3 variant repertoire participates in epigenetic reprogramming in Arabidopsis. *Current Biology, 20,* 2137–2143. https://doi.org/10.1016/j.cub.2010.11.012.

Ishii, T., Karimi-Ashtiyani, R., & Houben, A. (2016). Haploidization via chromosome elimination: Means and mechanisms. *Annual Review of Plant Biology, 67,* 421–438. https://doi.org/10.1146/annurev-arplant-043014-114714.

Iwata-Otsubo, A., Dawicki-McKenna, J. M., Akera, T., et al. (2017). Expanded satellite repeats amplify a discrete CENP-A nucleosome assembly site on chromosomes that drive in female meiosis. *Current Biology, 27*, 2365–2373, e8. https://doi.org/10.1016/j.cub.2017.06.069.

Jansen, L. E. T., Black, B. E., Foltz, D. R., & Cleveland, D. W. (2007). Propagation of centromeric chromatin requires exit from mitosis. *The Journal of Cell Biology, 176*, 795–805. https://doi.org/10.1083/jcb.200701066.

Karimi-Ashtiyani, R., Ishii, T., Niessen, M., et al. (2015). Point mutation impairs centromeric CENH3 loading and induces haploid plants. *Proceedings of the National Academy of Sciences of the United States of America, 112*, 11211–11216. https://doi.org/10.1073/pnas.1504333112.

Kato, H., Jiang, J., Zhou, B.-R., et al. (2013). A conserved mechanism for centromeric nucleosome recognition by centromere protein CENP-C. *Science, 340*, 1110–1113. https://doi.org/10.1126/science.1235532.

Kwenda, L., Collins, C. M., Dattoli, A. A., & Dunleavy, E. M. (2016). Nucleolar activity and CENP-C regulate CENP-A and CAL1 availability for centromere assembly in meiosis. *Development, 143*, 1400–1412. https://doi.org/10.1242/dev.130625.

Kynast, R. G., Riera-Lizarazu, O., Vales, M. I., et al. (2001). A complete set of maize individual chromosome additions to the oat genome. *Plant Physiology, 125*, 1216–1227. https://doi.org/10.1104/pp.125.3.1216.

Lagana, A., Dorn, J. F., De Rop, V., et al. (2010). A small GTPase molecular switch regulates epigenetic centromere maintenance by stabilizing newly incorporated CENP-A. *Nature Cell Biology, 12*, 1186–1193. https://doi.org/10.1038/ncb2129.

Lermontova, I., Koroleva, O., Rutten, T., et al. (2011). Knockdown of CENH3 in Arabidopsis reduces mitotic divisions and causes sterility by disturbed meiotic chromosome segregation. *The Plant Journal, 68*, 40–50. https://doi.org/10.1111/j.1365-313X.2011.04664.x.

Locke, D. P., Segraves, R., Carbone, L., et al. (2003). Large-scale variation among human and great ape genomes determined by array comparative genomic hybridization. *Genome Research, 13*, 347–357. https://doi.org/10.1101/gr.1003303.

Logsdon, G. A., Gambogi, C. W., Liskovykh, M. A., et al. (2019). Human artificial chromosomes that bypass centromeric DNA. *Cell, 178*, 624–639, e19. https://doi.org/10.1016/j.cell.2019.06.006.

Maloney, K. A., Sullivan, L. L., Matheny, J. E., et al. (2012). Functional epialleles at an endogenous human centromere. *Proceedings of the National Academy of Sciences of the United States of America, 109*, 13704–13709. https://doi.org/10.1073/pnas.1203126109.

McClintock, B. (1941). The stability of broken ends of chromosomes in Zea mays. *Genetics, 26*, 234–282.

Mendiburo, M. J., Padeken, J., Fulop, S., et al. (2011). Drosophila CENH3 is sufficient for centromere formation. *Science, 334*, 686–690. https://doi.org/10.1126/science.1206880.

Monen, J., Maddox, P. S., Hyndman, F., et al. (2005). Differential role of CENP-A in the segregation of holocentric *C. elegans* chromosomes during meiosis and mitosis. *Nature Cell Biology, 7*, 1248–1255. https://doi.org/10.1038/ncb1331.

Panzeri, L., & Philippsen, P. (1982). Centromeric DNA from chromosome VI in *Saccharomyces cerevisiae* strains. *The EMBO Journal, 1*, 1605–1611.

Pardo-Manuel de Villena, F., & Sapienza, C. (2001). Nonrandom segregation during meiosis: The unfairness of females. *Mammalian Genome, 12*, 331–339. https://doi.org/10.1007/s003350040003.

Pentakota, S., Zhou, K., Smith, C., et al. (2017). Decoding the centromeric nucleosome through CENP-N. *eLife, 6*, e33442. https://doi.org/10.7554/eLife.33442.

Perpelescu, M., & Fukagawa, T. (2011). The ABCs of CENPs. *Chromosoma, 120*, 425–446. https://doi.org/10.1007/s00412-011-0330-0.

Piras, F. M., Nergadze, S. G., Magnani, E., et al. (2010). Uncoupling of satellite DNA and centromeric function in the genus Equus. *PLoS Genetics*, *6*, e1000845. https://doi.org/10.1371/journal.pgen.1000845.

Rathke, C., Baarends, W. M., Awe, S., & Renkawitz-Pohl, R. (2014). Chromatin dynamics during spermiogenesis. *Biochimica et Biophysica Acta*, *1839*, 155–168. https://doi.org/10.1016/j.bbagrm.2013.08.004.

Ravi, M., & Chan, S. W. L. (2010). Haploid plants produced by centromere-mediated genome elimination. *Nature*, *464*, 615–618. https://doi.org/10.1038/nature08842.

Ravi, M., Marimuthu, M. P. A., Tan, E. H., et al. (2014). A haploid genetics toolbox for *Arabidopsis thaliana*. *Nature Communications*, *5*, 5334. https://doi.org/10.1038/ncomms6334.

Ravi, M., Shibata, F., Ramahi, J. S., et al. (2011). Meiosis-specific loading of the centromere-specific histone CENH3 in *Arabidopsis thaliana*. *PLoS Genetics*, 7, e1002121. https://doi.org/10.1371/journal.pgen.1002121.

Raychaudhuri, N., Dubruille, R., Orsi, G. A., et al. (2012). Transgenerational propagation and quantitative maintenance of paternal centromeres depends on Cid/Cenp-A presence in Drosophila sperm. *PLoS Biology*, *10*, e1001434. https://doi.org/10.1371/journal.pbio.1001434.

Rhoades, M. M. (1942). Preferential segregation in maize. *Genetics*, *27*, 395–407.

Riera-Lizarazu, O., Rines, H. W., & Phillips, R. L. (1996). Cytological and molecular characterization of oat x maize partial hybrids. *Theoretical and Applied Genetics*, *93*, 123–135. https://doi.org/10.1007/BF00225737.

Sandler, L., & Novitski, E. (1957). Meiotic drive as an evolutionary force. *The American Naturalist*, *91*, 105–110.

Sanei, M., Pickering, R., Kumke, K., et al. (2011). Loss of centromeric histone H3 (CENH3) from centromeres precedes uniparental chromosome elimination in interspecific barley hybrids. *Proceedings of the National Academy of Sciences of the United States of America*, *108*, E498–E505. https://doi.org/10.1073/pnas.1103190108.

Saunders, M., Fitzgerald-Hayes, M., & Bloom, K. (1988). Chromatin structure of altered yeast centromeres. *Proceedings of the National Academy of Sciences of the United States of America*, *85*, 175–179.

Schenk, R., Jenke, A., Zilbauer, M., et al. (2011). H3.5 is a novel hominid-specific histone H3 variant that is specifically expressed in the seminiferous tubules of human testes. *Chromosoma*, *120*, 275–285. https://doi.org/10.1007/s00412-011-0310-4.

Schubert, V., Lermontova, I., & Schubert, I. (2014). Loading of the centromeric histone H3 variant during meiosis—How does it differ from mitosis? *Chromosoma*, *123*, 491–497. https://doi.org/10.1007/s00412-014-0466-9.

Schuh, M., Lehner, C. F., & Heidmann, S. (2007). Incorporation of Drosophila CID/CENP-A and CENP-C into centromeres during early embryonic anaphase. *Current Biology*, *17*, 237–243. https://doi.org/10.1016/j.cub.2006.11.051.

Scott, K. C., & Sullivan, B. A. (2014). Neocentromeres: A place for everything and everything in its place. *Trends in Genetics*, *30*, 66–74. https://doi.org/10.1016/j.tig.2013.11.003.

Sekulic, N., Bassett, E. A., Rogers, D. J., & Black, B. E. (2010). The structure of (CENP-A–H4)2 reveals physical features that mark centromeres. *Nature*, *467*, 347–351. https://doi.org/10.1038/nature09323.

Shang, W.-H., Hori, T., Toyoda, A., et al. (2010). Chickens possess centromeres with both extended tandem repeats and short non-tandem-repetitive sequences. *Genome Research*, *20*, 1219–1228. https://doi.org/10.1101/gr.106245.110.

Shelby, R. D., Vafa, O., & Sullivan, K. F. (1997). Assembly of CENP-A into centromeric chromatin requires a cooperative array of nucleosomal DNA contact sites. *The Journal of Cell Biology*, *136*, 501–513. https://doi.org/10.1083/jcb.136.3.501.

Silva, M. C. C., Bodor, D. L., Stellfox, M. E., et al. (2012). Cdk activity couples epigenetic centromere inheritance to cell cycle progression. *Developmental Cell*, *22*, 52–63. https://doi.org/10.1016/j.devcel.2011.10.014.

Smoak, E. M., Stein, P., Schultz, R. M., et al. (2016). Long-term retention of CENP-A nucleosomes in mammalian oocytes underpins transgenerational inheritance of centromere identity. *Current Biology*, *26*, 1110–1116. https://doi.org/10.1016/j.cub.2016.02.061.

Stellfox, M. E., Bailey, A. O., & Foltz, D. R. (2013). Putting CENP-A in its place. *Cellular and Molecular Life Sciences*, *70*, 387–406. https://doi.org/10.1007/s00018-012-1048-8.

Swartz, S. Z., McKay, L. S., Su, K.-C., et al. (2018). *Quiescent cells actively replenish CENP-A nucleosomes to maintain centromere identity and proliferative potential.* bioRxiv, 433391. https://doi.org/10.1101/433391.

Tachiwana, H., Kagawa, W., Osakabe, A., et al. (2010). Structural basis of instability of the nucleosome containing a testis-specific histone variant, human H3T. *Proceedings of the National Academy of Sciences of the United States of America*, *107*, 10454–10459. https://doi.org/10.1073/pnas.1003064107.

Tyler-Smith, C., Gimelli, G., Giglio, S., et al. (1999). Transmission of a fully functional human neocentromere through three generations. *American Journal of Human Genetics*, *64*, 1440–1444. https://doi.org/10.1086/302380.

Urahama, T., Harada, A., Maehara, K., et al. (2016). Histone H3.5 forms an unstable nucleosome and accumulates around transcription start sites in human testis. *Epigenetics & Chromatin*, *9*, 2. https://doi.org/10.1186/s13072-016-0051-y.

Voullaire, L. E., Slater, H. R., Petrovic, V., & Choo, K. H. A. (1993). A functional marker centromere with no detectable alpha-satellite, satellite III, or CENP-B protein: Activation of a latent centromere? *American Journal of Human Genetics*, *52*, 1153–1163.

Wang, N., & Dawe, R. K. (2018). Centromere size and its relationship to haploid formation in plants. *Molecular Plant*, *11*, 398–406.

Wang, K., Wu, Y., Zhang, W., et al. (2014). Maize centromeres expand and adopt a uniform size in the genetic background of oat. *Genome Research*, *24*, 107–116. https://doi.org/10.1101/gr.160887.113.

Waye, J. S., & Willard, H. F. (1987). Nucleotide sequence heterogeneity of alpha satellite repetitive DNA: A survey of alphoid sequences from different human chromosomes. *Nucleic Acids Research*, *15*, 7549–7569. https://doi.org/10.1093/nar/15.18.7549.

White, T. A., Bordewich, M., & Searle, J. B. (2010). A network approach to study karyotypic evolution: The chromosomal races of the common shrew (*Sorex araneus*) and house mouse (*Mus musculus*) as model systems. *Systematic Biology*, *59*, 262–276. https://doi.org/10.1093/sysbio/syq004.

Wu, T., Lane, S. I. R., Morgan, S. L., & Jones, K. T. (2018). Spindle tubulin and MTOC asymmetries may explain meiotic drive in oocytes. *Nature Communications*, *9*, 2952. https://doi.org/10.1038/s41467-018-05338-7.

Zedek, F., & Bureš, P. (2016a). Absence of positive selection on CenH3 in Luzula suggests that holokinetic chromosomes may suppress centromere drive. *Annals of Botany*, *118*, 1347–1352. https://doi.org/10.1093/aob/mcw186.

Zedek, F., & Bureš, P. (2016b). CenH3 evolution reflects meiotic symmetry as predicted by the centromere drive model. *Scientific Reports*, *6*, 33308. https://doi.org/10.1038/srep33308.

CHAPTER THREE

Signaling between somatic follicle cells and the germline patterns the egg and embryo of *Drosophila*

Julie A. Merkle[a], Julia Wittes[b], Trudi Schüpbach[c,*]

[a]Department of Biology, University of Evansville, Evansville, IN, United States
[b]Department of Biological Sciences, Columbia University, New York, NY, United States
[c]Department of Molecular Biology, Princeton University, Princeton, NJ, United States
[*]Corresponding author: e-mail address: schupbac@princeton.edu

Contents

1. Introduction 56
2. Formation of the egg chamber 56
3. Selection of the oocyte 59
4. Germline-soma and soma-soma signaling pattern the polar and stalk cells and pre-polarize the newly formed oocyte 60
5. Anterior–posterior axis formation involves signaling between germline and follicle cells 62
6. Dorsal-ventral polarity establishment requires oocyte nuclear migration and signaling between the oocyte and follicle cells 67
7. Terminal patterning of the embryo requires follicle cell activity 72
8. Open questions in germline-follicle cell communication during *Drosophila* oogenesis and early development 74

Acknowledgments 75
References 76

Abstract

In *Drosophila*, specification of the embryonic body axes requires signaling between the germline and the somatic follicle cells. These signaling events are necessary to properly localize embryonic patterning determinants in the egg or eggshell during oogenesis. There are three maternal patterning systems that specify the anterior–posterior axis, and one that establishes the dorsal-ventral axis. We will first review oogenesis, focusing on the establishment of the oocyte and nurse cells and patterning of the follicle cells into different subpopulations. We then describe how two coordinated signaling events between the oocyte and follicle cells establish polarity of the oocyte and localize the anterior determinant *bicoid*, the posterior determinant *oskar*, and Gurken/epidermal growth factor (EGF), which breaks symmetry to initiate dorsal-ventral axis establishment. Next, we review how dorsal-ventral asymmetry of the follicle cells is transmitted to the embryo. This process also involves Gurken–EGF receptor (EGFR) signaling between

Current Topics in Developmental Biology, Volume 140
ISSN 0070-2153
https://doi.org/10.1016/bs.ctdb.2019.10.004

the oocyte and follicle cells, leading to ventrally-restricted expression of the sulfotransferase Pipe. These events promote the ventral processing of Spaetzle, a ligand for Toll, which ultimately sets up the embryonic dorsal-ventral axis. We then describe the activation of the terminal patterning system by specialized polar follicle cells. Finally, we present open questions regarding soma-germline signaling during *Drosophila* oogenesis required for cell identity and embryonic axis formation.

1. Introduction

The *Drosophila* embryo develops inside an egg which is covered by a finely patterned eggshell. The layers of the eggshell are secreted by somatic follicle cells (FCs) in the later stages of oogenesis. When the egg is laid, maternal determinants inside the egg specify the major axes of the embryo. As we will describe, the germline and FCs repeatedly signal to each other over the course of oogenesis to generate a complex array of FC fates and to pattern the oocyte. Mutations affecting these patterning processes generate characteristic defects in the structures of the eggshell and/or embryo; consequently, many key components and modulators of signaling pathways were initially identified and, in some cases, grouped together by screens for eggshell patterning and embryonic defects (Nüsslein-Volhard, Frohnhöfer, & Lehmann, 1987; Perrimon, Lanjuin, Arnold, & Noll, 1996; Schüpbach, 1987; Schüpbach & Wieschaus, 1986, 1991; reviewed in Berg, 2005). These screens revealed four maternal patterning systems; one for the dorsal-ventral (D/V) axis and three, with slightly overlapping domains, for the anterior–posterior (A/P) pattern of the developing *Drosophila* embryo. Among the latter three, the anterior system uses a gradient of Bicoid protein to specify anterior (head), thoracic and anterior abdominal segments. The posterior system relies on a gradient of Nanos protein to specify the abdominal segments; and finally, the terminal system regulates the activity of the receptor tyrosine kinase Torso. Disruptions to components of the terminal patterning system affect patterning at both the anterior and posterior ends of the embryo. Notably, all four pathways rely on signaling from the FCs, not only to coordinate the growth and development of the oocyte and FCs, but also to precisely align the final pattern of the embryo with that of the overlying eggshell.

2. Formation of the egg chamber

In *Drosophila*, oogenesis produces mature haploid eggs from immature multicellular structures called egg chambers. Females have paired ovaries, each containing approximately 18 strings of egg chambers called ovarioles. An ovariole is a chain of increasingly larger and more developed egg chambers. Each egg chamber produces a single egg and is made up of 16 germline cells (15 nurse cells and one oocyte) encased by somatic epithelial cells called follicle cells (FCs). Egg chamber development is subdivided into 14 distinct stages (for a more comprehensive review of oogenesis, see Spradling, 1993 and McLaughlin & Bratu, 2015). At the end of oogenesis, stage 14 (mature) oocytes pass through the oviduct, where they undergo egg activation and complete meiosis. Mature eggs are subsequently fertilized in the uterus by sperm stored in the spermatheca and then laid by the female, and embryonic development commences.

Egg chambers form in a structure called the germarium at the anterior end of the ovariole (Brown & King, 1962, 1964). The germarium is divided into four regions: 1, 2a, 2b, and 3. The germline stem cells (GSCs) are located at the anterior tip of the germarium in region 1 (Fig. 1A and B) and are in contact with somatic terminal filament and cap cells that secrete factors to establish and maintain the GSC niche, promoting self-renewal and preventing differentiation of the GSCs (Forbes, Lin, Ingham, & Spradling, 1996; Lin & Spradling, 1993, 1997; Song, Zhu, Doan, & Xie, 2002; Wieschaus & Szabad, 1979; Xie & Spradling, 1998, 2000). After GSC division, one daughter cell stays in contact with the niche and retains its potency as a stem cell. The other daughter cell, called the cystoblast, undergoes four more mitoses with incomplete cytokinesis in region 2a to form a germline cyst of 16 interconnected cells termed "cystocytes." All 16 cystocytes are connected by ring canals that form at each mitotic division, allowing transport of RNA and proteins within the germline cyst (Fig. 1B) (de Cuevas & Spradling, 1998; Deng & Lin, 1997). Somatic escort cells lie in close proximity to the cap cells and direct the differentiation and posterior movement of the dividing germline cyst (Decotto & Spradling, 2005). Once a 16-cell cyst has formed, escort cells direct it posteriorly to region 2b of the germarium, where the somatic follicle stem cells (FSCs) reside (Fig. 1A) (Decotto & Spradling, 2005; Kirilly, Wang, & Xie, 2011; Margolis & Spradling, 1995; Zhang & Kalderon, 2001). Immature FCs derived from the FSCs migrate to encapsulate the germline cyst. A newly encapsulated cyst, complete with a single uniform layer of FCs, constitutes a stage 1 egg chamber and also defines region 3 of the germarium. Once formed, the egg chambers of a given ovariole remain connected to each other throughout oogenesis by interfollicular stalk cells (discussed below). The FCs also produce and basally secrete a specialized extracellular matrix (ECM) called the basement membrane which surrounds each egg chamber

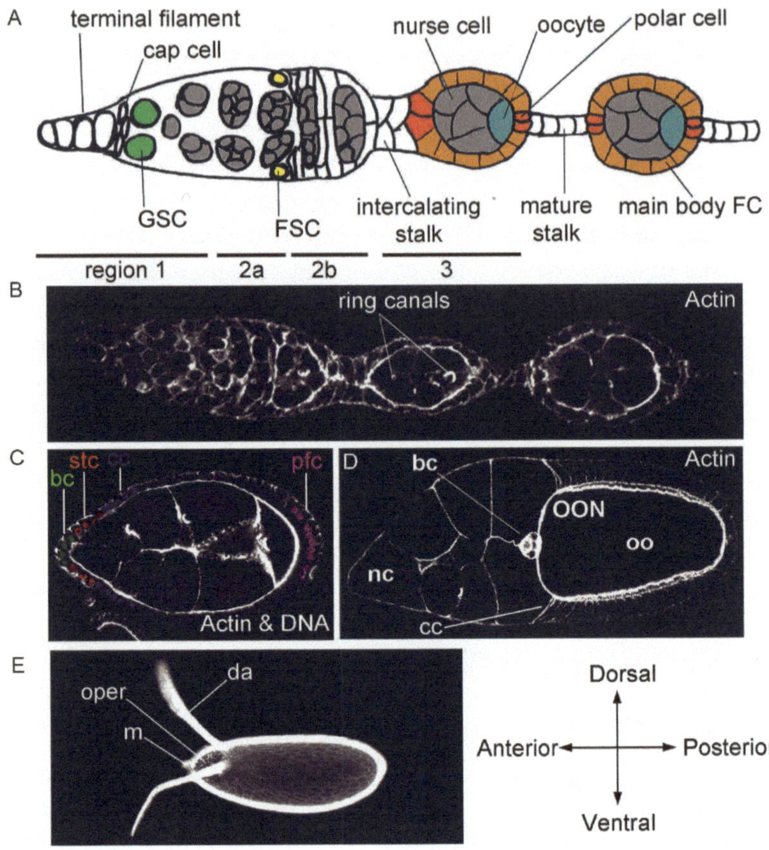

Fig. 1 An introduction to oogenesis. (A) A schematic of early oogenesis. Germline stem cells (GSCs, green) reside in region 1 at the anterior end of the germarium, in close association with the cap cells and terminal filament, which maintain the GSC niche. GSCs asymmetrically divide to generate a germline cyst (gray). Once specified, the oocyte (turquoise) moves to the posterior end of the cyst. Follicle stem cells (FSCs, yellow), found at the region 2a/2b boundary, divide to produce the somatic follicle cells of the egg chamber. Specialized follicle cells called the stalk cells intercalate to pinch off the region 3/stage 1 egg chamber from the germarium. Stalk cells intercalate further to form a 1-cell thick chain of cells separating each adjacent egg chamber in an ovariole (labeled "mature stalk"). Polar cells (red) function as important signaling hubs during oogenesis and are found at each terminus of an egg chamber. The main body follicle cells (FCs) are in orange. (B) An ovariole, consisting of a germarium and a stage 2 egg chamber, stained for F-actin. Ring canals are indicated. (C) Specification of the anterior and posterior terminal follicle cells. Graded JAK/STAT activity specifies three anterior follicle cell fates: the border cells (bc, green), stretch cells (stc, red) and centripetal cells (cc, blue). In the posterior, a combination of JAK/STAT and Gurken-EGFR signaling specifies the posterior follicle cells (pfc, magenta). The stage 7 egg chamber shown is stained for F-actin and DNA (both white). (D) A stage 10 egg chamber after completion of border

(Gutzeit, Eberhardt, & Gratwohl, 1991; Haigo & Bilder, 2011; Schneider et al., 2006). In addition, the entire ovariole is encased by a muscular sheath which contracts in waves to advance egg chambers toward the oviduct.

3. Selection of the oocyte

Via mechanisms that are only partially understood, 1 of the 16 cystocytes is selected as the oocyte and remains arrested in meiotic prophase I, while the other 15 cells become supportive cells called nurse cells (Fig. 1A and D) (Carpenter, 1975; Hammond & Laird, 1985; Theurkauf & Hawley, 1992; reviewed in Hughes, Miller, Miller, & Hawley, 2018). A precisely defined pattern of divisions gives rise to a 16-cell germline cyst and also causes some cystocytes to have more ring canals than others. In early region 2b, the two germline cells with the most (four) ring canals, the "oldest" cystocytes, are specified as pro-oocytes, each competent to become the oocyte (González-Reyes, Elliott, & St Johnston, 1997; Huynh & Johnston, 2000; Page & Hawley, 2001). Once specified, the oocyte can be distinguished from the nurse cells based on nuclear morphology, microtubule restructuring, entry into meiosis, and the accumulation of centrosomes and oocyte-specific proteins and mRNAs (Bolívar et al., 2001; Ephrussi, Dickinson, & Lehmann, 1991; Grieder, de Cuevas, & Spradling, 2000; Lantz & Schedl, 1994; Mach & Lehmann, 1997; Mahowald & Strassheim, 1970; Suter, Romberg, & Steward, 1989; Theurkauf, Smiley, Wong, & Alberts, 1992; Wharton & Struhl, 1989; reviewed in Huynh & St Johnston, 2004). Oocyte specification and determination are governed by intracellular processes and by communication between the germline cyst and the surrounding somatic cells (de Cuevas & Spradling, 1998; Lin, Yue, & Spradling, 1994; Theurkauf, 1994; Yue & Spradling, 1992).

cell migration (stained for F-actin; white). The border cells (bc) have migrated from the anterior tip of the egg chamber to the anterior end of the oocyte (oo). At this stage, the stretch cells (not shown) have elongated over the nurse cells (nc). The centripetal cells (cc) are at the boundary between the stretch and columnar main body follicle cells. Nurse cells (nc) and oocyte nucleus (OON) are indicated. (E) A wildtype eggshell, which is secreted by the follicle cells, is pictured. Key features of the eggshell: the micropyle (m), a pore through which sperm enter the egg; the dorsal appendages (da), with a respiratory function; and the operculum (oper), the opening through which the embryo will emerge upon hatching. (A–E) Anterior is to the left and posterior is to the right. (D–E) Dorsal is at the top and ventral is at the bottom.

4. Germline-soma and soma-soma signaling pattern the polar and stalk cells and pre-polarize the newly formed oocyte

Analysis of mosaic FCs established that two subpopulations of FC precursors are defined in the germarium: one population gives rise to the stalk and polar follicle cells and the other generates the rest of the epithelium surrounding the germline, referred to herein as the main body follicle cells (López-Schier, 2003; Margolis & Spradling, 1995; Tworoger, Larkin, Bryant, & Ruohola-Baker, 1999). Initially, the newly formed germline cyst is lens-shaped, with the two pro-oocytes at the center. In early region 2b, the precursor FCs shape and polarize the germline cyst, thereby contributing to the selection of a single oocyte from the two meiotic pro-oocytes (Margolis & Spradling, 1995). In late region 2b, as encapsulation proceeds, the oocyte moves posteriorly and adheres to a subset of FCs termed the terminal FCs. This anchoring is due to up-regulation of the adhesion protein E-cadherin in the oocyte and posterior terminal FCs (Godt & Tepass, 1998; González-Reyes & St Johnston, 1998b). E-cadherin accumulation at the posterior marks the first sign of anterior–posterior (A/P) patterning within the germline cyst. Disrupting E-cadherin in either the germline or FCs results in randomized positioning of the oocyte within the egg chamber. Additionally, genetic mosaic analysis demonstrated that each egg chamber helps to polarize the adjacent, anterior (younger) egg chamber to which it is connected in a "relay mechanism" mediated by Delta/Notch signaling (Torres, López-Schier, & St Johnston, 2003). Specifically, the anterior polar cells of the older cyst signal to induce the formation of the posterior polar cells of the younger, neighboring cyst by secreting Delta. It remains unknown how the first cyst in an ovariole, which lacks an older neighbor, acquires A/P polarity (Torres et al., 2003).

Both Hedgehog (Hh) and Notch/Delta signaling are required to specify the population of FCs that produces the polar and stalk cells. Genetic mosaic analysis has shown that both of these cell types derive from a common precursor population that differs early from the population that becomes the main body follicle cells. Castor (Cas), a nuclear protein initially expressed in all FC precursors, is repressed by the transcriptional activator Eyes absent (Eya) in the main body follicle cell precursors and becomes restricted to the polar cell/stalk cell population (Chang, Jang, Lin, & Montell, 2013). Hh signaling from the GSC niche appears to counteract Eya and maintain expression of Castor in the polar and stalk cell precursors. Overexpression of Hh increases the number of stalk and polar cells (Bai & Montell, 2002; Tworoger et al., 1999; Zhang & Kalderon, 2001). While the epithelial follicle cells that surround each germline cyst continue to divide mitotically until stage 5 or 6, the stalk and polar cell precursors exit the cell cycle before the egg chamber leaves the germarium.

In early oogenesis, 3–5 cells (Besse & Pret, 2003) express markers specific to the polar cells (Ruohola et al., 1991; Snow, Bieber, & Goodman, 1989). After apoptosis of the excess cells, exactly two polar FCs remain by stage 6 (McGregor, Xi, & Harrison, 2002). Although they are initially morphologically indistinguishable from the rest of the epithelium, the polar cells detach from the basement membrane and become round by stage 8 (Medioni & Noselli, 2005). Once specified by high levels of Delta/Notch signaling from adjacent cystocytes, the polar cells begin to express Unpaired (Upd), the ligand for the JAK/STAT pathway. In response, neighboring, still undifferentiated FCs that are not in contact with the germline differentiate into stalk cells (Chang et al., 2013; Keller Larkin et al., 1999; McGregor et al., 2002; Nystul & Spradling, 2010). Consequently, mutations in the JAK/STAT signaling pathway produce a loss of the stalk and ectopic JAK/STAT pathway activity increases the number of stalk cells (McGregor et al., 2002). The stalk cells facilitate egg chamber encapsulation and then maintain separation between the egg chambers throughout oogenesis. Stalk cells intercalate during approximately stages 1–5 to form a single column of cells connecting adjacent egg chambers in the ovariole (Fig. 1A and B) (Roth & Lynch, 2009).

The FCs that are in direct contact with the germline cyst form a polarized epithelial monolayer in which the apical edge faces the germline and the basal surface faces the exterior of the egg chamber (Bilder, Schober, & Perrimon, 2003; Tanentzapf & Tepass, 2003). At the anterior, FCs are further subdivided into terminal and main body follicle cells by the JAK/STAT ligand Upd that is secreted from the polar cells (Xi, McGregor, & Harrison, 2003). The 7–10 FCs closest to the polar cells, which are exposed to the highest levels of Upd, adopt the border cell fate (Fig. 1C). About 50 FCs slightly further away from the polar cells, and therefore exposed to a lower level of the Upd signal, assume the stretch cell fate. Lastly, 30–40 FCs even farther away from the source assume the centripetal cell fate (Beccari, Teixeira, & Rørth, 2002; Grammont & Irvine, 2002; Silver & Montell, 2001; Xi et al., 2003).

At the onset of stage 9 of oogenesis, these three distinct FC populations experience dramatic morphogenetic movements. The border cells and anterior polar cells delaminate from the epithelium and migrate as a group through the nurse cells to reach the anterior end of the oocyte (Fig. 1D) (Fulga & Rørth, 2002; Montell, Rørth, & Spradling, 1992; Montell, Yoon, & Starz-Gaiano, 2012). The main body FCs, however, contract over the oocyte and become columnar, while the stretch FCs flatten dramatically and envelop the nurse cells in a very thin, squamous epithelium (Fig. 1C and D) (Grammont & Irvine, 2002; Spradling, 1993). The centripetal FCs eventually become positioned at the precise boundary between the oocyte and the nurse cells. They later extend protrusions in between the nurse cells, contact the border cells and form a contiguous cover over the anterior end of the oocyte. Late in oogenesis, the border cells and centripetal FCs form the micropyle, a hollow tube through which sperm enter to fertilize the egg (Fig. 1E) (Karr, 1991; Montell et al., 1992; Rabinowitz, 1941). The specification and migration of the border cells possess hallmarks of an epithelial to mesenchymal transition (EMT), and the border cells have therefore been extensively studied as a model for EMT, cell migration, and tumor metastasis (reviewed in Montell, 2003 and Rosales-Nieves & González-Reyes, 2014).

5. Anterior–posterior axis formation involves signaling between germline and follicle cells

As described above, the movement of the oocyte and its cadherin-mediated attachment to the follicle cells at the pole opposite the germarium set up an initial oocyte anterior–posterior asymmetry. At around stage 6/7 of oogenesis, Delta is produced in the germline and signals to the FCs via the Notch receptor to initiate changes in the cell division cycle. This signaling event causes the main body follicle cells to switch from a canonical mitotic cycle to an endocycle in which DNA replication occurs in the absence of cell division (Fig. 2A) (Deng, Althauser, & Ruohola-Baker, 2001; López-Schier & St Johnston, 2001). At this stage Notch activity is essential for axis formation because it renders the posterior FCs competent to respond to JAK/STAT and EGFR inductive signaling.

Since two sets of polar cells are specified at each end of the egg chamber, secretion of Upd initially induces symmetrical terminal FC populations at both poles (Fig. 2B). In the anterior, graded JAK/STAT activity gives rise to the three anterior FC fates. At the posterior, these cell fates are blocked by Gurken (Grk), a TGFα-like signaling factor that is produced and released from the oocyte (Fig. 2C) (González-Reyes, Elliott, & St Johnston, 1995;

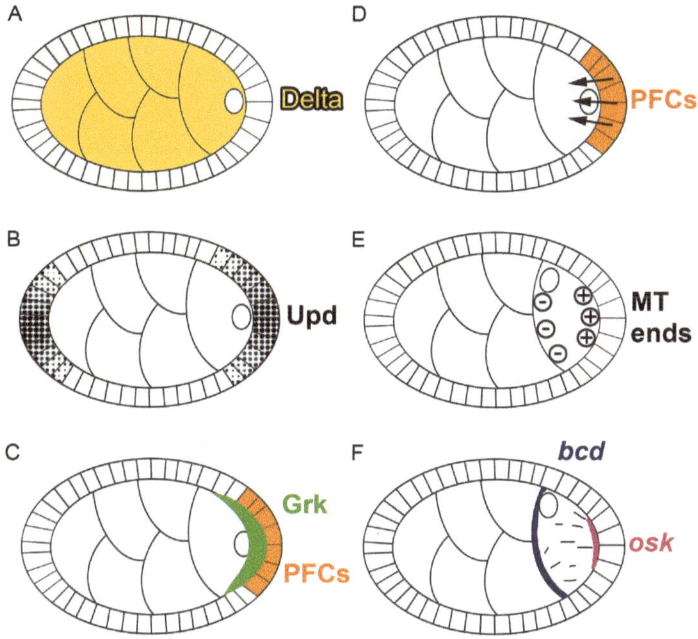

Fig. 2 Soma-germline signaling patterns the anterior–posterior axis. (A–F) The main body follicle cells are shown surrounding the germline (the oocyte and nurse cells). The oocyte nucleus is first anchored at the posterior of the oocyte (A–D) and then migrates to the dorsal-anterior corner (E–F). Anterior is to the left and posterior is to the right; dorsal is at the top and ventral is at the bottom. (A) Delta ligand (yellow) is produced by the germline and activates Notch signaling in all main body follicle cells. This causes the follicle cells to undergo endoreplication, a specialized cell cycle that replicates the DNA but lacks mitosis (also called an endocycle). (B) The polar cells secrete the JAK/STAT ligand Unpaired (Upd, black) in a gradient at each terminus of the egg chamber. (C) In the posterior, Gurken (Grk, green) signals from the oocyte to the overlying follicle cells to specify the posterior terminal follicle cell fate (PFCs, orange). (D) Once specified, the posterior terminal follicle cells (PFCs, orange) send an unknown secondary signal back to the oocyte (arrows). (E) The oocyte microtubules (MTs) reorganize such that the plus ends (+) are generally oriented toward the posterior and the minus ends (−) point toward the anterior. The oocyte nucleus migrates from the posterior to the dorsal-anterior corner. (F) *bicoid* RNA (*bcd*, blue) associates with the microtubule minus end-directed motor Dynein, and travels to the anterior end of the oocyte where it becomes anchored. *oskar* RNA (*osk*, pink) complexes with the protein Staufen (not shown) and migrates in association with the plus end-directed motor Kinesin to the posterior of the oocyte.

Roth, Neuman-Silberberg, Barcelo, & Schüpbach, 1995). *Gurken* is already expressed in the germline cells at stage 1 of oogenesis. *grk* mRNA is transported from the nurse cells into the oocyte along a microtubule network organized by a microtubule organizing center (MTOC) at the posterior pole of the oocyte (Januschke et al., 2006; Neuman-Silberberg & Schüpbach, 1993). Plus end-directed microtubules emanate through the ring canals between the oocyte and the nurse cells (Palacios & St Johnston, 2002). *grk* translation is blocked until it reaches the oocyte (Neuman-Silberberg & Schüpbach, 1993, 1996; Norvell, Kelley, Wehr, & Schüpbach, 1999). Grk protein accumulates in the oocyte during stages 1–6 of oogenesis. The Grk receptor, the epidermal growth factor receptor (EGFR) is expressed in all FCs, but at these early stages, Grk mainly reaches the posterior terminal FCs (Fig. 2C) (Van Buskirk & Schüpbach, 1999). Grk release from the oocyte membrane requires its cleavage by the intramembrane protease Brother of rhomboid (also called Stem cell tumor; Stet) (Ghiglione et al., 2002). After its secretion, Grk activates EGFR in the posterior terminal FCs. This activation prevents expression of genes that are characteristic of the anterior terminal FC fates. Instead, the posterior terminal FCs express a number of posterior-specific transcripts, such as *midline* (*mid*) and *H15*, as well as *pointed* (*pnt*) expression from the P1 promoter (*PntP1*) (Fregoso Lomas, Hails, Boisclair Lachance, & Nilson, 2013; Morimoto et al., 1996; Wittes & Schüpbach, 2019; Xi et al., 2003). In egg chambers mutant for *grk* or *Egfr*, the posterior terminal FCs assume the same cell fates (border, stretch and centripetal) as the anterior terminal cells. These experiments demonstrate that EGFR signaling suppresses anterior fates specifically at the posterior of the egg chamber (González-Reyes & St Johnston, 1998a; Roth et al., 1995). Thus, the localization of Grk and its activation of EGFR signaling at the posterior pole relays information about anterior versus posterior position within the egg chamber from the oocyte to the FCs.

Once specified, the posterior terminal FCs generate a new, so far molecularly undefined, secondary signal that is received by the oocyte and initiates the final anterior–posterior axis of the egg and future embryo (Fig. 2D). The evidence for this posterior polarizing signal derives from experiments in which disrupting EGFR or Notch signaling in the follicle cells produced a failure of oocyte nuclear migration and/or oocyte microtubule re-organization (González-Reyes et al., 1995; Larkin, Holder, Yost, Giniger, & Ruohola-Baker, 1996; Roth et al., 1995; Sun, Yan, Denef, & Schüpbach, 2011). This unknown signal restructures the oocyte microtubule network, such that the minus ends generally emanate from the anterior cortex of the oocyte and plus ends point toward the center and posterior of the oocyte (Fig. 2E) (Clark, Giniger, Ruohola-Baker, Jan, & Jan, 1994; Clark, Jan, & Jan, 1997; Nashchekin, Fernandes, & St Johnston, 2016; Theurkauf et al., 1992; Theurkauf, Alberts, Jan, & Jongens, 1993).

This microtubule reorganization is essential for the distribution of embryonic determinants (Fig. 2F). The mRNA encoding the anterior determinant Bicoid (Bcd) associates with Dynein, a minus end-directed microtubule motor, and therefore accumulates at the anterior end of the oocyte (Schnorrer, Bohmann, & Nüsslein-Volhard, 2000). Anteriorly-localized *bicoid* mRNA is translated in the early embryo after fertilization and forms a protein gradient that instructs much of the anterior–posterior pattern of the embryo (Driever & Nüsslein-Volhard, 1988a, 1988b; Ferrandon, Elphick, Nüsslein-Volhard, & St Johnston, 1994; Trovisco et al., 2016; Weil, Forrest, & Gavis, 2006). In contrast, the mRNA encoding the posterior pattern organizer Oskar (Osk) is bound by the RNA binding protein Staufen (Stau) and migrates posteriorly in association with plus end-directed microtubule motors (Brendza, Serbus, Duffy, & Saxton, 2000; St Johnston, Beuchle, & Nüsslein-Volhard, 1991). *Oskar* mRNA gradually accumulates at the posterior end of the oocyte and is translated during oogenesis; it is responsible for germ plasm formation (Ephrussi et al., 1991; Hachet & Ephrussi, 2004; Kim-Ha, Smith, & Macdonald, 1991; Lehmann, 2016). Oskar protein localization is also required for the later localization of *nanos* (*nos*) mRNA, which occurs by a trapping mechanism rather than directed microtubule transport (Becalska & Gavis, 2010; Forrest & Gavis, 2003). Localized *nanos* mRNA is translated at the posterior pole during early embryogenesis to form a protein gradient that patterns the abdominal structures of the developing embryo (Andrews, Snowflack, Clark, & Gavis, 2011; Crucs, Chatterjee, & Gavis, 2000; Gavis & Lehmann, 1992, 1994). In egg chambers where the FCs are mutant for components of Notch or EGFR signaling, the restructuring of the microtubule network fails, *bicoid* mRNA accumulates at both poles of the oocyte, and *oskar* mRNA and germ plasm components accumulate in the center of the oocyte (González-Reyes et al., 1995; Roth et al., 1995; Ruohola et al., 1991). Thus, the signal from the posterior terminal FCs is instrumental in setting up the embryonic anterior–posterior axis by defining a polarized microtubule network that allows the embryonic maternal determinants to be localized to the appropriate poles of the oocyte and egg.

In addition to promoting the localization of these important embryonic A/P axis determinants, the posterior polarizing signal from the FCs also instructs the oocyte nucleus to move from its original symmetric position to a cortical, asymmetric position within the developing oocyte (Fig. 3A and B) (González-Reyes et al., 1995; Roth et al., 1995; Zhao, Graham, Raposo, & St Johnston, 2012). This movement of the oocyte nucleus is

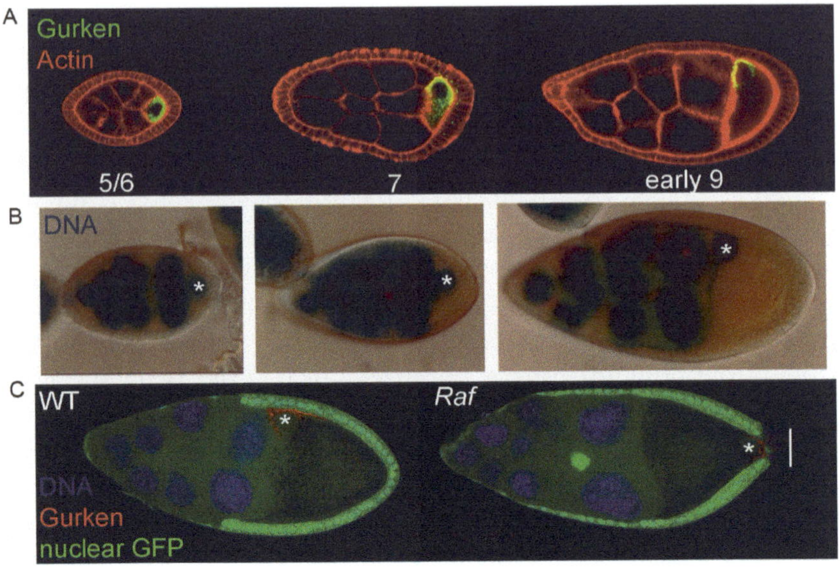

Fig. 3 Establishment of dorsal-ventral asymmetry through bidirectional signaling. (A) Distribution of Gurken protein at stages 5/6, 7 and 9 of oogenesis. Until stage 6, Gurken protein localizes throughout the oocyte and the nucleus is positioned centrally in the oocyte. At stage 6 or 7, Gurken from the oocyte binds to and activates EGFR from the follicle cells that are in contact with the oocyte and induces them to assume a posterior follicle cell fate. At stage 7 or 8, the posterior follicle cells send an unknown secondary signal back to the oocyte, causing the oocyte nucleus to move anteriorly. Gurken protein moves with the nucleus and becomes more concentrated on one side of the oocyte. This movement determines the dorsal side of the egg and future embryo. By stage 9, the oocyte nucleus is firmly anchored at the oocyte anterior and Gurken protein accumulates at the membrane close to the nucleus. Gurken then binds and activates EGFR in the overlying follicle cells, inducing a dorsal follicle cell fate. The actin cytoskeleton is in red and Gurken protein is in green. (B) Oocyte nuclear migration in stages 6, 7 and 9. The oocyte nucleus (asterisk) initially has a central position at stage 6. During stages 7/8, the oocyte nucleus breaks symmetry and moves anteriorly. At stage 9, it is anchored at the dorsal-anterior corner of the oocyte, in close proximity to the nurse cells and the oocyte cellular membrane. The germline nuclei are labeled in blue. (C) Egg chambers mosaic for a mutation in the MAPKKK *Raf*. In a wildtype (WT) egg chamber at late stage 9, the oocyte nucleus (asterisk) and Gurken protein (red) have moved to the dorsal side of the oocyte. In mosaic egg chambers in which the posterior follicle cells are mutant for *Raf* or any component of the EGFR pathway, the posterior follicle cells are not specified (white line indicates *Raf* mutant cells). The secondary polarizing signal is not produced and the nucleus and Gurken remain at the posterior of the oocyte. Nuclei are blue; wildtype follicle cells are green. (A–C) Anterior is to the left and posterior is to the right; dorsal is at the top and ventral is at the bottom. *Panel A: Images from Van Buskirk, C., & Schüpbach, T. (1999). Versatility in signalling: multiple responses to EGF receptor activation during* Drosophila *oogenesis.* Trends in Cell Biology, *9(1), 1–4. Panel C: Image courtesy of Yi Sun.*

coupled to microtubule reorganization, and it is essential for the establishment of dorsal-ventral polarity. In oocytes mutant for *grk* or Grk secretion (as in *cornichon* mutants) and in egg chambers where either the EGFR or Notch pathways are blocked in the FCs, the posterior polarizing signal fails (González-Reyes et al., 1995; Roth et al., 1995; Sun et al., 2011) (Larkin et al., 1996). As a result, the oocyte nucleus often remains at the posterior pole and no visible dorsal-ventral polarity is established in the developing oocyte and egg (Fig. 3C).

6. Dorsal-ventral polarity establishment requires oocyte nuclear migration and signaling between the oocyte and follicle cells

The asymmetric repositioning of the oocyte nucleus at the anterior end of the developing oocyte establishes the dorsal side of the egg and embryo. The movement of the nucleus is due in part to a pushing force exerted by microtubules (Tissot et al., 2017; Zhao et al., 2012), and the direction of nuclear migration toward the anterior cortex appears random (Tissot et al., 2017). It is not yet entirely clear how the nucleus becomes anchored to the anterior oocyte cortex following its migration (Guichet, Peri, & Roth, 2001; Tissot et al., 2017). This symmetry-breaking movement of the oocyte nucleus is assumed to be random in its orientation because in certain mutant cases where two nuclei become engulfed by the oocyte cytoplasm, both nuclei move anteriorly at the correct time, but their movement with respect to one another is random (Roth, Jordan, & Karess, 1999). This indicates that there is not a preformed path or anchor present in the oocyte that would predict the future dorsal side of egg and embryo. Recent live imaging of oocyte nuclear migration has confirmed that the oocyte nucleus can migrate along various paths to reach the anterior (Tissot et al., 2017).

grk mRNA always associates closely with the oocyte nucleus. If nuclear migration fails, *grk* mRNA remains at the posterior pole (Fig. 3C), and if there are two nuclei in the oocyte, *grk* mRNA associates with both (Roth et al., 1999). In wildtype egg chambers, *grk* mRNA and protein are symmetrically positioned around the oocyte nucleus with the protein mainly detected along the posterior cortex of the small oocyte until stage 6/7 (Neuman-Silberberg & Schüpbach, 1996). However, after the nucleus moves, *grk* mRNA and Grk protein localize asymmetrically within the oocyte, always in the vicinity of the oocyte nucleus, in the cytoplasmic space between the nucleus and the cellular membrane of the oocyte (Fig. 3A). Grk, which is secreted from the now dorsal side of the oocyte into the intracellular space between the oocyte and FCs, activates EGFR in the overlying FCs, and establishes dorsal-ventral asymmetry in the follicle cell epithelium (Neuman-Silberberg & Schüpbach, 1993, 1996). At this stage, the oocyte nucleus is firmly anchored at the anterior cortex of the oocyte, where *grk* mRNA and protein remain for the rest of oogenesis. As the oocyte grows anteriorly within the FC epithelium and as the FCs contract over the oocyte, Grk activates EGFR in a stripe of FCs. These FCs will later form the dorsal midline of the egg and future embryo (reviewed in Berg, 2005; Neuman-Silberberg & Schüpbach, 1993, 1996; Queenan, Barcelo, Van Buskirk, & Schüpbach, 1999; Roth & Schüpbach, 1994; Van Buskirk & Schüpbach, 1999). EGFR activity is graded in the FC epithelium with highest levels on the dorsal midline and lower levels along the lateral side of the epithelium (Boisclair Lachance, Fregoso Lomas, Eleiche, Bouchard Kerr, & Nilson, 2009; Goentoro et al., 2006; Pai, Barcelo, & Schüpbach, 2000). EGFR activity is transduced in the FCs through the Ras–Raf–Map Kinase pathway (Brand & Perrimon, 1994; Mantrova & Hsu, 2008; Schnorr & Berg, 1996), but in contrast to the earlier EGFR signaling event that induces posterior terminal FC fates, there is no simultaneous signal from Notch or JAK/STAT pathway activity in the lateral FCs (Fregoso Lomas, De Vito, Boisclair Lachance, Houde, & Nilson, 2016). This explains why these FCs adopt a distinct fate from the FCs at the posterior, even though both cell types are induced by Grk/EGFR signaling.

In response to EGFR activity, the FCs on the dorsal side express MAP Kinase target genes, such as *kekkon* (*kek*), *pointed* and *spitz* (*spi*) (Ghiglione et al., 1999; Peri, Bökel, & Roth, 1999; Ruohola-Baker et al., 1993; Sapir, Schweitzer, & Shilo, 1998; Wasserman & Freeman, 1998; Yakoby et al., 2008). EGFR activity also leads to the repression of genes with general activity in the FCs such as *Chorion factor 2* (*Cf2*) and *capicua* (*cic*) (Atkey, Boisclair Lachance, Walczak, Rebello, & Nilson, 2006; Goff, Nilson, & Morisato, 2001; Jiménez, Guichet, Ephrussi, & Casanova, 2000; Mantrova & Hsu, 2008). Importantly, repression of *capicua* on the dorsal side indirectly results in the restriction of *pipe* (*pip*) to the ventral side of the FC epithelium, which is crucial for dorsal-ventral patterning of the embryo (Sen, Goltz, Stevens, & Stein, 1998; Stein & Nüsslein-Volhard, 1992; see below).

At stage 10 of oogenesis at the anterior end of the developing oocyte, Decapentaplegic (Dpp; a *Drosophila* BMP2/4-like signaling molecule) is produced in the stretch and centripetal FCs (Deng & Bownes, 1997; Peri & Roth, 2000; Twombly et al., 1996) and activates the ubiquitously expressed BMP receptor Thickveins (Tkv) to initiate bone morphogenetic protein (BMP) signaling. The combination of EGFR and BMP signaling specifies an anterior dorsal domain in the main body FCs, which express markers such as *rhomboid* (*rho*) and *broad* (*br*) (Chen & Schüpbach, 2006; Dobens, Peterson, Treisman, & Raftery, 2000; Yakoby et al., 2008). Further interactions between the EGFR and BMP signaling pathways set up specific FC fates, in particular, the operculum and dorsal appendage precursor fates. These specialized follicle cell types secrete a finely patterned anterior eggshell toward the end of oogenesis (Fig. 1E) (Ruohola-Baker et al., 1993; Tzolovsky, Deng, Schlitt, & Bownes, 1999; Zartman, Kanodia, Cheung, & Shvartsman, 2009). During stages 8–14, the FCs secrete a three-layered proteinaceous eggshell. The innermost layer of the eggshell is called the vitelline membrane; we will next describe its key role in both dorsal-ventral and terminal patterning.

Dorsal-ventral patterning of the embryo requires EGFR signaling from the oocyte to the overlying FCs and involves a complex serine-protease cascade in the vitelline space surrounding the egg chamber (Fig. 4). EGFR activity on the dorsal side of the egg chamber represses *capicua* (Goff et al., 2001). Since Capicua represses *mirror* (*mirr*), EGFR signaling results in the expression of *mirror* in dorsal FCs and its exclusion from the ventral FCs (Atkey et al., 2006). Mirror, in turn, represses *pipe* expression dorsally (Fig. 4A and C) (Andreu et al., 2012). Pipe encodes a sulfotransferase which is active in the Golgi of the ventral FCs (Sen, Goltz, Konsolaki, Schüpbach, & Stein, 2000). It modifies vitelline membrane proteins as they are translated in the FCs and transported through the Golgi apparatus for secretion into the space between the oocyte and the FCs (Zhang, Stevens, & Stein, 2009). The vitelline membrane proteins that are sulfonated by Pipe protein on the ventral side of the egg will provide an assembly platform for a number of proteases, including Gastrulation-defective (Gd), Snake (Snk), and Easter (Ea), after the egg is fertilized (Cho, Stevens, & Stein, 2010). These proteases are present in an inactive form in the entire perivitelline space and only become active ventrally after they are cleaved. Gd localizes to the ventral side of the vitelline membrane where it recruits both Snake and Easter (Cho, Stevens, Sieverman, Nguyen, & Stein, 2012). Gd cleaves Snake, which subsequently cleaves and activates Easter, presumably by forming a functional protease complex (LeMosy, Tan, & Hashimoto, 2001; Smith & Delotto, 1994). Activated Easter then cleaves the signaling ligand Spaetzle (Spz) into its active form, rendering it capable of binding the Toll (Tl) receptor on the ventral side of the early embryo

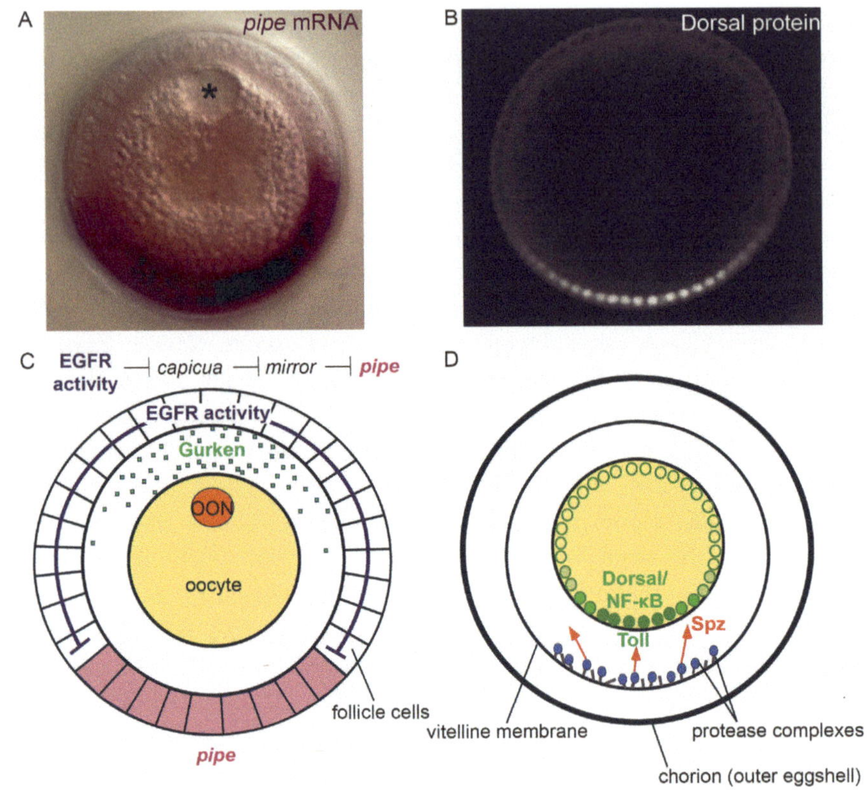

Fig. 4 Signaling from the ventral follicle cells sets up the dorsal-ventral pattern of the embryo. (A) Cross-section through an oocyte stained for *pipe* mRNA (purple) at late stage 9. *pipe* is expressed in the ventral follicle cells surrounding the oocyte and shows a sharp on/off transition in the epithelium. The oocyte nucleus (asterisk) is shown in its dorsal location. (B) Cross-section through an early *Drosophila* embryo at the syncytial blastoderm stage. Dorsal protein (white) shows a graded distribution with the highest concentration in the nuclei on the ventral side. (C) Schematic of dorsal signaling by Gurken. Gurken (green squares) secreted by the oocyte (yellow) activates EGFR in the overlying follicle cells. EGFR activity (blue line) ultimately leads to the repression of *pipe* (pink), which is only expressed in ventral follicle cells. The oocyte nucleus (OON) is localized dorsally. (D) Schematic of ventral signaling in response to Pipe activity. On the ventral side, where Pipe is active, vitelline membrane proteins (gray lines) are sulfonated, which promotes the formation and anchoring of active protease complexes (blue), such as Gastrulation defective (Gd), Snake and Easter. This leads to the cleavage and activation of Spaetzle (Spz, red) which binds to Toll on the embryonic cell membrane. Toll activity leads to nuclear import of Dorsal/NF-κB protein (green) into the embryonic nuclei, with the highest levels in the ventral-most region. (A–D) Dorsal is at the top and ventral is at the bottom. *Panel A: Image courtesy of Lea Goentoro. Panel B: Image courtesy of Eric Wieschaus.*

(Fig. 4D) (Anderson, Bokla, & Nüsslein-Volhard, 1985; Chasan & Anderson, 1989; Cho et al., 2010; DeLotto & DeLotto, 1998; Hashimoto et al., 1991; LeMosy et al., 2001; Morisato, 2001; Morisato & Anderson, 1994; Stein & Stevens, 2014, for review). Negative feedback mechanisms restrict production and diffusion of active Spaetzle and ensure that Toll activation is restricted to a stripe on the ventral side, roughly reflecting the earlier ventral *pipe* domain in the FCs (Hashimoto, Kim, Weiss, Miller, & Morisato, 2003; Morisato, 2001). Toll activation is transduced to the egg/embryo cytoplasm via the NF-κB homolog Dorsal (Dl). Dorsal is initially bound by the IκB homolog Cactus (Cact), but Toll activity results in the activation of a series of phosphorylation events that free Dorsal from Cactus and allow Dorsal's nuclear entry (Belvin, Jin, & Anderson, 1995; Bergmann et al., 1996; Großhans, Bergmann, Haffter, & Nüsslein-Volhard, 1994; Roth, Stein, & Nüsslein-Volhard, 1989; Rushlow, Han, Manley, & Levine, 1989; Steward, 1989). Even though *pipe* expression in the FCs appears to be uniform within the ventral domain, and shows a sharp on/off margin along the lateral side of the FC epithelium (Fig. 4A), the activation of Dorsal via Spaetzle and Toll is graded (Fig. 4B) (Nilson & Schüpbach, 1998; Roth & Schüpbach, 1994). The nuclear concentration of Dorsal in the embryo forms a clear gradient of activity that allows different target genes to be expressed in varying ventral to dorsal domains. At the highest concentrations of Dorsal, genes responsible for the most ventral embryonic fates are expressed, notably *twist* (*twi*) and *snail* (*sna*) (Ray, Arora, Nüsslein-Volhard, & Gelbart, 1991). *short gastrulation* (*sog*), *rhomboid* and *ventral nervous system defective* (*vnd*) are expressed in more lateral domains (Ip, Park, Kosman, Bier, & Levine, 1992; Stathopoulos & Levine, 2005; Stathopoulos, Van Drenth, Erives, Markstein, & Levine, 2002). Dorsal also restricts *dpp* and *zerknullt* (*zen*) to the dorsal domain of the early embryo, which in turn, determines the dorsal cell fates of the developing embryo (Ferguson & Anderson, 1992; Ip, Kraut, Levine, & Rushlow, 1991; Kirov, Childs, O'Connor, & Rushlow, 2015; Ray et al., 1991; Reeves & Stathopoulos, 2009).

In summary, formation of the dorsal-ventral (D/V) axis in *Drosophila* depends on the prior establishment of the A/P axis and involves a series of signals between the oocyte and FCs (Fig. 4C and D). As described previously, Grk-EGFR signaling first establishes the posterior terminal FC fate. This results in a secondary signal from the FCs back to the oocyte that causes the oocyte nucleus to move and thus breaks the radial symmetry of the egg chamber. Once the nucleus has assumed an asymmetric position, another Grk-EGFR signal from the oocyte to the FCs is initiated to establish the D/V axis. Dorsal EGFR and anterior TGFβ signaling establish a more finely compartmentalized anterior–posterior and dorsal-ventral pattern that is eventually reflected in the eggshell and affects the morphology of the operculum and dorsal appendages. The FCs also use the Grk-EGFR signal to repress expression of *pipe*, which is restricted to the ventral side and initiates a new cascade of signaling in the early embryo from the vitelline membrane via a protease cascade that ultimately activates Toll and Dorsal. The continuous back and forth signaling between the oocyte and FCs serves to coordinate the development of those two tissues and guarantees the precise alignment of the developing embryo with the eggshell, from which the fully developed embryo will hatch as a first instar larva through the preformed operculum.

7. Terminal patterning of the embryo requires follicle cell activity

The third patterning system that participates in specifying the A/P pattern of the embryo is called the terminal system. It is governed by the receptor tyrosine kinase Torso (Tor), and is responsible for assigning cell fates in the anterior-most and posterior-most regions of the embryo (Klingler, Erdélyi, Szabad, & Nüsslein-Volhard, 1988; Schüpbach & Wieschaus, 1986; Sprenger, Stevens, & Nüsslein-Volhard, 1989; Strecker, Halsell, Fisher, & Lipshitz, 1989). Torso signaling is activated when the Torso receptor is bound by its ligand, Trunk (Trk) (Casanova, Furriols, McCormick, & Struhl, 1995; Johnson et al., 2015). Torso is expressed in the germline and is present on the entire membrane of the syncytial blastoderm embryo (Casanova & Struhl, 1989; Sprenger et al., 1989). Its activity, however, is spatially limited to the poles by the expression of *torso-like* (*tsl*) in the polar FCs during oogenesis (Martin, Raibaud, & Ollo, 1994; Savant-Bhonsale & Montell, 1993; Stevens, Frohnhöfer, Klingler, & Nüsslein-Volhard, 1990). Tsl is a secreted protein that becomes specifically localized to the vitelline membrane at the poles of the mature egg (Stevens, Beuchle, Jurcsak, Tong, & Stein, 2003). Expression of *tsl* in all FCs results in a dramatic expansion of the terminal structures of the embryo at the expense of central segments (Martin et al., 1994; Savant-Bhonsale & Montell, 1993), illustrating that Tsl is a limiting factor in the terminal patterning system.

While Tsl and Trunk are required for the terminal patterning process in the embryo, the relationship between the two is not entirely understood.

Tsl encodes a membrane attack complex/perforin (MACPF) protein (Johnson, Henstridge, & Warr, 2017; Ponting, 1999). Trunk ligand is produced in the early embryo from maternally deposited mRNA, which is not spatially localized (Casali & Casanova, 2001; Casanova et al., 1995; Johnson et al., 2015). Instead, Trunk protein appears to be activated specifically at the poles in a process mediated by Tsl. Initially, it was thought that Trunk might be secreted in a precursor form and activated by proteolytic cleavage (Casali & Casanova, 2001; Casanova et al., 1995; Henstridge, Johnson, Warr, & Whisstock, 2014). However, since Tsl does not encode a protease, the activation of Trunk by Tsl must be more indirect. More recently, Trunk cleavage was shown to be promoted by Furin proteases (Johnson et al., 2015), which suggests that Trunk is cleaved intracellularly before it is secreted. Johnson and colleagues suggest that Tsl may stimulate secretion of Trunk at the poles. In a recent study in cultured cells, Amarnath, Stevens, and Stein (2017) found that Trunk and Tsl act synergistically and may in fact both be required to allow stable dimerization of Torso, which is necessary to activate its tyrosine kinase function (Amarnath et al., 2017). However, these interpretations do not fully explain the finding that when Torso is only expressed in the central portion of the embryo, it becomes activated, possibly by diffusing ligand (Trunk) from the poles (Casanova & Struhl, 1993; Sprenger & Nüsslein-Volhard, 1992). Further biochemical experiments are warranted to precisely determine the interactions of Trunk, Tsl and Torso in terminal patterning (for further discussion, see Johnson et al., 2017; Mineo, Furriols, & Casanova, 2018).

Ultimately, Torso activation triggers a Ras/Raf/Map Kinase cascade that represses the transcriptional regulator Capicua at the poles of the embryo (Furriols & Casanova, 2003; Grimm et al., 2012; Jiménez et al., 2000). This allows expression of the terminal patterning genes *tailless* (*tll*) and *huckebein* (*hkb*) (Jiménez et al., 2000), which are the main pattern organizers of the anterior- and posterior-most regions of the embryo.

In summary, all four major patterning systems of the embryo, the anterior, posterior, terminal and dorsal-ventral pathways, rely on communication between the germline and the FCs that surround the oocyte during oogenesis. This is an unusual situation in animal development; more commonly, only one of the major embryonic axes (usually the animal-vegetal axis) is laid down during oogenesis (Boettger, Knoetgen, Wittler, & Kessel, 2001; Elkouby & Mullins, 2017; Momose & Houliston, 2007; Rebagliati, Weeks, Harvey, & Melton, 1985; Wilt, 1987). In mammalian eggs, none of the body axes are stably preformed at the time of fertilization. In *Drosophila*, however, rapid embryonic development is facilitated by the follicle cell epithelium to pre-pattern the egg.

8. Open questions in germline-follicle cell communication during *Drosophila* oogenesis and early development

One open question is whether small molecules or ions are involved in germline-follicle cell signaling. Gap junctions allow small molecules, ions and electrical signals to pass between adjoining cells, and are physically composed of protein subunits called innexins. Immunostaining suggests that the gap junction protein Innexin-2 (Inx-2) is expressed in both the germline and FCs, and that it localizes to the plasma membranes where the oocyte and FCs meet. Furthermore, experiments show that small fluorescent dyes injected into the oocyte can pass into the FCs, suggesting that channels exist through which small molecules can pass between the germline and FCs (Bohrmann & Haas-Assenbaum, 1993; Bohrmann & Zimmermann, 2008). Injecting antibodies against Inx-2 into the oocyte interferes with this movement, suggesting that Inx-2 may function in the oocyte as a component of one such channel (Bohrmann & Zimmermann, 2008). Other innexins have also been found in the germline and FCs and could potentially be involved in signaling. However, as of yet, gap junctions and innexins have not been reported to play a functional role in patterning.

Another open question is the identity of the posterior polarizing signal, which is generated in the posterior terminal FCs and triggers oocyte microtubule reorganization (González-Reyes et al., 1995; Larkin et al., 1996; Nashchekin et al., 2016; Roth et al., 1995; Sun et al., 2011; Theurkauf et al., 1992). The Schüpbach laboratory and others have used forward genetic screens to attempt to identify this signal, however, while these screens have been fruitful, they have not identified the signal itself (Chen & Schüpbach, 2006; Laplante & Nilson, 2006; Pai et al., 2000; Wittes & Schüpbach, 2019; Yan, Denef, & Schüpbach, 2009). There are a number of potential reasons why the posterior signal has not been identified to date. The polarizing signal may be encoded by redundant proteins, or it could require two pathways functioning in parallel. In either case, both would need to be disrupted to yield an oocyte polarity defect. It has also been proposed that the signal may be mechanical in nature; however, mutations affecting mechanical properties of the egg chamber have not been reported to disrupt oocyte polarity (Cetera et al., 2014;

Crest, Diz-Muñoz, Chen, Fletcher, & Bilder, 2017; Haigo & Bilder, 2011). Finally, the posterior signal may be encoded by a protein that is required for earlier stages of oogenesis, such that mutants disrupt cyst production or cause oogenesis to arrest before a polarity phenotype can be identified.

Another question in the field surrounds the precise molecular mechanisms that specify and maintain the oocyte fate. Although many distinctive features of oocyte identity are well-described, the exact sequence of events required for oocyte fate are unclear. Two prevailing models for the mechanism of oocyte specification have been proposed (reviewed in Huynh & St Johnston, 2004). These models differ mainly in the timing of oocyte specification. One model suggests that the oocyte is specified in the germarium after the last cyst division, while the other suggests that the determinants specifying the oocyte are already asymmetrically distributed at the first mitotic division from the GSC. Both models, however, propose that an asymmetric accumulation of specific mRNAs, proteins, and/or organelles are responsible for oocyte identity establishment. Interestingly, once oocyte-associated proteins and RNAs are restricted to a single cell, oocyte identity may still be lost in early oogenesis. Specifically, if polarity is lost and/or the meiotic prophase I arrest is not maintained, the oocyte nucleus initiates endoreplication, thereby reverting to a nurse cell-like program (Hong, Lee-Kong, Iida, Sugimura, & Lilly, 2003; Huynh, Petronczki, Knoblich, & St Johnston, 2001). Although many molecules and organelles have been identified as oocyte determining factors based on localization patterns and mutant phenotypes, it has been difficult to elucidate the precise timing at which each of these components are required for oocyte fate.

A further open question concerns the activation of the receptor tyrosine kinase Torso. The expression of *torso-like* in the polar cells is instrumental in limiting the activation of Torso to the anterior and posterior ends of the developing embryo. How spatial information is relayed from the follicle cells to the embryo, however, is unclear. Further biochemical and genetic experiments are required to determine whether the MACPF-encoding protein Torso-like promotes the secretion of Trunk at the poles of the egg, and/or Trunk and Torso binding.

Acknowledgments

We would like to acknowledge grant support from the National Institutes of Health: R01GM077620 to T.S. and 5T32DK007328-40 to J.W.

References

Amarnath, S., Stevens, L. M., & Stein, D. S. (2017). Reconstitution of Torso signaling in cultured cells suggests a role for both Trunk and Torso-like in receptor activation. *Development, 144*(4), 677–686.

Anderson, K. V., Bokla, L., & Nüsslein-Volhard, C. (1985). Establishment of dorsal-ventral polarity in the Drosophila embryo: The induction of polarity by the Toll gene product. *Cell, 42*(3), 791–798.

Andreu, M. J., Gonzalez-Perez, E., Ajuria, L., Samper, N., Gonzalez-Crespo, S., Campuzano, S., et al. (2012). Mirror represses *pipe* expression in follicle cells to initiate dorsoventral axis formation in *Drosophila*. *Development, 139*(6), 1110–1114.

Andrews, S., Snowflack, D. R., Clark, I. E., & Gavis, E. R. (2011). Multiple mechanisms collaborate to repress *nanos* translation in the *Drosophila* ovary and embryo. *RNA, 17*(5), 967–977.

Atkey, M. R., Boisclair Lachance, J.-F., Walczak, M., Rebello, T., & Nilson, L. A. (2006). Capicua regulates follicle cell fate in the *Drosophila* ovary through repression of *mirror*. *Development, 133*(11), 2115–2123.

Bai, J., & Montell, D. (2002). Eyes absent, a key repressor of polar cell fate during *Drosophila* oogenesis. *Development, 129*(23), 5377–5388.

Becalska, A. N., & Gavis, E. R. (2010). Bazooka regulates microtubule organization and spatial restriction of germ plasm assembly in the *Drosophila* oocyte. *Developmental Biology, 340*(2), 528–538.

Beccari, S., Teixeira, L., & Rørth, P. (2002). The JAK/STAT pathway is required for border cell migration during *Drosophila* oogenesis. *Mechanisms of Development, 111*(1–2), 115–123.

Belvin, M. P., Jin, Y., & Anderson, K. V. (1995). Cactus protein degradation mediates *Drosophila* dorsal-ventral signaling. *Genes & Development, 9*(7), 783–793.

Berg, C. A. (2005). The *Drosophila* shell game: Patterning genes and morphological change. *Trends in Genetics, 21*(6), 346–355.

Bergmann, A., Stein, D., Geisler, R., Hagenmaier, S., Schmid, B., Fernandez, N., et al. (1996). A gradient of cytoplasmic Cactus degradation establishes the nuclear localization gradient of the *dorsal* morphogen in *Drosophila*. *Mechanisms of Development, 60*(1), 109–123.

Besse, F., & Pret, A.-M. (2003). Apoptosis-mediated cell death within the ovarian polar cell lineage of *Drosophila melanogaster*. *Development, 130*(5), 1017–1027.

Bilder, D., Schober, M., & Perrimon, N. (2003). Integrated activity of PDZ protein complexes regulates epithelial polarity. *Nature Cell Biology, 5*(1), 53–58.

Boettger, T., Knoetgen, H., Wittler, L., & Kessel, M. (2001). The avian organizer. *The International Journal of Developmental Biology, 45*(1), 281–287.

Bohrmann, J., & Haas-Assenbaum, A. (1993). Gap junctions in ovarian follicles of *Drosophila melanogaster*: Inhibition and promotion of dye-coupling between oocyte and follicle cells. *Cell and Tissue Research, 273*(1), 163–173.

Bohrmann, J., & Zimmermann, J. (2008). Gap junctions in the ovary of *Drosophila melanogaster*: Localization of innexins 1, 2, 3 and 4 and evidence for intercellular communication via innexin-2 containing channels. *BMC Developmental Biology, 8*(1), 111.

Boisclair Lachance, J.-F., Fregoso Lomas, M., Eleiche, A., Bouchard Kerr, P., & Nilson, L. A. (2009). Graded EGFR activity patterns the *Drosophila* eggshell independently of autocrine feedback. *Development, 136*(17), 2893–2902.

Bolívar, J., Huynh, J.-R., López-Schier, H., González, C., St Johnston, D., & González-Reyes, A. (2001). Centrosome migration into the *Drosophila* oocyte is independent of *BicD* and *egl*, and of the organisation of the microtubule cytoskeleton. *Development, 128*(10), 1889–1897.

Brand, A. H., & Perrimon, N. (1994). Raf acts downstream of the EGF receptor to determine dorsoventral polarity during *Drosophila* oogenesis. *Genes & Development, 8*(5), 629–639.
Brendza, R. P., Serbus, L. R., Duffy, J. B., & Saxton, W. M. (2000). A function for kinesin I in the posterior transport of *oskar* mRNA and Staufen protein. *Science, 289*(5487), 2120–2122.
Brown, E. H., & King, R. C. (1962). Oogonial and spermatogonial differentiation within a mosaic gonad of *Drosophila melanogaster. Growth, 26*, 53–69.
Brown, E. H., & King, R. C. (1964). Studies on the events resulting in the formation of an egg chamber in *Drosophila melanogaster. Growth, 28*, 41–81.
Carpenter, A. T. (1975). Electron microscopy of meiosis in Drosophila melanogaster females. I. Structure, arrangement, and temporal change of the synaptonemal complex in wild-type. *Chromosoma, 51*(2), 157–182.
Casali, A., & Casanova, J. (2001). The spatial control of Torso RTK activation: A C-terminal fragment of the Trunk protein acts as a signal for Torso receptor in the *Drosophila* embryo. *Development, 128*(9), 1709–1715.
Casanova, J., Furriols, M., McCormick, C. A., & Struhl, G. (1995). Similarities between trunk and spätzle, putative extracellular ligands specifying body pattern in Drosophila. *Genes & Development, 9*(20), 2539–2544.
Casanova, J., & Struhl, G. (1989). Localized surface activity of *torso*, a receptor tyrosine kinase, specifies terminal body pattern in Drosophila. *Genes & Development, 3*(12 B), 2025–2038.
Casanova, J., & Struhl, G. (1993). The *torso* receptor localizes as well as transduces the spatial signal specifying terminal body pattern in *Drosophila. Nature, 362*(6416), 152–155.
Cetera, M., Ramirez-San Juan, G. R., Oakes, P. W., Lewellyn, L., Fairchild, M. J., Tanentzapf, G., et al. (2014). Epithelial rotation promotes the global alignment of contractile actin bundles during *Drosophila* egg chamber elongation. *Nature Communications, 5*(1), 5511.
Chang, Y.-C., Jang, A. C., Lin, C.-H., & Montell, D. J. (2013). Castor is required for hedgehog-dependent cell-fate specification and follicle stem cell maintenance in *Drosophila* oogenesis. *Proceedings of the National Academy of Sciences of the United States of America, 110*(19), E1734–E1742.
Chasan, R., & Anderson, K. V. (1989). The role of *easter*, an apparent serine protease, in organizing the dorsal-ventral pattern of the Drosophila embryo. *Cell, 56*(3), 391–400.
Chen, Y., & Schüpbach, T. (2006). The role of brinker in eggshell patterning. *Mechanisms of Development, 123*(5), 395–406.
Cho, Y. S., Stevens, L. M., Sieverman, K. J., Nguyen, J., & Stein, D. (2012). A ventrally localized protease in the *Drosophila* egg controls embryo dorsoventral polarity. *Current Biology, 22*(11), 1013–1018.
Cho, Y. S., Stevens, L. M., & Stein, D. (2010). Pipe-dependent ventral processing of Easter by Snake is the defining step in *Drosophila* embryo DV axis formation. *Current Biology, 20*(12), 1133–1137.
Clark, I., Giniger, E., Ruohola-Baker, H., Jan, L. Y., & Jan, Y. N. (1994). Transient posterior localization of a kinesin fusion protein reflects anteroposterior polarity of the *Drosophila* oocyte. *Current Biology, 4*(4), 289–300.
Clark, I. E., Jan, L. Y., & Jan, Y. N. (1997). Reciprocal localization of Nod and kinesin fusion proteins indicates microtubule polarity in the *Drosophila* oocyte, epithelium, neuron and muscle. *Development, 124*(2), 461–470.
Crest, J., Diz-Muñoz, A., Chen, D. Y., Fletcher, D. A., & Bilder, D. (2017). Organ sculpting by patterned extracellular matrix stiffness. *eLife, 6*, e25958.
Crucs, S., Chatterjee, S., & Gavis, E. R. (2000). Overlapping but distinct RNA elements control repression and activation of *nanos* translation. *Molecular Cell, 5*(3), 457–467.

de Cuevas, M., & Spradling, A. C. (1998). Morphogenesis of the *Drosophila* fusome and its implications for oocyte specification. *Development, 125*(15), 2781–2789.

Decotto, E., & Spradling, A. C. (2005). The *Drosophila* ovarian and testis stem cell niches: Similar somatic stem cells and signals. *Developmental Cell, 9*(4), 501–510.

DeLotto, Y., & DeLotto, R. (1998). Proteolytic processing of the Drosophila Spätzle protein by Easter generates a dimeric NGF-like molecule with ventralising activity. *Mechanisms of Development, 72*(1–2), 141–148.

Deng, W.-M., Althauser, C., & Ruohola-Baker, H. (2001). Notch-Delta signaling induces a transition from mitotic cell cycle to endocycle in *Drosophila* follicle cells. *Development, 4746*(23), 4737–4746.

Deng, W.-M., & Bownes, M. (1997). Two signalling pathways specify localised expression of the *Broad-Complex* in *Drosophila* eggshell patterning and morphogenesis. *Development, 124*(22), 4639–4647.

Deng, W.-M., & Lin, H. (1997). Spectrosomes and fusomes anchor mitotic spindles during asymmetric germ cell divisions and facilitate the formation of a polarized microtubule array for oocyte specification in *Drosophila. Developmental Biology, 189*(1), 79–94.

Dobens, L. L., Peterson, J. S., Treisman, J., & Raftery, L. A. (2000). *Drosophila bunched* integrates opposing DPP and EGF signals to set the operculum boundary. *Development, 127*(4), 745–754.

Driever, W., & Nüsslein-Volhard, C. (1988a). A gradient of *bicoid* protein in Drosophila embryos. *Cell, 54*(1), 83–93.

Driever, W., & Nüsslein-Volhard, C. (1988b). The *bicoid* protein determines position in the Drosophila embryo in a concentration-dependent manner. *Cell, 54*(1), 95–104.

Elkouby, Y. M., & Mullins, M. C. (2017). Coordination of cellular differentiation, polarity, mitosis and meiosis—New findings from early vertebrate oogenesis. *Developmental Biology, 430*(2), 275–287.

Ephrussi, A., Dickinson, L. K., & Lehmann, R. (1991). Oskar organizes the germ plasm and directs localization of the posterior determinant nanos. *Cell, 66*(1), 37–50.

Ferguson, E. L., & Anderson, K. V. (1992). Decapentaplegic acts as a morphogen to organize dorsal-ventral pattern in the Drosophila embryo. *Cell, 71*(3), 451–461.

Ferrandon, D., Elphick, L., Nüsslein-Volhard, C., & St Johnston, D. (1994). Staufen protein associates with the 3′UTR of *bicoid* mRNA to form particles that move in a microtubule-dependent manner. *Cell, 79*(7), 1221–1232.

Forbes, A. J., Lin, H., Ingham, P. W., & Spradling, A. C. (1996). *hedgehog* is required for the proliferation and specification of ovarian somatic cells prior to egg chamber formation in *Drosophila. Development, 122*(4), 1125–1135.

Forrest, K. M., & Gavis, E. R. (2003). Live imaging of endogenous RNA reveals a diffusion and entrapment mechanism for *nanos* mRNA localization in *Drosophila. Current Biology, 13*(14), 1159–1168.

Fregoso Lomas, M., De Vito, S., Boisclair Lachance, J.-F., Houde, J., & Nilson, L. A. (2016). Determination of EGFR signaling output by opposing gradients of BMP and JAK/STAT activity. *Current Biology, 26*(19), 2572–2582.

Fregoso Lomas, M., Hails, F., Boisclair Lachance, J.-F., & Nilson, L. A. (2013). Response to the dorsal anterior gradient of EGFR signaling in *Drosophila* oogenesis is prepatterned by earlier posterior EGFR activation. *Cell Reports, 4*(4), 791–802.

Fulga, T. A., & Rørth, P. (2002). Invasive cell migration is initiated by guided growth of long cellular extensions. *Nature Cell Biology, 4*(9), 715–719.

Furriols, M., & Casanova, J. (2003). In and out of Torso RTK signalling. *EMBO Journal, 22*(9), 1947–1952.

Gavis, E. R., & Lehmann, R. (1992). Localization of *nanos* RNA controls embryonic polarity. *Cell, 71*(2), 301–313.

Gavis, E. R., & Lehmann, R. (1994). Translational regulation of *nanos* by RNA localization. *Nature, 369*(6478), 315–318.
Ghiglione, C., Bach, E. A., Paraiso, Y., Carraway, K. L., Noselli, S., & Perrimon, N. (2002). Mechanism of activation of the *Drosophila* EGF receptor by the TGFalpha ligand Gurken during oogenesis. *Development, 129*(1), 175–186.
Ghiglione, C., Carraway, K. L., Amundadottir, L. T., Boswell, R. E., Perrimon, N., & Duffy, J. B. (1999). The transmembrane molecule kekkon 1 acts in a feedback loop to negatively regulate the activity of the Drosophila EGF receptor during oogenesis. *Cell, 96*(6), 847–856.
Godt, D., & Tepass, U. (1998). *Drosophila* oocyte localization is mediated by differential cadherin-based adhesion. *Nature, 395*(6700), 387–391.
Goentoro, L. A., Reeves, G. T., Kowal, C. P., Martinelli, L., Schüpbach, T., & Shvartsman, S. Y. (2006). Quantifying the Gurken morphogen gradient in *Drosophila* oogenesis. *Developmental Cell, 11*(2), 263–272.
Goff, D. J., Nilson, L. A., & Morisato, D. (2001). Establishment of dorsal-ventral polarity of the *Drosophila* egg requires *capicua* action in ovarian follicle cells. *Development, 128*(22), 4553–4562.
González-Reyes, A., Elliott, H., & St Johnston, D. (1995). Polarization of both major body axes in *Drosophila* by *gurken-torpedo* signalling. *Nature, 375*(6533), 654–658.
González-Reyes, A., Elliott, H., & St Johnston, D. (1997). Oocyte determination and the origin of polarity in *Drosophila*: The role of the *spindle* genes. *Development, 124*(24), 4927–4937.
González-Reyes, A., & St Johnston, D. (1998a). Patterning of the follicle cell epithelium along the anterior-posterior axis during *Drosophila* oogenesis. *Development, 125*(15), 2837–2846.
González-Reyes, A., & St Johnston, D. (1998b). The *Drosophila* AP axis is polarised by the cadherin-mediated positioning of the oocyte. *Development, 125*(18), 3635–3644.
Grammont, M., & Irvine, K. D. (2002). Organizer activity of the polar cells during *Drosophila* oogenesis. *Development, 129*(22), 5131–5140.
Grieder, N. C., de Cuevas, M., & Spradling, A. C. (2000). The fusome organizes the microtubule network during oocyte differentiation in *Drosophila*. *Development, 127*(19), 4253–4264.
Grimm, O., Zini, V. S., Kim, Y., Casanova, J., Shvartsman, S. Y., & Wieschaus, E. (2012). Torso RTK controls Capicua degradation by changing its subcellular localization. *Journal of Cell Science, 125*(21), 3962–3968.
Großhans, J., Bergmann, A., Haffter, P., & Nüsslein-Volhard, C. (1994). Activation of the kinase Pelle by Tube in the dorsoventral signal transduction pathway of *Drosophila* embryo. *Nature, 372*(6506), 563–566.
Guichet, A., Peri, F., & Roth, S. (2001). Stable anterior anchoring of the oocyte nucleus is required to establish dorsoventral polarity of the *Drosophila* egg. *Developmental Biology, 237*(1), 93–106.
Gutzeit, H. O., Eberhardt, W., & Gratwohl, E. (1991). Laminin and basement membrane-associated microfilaments in wild-type and mutant *Drosophila* ovarian follicles. *Journal of Cell Science, 100*, 781–788.
Hachet, O., & Ephrussi, A. (2004). Splicing of *oskar* RNA in the nucleus is coupled to its cytoplasmic localization. *Nature, 428*(6986), 959–963.
Haigo, S. L., & Bilder, D. (2011). Global tissue revolutions in a morphogenetic movement controlling elongation. *Science, 331*(6020), 1071–1074.
Hammond, M. P., & Laird, C. D. (1985). Chromosome structure and DNA replication in nurse and follicle cells of *Drosophila melanogaster*. *Chromosoma, 91*(3–4), 267–278.

Hashimoto, C., Gerttula, S., Anderson, K. V., Duman, M., Lawinger, P., Botas, J., et al. (1991). Plasma membrane localization of the *Toll* protein in the syncytial *Drosophila* embryo: Importance of transmembrane signaling for dorsal-ventral pattern formation. *Development, 111*(4), 1021–1028.

Hashimoto, C., Kim, D. R., Weiss, L. A., Miller, J. W., & Morisato, D. (2003). Spatial regulation of developmental signaling by a serpin. *Developmental Cell, 5*(6), 945–950.

Henstridge, M. A., Johnson, T. K., Warr, C. G., & Whisstock, J. C. (2014). Trunk cleavage is essential for Drosophila terminal patterning and can occur independently of Torso-like. *Nature Communications, 5*(1), 1–7.

Hong, A., Lee-Kong, S., Iida, T., Sugimura, I., & Lilly, M. A. (2003). The p27cip/kip ortholog *dacapo* maintains the *Drosophila* oocyte in prophase of meiosis I. *Development, 130*(7), 1235–1242.

Hughes, S. E., Miller, D. E., Miller, A. L., & Hawley, R. S. (2018). Female meiosis: Synapsis, recombination, and segregation in *Drosophila melanogaster. Genetics, 208*(3), 875–908.

Huynh, J.-R., & Johnston, D. S. (2000). The role of BicD, Egl, Orb and the microtubules in the restriction of meiosis to the *Drosophila* oocyte. *Development, 127*(13), 2785–2794.

Huynh, J.-R., Petronczki, M., Knoblich, J. A., & St Johnston, D. (2001). Bazooka and PAR-6 are required with PAR-1 for the maintenance of oocyte fate in *Drosophila. Current Biology, 11*(11), 901–906.

Huynh, J.-R., & St Johnston, D. (2004). The origin of asymmetry: Early polarisation of the *Drosophila* germline cyst and oocyte. *Current Biology, 14*(11), 438–449.

Ip, Y. T., Kraut, R., Levine, M., & Rushlow, C. A. (1991). The *dorsal* morphogen is a sequence-specific DNA-binding protein that interacts with a long-range repression element in *Drosophila. Cell, 64*(2), 439–446.

Ip, Y. T., Park, R. E., Kosman, D., Bier, E., & Levine, M. (1992). The *dorsal* gradient morphogen regulates stripes of *rhomboid* expression in the presumptive neuroectoderm of the *Drosophila* embryo. *Genes & Development, 6*(9), 1728–1739.

Januschke, J., Gervais, L., Gillet, L., Keryer, G., Bornens, M., & Guichet, A. (2006). The centrosome-nucleus complex and microtubule organization in the *Drosophila* oocyte. *Development, 133*(1), 129–139.

Jiménez, G., Guichet, A., Ephrussi, A., & Casanova, J. (2000). Relief of gene repression by Torso RTK signaling: Role of *capicua* in *Drosophila* terminal and dorsoventral patterning. *Genes & Development, 14*(2), 224–231.

Johnson, T. K., Henstridge, M. A., Herr, A., Moore, K. A., Whisstock, J. C., & Warr, C. G. (2015). Torso-like mediates extracellular accumulation of Furin-cleaved Trunk to pattern the Drosophila embryo termini. *Nature Communications, 6*(1), 1–6.

Johnson, T. K., Henstridge, M. A., & Warr, C. G. (2017). MACPF/CDC proteins in development: Insights from *Drosophila* torso-like. *Seminars in Cell and Developmental Biology, 72*, 163–170.

Karr, T. L. (1991). Intracellular sperm/egg interactions in *Drosophila*: A three-dimensional structural analysis of a paternal product in the developing egg. *Mechanisms of Development, 34*(2–3), 101–111.

Keller Larkin, M., Deng, W.-M., Holder, K., Tworoger, M., Clegg, N., & Ruohola-Baker, H. (1999). Role of Notch pathway in terminal follicle cell differentiation during *Drosophila* oogenesis. *Development Genes and Evolution, 209*(5), 301–311.

Kim-Ha, J., Smith, J. L., & Macdonald, P. M. (1991). *oskar* mRNA is localized to the posterior pole of the Drosophila oocyte. *Cell, 66*(1), 23–35.

Kirilly, D., Wang, S., & Xie, T. (2011). Self-maintained escort cells form a germline stem cell differentiation niche. *Development, 138*(23), 5087–5097.

Kirov, N., Childs, S., O'Connor, M., & Rushlow, C. (2015). The *Drosophila dorsal* morphogen represses the *tolloid* gene by interacting with a silencer element. *Molecular and Cellular Biology, 14*(1), 713–722.

Klingler, M., Erdélyi, M., Szabad, J., & Nüsslein-Volhard, C. (1988). Function of torso in determining the terminal anlagen of the *Drosophila* embryo. *Nature*, *335*(6187), 275–277.

Lantz, V., & Schedl, P. (1994). Multiple *cis*-acting targeting sequences are required for *orb* mRNA localization during *Drosophila* oogenesis. *Molecular and Cellular Biology*, *14*(4), 2235–2242.

Laplante, C., & Nilson, L. A. (2006). Differential expression of the adhesion molecule Echinoid drives epithelial morphogenesis in *Drosophila*. *Development*, *133*(16), 3255–3264.

Larkin, M. K., Holder, K., Yost, C., Giniger, E., & Ruohola-Baker, H. (1996). Expression of constitutively active Notch arrests follicle cells at a precursor stage during *Drosophila* oogenesis and disrupts the anterior-posterior axis of the oocyte. *Development*, *122*(11), 3639–3650.

Lehmann, R. (2016). Germ plasm biogenesis—An Oskar-centric perspective. *Current Topics in Developmental Biology*, *116*, 679–707.

LeMosy, E. K., Tan, Y.-Q., & Hashimoto, C. (2001). Activation of a protease cascade involved in patterning the *Drosophila* embryo. *Proceedings of the National Academy of Sciences of the United States of America*, *98*(9), 5055–5060.

Lin, H., & Spradling, A. C. (1993). Germline stem cell division and egg chamber development in transplanted *Drosophila* germaria. *Developmental Biology*, *159*(1), 140–152.

Lin, H., & Spradling, A. C. (1997). A novel group of pumilio mutations affects the asymmetric division of germline stem cells in the Drosophila ovary. *Development*, *124*(12), 2463–2476.

Lin, H., Yue, L., & Spradling, A. C. (1994). The *Drosophila* fusome, a germline-specific organelle, contains membrane skeletal proteins and functions in cyst formation. *Development*, *120*(4), 947–956.

López-Schier, H. (2003). The polarisation of the anteroposterior axis in *Drosophila*. *BioEssays*, *25*(8), 781–791.

López-Schier, H., & St Johnston, D. (2001). Delta signaling from the germ line controls the proliferation and differentiation of the somatic follicle cells during *Drosophila* oogenesis. *Genes & Development*, *15*(11), 1393–1405.

Mach, J. M., & Lehmann, R. (1997). An Egalitarian-BicaudalD complex is essential for oocyte specification and axis determination in *Drosophila*. *Genes & Development*, *11*(4), 423–435.

Mahowald, A. P., & Strassheim, J. M. (1970). Intercellular migration of centrioles in the germarium of *Drosophila melanogaster*. An electron microscopic study. *The Journal of Cell Biology*, *45*(2), 306–320.

Mantrova, E. Y., & Hsu, T. (2008). Down-regulation of transcription factor CF2 by *Drosophila* Ras/MAP kinase signaling in oogenesis: Cytoplasmic retention and degradation. *Genes & Development*, *12*(8), 1166–1175.

Margolis, J., & Spradling, A. (1995). Identification and behavior of epithelial stem cells in the *Drosophila* ovary. *Development*, *121*(11), 3797–3807.

Martin, J.-R., Raibaud, A., & Ollo, R. (1994). Terminal pattern elements in *Drosophila* embryo induced by the *torso-like* protein. *Nature*, *367*(6465), 741–745.

McGregor, J. R., Xi, R., & Harrison, D. A. (2002). JAK signaling is somatically required for follicle cell differentiation in *Drosophila*. *Development*, *129*(3), 705–717.

McLaughlin, J. M., & Bratu, D. P. (2015). *Drosophila melanogaster* oogenesis: An overview. In *Drosophila oogenesis: Methods and protocols* (pp. 1–20). New York, NY: Humana Press.

Medioni, C., & Noselli, S. (2005). Dynamics of the basement membrane in invasive epithelial clusters in *Drosophila*. *Development*, *132*(13), 3069–3077.

Mineo, A., Furriols, M., & Casanova, J. (2018). The trigger (and the restriction) of Torso RTK activation. *Open Biology*, *8*(12), 1–6.

Momose, T., & Houliston, E. (2007). Two oppositely localised frizzled RNAs as axis determinants in a Cnidarian embryo. *PLoS Biology, 5*(4), 889–899.
Montell, D. J. (2003). Border-cell migration: The race is on. *Nature Reviews. Molecular Cell Biology, 4*(1), 13–24.
Montell, D. J., Rørth, P., & Spradling, A. C. (1992). *Slow border cells*, a locus required for a developmentally regulated cell migration during oogenesis, encodes Drosophila CEBP. *Cell, 71*(1), 51–62.
Montell, D. J., Yoon, W. H., & Starz-Gaiano, M. (2012). Group choreography: Mechanisms orchestrating the collective movement of border cells. *Nature Reviews Molecular Cell Biology, 13*(10), 631–645.
Morimoto, A. M., Jordan, K. C., Tietze, K., Britton, J. S., O'Neill, E. M., & Ruohola-Baker, H. (1996). Pointed, an ETS domain transcription factor, negatively regulates the EGF receptor pathway in *Drosophila* oogenesis. *Development, 122*(12), 3745–3754.
Morisato, D. (2001). Spätzle regulates the shape of the Dorsal gradient in the *Drosophila* embryo. *Development, 128*(12), 2309–2319.
Morisato, D., & Anderson, K. V. (1994). The *spätzle* gene encodes a component of the extracellular signaling pathway establishing the dorsal-ventral pattern of the Drosophila embryo. *Cell, 76*(4), 677–688.
Nashchekin, D., Fernandes, A. R., & St Johnston, D. (2016). Patronin/shot cortical foci assemble the noncentrosomal microtubule array that specifies the *Drosophila* anterior-posterior axis. *Developmental Cell, 38*(1), 61–72.
Neuman-Silberberg, F. S., & Schüpbach, T. (1993). The Drosophila dorsoventral patterning gene *gurken* produces a dorsally localized RNA and encodes a TGF alpha-like protein. *Cell, 75*(1), 165–174.
Neuman-Silberberg, F. S., & Schüpbach, T. (1996). The *Drosophila* TGF-alpha-like protein Gurken: Expression and cellular localization during *Drosophila* oogenesis. *Mechanisms of Development, 59*(2), 105–113.
Nilson, L. A., & Schüpbach, T. (1998). Localized requirements for *windbeutel* and *pipe* reveal a dorsoventral prepattern within the follicular epithelium of the *Drosophila* ovary. *Cell, 93*(2), 253–262.
Norvell, A., Kelley, R. L., Wehr, K., & Schüpbach, T. (1999). Specific isoforms of Squid, a *Drosophila* hnRNP, perform distinct roles in Gurken localization during oogenesis. *Genes and Development, 13*(7), 864–876.
Nüsslein-Volhard, C., Frohnhöfer, H. G., & Lehmann, R. (1987). Determination of anteroposterior polarity in Drosophila. *Science, 238*(4834), 1675–1681.
Nystul, T., & Spradling, A. (2010). Regulation of epithelial stem cell replacement and follicle formation in the Drosophila ovary. *Genetics, 184*(2), 503–515.
Page, S. L., & Hawley, R. S. (2001). *c(3)G* encodes a *Drosophila* synaptonemal complex protein. *Genes & Development, 15*(23), 3130–3143.
Pai, L. M., Barcelo, G., & Schüpbach, T. (2000). *D-cbl*, a negative regulator of the EGFR pathway, is required for dorsoventral patterning in *Drosophila* oogenesis. *Cell, 103*(1), 51–61.
Palacios, I. M., & St Johnston, D. (2002). Kinesin light chain-independent function of the Kinesin heavy chain in cytoplasmic streaming and posterior localisation in the *Drosophila* oocyte. *Development, 129*(23), 5473–5485.
Peri, F., Bökel, C., & Roth, S. (1999). Local Gurken signaling and dynamic MAPK activation during Drosophila oogenesis. *Mechanisms of Development, 81*(1–2), 75–88.
Peri, F., & Roth, S. (2000). Combined activities of Gurken and Decapentaplegic specify dorsal chorion structures of the *Drosophila* egg. *Development, 127*(4), 841–850.
Perrimon, N., Lanjuin, A., Arnold, C., & Noll, E. (1996). Zygotic lethal mutations with maternal effect phenotypes in *Drosophila melanogaster*. II. Loci on the second and third chromosomes identified by *P*-element-induced mutations. *Genetics, 144*(4), 1681–1692.

Ponting, C. P. (1999). Chlamydial homologues of the MACPF (MAC/perforin) domain. *Current Biology*, *9*(24), R911–R913.

Queenan, A. M., Barcelo, G., Van Buskirk, C., & Schüpbach, T. (1999). The transmembrane region of Gurken is not required for biological activity, but is necessary for transport to the oocyte membrane in Drosophila. *Mechanisms of Development*, *89*(1–2), 35–42.

Rabinowitz, M. (1941). Studies on the cytology and early embryology of the egg of Drosophila melanogaster. *Journal of Morphology*, *69*(1), 1–49.

Ray, R. P., Arora, K., Nüsslein-Volhard, C., & Gelbart, W. M. (1991). The control of cell fate along the dorsal-ventral axis of the *Drosophila* embryo. *Development*, *113*(1), 35–54.

Rebagliati, M. R., Weeks, D. L., Harvey, R. P., & Melton, D. A. (1985). Identification and cloning of localized maternal RNAs from Xenopus eggs. *Cell*, *42*(3), 769–777.

Reeves, G. T., & Stathopoulos, A. (2009). Graded Dorsal and differential gene regulation in the *Drosophila* embryo. *Cold Spring Harbor Perspectives in Biology*, *1*(4), 1–16.

Rosales-Nieves, A. E., & González-Reyes, A. (2014). Genetics and mechanisms of ovarian cancer: Parallels between *Drosophila* and humans. *Seminars in Cell and Developmental Biology*, *28*, 104–109.

Roth, S., Jordan, P., & Karess, R. (1999). Binuclear *Drosophila* oocytes: Consequences and implications for dorsal-ventral patterning in oogenesis and embryogenesis. *Development*, *126*(5), 927–934.

Roth, S., & Lynch, J. A. (2009). Symmetry breaking during *Drosophila* oogenesis. *Cold Spring Harbor Perspectives in Biology*, *1*(2), 1–21.

Roth, S., Neuman-Silberberg, F. S., Barcelo, G., & Schüpbach, T. (1995). *Cornichon* and the EGF receptor signaling process are necessary for both anterior-posterior and dorsal-ventral pattern formation in Drosophila. *Cell*, *81*(6), 967–978.

Roth, S., & Schüpbach, T. (1994). The relationship between ovarian and embryonic dorsoventral patterning in *Drosophila*. *Development*, *120*(8), 2245–2257.

Roth, S., Stein, D., & Nüsslein-Volhard, C. (1989). A gradient of nuclear localization of the *dorsal* protein determines dorsoventral pattern in the Drosophila embryo. *Cell*, *59*(6), 1189–1202.

Ruohola, H., Bremer, K. A., Baker, D., Swedlow, J. R., Jan, L. Y., & Jan, Y. N. (1991). Role of neurogenic genes in establishment of follicle cell fate and oocyte polarity during oogenesis in Drosophila. *Cell*, *66*(3), 433–449.

Ruohola-Baker, H., Grell, E., Chou, T. B., Baker, D., Jan, L. Y., & Jan, Y. N. (1993). Spatially localized rhomboid is required for establishment of the dorsal-ventral axis in Drosophila oogenesis. *Cell*, *73*(5), 953–965.

Rushlow, C. A., Han, K., Manley, J. L., & Levine, M. (1989). The graded distribution of the *dorsal* morphogen is initiated by selective nuclear transport in Drosophila. *Cell*, *59*(6), 1165–1177.

Sapir, A., Schweitzer, R., & Shilo, B.-Z. (1998). Sequential activation of the EGF receptor pathway during *Drosophila* oogenesis establishes the dorsoventral axis. *Development*, *125*(2), 191–200.

Savant-Bhonsale, S., & Montell, D. J. (1993). *Torso-like* encodes the localized determinant of *Drosophila* terminal pattern formation. *Genes & Development*, 7(12 B), 2548–2555.

Schneider, M., Khalil, A. A., Poulton, J., Castillejo-Lopez, C., Egger-Adam, D., Wodarz, A., et al. (2006). Perlecan and Dystroglycan act at the basal side of the *Drosophila* follicular epithelium to maintain epithelial organization. *Development*, *133*(19), 3805–3815.

Schnorr, J. D., & Berg, C. A. (1996). Differential activity of *Ras1* during patterning of the Drosophila dorsoventral axis. *Genetics*, *144*(4), 1545–1557.

Schnorrer, F., Bohmann, K., & Nüsslein-Volhard, C. (2000). The molecular motor dynein is involved in targeting Swallow and *bicoid* RNA to the anterior pole of *Drosophila* oocytes. *Nature Cell Biology*, *2*(4), 185–190.

Schüpbach, T. (1987). Germ line and soma cooperate during oogenesis to establish the dorsoventral pattern of egg shell and embryo in Drosophila melanogaster. *Cell*, *49*(5), 699–707.

Schüpbach, T., & Wieschaus, E. (1986). Maternal-effect mutations altering the anterior-posterior pattern of the Drosophila embryo. *Roux's Archives of Developmental Biology*, *195*(5), 302–317.

Schüpbach, T., & Wieschaus, E. (1991). Female sterile mutations on the second chromosome of *Drosophila melanogaster*. II. Mutations blocking oogenesis or altering egg morphology. *Genetics*, *129*(4), 1119–1136.

Sen, J., Goltz, J. S., Konsolaki, M., Schüpbach, T., & Stein, D. (2000). Windbeutel is required for function and correct subcellular localization of the *Drosophila* patterning protein pipe. *Development*, *127*(24), 5541–5550.

Sen, J., Goltz, J. S., Stevens, L., & Stein, D. (1998). Spatially restricted expression of *pipe* in the *Drosophila* egg chamber defines embryonic dorsal-ventral polarity. *Cell*, *95*(4), 471–481.

Silver, D. L., & Montell, D. J. (2001). Paracrine signaling through the JAK/STAT pathway activates invasive behavior of ovarian epithelial cells in *Drosophila*. *Cell*, *107*(7), 831–841.

Smith, C. L., & Delotto, R. (1994). Ventralizing signal determined by protease activation in *Drosophila* embryogenesis. *Nature*, *368*(6471), 548–551.

Snow, P. M., Bieber, A. J., & Goodman, C. S. (1989). Fasciclin III: A novel homophilic adhesion molecule in Drosophila. *Cell*, *59*(2), 313–323.

Song, X., Zhu, C. H., Doan, C., & Xie, T. (2002). Germline stem cells anchored by adherens junctions in the *Drosophila* ovary niches. *Science*, *296*(5574), 1855–1857.

Spradling, A. C. (1993). Developmental genetics of oogenesis. In M. Bate & A. Martinez-Arias (Eds.), *The development of* Drosophila melanogaster*: Vol. 1* (pp. 1–70). Cold Spring Harbor, NY: Cold Spring Harbor Press.

Sprenger, F., & Nüsslein-Volhard, C. (1992). Torso receptor activity is regulated by a diffusible ligand produced at the extracellular terminal regions of the Drosophila egg. *Cell*, *71*(6), 987–1001.

Sprenger, F., Stevens, L. M., & Nüsslein-Volhard, C. (1989). The *Drosophila* gene *torso* encodes a putative receptor tyrosine kinase. *Nature*, *338*(6215), 478–483.

St Johnston, D., Beuchle, D., & Nüsslein-Volhard, C. (1991). *Staufen*, a gene required to localize maternal RNAs in the Drosophila egg. *Cell*, *66*(1), 51–63.

Stathopoulos, A., & Levine, M. (2005). Genomic regulatory networks and animal development. *Developmental Cell*, *9*(4), 449–462.

Stathopoulos, A., Van Drenth, M., Erives, A., Markstein, M., & Levine, M. (2002). Whole-genome analysis of Dorsal-ventral patterning in the *Drosophila* embryo. *Cell*, *111*(5), 687–701.

Stein, D., & Nüsslein-Volhard, C. (1992). Multiple extracellular activities in Drosophila egg perivitelline fluid are required for establishment of embryonic dorsal-ventral polarity. *Cell*, *68*(3), 429–440.

Stein, D. S., & Stevens, L. M. (2014). Maternal control of the *Drosophila* dorsal-ventral body axis. *Wiley Interdisciplinary Reviews: Developmental Biology*, *3*(5), 301–330.

Stevens, L. M., Beuchle, D., Jurcsak, J., Tong, X., & Stein, D. (2003). The *Drosophila* embryonic patterning determinant torsolike is a component of the eggshell. *Current Biology*, *13*(12), 1058–1063.

Stevens, L. M., Frohnhöfer, H. G., Klingler, M., & Nüsslein-Volhard, C. (1990). Localized requirement for *torso-like* expression in follicle cells for development of terminal anlagen of the *Drosophila* embryo. *Nature*, *346*(6285), 660–663.

Steward, R. (1989). Relocalization of the *dorsal* protein from the cytoplasm to the nucleus correlates with its function. *Cell*, *59*(6), 1179–1188.

Strecker, T. R., Halsell, S. R., Fisher, W. W., & Lipshitz, H. D. (1989). Reciprocal effects of hyper- and hypoactivity mutations in the *Drosophila* pattern gene *torso*. *Science*, *243*(4894), 1062–1066.

Sun, Y., Yan, Y., Denef, N., & Schüpbach, T. (2011). Regulation of somatic myosin activity by protein phosphatase 1β controls *Drosophila* oocyte polarization. *Development*, *138*(10), 1991–2001.

Suter, B., Romberg, L. M., & Steward, R. (1989). *Bicaudal-D*, a *Drosophila* gene involved in developmental asymmetry: Localized transcript accumulation in ovaries and sequence similarity to myosin heavy chain tail domains. *Genes & Development*, *3*(12A), 1957–1968.

Tanentzapf, G., & Tepass, U. (2003). Interactions between the *crumbs*, *lethal giant larvae* and *bazooka* pathways in epithelial polarization. *Nature Cell Biology*, *5*(1), 46–52.

Theurkauf, W. E. (1994). Microtubules and cytoplasm organization during Drosophila oogenesis. *Developmental Biology*, *165*(2), 352–360.

Theurkauf, W. E., Alberts, B. M., Jan, Y. N., & Jongens, T. A. (1993). A central role for microtubules in the differentiation of *Drosophila* oocytes. *Development*, *118*(4), 1169–1180.

Theurkauf, W. E., & Hawley, R. S. (1992). Meiotic spindle assembly in *Drosophila* females: Behavior of nonexchange chromosomes and the effects of mutations in the *nod* kinesin-like protein. *Journal of Cell Biology*, *116*(5), 1167–1180.

Theurkauf, W. E., Smiley, S., Wong, M. L., & Alberts, B. M. (1992). Reorganization of the cytoskeleton during *Drosophila* oogenesis: Implications for axis specification and intercellular transport. *Development*, *115*(4), 923–936.

Tissot, N., Lepesant, J. A., Bernard, F., Legent, K., Bosveld, F., Martin, C., et al. (2017). Distinct molecular cues ensure a robust microtubule-dependent nuclear positioning in the *Drosophila* oocyte. *Nature Communications*, *8*(1), 1–13.

Torres, I. L., López-Schier, H., & St Johnston, D. (2003). A Notch/Delta-dependent relay mechanism establishes anterior-posterior polarity in *Drosophila*. *Developmental Cell*, *5*(4), 547–558.

Trovisco, V., Belaya, K., Nashchekin, D., Irion, U., Sirinakis, G., Butler, R., et al. (2016). *Bicoid* mRNA localises to the *Drosophila* oocyte anterior by random Dynein-mediated transport and anchoring. *eLife*, *5*, e17537.

Twombly, V., Blackman, R. K., Jin, H., Graff, J. M., Padgett, R. W., & Gelbart, W. M. (1996). The TGF-β signaling pathway is essential for *Drosophila* oogenesis. *Development*, *122*(5), 1555–1565.

Tworoger, M., Larkin, M. K., Bryant, Z., & Ruohola-Baker, H. (1999). Mosaic analysis in the Drosophila ovary reveals a common hedgehog-inducible precursor stage for stalk and polar cells. *Genetics*, *151*(2), 739–748.

Tzolovsky, G., Deng, W.-M., Schlitt, T., & Bownes, M. (1999). The function of the broad-complex during *Drosophila* oogenesis. *Genetics*, *153*(3), 1–13.

Van Buskirk, C., & Schüpbach, T. (1999). Versatility in signalling: Multiple responses to EGF receptor activation during *Drosophila* oogenesis. *Trends in Cell Biology*, *9*(1), 1–4.

Wasserman, J. D., & Freeman, M. (1998). An autoregulatory cascade of EGF receptor signaling patterns the *Drosophila* egg. *Cell*, *95*(3), 355–364.

Weil, T. T., Forrest, K. M., & Gavis, E. R. (2006). Localization of *bicoid* mRNA in late oocytes is maintained by continual active transport. *Developmental Cell*, *11*(2), 251–262.

Wharton, R. P., & Struhl, G. (1989). Structure of the Drosophila *BicaudalD* protein and its role in localizing the posterior determinant *nanos*. *Cell*, *59*(5), 881–892.

Wieschaus, E., & Szabad, J. (1979). The development and function of the female germ line in *Drosophila melanogaster*: A cell lineage study. *Developmental Biology*, *68*(1), 29–46.

Wilt, F. H. (1987). Determination and morphogenesis in the sea urchin embryo. *Development*, *100*(4), 559–576.

Wittes, J., & Schüpbach, T. (2019). A gene expression screen in *Drosophila melanogaster* identifies novel JAK/STAT and EGFR targets during oogenesis. *G3 (Bethesda)*, *9*(1), 47–60.

Xi, R., McGregor, J. R., & Harrison, D. A. (2003). A gradient of JAK pathway activity patterns the anterior-posterior axis of the follicular epithelium. *Developmental Cell*, *4*(2), 167–177.

Xie, T., & Spradling, A. (1998). *Decapentaplegic* is essential for the maintenance and division of germline stem cells in the *Drosophila* ovary. *Cell*, *94*(2), 251–260.

Xie, T., & Spradling, A. C. (2000). A niche maintaining germ line stem cells in the *Drosophila* ovary. *Science*, *290*(5490), 328–330.

Yakoby, N., Bristow, C. A., Gong, D., Schafer, X., Lembong, J., Zartman, J. J., et al. (2008). A combinatorial code for pattern formation in *Drosophila* oogenesis. *Developmental Cell*, *15*(5), 725–737.

Yan, Y., Denef, N., & Schüpbach, T. (2009). The vacuolar proton pump, V-ATPase, is required for notch signaling and endosomal trafficking in *Drosophila*. *Developmental Cell*, *17*(3), 387–402.

Yue, L., & Spradling, A. C. (1992). *hu-li tai shao*, a gene required for ring canal formation during *Drosophila* oogenesis, encodes a homolog of adducin. *Genes & Development*, *6*(12B), 2443–2454.

Zartman, J. J., Kanodia, J. S., Cheung, L. S., & Shvartsman, S. Y. (2009). Feedback control of the EGFR signaling gradient: Superposition of domain-splitting events in *Drosophila* oogenesis. *Development*, *136*(17), 2903–2911.

Zhang, Y., & Kalderon, D. (2001). Hedgehog acts as a somatic stem cell factor in the *Drosophila* ovary. *Nature*, *410*(6828), 599–604.

Zhang, Z., Stevens, L. M., & Stein, D. (2009). Sulfation of eggshell components by pipe defines dorsal-ventral polarity in the *Drosophila* embryo. *Current Biology*, *19*(14), 1200–1205.

Zhao, T., Graham, O. S., Raposo, A., & St Johnston, D. (2012). Growing microtubules push the oocyte nucleus to polarize the *Drosophila* dorsal-ventral axis. *Science*, *336*(6084), 999–1003.

CHAPTER FOUR

Organizing the oocyte: RNA localization meets phase separation

Sarah E. Cabral, Kimberly L. Mowry*
Department of Molecular Biology, Cell Biology, and Biochemistry, Brown University, Providence, RI, United States
*Corresponding author: e-mail address: kimberly_mowry@brown.edu

Contents

1. Introduction 88
2. Functional advantages of RNA localization 89
3. Mechanisms of RNA localization 91
4. RNA localization in *Xenopus* oocytes 94
   4.1 *Xenopus* oogenesis 95
   4.2 *Xenopus* RNA localization pathways 96
5. RNPs and localization 100
   5.1 *Cis* sequences and "zip codes" 100
   5.2 *Trans*-factors 101
6. Biomolecular condensates and phase separation 102
   6.1 General principles of phase separation 103
   6.2 RNA localization and phase separation 105
7. Biomolecular condensates in oocytes and embryos 106
   7.1 Nucleoli 106
   7.2 P granules in *C. elegans* 106
   7.3 Germ granules in *Drosophila* 107
   7.4 The Balbiani body in *Xenopus* oocytes 108
8. Conclusions 109
Acknowledgments 110
References 110

Abstract

RNA localization is a key biological strategy for organizing the cytoplasm and generating both cellular and developmental polarity. During RNA localization, RNAs are targeted asymmetrically to specific subcellular destinations, resulting in spatially and temporally restricted gene expression through local protein synthesis. First discovered in oocytes and embryos, RNA localization is now recognized as a significant regulatory strategy for diverse RNAs, both coding and non-coding, in a wide range of cell types. Yet, the highly polarized cytoplasm of the oocyte remains a leading model to understand not only the

Current Topics in Developmental Biology, Volume 140
ISSN 0070-2153
https://doi.org/10.1016/bs.ctdb.2020.02.007

principles and mechanisms underlying RNA localization, but also links to the formation of biomolecular condensates through phase separation. Here, we discuss both RNA localization and biomolecular condensates in oocytes with a particular focus on the oocyte of the frog, *Xenopus laevis*.

1. Introduction

In 1957, Francis Crick formulated the modern framework for molecular biology by coining the term "central dogma" and describing the flow of biological information from DNA to RNA to protein (Crick, 1958). Since that time, regulatory paradigms have been described for each of the steps of the central dogma. In particular, RNA localization has been found to exert precise control of gene expression both temporally and spatially. The first examples of localized RNAs were identified in the 1980s, and in the intervening decades RNA localization has been described for a stunning number of RNAs in a wide range of cells types (Jeffery, Tomlinson, & Brodeur, 1983; Lawrence & Singer, 1986; Rebagliati, Weeks, Harvey, & Melton, 1985). Notable examples are found in evolutionarily distant organisms, with functions that include regulation of mating type switching in yeast, facilitation of motility in fibroblasts, and control of synaptic plasticity and axonal guidance in neurons (Mingle et al., 2005; Puthanveettil, 2013; reviewed in Singer-Krüger & Jansen, 2014). Although nearly ubiquitous in cells, RNA localization is perhaps best-characterized as a regulator of developmental patterning in vertebrate and invertebrate oocytes (reviewed in Houston, 2013; Medioni, Mowry, & Besse, 2012). However, asymmetric RNA localization is not restricted to mRNAs and is used to localize both coding and non-coding RNAs in a wide variety of cell types, suggesting that RNA localization provides a paradigm for organization of the cell throughout biology (reviewed in Cabili et al., 2015; Lécuyer et al., 2007; Weiß, Antoniou, & Schratt, 2015; Wilk, Hu, Blotsky, & Krause, 2016).

Organization of the cytoplasm is also achieved through formation of biomolecular condensates, which are cellular subcompartments that form through phase separation and function to concentrate proteins and nucleic acids without the use of a membrane (reviewed in Banani, Lee, Hyman, & Rosen, 2017). While now identified in an ever-growing set of cell types and subcellular locations, the principles of phase separation were first described in oocytes and condensates have been found across many species in these cells (Brangwynne et al., 2009). Importantly, new insights into the biophysical basis for phase separation in cells coupled with intriguing links to RNA localization have reframed our view of both processes. Here, we review how studies in the oocyte model have established a basis for our understanding of the mechanisms directing RNA localization and the functional advantages it can confer. We then discuss recent advances in the field of phase separation and the range of biomolecular condensates found in oocytes, highlighting the intrinsic links between RNA localization and phase separation.

2. Functional advantages of RNA localization

In an mRNA-centric view, RNA localization provides numerous advantages to the cell over transport of protein products. Particularly in large cells such as oocytes, it has been hypothesized that it would be impossible to establish the gradients of factors and subcellular localization patterns required by the cell by merely transporting translated protein products (reviewed in Blower, 2013). However, as noted above, RNA localization is not only utilized in large, highly polarized cells like oocytes, but has been shown to be a general mechanism used in a wide variety of non-germline cells, including somatic cells, plant cells, and even unicellular organisms (Buskilay, Kannaiahy, & Amster-Choder, 2014; Cajigas et al., 2012; Okita & Choi, 2002; Singer-Krüger & Jansen, 2014; Sundell & Singer, 1991). In some of these examples, the advantages of RNA localization to the cell have not yet been defined. However, well-characterized examples of RNA localization have demonstrated a number of non-mutually exclusive advantages to this process, as detailed below (reviewed in Blower, 2013; Buxbaum, Haimovich, & Singer, 2015; Du, Schmid, & Jansen, 2007; Martin & Ephrussi, 2009; St Johnston, 2005).

First, asymmetric enrichment of mRNAs in a particular subcellular domain through RNA localization provides an energetically favorable mechanism to spatially and temporally decouple protein translation from mRNA transcription. Thus, cells can transport translation machinery and only a few copies of an mRNA to a particular site, often far from the nucleus, where multiple rounds of local translation can produce strong enrichment of the encoded protein. This energetic favorability is clearly advantageous in large cells such as oocytes and neurons, as well as in other polarized cells. For example, in fibroblasts mRNAs are localized to lamellipodia to allow for the locally high concentrations of cytoskeletal elements necessary for rapid motility (Kislauskis, Zhu, & Singer, 1997; Lawrence & Singer, 1986; Mardakheh et al., 2015). Second, mRNA localization, which generally transports translationally silenced mRNAs, can prevent ectopic protein activity during localization which may be toxic or deleterious to the cell. This is best-illustrated in germ cells, where developmental determinants are often localized as mRNAs rather than proteins, preventing errors in embryonic patterning that can arise through spatially inappropriate protein activity (Ephrussi, Dickinson, & Lehmann, 1991; Gavis & Lehmann, 1992). Third, mRNA localization and local translation allow for rapid response to stimuli by bypassing the need for signaling from the cytoplasm back to the nucleus, waiting for a transcriptional and translational response, and then transporting the protein product back to the appropriate site. This rapid response to external stimuli is particularly important in neurons where rapid translational responses far from the cell body are important for synaptic plasticity (reviewed in Martin, Barad, & Kandel, 2000). However, RNA localization is not restricted to protein-coding mRNAs. Localization of non-coding RNAs has been proposed to play a role in long-term memory formation by regulating the local translation and stability of these synapse localized mRNAs (reviewed in Mercer et al., 2008). Finally, an underappreciated advantage of localizing RNAs is that this process can bias macromolecular complex formation by creating intracellular regions of high RNA and protein concentration. Traditionally, this process has been thought to facilitate the incorporation of nascent proteins into complexes with particular stoichiometries or differential binding partners depending on the site of translation (reviewed in Rodriguez, Czaplinski, Condeelis, & Singer, 2008). However, this inherent advantage of RNA localization can also be reconsidered in the context of biomolecular condensate formation. In this view, locally high concentrations of RNA and protein factors established by RNA localization can facilitate phase separation. This advantage may underlie the localization of non-coding RNAs where the RNA itself could be acting as a scaffold for macromolecular complex formation or as a regulator at the site of enrichment (Clemson et al., 2009; Kloc et al., 2005; Weiß et al., 2015).

While many advantages of mRNA localization coupled with local translation have been well described in the literature, the functions of localized non-coding RNAs are only recently emerging. In particular, functions for localized long non-coding RNAs (lncRNAs) have been identified both in the nucleus where lncRNAs can perform structural or gene regulatory roles, and in the cytoplasm, where localized lncRNAs can also function to regulate gene expression (Chen, 2016; Clemson et al., 2009; Noh, Kim, McClusky, Abdelmohsen, & Gorospe, 2018). For example, the *neat1* lncRNA functions in the nucleus of mammalian cells as a necessary RNA scaffold for paraspeckle formation (Clemson et al., 2009). However, the functional advantages for localizing most lncRNAs, as well as many other classes of non-coding RNAs, remain to be determined. An exciting challenge for the field will be to not only identify the particular non-coding RNAs that are localized asymmetrically, but also to define how those RNAs function at their destinations.

3. Mechanisms of RNA localization

Traditionally, three primary mechanisms have been described to generate asymmetric RNA localization: diffusion and local entrapment, local stabilization and regulated degradation, and, most prominently in the literature, active transport by molecular motors (reviewed in Gagnon & Mowry, 2011; Martin & Ephrussi, 2009). More recently, formation of biomolecular condensates has emerged as a fourth mechanism for generating asymmetry in RNA distribution (reviewed in Langdon & Gladfelter, 2018). These mechanisms, schematized in Fig. 1, have been observed in a wide array of somatic cells, germ cells, and even unicellular organisms and facilitate the localization of diverse RNAs.

The first mechanism of RNA localization, diffusion and local entrapment, is characterized by the accumulation of RNA at a subcellular destination through the capture of RNA molecules which are freely diffusing through the cell. For example, in *Drosophila*, *nanos* mRNA is maternally deposited at the anterior of the developing oocyte and freely diffuses until it is captured and stably anchored at the posterior pole of the oocyte (Forrest & Gavis, 2003). As shown in Fig. 1A, the *Xenopus* homolog of *nanos* is enriched in the Balbiani body during early oogenesis via a diffusion and local entrapment mechanism (Chang et al., 2004). Interestingly, recent evidence has shown that both *Drosophila* germ granules and the *Xenopus* Balbiani body are biomolecular condensates, illustrating not only that many RNAs are localized in phase separated intermediates, but also suggesting that incorporation into phase separated structures may be a key mechanism for entrapment of diffusing RNAs (Boke et al., 2016; Kistler et al., 2018).

The second mechanism of RNA localization, local stabilization and regulated degradation, is characterized by the enrichment of RNAs in a subcellular destination through global degradation of non-localized RNA

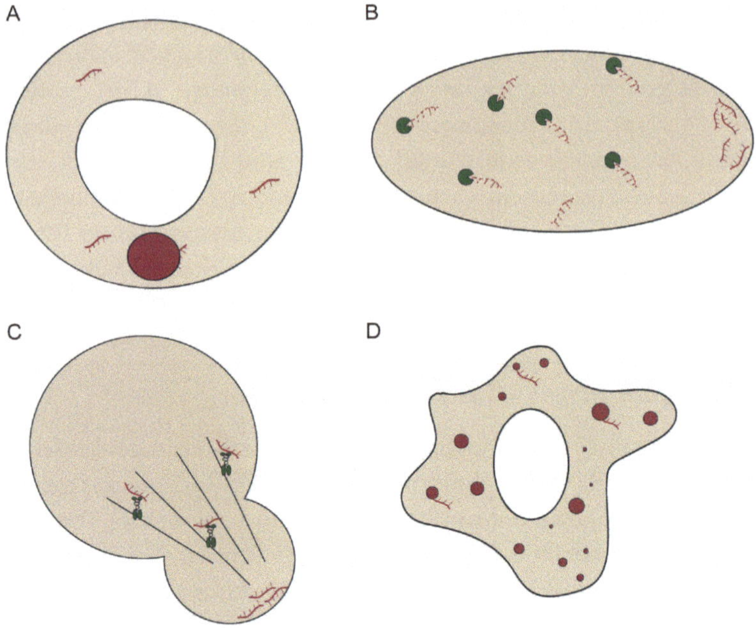

Fig. 1 Mechanisms of RNA localization. Asymmetric RNA localization is achieved through four mechanisms. (A) Diffusion and local entrapment leads to the enrichment of *nanos1* mRNA (magenta) in the Balbiani body in *Xenopus* oocytes. The Balbiani body is shown below the oocyte nucleus (GV), which is in the center of the oocyte. (B) Local stabilization and regulated degradation leads to the enrichment of *hsp83* mRNA (magenta) in the posterior pole (right) of the *Drosophila* embryo. Degradation outside the posterior pole is directed by Smaug-RNA binding and recruitment of the CCR4-NOT complex (green). (C) Active motor-based transport leads to the enrichment of *ash1* mRNA (magenta) at the tip of the daughter cell (lower right) in budding yeast. Transport is dependent on myosin (green) along actin filaments (black). (D) Incorporation into biomolecular condensates enriches *actin* mRNAs (magenta) in stress granules in mammalian cells.

and selective protection of localized RNA, rather than the entrapment or transport of the RNA. For *nanos* mRNA in the *Drosophila* embryo, the diffusion and local entrapment process described above is highly inefficient and only ~4% of the total *nanos* mRNA is enriched at the posterior pole (Gavis & Lehmann, 1992). To create the robust localization necessary for proper anterior-posterior patterning, the fraction of the mRNA enriched at the posterior pole is stabilized, whereas the non-localized transcripts are targeted for degradation through recruitment of the CCR4-NOT deadenylase complex by Smaug, an RNA binding protein (Zaessinger, Busseau, & Simonelig, 2006). The dual mechanisms for localization of *nanos* mRNA, coupling diffusion and local entrapment with local stabilization and selective degradation, highlight how a single transcript can utilize multiple mechanisms of RNA localization to achieve stringent enrichment at a particular location. As shown in Fig. 1B, local stabilization and regulated degradation is also used in the *Drosophila* embryo to enrich *heat shock protein 83* (*hsp83*) mRNA in the posterior pole (Ding, Parkhurst, Halsell, & Lipshitz, 1993). Similarly to *nanos* mRNA, *hsp83* mRNA is degraded outside of the posterior pole by the binding of Smaug to elements in the 3′ untranslated region (UTR) of the mRNA (Bashirullah, Cooperstock, & Lipshitz, 2001; Chen et al., 2014; Semotok et al., 2008; Tadros et al., 2007).

The third mechanism of RNA localization, and most predominant in the literature, is active, motor-based transport of RNA along the cytoskeleton. Active transport allows for faster and more long-range RNA enrichment than simple diffusion and is essential for the localization of many RNAs. For example, as shown in Fig. 1C, *ash1* mRNA, a regulator of mating type switching in budding yeast, is localized to the tip of the daughter cell along actin filaments in a myosin-dependent manner (Long et al., 1997; Takizawa & Vale, 2000). Motor-based transport is also important for the localization of RNAs in neurons, with numerous examples including the kinesin-based transport of β-actin mRNA to growth cones in immature neurons (reviewed in Das, Singer, & Yoon, 2019; Kiebler & Bassell, 2006; Zhang, Singer, & Bassell, 1999). Active transport of RNAs allows not only for rapid localization, but for exquisite control over the site of enrichment based on both biases in the orientation of cytoskeletal elements and by the activity of different molecular motors (reviewed in Gagnon & Mowry, 2011). This tight control of localization is critical for embryonic patterning, as illustrated by the distinct localization patterns adopted by *bicoid*, *oskar*, and *gurken* RNAs in the *Drosophila* oocyte (reviewed in Weil, 2014). In this system, many mRNAs are synthesized in nurse cells and are transported through ring canals, or cytoplasmic bridges, into the oocyte by dynein (Clark, Meignin, & Davis, 2007). Once in the oocyte, continued motor-based transport of these RNAs allows them to adopt specific localization patterns (reviewed in Kugler & Lasko, 2009; St Johnston, 2005). For example, *bicoid* mRNA localizes to the anterior of the oocyte via active transport by the minus end directed motor, dynein, along microtubules nucleated at the anterior cortex (Duncan & Warrior, 2002). Also transported by dynein is *gurken* mRNA, which is first transported, like *bicoid* mRNA, to the anterior of the oocyte (MacDougall, Clark, MacDougall, & Davis, 2003). In a second step, *gurken* RNA is subsequently transported to the dorsal-anterior corner of the oocyte along microtubules nucleated by the oocyte nucleus (MacDougall et al., 2003). Localization of *oskar* mRNA to the posterior of the oocyte relies instead on the plus end directed microtubule motor, kinesin (Brendza, Serbus, Duffy, & Saxton, 2000; Cha, Serbus, Koppetsch, & Theurkauf, 2002; Gáspár, Sysoev, Komissarov, & Ephrussi, 2017; Januschke et al., 2002; Zimyanin et al., 2008). As will be discussed in detail below, RNAs are also transported by molecular motors along microtubule arrays to the vegetal pole of the *Xenopus* oocyte (Gagnon, Kreiling, Powrie, Wood, & Mowry, 2013; Messitt et al., 2008).

Finally, the most recently described mechanism of RNA localization is incorporation into biomolecular condensates (reviewed in Langdon & Gladfelter, 2018). Through this mechanism, RNAs are enriched into non-membrane bound substructures based on the thermodynamic properties of the RNA and surrounding proteins, often in response to cellular cues. For example, as shown in Fig. 1D, polyA$^+$ mRNAs are sequestered during times of cellular distress to stress granules, which form by phase separation (reviewed in Kedersha, Ivanov, & Anderson, 2013). This mechanism of RNA localization allows for specific RNAs to be enriched within biomolecular condensates depending on the cellular environment and type of condensate. For example, as a means to reduce energy expenditure while allowing cells to repair damage, stress granules recruit RNAs encoding house-keeping genes, such as β-actin and GAPDH, restricting them from being actively translated, while excluding RNAs encoding proteins involved in stress-related repair such as heat shock proteins (reviewed in Fay & Anderson, 2018) While the mechanisms by which RNAs are incorporated into these structures is less well characterized, RNA has been shown to play a variety of roles in forming the biomolecular condensates in which they are enriched (reviewed in Van Treeck & Parker, 2018). Such roles include active structural roles in condensates, buffering phase separation, and regulation of the physical state of the condensates (Clemson et al., 2009; Jain & Vale, 2017; Langdon et al., 2018; Maharana et al., 2018; Van Treeck et al., 2018).

4. RNA localization in *Xenopus* oocytes

While RNA localization has been observed nearly universally across cell types and organisms, the developing oocyte of the African clawed frog, *Xenopus laevis*, has proven to be an ideal model system for the study of RNA localization in vertebrates. The *Xenopus* oocyte is an extremely large, highly polarized cell that uses a variety of mechanisms to localize both maternal mRNAs and non-coding RNAs during oogenesis, many of which are required for proper patterning of the embryo and specification of the germline (reviewed in Holt & Bullock, 2010; King, Messitt, & Mowry, 2005; Medioni et al., 2012). The localization of these RNAs is developmentally coordinated during oogenesis, the stages of which are briefly described below and provide a framework for the pathways of RNA localization that operate during *Xenopus* oogenesis.

4.1 *Xenopus* oogenesis

As depicted in Fig. 2, *Xenopus* oogenesis proceeds through six morphologically distinct stages, designated I–VI, which are defined by increasing size, yolk accumulation, and changes in external pigmentation (Dumont, 1972). During stage I of oogenesis, the oocyte is transparent with the germinal vesicle (GV) and Balbiani body, also termed the mitochondrial cloud, visible in the cytoplasm (Heasman, Quarmby, & Wylie, 1984). The Balbiani body is closely associated with the GV through a network of cytokeratin filaments and is the primary site of mitochondrial replication in the pre-vitellogenic oocyte. It also contains many classic germ line determinants, such as the *Xenopus* homolog of *nanos* mRNA and Vasa protein (reviewed in King et al., 2005). During stage II, the oocytes grow and become increasingly opaque, turning white in color and obscuring the GV from view (Dumont, 1972). In stage III, vitellogenesis, or the production of yolk, and pigmentation of the oocyte begin (Wallace & Dumont, 1968). These oocytes have nonpolarized dark pigment, turning the oocyte light brown

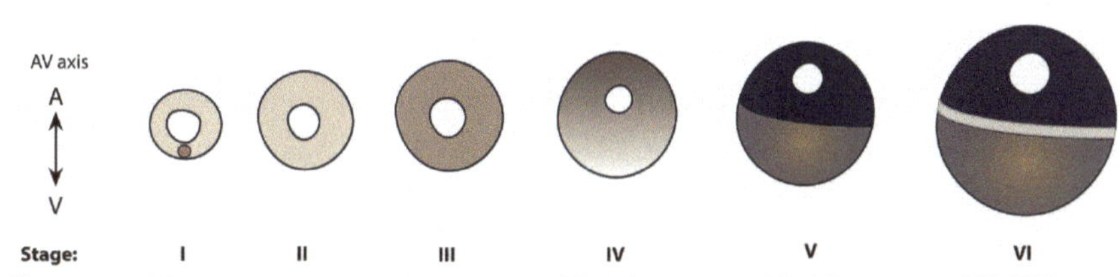

Fig. 2 Schematic of *Xenopus* oogenesis. *Xenopus* oogenesis proceeds through six stages designated I–VI. Oocyte stages are defined by the increase in oocyte diameter, accumulation of yolk, and the onset and polarization of pigmentation. Oocytes are oriented along the animal-vegetal (AV) axis with the vegetal pole at the bottom. The white circle is the oocyte nucleus (GV) and the tan circle in stage I is the Balbiani body. Ranges in oocyte diameter are denoted below each stage, but individual oocytes are not to scale.

in early stage III and black in late stage III. In stage IV, vitellogenesis continues and oocytes continue to rapidly expand in size; the largest yolk platelets accumulate in the vegetal hemisphere of the oocyte and the GV is displaced toward the animal hemisphere. Stage IV cells are marked by the beginning of pigment polarization, with the dark pigment enriched in the animal hemisphere of the oocyte. In stage V, vitellogenesis slows and distinct hemispheres are clearly visible with the pigment highly enriched in the animal hemisphere. Finally, stage VI oocytes have reached their full size and exhibit an unpigmented equatorial band, approximately 0.2 mm in diameter, dividing the animal and vegetal hemispheres of the oocyte; stage VI oocytes are fully mature and are ready for oviposition.

4.2 *Xenopus* RNA localization pathways

In addition to changes in size, pigmentation, and vitellogenesis, the distinct stages of *Xenopus* oogenesis are also marked by the developmentally staged localization of maternal RNAs. Localization of RNAs in the vegetal cytoplasm occurs through two different RNA localization pathways, the early, or METRO (*me*ssenger *tr*ansport *o*rganizing center) pathway and the late pathway, with some RNAs displaying an intermediate phenotype (Forristall, Pondel, Chen, & King, 1995; King et al., 2005; Kloc & Etkin, 1995). Several other RNAs, including the *An1, 2, 3* and *4* RNAs, enrich in the animal cytoplasm of the oocyte through unknown mechanisms (Hudson, Alarcón, & Elinson, 1996; Rebagliati et al., 1985). The polarized distribution of RNAs along the animal-vegetal axis of the oocyte is critical for embryonic patterning, with the animal hemisphere giving rise to the ectoderm, the vegetal hemisphere giving rise to the endoderm, and the mesoderm induced via signaling from the vegetal blastomeres (Nieuwkoop, 1969; Sudarwati & Nieuwkoop, 1971). For a comprehensive list of known localized RNAs in *Xenopus* oocytes as well as the techniques used to identify them, please see Houston (2013).

As shown in Fig. 3A, during stages I–II of oogenesis, several RNAs, including *nanos1/xcat-2* mRNA, the *Xenopus* homolog of *nanos*, are localized via the early localization pathway. Generally, the early pathway is used to localize RNAs that function in germ cell determination, and RNAs localized by this pathway include mRNAs encoding germ plasm components and signaling molecules, such as *Xdazl* and *Xwnt11* mRNAs, as well as a family of non-coding RNAs called *Xlsirts* (Forristall et al., 1995; Heasman et al., 1984; Houston, Zhang, Maines, Wasserman, & King, 1998;

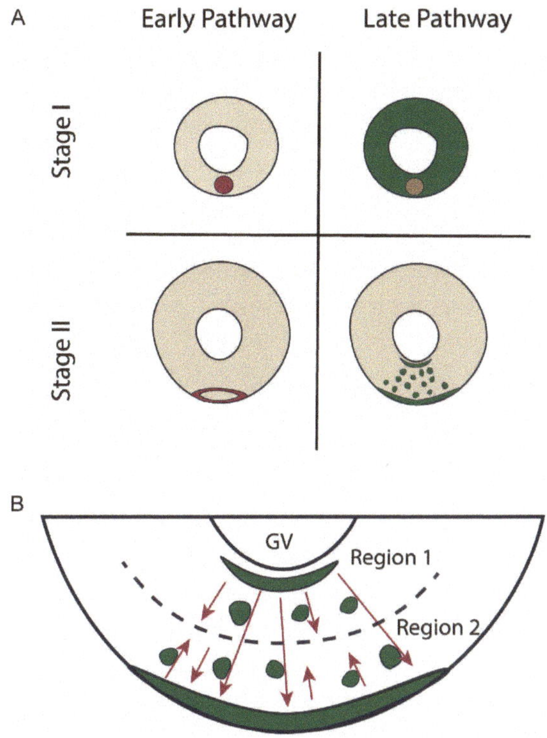

Fig. 3 RNA localization in *Xenopus* oocytes. (A) RNA localization in *Xenopus* oocytes proceeds through two primary pathways: early (or METRO) and late. Early pathway RNAs (magenta) are enriched in the Balbiani body in stage I oocytes and anchored in a ring-like pattern at the vegetal cortex in stage II. Late pathway RNAs (green) are ubiquitous throughout the cytoplasm in stage I and become enriched in the vegetal cytoplasm in stage II in a characteristic pattern: RNAs enrich in the perinuclear cup, cytoplasmic islands, and at the cortex. (B) Late pathway RNAs (green) are transported (arrows) along microtubules in two discrete steps: unidirectional transport by dynein from the perinuclear cup toward to cortex in the upper vegetal cytoplasm (region 1) and bidirectional transport by kinesin motors in the lower vegetal cytoplasm (region 2).

Kloc, Spohr, & Etkin, 1993; Mosquera, Forristall, Zhou, & King, 1993; Zearfoss, Chan, Kloc, Allen, & Etkin, 2003). This pathway proceeds via a three step process in which RNAs are first exported from the GV into the cytoplasm, then enriched in the Balbiani body in stage I, and finally deposited onto the vegetal cortex in a ring-like pattern in stage II (Kloc & Etkin, 1995). As shown in Fig. 1A, early localizing RNAs are enriched into the Balbiani body by the diffusion and local entrapment mechanism, as evidenced by the microtubule and microfilament independent, linear enrichment of microinjected, fluorescently-labeled RNA in the Balbiani body, with no degradation of the RNA in the surrounding cytoplasm (Chang et al., 2004). The mechanistic conservation between *Drosophila* and *Xenopus* oocytes for localization of germ line determining mRNAs highlights both the importance of the diffusion and local entrapment mechanism and the overall conservation of germplasm establishment from invertebrates to vertebrates (Kloc, Bilinski, & Etkin, 2004).

During stages II–IV of oogenesis, a second group of RNAs are enriched asymmetrically via the late localization pathway. Generally, the late pathway is used to localize RNAs, such as *vg1* and *vegT*, which are necessary for germ-layer patterning in the embryo (Birsoy, Kofron, Schaible, Wylie, & Heasman, 2006; Joseph & Melton, 1998; Kofron et al., 1999; White & Heasman, 2008; Zhang et al., 1998). Importantly, spatially inappropriate expression of either *vg1* and *vegT* mRNAs, which encode, respectively, a TGFβ growth factor family member and a T-box transcription factor, cause lethal embryonic phenotypes (Thomsen & Melton, 1993; Wallace & Dumont, 1968; Zhang & King, 1996). In stage I oocytes, while early localizing RNAs are enriched within the Balbiani body, late localizing RNAs are distributed ubiquitously in the cytoplasm, indicating a clear mechanistic distinction between the two pathways. During stages II–III, late pathway RNAs become restricted within the vegetal cytoplasm in a characteristic pattern: in the perinuclear cup region, in distinct, non-spherical cytoplasmic islands, and at the vegetal cortex (Fig. 3A). By stage IV, the RNAs are restricted to the vegetal cortex, where they remain, translationally silenced, until expression is needed to pattern the embryo (Dale, Matthews, Tabe, & Colman, 1989; Tannahill & Melton, 1989).

Like early pathway RNAs, initial enrichment of late pathway RNAs in the vegetal cytoplasm is independent of the cytoskeleton and may proceed through a diffusion and local entrapment mechanism (Yisraeli, Sokol, & Melton, 1990). However, the next step in late pathway RNA localization, transport to the vegetal cortex, relies on the cytoskeletal network and molecular motors (Gagnon et al., 2013; Messitt et al., 2008; Yisraeli et al., 1990; Yoon & Mowry, 2004). Interestingly, as in *Drosophila* oocytes, both plus end and minus end directed motors are employed by late localizing RNAs to reach their ultimate site of enrichment (Gagnon et al., 2013). The transport process has been best described for the *vg1* mRNA, which is transported in a multistep process from the perinuclear cup to the vegetal cortex using a mixed microtubule array (Messitt et al., 2008). As shown in Fig. 3B, in the upper vegetal cytoplasm of the oocyte, *vg1* is transported unidirectionally from the perinuclear cup at the vegetal side of the GV toward microtubule minus ends by the molecular motor dynein (Gagnon et al., 2013). In the lower vegetal cytoplasm, *vg1* is transported bi-directionally both toward the cortex and back toward microtubule plus ends in the upper vegetal cytoplasm by the molecular motor kinesin. This bidirectional transport continues until the *vg1* mRNA has been stably anchored at the vegetal cortex by an unknown mechanism likely involving interaction with the cytoskeletal network (Gagnon et al., 2013; Yisraeli et al., 1990). It is proposed that this bidirectional transport may provide a mechanism in which multiple iterations of transport enriches the RNA at the vegetal cortex even if the RNA is inefficiently captured by the anchoring process.

Although the early and late RNA localization pathways are clearly distinct in timing, mechanisms of RNA localization, and localization patterns, some RNAs exhibit characteristics of both pathways. These RNAs, which include *hermes* and *fatvg* mRNAs, are classified as localizing via an intermediate pathway as they are enriched in the Balbiani body during early oogenesis and to vegetal cytoplasm islands during mid-oogenesis (Chan, Kloc, Bilinski, & Etkin, 2001; Chan, Kloc, & Etkin, 1999; Zearfoss, Chan, Wu, Kloc, & Etkin, 2004). However, it remains to be determined whether this represents a distinct pathway of localization or if these RNAs are using the established early and late pathways sequentially.

Interestingly, while these localization pathways have been viewed as important for localizing mRNAs for subsequent local translation, both *xlsirts* non-coding RNAs, which localize via the early pathway, and *vegT* mRNA, which localizes via the late pathway, have been shown to play structural roles in organizing the vegetal cytoplasm of the *Xenopus* oocyte. In addition to encoding a protein necessary for germ-layer patterning, *vegT* mRNA is also thought to play a role as a RNA scaffold to nucleate cytokeratin networks during oogenesis (King, et al., 2005; Kloc, Bilinski, & Dougherty, 2007; Kloc, Foreman, & Reddy, 2011). This role in cytoskeletal organizational may explain why depletion of *vegT* mRNA, independent of its protein product, causes mislocalization of both early (*bicaudal-C* and *Xwnt11*) and late (*vg1*) mRNAs in the oocyte (Heasman, Wessely, Langland, Craig, & Kessler, 2001). Vegetal cytoskeletal arrangement also relies on the localization of the *xlsirt* non-coding RNAs, as depletion of these transcripts also causes disruption of the cytokeratin network, demonstrating a potential function of localizing these non-coding RNAs (Kloc et al., 2005).

5. RNPs and localization

In *Xenopus* oocytes, as well as all other cells, RNAs do not localize independently. Rather, RNAs are packaged with proteins to form ribonucleoprotein complexes (RNPs). These RNA-protein interactions, which are often required for the localization, translational silencing, and stability of the RNA are initiated by *cis*-acting sequences in the RNA called zip codes or localization elements (Kislauskis, Zhu, & Singer, 1994; reviewed in Singh, Pratt, Yeo, & Moore, 2015). These *cis*-sequences can bind proteins based on their primary sequence, secondary structure, or both, and are most often found in the 3′ UTR of mRNAs (reviewed in Jambhekar & Derisi, 2007). *cis*-Sequences are subsequently bound by *trans*-acting proteins to create a localization competent RNP (reviewed in Singh et al., 2015). A growing body of evidence suggests that perturbations to RNP assembly and subsequent RNA localization can lead to pathological phenotypes, particularly in neurodegenerative diseases (reviewed in Bovaird, Patel, Padilla, & Lécuyer, 2018; Cody, Iampietro, & Lécuyer, 2013).

5.1 *Cis* sequences and "zip codes"

Not long after the asymmetric distribution of RNA was first observed, researchers began to identify *cis*-acting sequences that were required for proper localization of RNAs in a variety of organisms, including *Drosophila* oocytes, *Xenopus* oocytes, and fibroblasts (Kislauskis et al., 1994; Macdonald & Struhl, 1988; Mowry & Melton, 1992). Many of these early experiments identified minimal sequences which were sufficient for localization to their respective subcellular destinations, but those elements showed few shared features at the level of either primary sequence or secondary structure and varied in dramatically in length. For example, the localization element for *bicoid* mRNA in *Drosophila* oocytes was mapped to a 625 nucleotide (nt.) region of the 3′ UTR which included extensive regions of predicted secondary structure (Macdonald & Struhl, 1988). Mutational analysis showed that both a non-sequence specific helical structure and an adjacent sequence specific recognition domain were both required for *bicoid* localization, highlighting the diverse ways in which *cis*-sequences can interact with *trans*-factors (Macdonald & Kerr, 1998). In contrast, the localization element for β-actin mRNA in chick embryonic fibroblasts was mapped to a bipartite RNA sequence: a primary 54-nt. region (termed the "zip code") with a weaker 43-nt. element downstream in the 3′ UTR (Kislauskis et al., 1993, 1994).

cis-Sequences directing localization have also been well characterized for *vg1* mRNA, a late pathway localizing RNA in *Xenopus* oocytes. A 340-nt. region of the 3′ UTR of *vg1* mRNA, termed the *v*egetal *l*ocalization *e*lement (*VLE*), is sufficient to drive late pathway-like RNA localization in *Xenopus* oocytes (Mowry & Melton, 1992). A second, distinct 250-nt. region of the *vg1* 3′ UTR, termed the *vg1* translational *e*lement (*VTE*) has been identified as important for the maintenance of translational repression during transport (Otero, Devaux, & Standart, 2001). This AU-rich region is 118-nt. downstream of the *VLE* and binds to Elavl1 and Elavl2 to maintain translational repression (Colegrove-Otero, Devaux, & Standart, 2005). Similarly, the localization element of a second late pathway RNA, *vegT*, has also been identified and, while there is no overall primary sequence homology with the *vg1 VLE*, repeated clusters of redundant *cis*-sequences, which bind *trans*-factors, does appear to be a conserved localization signal in *Xenopus* oocytes (Bubunenko, Kress, Vempati, Mowry, & King, 2002). Subsequent studies have identified a number of *cis*-elements important for RNA localization in a wide range of other organisms, ranging in size from tens of nucleotides to 1 kb, but no universal patterns in sequence or structure have emerged (reviewed in Jambhekar & Derisi, 2007; Marchand, Gaspar, & Ephrussi, 2012; Van De Bor & Davis, 2004).

5.2 *Trans*-factors

In most cases where the *cis*-acting localization elements have been identified, the full complement of RNA binding proteins (RBPs), protein adapters, and molecular motors acting as *trans*-factors are not yet known. One exception is the well-characterized RNPs of pair-rule RNAs, such as *wingless, hairy*, and *ftz* mRNAs, which localize apically in the *Drosophila* blastoderm embryo (Davis & Ish-Horowicz, 1991). The localization of these RNPs, which is required for coordinating segmentation in the embryo, is dynein-dependent and directed toward the minus ends of microtubules (Bullock & Ish-Horowicz, 2001; Wilkie & Davis, 2001). Extensive studies have revealed both the *cis*-sequences required for localization of these mRNAs and the minimal set of *trans*-factors which are required to create a localization competent RNP. mRNAs are directly bound by Egalitarian (Egl), which in turn binds to the adapter protein Bicaudal D (BicD), which then licenses recruitment of the dynein/dynactin motor complex (Bullock & Ish-Horowicz, 2001; Dienstbier, Boehl, Li, & Bullock, 2009; Wilkie & Davis, 2001). Interestingly, *cis*-acting sequences in the RNA cargoes promote the interaction of Egl with BicD, providing a mechanism for RNAs to regulate their own intracellular transport (McClintock et al., 2018).

Another well-characterized example of *trans*-acting factors interacting with *cis*-sequences is found in the *vg1* RNP in *Xenopus* oocytes. Here, the *vg1 VLE* interacts with several RBPs, including heterogeneous nuclear ribonucleoprotein AB (hnRNPAB/40LoVe), polypyrimidine tract binding protein (PTB/hnRNPI), insulin-like growth factor 2 mRNA binding protein 3 (Igf2bp3/Vg1 RBP/Vera), and Staufen1 (Cote et al., 1999; Czaplinski et al., 2005; Deshler, Highett, Abramson, & Schnapp, 1998; Havin et al., 1998; Mowry, 1996; Yoon & Mowry, 2004). Binding of PTB and Vera to the *VLE* is required for *VLE* localization, as point mutations in their *cis*-binding sites abolish *VLE* localization (Cote et al., 1999; Deshler et al., 1998; Lewis et al., 2004; Lewis, Gagnon, & Mowry, 2008). In the *vg1* RNP, both kinesin and dynein molecular motors drive vegetal localization, but the *trans*-factor adapters linking the motors to the RNP have not yet been identified (Gagnon et al., 2013; Messitt et al., 2008). Many *trans*-factors, including each of the above-mentioned *vg1 VLE* RBPs, contain multiple RNA binding domains and are therefore capable of interactions either with a single RNA molecule at multiple sites or multiple RNA molecules at the same time (Cote et al., 1999; Deshler et al., 1998; Havin et al., 1998; Yoon & Mowry, 2004). This multivalency of interactions is important not only for RNP formation and subsequent RNA localization, but has also recently been characterized as a driver of phase separation (Lin, Protter, Rosen, & Parker, 2015).

6. Biomolecular condensates and phase separation

In 2009, the paradigm for understanding RNP complex structure, formation, and biology shifted with the first description of liquid-liquid phase separation in cells (Brangwynne et al., 2009). The observation of non-membrane bound compartments within the cell was not novel, as the membraneless nucleolus was first observed in the 1800s using the earliest forms of light microscopy (reviewed in Pederson, 2011). Rather, phase separation provided a new biophysical framework to describe how these compartments regulate their composition, concentrate specific biomolecules above their surroundings, and modulate internal biochemical activity without the use of a membrane. In the years since the first description of liquid-liquid phase separation, RNP granules (many of which have been revealed to be biomolecular condensates—also termed membraneless organelles) have been described in diverse cell types, including oocytes, and in a multitude of subcellular locations (reviewed in Banani et al., 2017). Notable examples include stress granules, processing bodies, and germ granules in several species (Brangwynne et al., 2009; reviewed in Decker & Parker, 2012; Saha et al., 2016; Voronina, Seydoux, Sassone-Corsi, & Nagamori, 2011).

6.1 General principles of phase separation

The understanding of phase separation in biology is based on principles of thermodynamics applied within the cellular context (reviewed in Hyman, Weber, & Jülicher, 2014). In this model, molecules overcome entropy and form a dilute solution phase and a concentrated condensate phase based on the free energy of the solution and the chemical potential (reviewed in Banani et al., 2017). Thus, molecules are miscible in the cell until they reach a threshold concentration at which they undergo liquid-liquid demixing and phase separate. A defining feature of biomolecular condensates is their ability to exert control over their contents without a membrane, concentrating certain molecules and excluding others. One framework to understand this phenomenon is to define condensate components as either scaffolds or clients (Banani et al., 2016). Scaffold components, which are often only a small portion of the condensate components, drive condensate formation, while clients, which are often the majority of the condensate components, are selectively recruited into the scaffold complex without playing a significant role in condensate formation (reviewed in Ditlev, Case, & Rosen, 2018).

While examples of biomolecular condensates vary in their biological functions, compositions, subcellular localizations, and size, the physical principles that drive their formation dictate certain similarities between them. As depicted in Fig. 4, biomolecular condensates are often enriched for multivalent macromolecules that can drive phase separation through a multiplicity of interactions between protein-protein or protein-RNA interaction domains (Fig. 4A), through transient interactions between protein intrinsically disordered regions (IDRs) with "sticker" domains (Fig. 4B), or a combination of both types of interactions (Lin et al., 2015; Mittag & Parker, 2018; Van Treeck & Parker, 2018). Perhaps for this reason, many

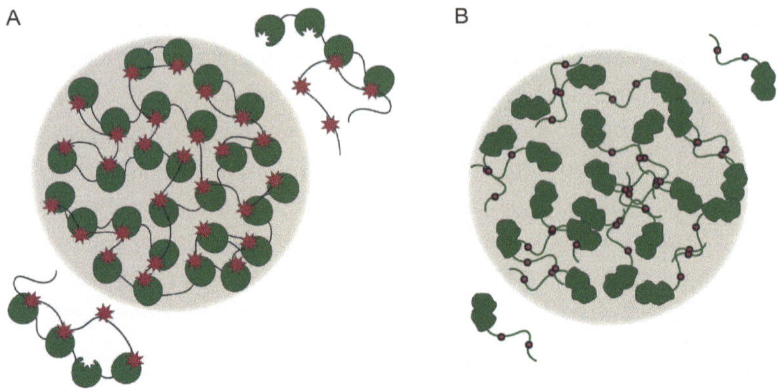

Fig. 4 Multivalent interactions mediate phase separation. (A) Phase separation can be mediated by multivalent interactions between modular binding domains in proteins (green) interacting with protein or RNA binding partners (magenta). These interactions mediate condensate formation and form a dense phase (gray shaded circle), with networked complexes of interactions, surrounded by a dilute phase. (B) Phase separation can also be mediated through multivalent interactions between "sticker" domains (magenta) in IDR-containing proteins (green) which interact weakly and transiently with one another to drive biocondensate formation (gray shaded circle).

proteins capable of multivalent interactions are conserved components of biomolecular condensates, with diverse types of RNP granules exhibiting broad compositional overlap (Buchan, 2014; Cumberworth, Lamour, Babu, & Gsponer, 2013). In addition, RNA itself is multivalent, capable of interacting with multiple RBPs (reviewed in Van Treeck & Parker, 2018). Other similarities shared by biomolecular condensates include the reversibility of condensate formation and, in their liquid state, the capability of droplets to undergo fusion and fission, as well as distorting their spherical shape in response to shear forces (Brangwynne et al., 2009; Brangwynne, Mitchison, & Hyman, 2011; Elbaum-Garfinkle et al., 2015; Weber & Brangwynne, 2015).

Biomolecular condensates do not exist solely in liquid-like states (reviewed in Alberti & Hyman, 2016). In a process thought to be driven by IDRs, biomolecular condensates can exist in a reversible continuum of decreasing dynamics from mixed liquids, to demixed liquid droplets, to gels, and finally to solid or fibrous aggregates (Han et al., 2012; Kato et al., 2012; Lin et al., 2015). The final transition to a solid aggregate is often irreversible and associated with pathology (Elbaum-Garfinkle & Brangwynne, 2015; Murakami et al., 2015). This continuum of phase transitions can be recapitulated in vitro, as many IDR-containing proteins form droplets which are initially liquid-like, but mature over time to a more solid-like state (Lin et al., 2015; Molliex et al., 2015; Patel et al., 2015; Xiang et al., 2015; Zhang et al., 2015). Recent evidence in oocytes and other model systems demonstrates that in vivo many non-pathological phase separated bodies are heterogeneous in phase, with particular elements or regions being more or less liquid-like (Feric et al., 2016; Putnam, Cassani, Smith, & Seydoux, 2019).

6.2 RNA localization and phase separation

As our understanding of both RNA localization and phase separation expands, several mechanisms by which the processes facilitate one another have emerged. For example, RNA localization and local translation can provide a mechanism for spatial and temporal control of phase separation in the cell. A recent study comparing the properties of proteins which are either translated upon nuclear export and localized as proteins or transported as mRNAs and translated distally, demonstrated that the latter class of mRNAs is significantly enriched for transcripts encoding proteins with IDRs (Weatheritt, Gibson, & Babu, 2014). Thus, the advantage of RNA localization as a means to prevent ectopic protein activity can also be viewed in the context of preventing ectopic phase separation by local translation of protein IDRs. RNA localization and local translation of IDR-containing proteins may also be advantageous to cells as it provides a mechanism for rapid phase separation in response to external stimuli. Furthermore, RNA localization may facilitate local phase separation by creating concentrations of both RNAs and proteins above the threshold concentration for phase separation of scaffolds. In this way, RNA localization can serve as an energetically favorable biological means to regulate phase separation, as with phase separation of *Drosophila* germ granules, described below (Niepielko, Eagle, & Gavis, 2018). Conversely, phase separation can facilitate RNA localization, as some biomolecular condensates, such as the *Xenopus* Balbiani body, contain localized RNAs (Boke et al., 2016). Indeed, RNA incorporation into biomolecular condensates itself can be described as a form of RNA localization. Finally, many of the *trans*-factors important for RNA localization are also highly enriched in biomolecular condensates, suggesting additional links between RNA localization and phase separation (Lin et al., 2015; reviewed in Mittag & Parker, 2018; Van Treeck & Parker, 2018).

7. Biomolecular condensates in oocytes and embryos

Just as the large, highly polarized oocyte was the source of many of the first examples of RNA localization, so too it has proved to be an excellent model for the identification and study of biomolecular condensates. Examples include phase separated germ granules, such as P granules in *C. elegans*, the germ granules of *Drosophila* oocytes, and the Balbiani body in *Xenopus* oocytes, as well as other non-germ granule structures such as nucleoli, which have been studied in *Xenopus* oocytes. These examples of biomolecular condensates, which are detailed below, demonstrate the diversity of phase separated structures and the principles of phase separation that have been gleaned from work in these systems.

7.1 Nucleoli

Nucleoli are subdomains of the nucleus dedicated to ribosome biogenesis and, as mentioned previously, were the first membraneless compartment observed in the cell (reviewed in Pederson, 2011). Using the thermodynamic principles of liquid-liquid phase separation and the enormous nuclei of the stage V *Xenopus* oocyte (~0.4–0.5 mm), researchers quickly characterized the nucleolus as a biomolecular condensate with liquid-like properties (Brangwynne et al., 2011). Nucleoli contain subcompartments important for different stages of ribosome production, but until recently the mechanisms of subdomain formation were unclear (reviewed in Thiry & Lafontaine, 2005). Interestingly, the subcompartments of the nucleolus were found to be co-existing, immiscible liquid phases within the larger biomolecular condensate, with differential biophysical properties and primary surface tension driving organization of the subcompartments (Feric et al., 2016). The large size of *Xenopus* oocyte nucleoli has made them the ideal model to study additional questions in phase separation, such as the role of ATP in phase separation and the role of transcription within condensates (Berry et al., 2015; Hayes, Peuchen, Dovichi, & Weeks, 2018).

7.2 P granules in *C. elegans*

P granules are germ granules in *C. elegans* that form perinuclearly in germ cells, become cytoplasmic in growing oocytes, and are subsequently enriched in the posterior of the one-cell embryo (reviewed in Seydoux, 2018). During cell division, P granules are ultimately asymmetrically inherited by the cells that will become the germ line. While originally observed in the 1980s, several decades later, P granules were the first biomolecular condensate to be described as forming via liquid-liquid phase separation (Brangwynne et al., 2009; Strome & Wood, 1982). Two scaffold proteins of the P granule, PGL-1 and PGL-3, are RBPs which bind RNA through RGG boxes and have dimerization domains that are required for condensate formation (Hanazawa, Yonetani, & Sugimoto, 2011; Kawasaki et al., 1998). The asymmetrical distribution of P granules in the posterior of the embryo depends not on these scaffold proteins, but on the anticorrelated gradients of MEG-3, an IDR-containing protein enriched in the posterior, and MEX-5, an RBP enriched in the anterior of the cell (Smith et al., 2016). This model for the spatial control of phase separation hypothesizes that phase transition is based on the availability of RNA, whereby MEX-5 RNA binding prevents MEG-3 from interacting with RNA, driving it to locally phase separate and leading to P granule assembly (Saha et al., 2016).

Much like nucleoli, individual P granules are surprisingly heterogeneous, as the proteins in assembled P granules are not uniformly distributed throughout the individual condensate and do not exhibit the same dynamics (reviewed in Updike & Strome, 2010). MEG-3 forms a gel-like assembly primarily at the periphery of the structure while PGL-3 is in a more dynamic phase primarily in the core, once again highlighting the diversity of biophysical states possible within a single condensate (Putnam et al., 2019). Although many of the biological functions of forming subcompartments within a larger condensate remain unclear, regulating the biophysical state of distinct subcompartments provides an attractive model for both stabilizing the structure against mechanical forces in the cell and tuning reaction kinetics (reviewed in Banani et al., 2017).

7.3 Germ granules in *Drosophila*

Similar to *C. elegans*, *Drosophila* germ line establishment occurs through asymmetrical inheritance of biomolecular condensates in the embryo (reviewed in Mahowald, 2001; Trcek & Lehmann, 2019). These condensates, termed germ granules, contain several localized RNAs, including *nanos* and *pgc*, and proteins required for germ line specification such as Vasa (reviewed in Trcek & Lehmann, 2019). The formation of germ granules is intrinsically linked to the polarization created by RNA localization and local translation of their protein products in the oocyte.

Specifically, germ granule formation is nucleated by translation of the *oskar* mRNA, which is localized to the posterior pole by active transport (Ephrussi et al., 1991; Niepielko et al., 2018). *Nanos* mRNA, which was localized to the posterior pole via a combination of diffusion and local entrapment and local stabilization and selective degradation methods, is then incorporated into nascent germ granules, where *nanos* RNA acts to recruit more of the locally enriched *nanos* mRNA (Niepielko et al., 2018). In this way, three RNA localization mechanisms work in concert to produce the conditions and macromolecular gradients necessary for seeding and growth of a biomolecular condensate. Like nucleoli, P granules, and other biomolecular condensates, *Drosophila* germ granules display a range of biophysical states with both liquid-like and more solid or gel-like properties (Kistler et al., 2018).

7.4 The Balbiani body in *Xenopus* oocytes

Finally, the Balbiani body, the site of early pathway RNA localization during stage I of *Xenopus* oogenesis, and the site of germ granule formation has also been shown to be a biomolecular condensate (Boke et al., 2016). The Balbiani body is a non-membrane bound compartment which, in addition to localized RNAs and proteins, contains high numbers of membrane bound organelles, such as mitochondria and endoplasmic reticulum (Kloc, Jedrzejowska, Tworzydlo, & Bilinski, 2014). Self-assembly of this structure is driven by the N-terminal prion-like domain of Xvelo protein, which is highly enriched in the Balbiani body in vivo and able to form amyloid-like structures in vitro (Boke et al., 2016) Intriguingly, Xvelo displays characteristics of a scaffold in vitro, as phase separating Xvelo is able to recruit both mitochondria, through interactions with the N-terminal domain, and *nanos1* mRNA, through a putative C-terminal RNA binding domain, as clients (Boke et al., 2016). Unlike many other biomolecular condensates, the Balbiani body does not appear to be liquid in vivo (Boke et al., 2016). Instead, the Balbiani body likely represents a more solid-like phase separated structure, although it is not known if the initial formation proceeds through a liquid-like intermediate that eventually matures to a more solid-like state.

As *Xenopus* oocyte maturation occurs over months, a less dynamic condensate is an attractive model for stably maintaining factors in the proper subcellular location on an extended developmental timescale. The mechanism by which this solid-like structure disassembles and deposits the early pathway RNAs at the vegetal cortex during the transition to stage II of oogenesis is currently unclear. Nonetheless, the Balbani body is an intriguing example of a solid-like biomolecular condensate that is not associated with a pathological state and the mechanisms by which it disassembles, once elucidated, may have important implications for disease-associated amyloid-like condensates. Additionally, while the biophysical state of the cytoplasmic islands in the *Xenopus* late RNA localization pathway has not yet been described, based on the similarities in function and timescale between the Balbiani body and the late pathway vegetal RNA islands and the emergence of biomolecular condensates in RNA transport, it is an attractive hypothesis that late pathway RNAs also enrich within phase separated structures.

8. Conclusions

Over the decades since RNA localization was first described, it has become clear that this process plays a crucial role in subcellular organization. Recent technological advances have enabled global studies highlighting the astonishingly complex combinations of *cis*-sequences and *trans*-factors that facilitate localization of RNAs. As the list of localized RNAs rapidly expands, it remains a challenge in the field to understand the mechanisms by which newly identified RNPs are localized and the cellular functions linked to their localization. This is particularly true of non-coding RNAs, as the mechanisms by which they are transported, the advantages this provides to the cell, and the functions of the localized transcripts are all much less well understood than for many mRNAs.

An exciting and new avenue for novel insights into RNA localization comes through appreciating the intrinsic linkage between RNA localization and phase separation. As both processes are emerging as means of creating asymmetry in a broad variety of cell types and organisms, it is becoming clear that these processes may be inherently linked and facilitate one another. This seems to be particularly true in oocytes as invertebrate and vertebrate germ granules alike have been shown to be structures in which both phase separation and RNA localization occur in concert. RNA localization can establish the local enrichment of factors necessary for phase separation, as in the case of *Drosophila* germ granules. Additionally, RNA localization, coupled with local translation can also allow for proteins with IDRs to phase separate only at particular subcellular destinations and in response to external cues, providing a mechanism for regulating this biophysical process within cells. In the coming years, an exciting challenge for the field will be to utilize many of the new techniques for studying both biomolecular condensates and RNA localization to inform our understanding of these processes.

Acknowledgments

We thank J. Otis and L. O'Connell for careful reading of the manuscript. Work in our laboratory is supported by the NIH (GM071049).

References

Alberti, S., & Hyman, A. A. (2016). Are aberrant phase transitions a driver of cellular aging? *BioEssays, 38*, 959–968.

Banani, S. F., Lee, H. O., Hyman, A. A., & Rosen, M. K. (2017). Biomolecular condensates: Organizers of cellular biochemistry. *Nature Reviews Molecular Cell Biology, 18*, 285–298.

Banani, S. F., Rice, A. M., Peeples, W. B., Lin, Y., Jain, S., Parker, R., et al. (2016). Compositional control of phase-separated cellular bodies. *Cell, 166*, 651–663.

Bashirullah, A., Cooperstock, R. L., & Lipshitz, H. D. (2001). Spatial and temporal control of RNA stability. *Proceedings of the National Academy of Sciences of the United States of America, 98*, 7025–7028.

Berry, J., Weber, S. C., Vaidya, N., Haataja, M., Brangwynne, C. P., & Weitz, D. A. (2015). RNA transcription modulates phase transition-driven nuclear body assembly. *Proceedings of the National Academy of Sciences of the United States of America, 112*, E5237–E5245.

Birsoy, B., Kofron, M., Schaible, K., Wylie, C., & Heasman, J. (2006). Vg1 is an essential signaling molecule in Xenopus development. *Development, 133*, 15–20.

Blower, M. D. (2013). Molecular insights into intracellular RNA localization. *International Review of Cell and Molecular Biology, 302*, 1–39. Elsevier.

Boke, E., Ruer, M., Wühr, M., Coughlin, M., Lemaitre, R., Gygi, S. P., et al. (2016). Amyloid-like self-assembly of a cellular compartment. *Cell, 166*, 637–650.

Bovaird, S., Patel, D., Padilla, J. C. A., & Lécuyer, E. (2018). Biological functions, regulatory mechanisms, and disease relevance of RNA localization pathways. *FEBS Letters, 592*, 2948–2972.

Brangwynne, C. P., Eckmann, C. R., Courson, D. S., Rybarska, A., Hoege, C., Gharakhani, J., et al. (2009). Germline P granules are liquid droplets that localize by controlled dissolution/condensation. *Science, 324*, 1729–1732.

Brangwynne, C. P., Mitchison, T. J., & Hyman, A. A. (2011). Active liquid-like behavior of nucleoli determines their size and shape in Xenopus laevis oocytes. *Proceedings of the National Academy of Sciences of the United States of America, 108*, 4334–4339.

Brendza, R. P., Serbus, L. R., Duffy, J. B., & Saxton, W. M. (2000). A function for kinesin I in the posterior transport of oskar mRNA and staufen protein. *Science, 289*, 2120–2122.

Bubunenko, M., Kress, T. L., Vempati, U. D., Mowry, K. L., & King, M. L. (2002). A consensus RNA signal that directs germ layer determinants to the vegetal cortex of Xenopus oocytes. *Developmental Biology, 248*, 82–92.

Buchan, J. R. (2014). mRNP granules assembly, function, and connections with disease. *RNA Biology, 11*, 1019–1030.

Bullock, S. L., & Ish-Horowicz, D. (2001). Conserved signals and machinery for RNA transport in Drosophila oogenesis and embryogenesis. *Nature, 414*, 611–616.

Buskilay, A. A. A., Kannaiahy, S., & Amster-Choder, O. (2014). RNA localization in bacteria. *RNA Biology, 11*, 1051–1060.

Buxbaum, A. R., Haimovich, G., & Singer, R. H. (2015). In the right place at the right time: Visualizing and understanding mRNA localization. *Nature Reviews Molecular Cell Biology*, *16*, 95–109.

Cabili, M. N., Dunagin, M. C., McClanahan, P. D., Biaesch, A., Padovan-Merhar, O., Regev, A., et al. (2015). Localization and abundance analysis of human lncRNAs at single-cell and single-molecule resolution. *Genome Biology*, *16*, 1–16.

Cajigas, I. J., Tushev, G., Will, T. J., Tom Dieck, S., Fuerst, N., & Schuman, E. M. (2012). The local transcriptome in the synaptic neuropil revealed by deep sequencing and high-resolution imaging. *Neuron*, *74*, 453–466.

Cha, B. J., Serbus, L. R., Koppetsch, B. S., & Theurkauf, W. E. (2002). Kinesin I-dependent cortical exclusion restricts pole plasm to the oocyte posterior. *Nature Cell Biology*, *4*, 592–598.

Chan, A. P., Kloc, M., Bilinski, S., & Etkin, L. D. (2001). The vegetally localized mRNA fatvg is associated with the germ plasm in the early embryo and is later expressed in the fat body. *Mechanisms of Development*, *100*, 137–140.

Chan, A. P., Kloc, M., & Etkin, L. D. (1999). fatvg encodes a new localized RNA that uses a 25-nucleotide element (FVLE1) to localize to the vegetal cortex of Xenopus oocytes. *Development*, *126*, 4943–4953.

Chang, P., Torres, J., Lewis, R. A., Mowry, K. L., Houliston, E., & King, M. L. (2004). Localization of RNAs to the mitochondrial cloud in Xenopus oocytes through entrapment and association with endoplasmic reticulum. *Molecular Biology of the Cell*, *15*, 4669–4681.

Chen, L. L. (2016). Linking long noncoding RNA localization and function. *Trends in Biochemical Sciences*, *41*, 761–772.

Chen, L., Dumelie, J. G., Li, X., Cheng, M. H. K., Yang, Z., Laver, J. D., et al. (2014). Global regulation of mRNA translation and stability in the early Drosophila embryo by the Smaug RNA-binding protein. *Genome Biology*, *15*, 19–22.

Clark, A., Meignin, C., & Davis, I. (2007). A dynein-dependent shortcut rapidly delivers axis determination transcripts into the Drosophila oocyte. *Development*, *134*, 1955–1965.

Clemson, C. M., Hutchinson, J. N., Sara, S. A., Ensminger, A. W., Fox, A. H., Chess, A., et al. (2009). An architectural role for a nuclear noncoding RNA: NEAT1 RNA Is essential for the structure of paraspeckles. *Molecular Cell*, *33*, 717–726.

Cody, N. A. L., Iampietro, C., & Lécuyer, E. (2013). The many functions of mRNA localization during normal development and disease: From pillar to post. *Wiley Interdisciplinary Reviews: Developmental Biology*, *2*, 781–796.

Colegrove-Otero, L. J., Devaux, A., & Standart, N. (2005). The Xenopus ELAV protein ElrB represses Vg1 mRNA translation during oogenesis. *Molecular and Cellular Biology*, *25*, 9028–9039.

Cote, C. A., Gautreau, D., Denegre, J. M., Kress, T. L., Terry, N. A., & Mowry, K. L. (1999). A Xenopus protein related to hnRNP I has a role in cytoplasmic RNA localization. *Molecular Cell*, *4*, 431–437.

Crick, F. (1958). On protein synthesis. In *Symposia of the Society for Experimental Biology* (pp. 138–163).

Cumberworth, A., Lamour, G., Babu, M. M., & Gsponer, J. (2013). Promiscuity as a functional trait: Intrinsically disordered regions as central players of interactomes. *The Biochemical Journal*, *454*, 361–369.

Czaplinski, K., Köcher, T., Schelder, M., Segref, A., Wilm, M., & Mattaj, I. W. (2005). Identification of 40LoVe, a Xenopus hnRNP D family protein involved in localizing a TGF-β-related mRNA during oogenesis. *Developmental Cell*, *8*, 505–515.

Dale, L., Matthews, G., Tabe, L., & Colman, A. (1989). Developmental expression of the protein product of Vg1, a localized maternal mRNA in the frog Xenopus laevis. *The EMBO Journal*, *8*, 1057–1065.

Das, S., Singer, R. H., & Yoon, Y. J. (2019). The travels of mRNAs in neurons: do they know where they are going? *Current Opinion in Neurobiology*, *57*, 110–116.

Davis, I., & Ish-Horowicz, D. (1991). Apical localization of pair-rule transcripts requires 3′ sequences and limits protein diffusion in the Drosophila blastoderm embryo. *Cell*, *67*, 927–940.

Decker, C. J., & Parker, R. (2012). P-bodies and stress granules: Possible roles in the control of translation and mRNA degradation. *Cold Spring Harbor Perspectives in Biology*, *4*, a012286.

Deshler, J. O., Highett, M. I., Abramson, T., & Schnapp, B. J. (1998). A highly conserved RNA-binding protein for cytoplasmic mRNA localization in vertebrates. *Current Biology*, *8*, 489–496.

Dienstbier, M., Boehl, F., Li, X., & Bullock, S. L. (2009). Egalitarian is a selective RNA-binding protein linking mRNA localization signals to the dynein motor. *Genes & Development*, *23*, 1546–1558.

Ding, D., Parkhurst, S. M., Halsell, S. R., & Lipshitz, H. D. (1993). Dynamic Hsp83 RNA localization during Drosophila oogenesis and embryogenesis. *Molecular and Cellular Biology*, *13*, 3773–3781.

Ditlev, J. A., Case, L. B., & Rosen, M. K. (2018). Who's in and who's out—Compositional control of biomolecular condensates. *Journal of Molecular Biology*, *430*, 4666–4684.

Du, T. G., Schmid, M., & Jansen, R. P. (2007). Why cells move messages: The biological functions of mRNA localization. *Seminars in Cell & Developmental Biology*, *18*, 171–177.

Dumont, J. N. (1972). Oogenesis in Xenopus laevis (Daudin). I. Stages of oocyte development in laboratory maintained animals. *Journal of Morphology*, *136*, 153–179.

Duncan, J. E., & Warrior, R. (2002). The cytoplasmic dynein and kinesin motors have interdependent roles in patterning the Drosophila oocyte. *Current Biology*, *12*, 1982–1991.

Elbaum-Garfinkle, S., & Brangwynne, C. P. (2015). Liquids, fibers, and gels: The many phases of neurodegeneration. *Developmental Cell*, *35*, 531–532.

Elbaum-Garfinkle, S., Kim, Y., Szczepaniak, K., Chen, C. C. H., Eckmann, C. R., Myong, S., et al. (2015). The disordered P granule protein LAF-1 drives phase separation into droplets with tunable viscosity and dynamics. *Proceedings of the National Academy of Sciences of the United States of America*, *112*, 7189–7194.

Ephrussi, A., Dickinson, L. K., & Lehmann, R. (1991). Oskar organizes the germ plasm and directs localization of the posterior determinant nanos. *Cell*, *66*, 37–50.

Fay, M. M., & Anderson, P. J. (2018). The role of RNA in biological phase separations. *Journal of Molecular Biology*, *430*, 4685–4701.

Feric, M., Vaidya, N., Harmon, T. S., Mitrea, D. M., Zhu, L., Richardson, T. M., et al. (2016). Coexisting liquid phases underlie nucleolar subcompartments. *Cell*, *165*, 1686–1697.

Forrest, K. M., & Gavis, E. R. (2003). Live imaging of endogenous RNA reveals a diffusion and entrapment mechanism for nanos mRNA localization in Drosophila. *Current Biology*, *13*, 1159–1168.

Forristall, C., Pondel, M., Chen, L., & King, M. L. (1995). Patterns of localization and cytoskeletal association of two vegetally localized RNAs, Vg1 and Xcat-2. *Development*, *121*, 201–208.

Gagnon, J. A., Kreiling, J. A., Powrie, E. A., Wood, T. R., & Mowry, K. L. (2013). Directional transport Is mediated by a dynein-dependent step in an RNA localization pathway. *PLoS Biology*, *11*, e1001551.

Gagnon, J. A., & Mowry, K. L. (2011). Molecular motors: Directing traffic during RNA localization. *Critical Reviews in Biochemistry and Molecular Biology*, *46*, 229–239.

Gáspár, I., Sysoev, V., Komissarov, A., & Ephrussi, A. (2017). An RNA-binding atypical tropomyosin recruits kinesin-1 dynamically to oskar mRNPs. *The EMBO Journal*, *36*, 319–333.

Gavis, E. R., & Lehmann, R. (1992). Localization of nanos RNA controls embryonic polarity. *Cell*, *71*, 301–313.

Han, T. W., Kato, M., Xie, S., Wu, L. C., Mirzaei, H., Pei, J., et al. (2012). Cell-free formation of RNA granules: Bound RNAs identify features and components of cellular assemblies. *Cell*, *149*, 768–779.

Hanazawa, M., Yonetani, M., & Sugimoto, A. (2011). PGL proteins self associate and bind RNPs to mediate germ granule assembly in C. elegans. *The Journal of Cell Biology*, *192*, 929–937.

Havin, L., Git, A., Elisha, Z., Oberman, F., Yaniv, K., Schwartz, S. P., et al. (1998). RNA-binding protein conserved in both microtubule- and microfilament-based RNA localization. *Genes & Development*, *12*, 1593–1598.

Hayes, M. H., Peuchen, E. H., Dovichi, N. J., & Weeks, D. L. (2018). Dual roles for ATP in the regulation of phase separated protein aggregates in Xenopus oocyte nucleoli. *eLife*, *7*, 1–24.

Heasman, J., Quarmby, J., & Wylie, C. C. (1984). The mitochondrial cloud of Xenopus oocytes: The source of germinal granule material. *Developmental Biology*, *105*, 458–469.

Heasman, J., Wessely, O., Langland, R., Craig, E. J., & Kessler, D. S. (2001). Vegetal localization of maternal mRNAs is disrupted by VegT depletion. *Developmental Biology*, *240*, 377–386.

Holt, C. E., & Bullock, S. L. (2010). Subcellular mRNA localization in animal cells and why it matters. *Science*, *1212*, 1212–1216.

Houston, D. W. (2013). *Regulation of cell polarity and RNA localization in vertebrate oocytes* (1st ed.). Elsevier Inc: International Review of Cell and Molecular Biology.

Houston, D. W., Zhang, J., Maines, J. Z., Wasserman, S. A., & King, M. L. (1998). A Xenopus DAZ-like gene encodes an RNA component of germ plasm and is a functional homologue of Drosophila boule. *Development*, *125*, 171–180.

Hudson, J. W., Alarcón, V. B., & Elinson, R. P. (1996). Identification of new localized RNAs in the Xenopus oocyte by differential display PCR. *Developmental Genetics*, *19*, 190–198.

Hyman, A. A., Weber, C. A., & Jülicher, F. (2014). Liquid-liquid phase separation in biology. *Annual Review of Cell and Developmental Biology*, *30*, 39–58.

Jain, A., & Vale, R. D. (2017). RNA phase transitions in repeat expansion disorders. *Nature*, *546*, 243–247.

Jambhekar, A., & Derisi, J. L. (2007). Cis-acting determinants of asymmetric, cytoplasmic RNA transport. *RNA*, *13*, 625–642.

Januschke, J., Gervais, L., Dass, S., Kaltschmidt, J. A., Lopez-Schier, H., St. Johnston, D., et al. (2002). Polar transport in the Drosophila oocyte requires Dynein and Kinesin I cooperation. *Current Biology*, *12*, 1971–1981.

Jeffery, W. R., Tomlinson, C. R., & Brodeur, R. D. (1983). Localization of actin messenger RNA during early ascidian development. *Developmental Biology*, *99*, 408–417.

Joseph, E. M., & Melton, D. A. (1998). Mutant Vg1 ligands disrupt endoderm and mesoderm formation in Xenopus embryos. *Development*, *125*, 2677–2685.

Kato, M., Han, T. W., Xie, S., Shi, K., Du, X., Wu, L. C., et al. (2012). Cell-free formation of RNA granules: Low complexity sequence domains form dynamic fibers within hydrogels. *Cell*, *149*, 753–767.

Kawasaki, I., Shim, Y. H., Kirchner, J., Kaminker, J., Wood, W. B., & Strome, S. (1998). PGL-1, a predicted RNA-binding component of germ granules, is essential for fertility in *C. elegans*. *Cell*, *94*, 635–645.

Kedersha, N., Ivanov, P., & Anderson, P. (2013). Stress granules and cell signaling: More than just a passing phase? *Trends in Biochemical Sciences*, *38*, 494–506.

Kiebler, M. A., & Bassell, G. J. (2006). Neuronal RNA granules: Movers and makers. *Neuron*, *51*, 685–690.

King, M. L., Messitt, T. J., & Mowry, K. L. (2005). Putting RNAs in the right place at the right time: RNA localization in the frog oocyte. *Biology of the Cell*, *97*, 19–33.

Kislauskis, E. H., Li, Z., Singer, R. H., & Taneja, K. L. (1993). Isoform-specific 3'-untranslated sequences sort α-cardiac and β-cytoplasmic actin messenger RNAs to different cytoplasmic compartments. *The Journal of Cell Biology*, *123*, 165–172.

Kislauskis, E. H., Zhu, X., & Singer, R. H. (1994). Sequences responsible for intracellular localization of β-actin messenger RNA also affect cell phenotype. *The Journal of Cell Biology*, *127*, 441–451.

Kislauskis, E. H., Zhu, X. C., & Singer, R. H. (1997). β-Actin messenger RNA localization and protein synthesis augment cell motility. *The Journal of Cell Biology*, *136*, 1263–1270.

Kistler, K. E., Trcek, T., Hurd, T. R., Chen, R., Liang, F. X., Sall, J., et al. (2018). Phase transitioned nuclear oskar promotes cell division of Drosophila primordial germ cells. *eLife*, *7*, 1–35.

Kloc, M., Bilinski, S., & Dougherty, M. T. (2007). Organization of cytokeratin cytoskeleton and germ plasm in the vegetal cortex of Xenopus laevis oocytes depends on coding and non-coding RNAs: Three-dimensional and ultrastructural analysis. *Experimental Cell Research*, *313*, 1639–1651.

Kloc, M., Bilinski, S., & Etkin, L. D. (2004). The balbiani body and germ cell determinants: 150 years later. *Current Topics in Developmental Biology*, *59*, 1–36.

Kloc, M., & Etkin, L. D. (1995). Two distinct pathways for the localization of RNAs at the vegetal cortex in Xenopus oocytes. *Development*, *121*, 287–297.

Kloc, M., Foreman, V., & Reddy, S. A. (2011). Binary function of mRNA. *Biochimie*, *93*, 1955–1961.

Kloc, M., Jedrzejowska, I., Tworzydlo, W., & Bilinski, S. (2014). Balbiani body, nuage and sponge bodies—The germ plasm pathway players. *Arthropod Structure & Development*, *43*, 341–348.

Kloc, M., Spohr, G., & Etkin, L. D. (1993). Translocation of repetitive RNA sequences with the germ plasm in Xenopus oocytes. *Science*, *262*, 1712–1714.

Kloc, M., Wilk, K., Vargas, D., Shirato, Y., Bilinski, S., & Etkin, L. D. (2005). Potential structural role of non-coding and coding RNAs in the organization of the cytoskeleton at the vegetal cortex of Xenopus oocytes. *Development*, *132*, 3445–3457.

Kofron, M., Demel, T., Xanthos, J., Lohr, J., Sun, B., Sive, H., et al. (1999). Mesoderm induction in Xenopus is a zygotic event regulated by maternal VegT via TGFβ growth factors. *Development*, *126*, 5759–5770.

Kugler, J. M., & Lasko, P. (2009). Localization, anchoring and translational control of oskar, gurken, bicoid and nanos mRNA during drosophila oogenesis. *Fly*, *3*, 15–28.

Langdon, E. M., & Gladfelter, A. S. (2018). A new lens for RNA localization: Liquid-liquid phase separation. *Annual Review of Microbiology*, *72*, 255–271.

Langdon, E. M., Qiu, Y., Niaki, A. G., Mclaughlin, G. A., Weidmann, C. A., Gerbich, T. M., et al. (2018). mRNA structure determines specificity ofa polyQ-driven phase separation. *Science*, *360*, 922–927.

Lawrence, J. B., & Singer, R. H. (1986). Intracellular localization of messenger RNAs for cytoskeletal proteins. *Cell*, *45*, 407–415.

Lécuyer, E., Yoshida, H., Parthasarathy, N., Alm, C., Babak, T., Cerovina, T., et al. (2007). Global analysis of mRNA localization reveals a prominent role in organizing cellular architecture and function. *Cell*, *131*, 174–187.

Lewis, R. A., Gagnon, J. A., & Mowry, K. L. (2008). PTB/hnRNP I Is required for RNP remodeling during RNA localization in Xenopus oocytes. *Molecular and Cellular Biology*, *28*, 678–686.
Lewis, R. A., Kress, T. L., Cote, C. A., Gautreau, D., Rokop, M. E., & Mowry, K. L. (2004). Conserved and clustered RNA recognition sequences are a critical feature of signals directing RNA localization in Xenopus oocytes. *Mechanisms of Development*, *121*, 101–109.
Lin, Y., Protter, D. S. W., Rosen, M. K., & Parker, R. (2015). Formation and maturation of phase-separated liquid droplets by RNA-binding proteins. *Molecular Cell*, *60*, 208–219.
Long, R. M., Singer, R. H., Meng, X., Gonzalez, I., Nasmyth, K., & Jansen, R. P. (1997). Mating type switching in yeast controlled by asymmetric localization of ASH1 mRNA. *Science*, *277*, 383–387.
Macdonald, P. M., & Kerr, K. (1998). Mutational analysis of an RNA recognition element that mediates localization of bicoid mRNA. *Molecular and Cellular Biology*, *18*, 3788–3795.
Macdonald, P., & Struhl, G. (1988). Cis-acting sequences responsible for localizing bicoid mRNA at the anterior pole of Drosophila embryos. *Nature*, *336*, 595–598.
MacDougall, N., Clark, A., MacDougall, E., & Davis, I. (2003). Drosophila gurken (TGFα) mRNA localizes as particles that move within the oocyte in two dynein-dependent steps. *Developmental Cell*, *4*, 307–319.
Maharana, S., Wang, J., Papadopoulos, D. K., Richter, D., Pozniakovsky, A., Poser, I., et al. (2018). RNA buffers the phase separation behavior of prion-like RNA binding proteins. *Science*, *360*, 918–921.
Mahowald, A. P. (2001). Assembly of the Drosophila germ plasm. *International Review of Cytology*, *203*, 187–213.
Marchand, V., Gaspar, I., & Ephrussi, A. (2012). An intracellular transmission control protocol: Assembly and transport of ribonucleoprotein complexes. *Current Opinion in Cell Biology*, *24*, 202–210.
Mardakheh, F. K., Paul, A., Kümper, S., Sadok, A., Paterson, H., Mccarthy, A., et al. (2015). Global analysis of mRNA, translation, and protein localization: Local translation is a key regulator of cell protrusions. *Developmental Cell*, *35*, 344–357.
Martin, K. C., Barad, M., & Kandel, E. R. (2000). Local protein synthesis and its role in synapse-specific plasticity. *Current Opinion in Neurobiology*, *10*, 587–592.
Martin, K. C., & Ephrussi, A. (2009). mRNA localization: Gene expression in the spatial dimension. *Cell*, *136*, 719–730.
McClintock, M. A., Dix, C. I., Johnson, C. M., McLaughlin, S. H., Maizels, R. J., Hoang, H. T., et al. (2018). RNA-directed activation of cytoplasmic dynein-1 in reconstituted transport RNPs. *eLife*, *7*, 1–29.
Medioni, C., Mowry, K., & Besse, F. (2012). Principles and roles of mRNA localization in animal development. *Development*, *139*, 3263–3276.
Mercer, T. R., Dinger, M. E., Mariani, J., Kosik, K. S., Mehler, M. F., & Mattick, J. S. (2008). Noncoding RNAs in long-term memory formation. *The Neuroscientist*, *14*, 434–445.
Messitt, T. J., Gagnon, J. A., Kreiling, J. A., Pratt, C. A., Yoon, Y. J., & Mowry, K. L. (2008). Multiple kinesin motors coordinate cytoplasmic RNA transport on a subpopulation of microtubules in Xenopus oocytes. *Developmental Cell*, *15*, 426–436.
Mingle, L. A., Okuhama, N. N., Shi, J., Singer, R. H., Condeelis, J., & Liu, G. (2005). Localization of all seven messenger RNAs for the actin-polymerization nucleator Arp2/3 complex in the protrusions of fibroblasts. *Journal of Cell Science*, *118*, 2425–2433.
Mittag, T., & Parker, R. (2018). Multiple modes of protein–protein interactions promote RNP granule assembly. *Journal of Molecular Biology*, *430*, 4636–4649.

Molliex, A., Temirov, J., Lee, J., Coughlin, M., Kanagaraj, A. P., Kim, H. J., et al. (2015). Phase separation by low complexity domains promotes stress granule assembly and drives pathological fibrillization. *Cell*, *163*, 123–133.

Mosquera, L., Forristall, C., Zhou, Y., & King, M. L. (1993). A mRNA localized to the vegetal cortex of Xenopus oocytes encodes a protein with a nanos-like zinc finger domain. *Development*, *117*, 377–386.

Mowry, K. L. (1996). Complex formation between stage-specific oocyte factors and a Xenopus mRNA localization element. *Proceedings of the National Academy of Sciences of the United States of America*, *93*, 14608–14613.

Mowry, K. L., & Melton, D. A. (1992). Vegetal messenger RNA localization directed by a 340-nt RNA sequence element in Xenopus oocytes. *Science*, *255*, 991–994.

Murakami, T., Qamar, S., Lin, J. Q., Schierle, G. S. K., Rees, E., Miyashita, A., et al. (2015). ALS/FTD mutation-induced phase transition of FUS liquid droplets and reversible hydrogels into irreversible hydrogels impairs RNP granule function. *Neuron*, *88*, 678–690.

Niepielko, M. G., Eagle, W. V. I., & Gavis, E. R. (2018). Stochastic seeding coupled with mRNA self-recruitment generates heterogeneous Drosophila germ granules. *Current Biology*, *28*, 1872–1881.e3.

Nieuwkoop, P. D. (1969). The formation of the mesoderm in Urodelean amphibians—II. The origin of the dorso-ventral polarity of the mesoderm. *Wilhelm Roux' Archiv für Entwicklungsmechanik der Organismen*, *163*, 298–315.

Noh, J. H., Kim, K. M., McClusky, W. G., Abdelmohsen, K., & Gorospe, M. (2018). Cytoplasmic functions of long noncoding RNAs. *Wiley Interdisciplinary Reviews RNA*, *9*, 1–15.

Okita, T. W., & Choi, S. B. (2002). mRNA localization in plants: Targeting to the cell's cortical region and beyond. *Current Opinion in Plant Biology*, *5*, 553–559.

Otero, L. J., Devaux, A., & Standart, N. (2001). A 250-nucleotide UA-rich element in the 3′ untranslated region of Xenopus laevis Vg1 mRNA represses translation both in vivo and in vitro. *RNA*, *7*, 1753–1767.

Patel, A., Lee, H. O., Jawerth, L., Maharana, S., Jahnel, M., Hein, M. Y., et al. (2015). A liquid-to-solid phase transition of the ALS protein FUS accelerated by disease mutation. *Cell*, *162*, 1066–1077.

Pederson, T. (2011). The nucleolus. *Cold Spring Harbor Perspectives in Biology*, *3*, 1–15. a000638.

Puthanveettil, S. V. (2013). RNA transport and long-term memory storage. *RNA Biology*, *10*, 1765–1770.

Putnam, A., Cassani, M., Smith, J., & Seydoux, G. (2019). A gel phase promotes condensation of liquid P granules in *Caenorhabditis elegans* embryos. *Nature Structural & Molecular Biology*, *26*, 1.

Rebagliati, M. R., Weeks, D. L., Harvey, R. P., & Melton, D. A. (1985). Identification and cloning of localized maternal RNAs from Xenopus eggs. *Cell*, *42*, 769–777.

Rodriguez, A. J., Czaplinski, K., Condeelis, J. S., & Singer, R. H. (2008). Mechanisms and cellular roles of local protein synthesis in mammalian cells. *Current Opinion in Cell Biology*, *20*, 144–149.

Saha, S., Weber, C. A., Nousch, M., Adame-Arana, O., Hoege, C., Hein, M. Y., et al. (2016). Polar positioning of phase-separated liquid compartments in cells regulated by an mRNA competition mechanism. *Cell*, *166*, 1572–1584.e16.

Semotok, J. L., Luo, H., Cooperstock, R. L., Karaiskakis, A., Lipshitz, H. D., Vari, H. K., et al. (2008). Drosophila maternal Hsp83 mRNA destabilization is directed by multiple SMAUG recognition elements in the open reading frame. *Molecular and Cellular Biology*, *28*, 6757–6772.

Seydoux, G. (2018). The P granules of *C. elegans*: A genetic model for the study of RNA–protein condensates. *Journal of Molecular Biology, 430*, 4702–4710.
Singer-Krüger, B., & Jansen, R. P. (2014). Here, there, everywhere: mRNA localization in budding yeast. *RNA Biology, 11*, 1031–1039.
Singh, G., Pratt, G., Yeo, G. W., & Moore, M. J. (2015). The clothes make the mRNA: Past and present trends in mRNP fashion. *Annual Review of Biochemistry, 84*, 325–354.
Smith, J., Calidas, D., Schmidt, H., Lu, T., Rasoloson, D., & Seydoux, G. (2016). Spatial patterning of P granules by RNA-induced phase separation of the intrinsically-disordered protein MEG-3. *eLife, 5*, 1–18.
St Johnston, D. (2005). Moving messages: The intracellular localization of mRNAs. *Nature Reviews Molecular Cell Biology, 6*, 363–375.
Strome, S., & Wood, W. B. (1982). Immunofluorescence visualization of germ-line-specific cytoplasmic granules in embryos, larvae, and adults of Caenorhabditis elegans. *Proceedings of the National Academy of Sciences of the United States of America, 79*, 1558–1562.
Sudarwati, S., & Nieuwkoop, P. D. (1971). Mesoderm formation in the anuran *Xenopus laevis* (Daudin). *Wilhelm Roux' Archiv für Entwicklungsmechanik der Organismen, 166*, 189–204.
Sundell, C. L., & Singer, R. H. (1991). Requirement of microfilaments in sorting of actin messenger RNA. *Science, 252*, 1275–1277.
Tadros, W., Goldman, A. L., Babak, T., Menzies, F., Vardy, L., Orr-Weaver, T., et al. (2007). SMAUG Is a major regulator of maternal mRNA destabilization in Drosophila and Its translation Is activated by the PAN GU kinase. *Developmental Cell, 12*, 143–155.
Takizawa, P. A., & Vale, R. D. (2000). The myosin motor, Myo4p, binds Ash1 mRNA via the adapter protein, She3p. *Proceedings of the National Academy of Sciences of the United States of America, 97*, 5273–5278.
Tannahill, D., & Melton, D. A. (1989). Localized synthesis of the Vg1 protein during early Xenopus development. *Development, 106*, 775–785.
Thiry, M., & Lafontaine, D. L. J. (2005). Birth of a nucleolus: The evolution of nucleolar compartments. *Trends in Cell Biology, 15*, 194–199.
Thomsen, G. H., & Melton, D. A. (1993). Processed Vg1 protein is an axial mesoderm inducer in Xenopus. *Cell, 74*, 433–441.
Trcek, T., & Lehmann, R. (2019). Germ granules in Drosophila. *Traffic, 20*, 650–660.
Updike, D., & Strome, S. (2010). P granule assembly and function in *Caenorhabditis elegans* germ cells. *Journal of Andrology, 31*, 53–60.
Van De Bor, V., & Davis, I. (2004). mRNA localisation gets more complex. *Current Opinion in Cell Biology, 16*, 300–307.
Van Treeck, B., & Parker, R. (2018). Emerging roles for intermolecular RNA-RNA interactions in RNP assemblies. *Cell, 174*, 791–802.
Van Treeck, B., Protter, D. S. W., Matheny, T., Khong, A., Link, C. D., & Parker, R. (2018). RNA self-assembly contributes to stress granule formation and defining the stress granule transcriptome. *Proceedings of the National Academy of Sciences of the United States of America, 115*, 2734–2739.
Voronina, E., Seydoux, G., Sassone-Corsi, P., & Nagamori, I. (2011). RNA granules in germ cells. *Cold Spring Harbor Perspectives in Biology, 3*, 1–28.
Wallace, R. A., & Dumont, J. N. (1968). The induced synthesis and transport of yolk proteins and their accumulation by the oocyte in *Xenopus laevis*. *Journal of Cellular Physiology, 72*, 73–89.
Weatheritt, R. J., Gibson, T. J., & Babu, M. M. (2014). Asymmetric mRNA localization contributes to fidelity and sensitivity of spatially localized systems. *Nature Structural & Molecular Biology, 21*, 833–839.
Weber, S. C., & Brangwynne, C. P. (2015). Inverse size scaling of the nucleolus by a concentration-dependent phase transition. *Current Biology, 25*, 641–646.

Weil, T. T. (2014). mRNA localization in the Drosophila germline. *RNA Biology*, *11*, 1010–1018.

Weiß, K., Antoniou, A., & Schratt, G. (2015). Non-coding mechanisms of local mRNA translation in neuronal dendrites. *European Journal of Cell Biology*, *94*, 363–367.

White, J., & Heasman, J. (2008). Maternal control of pattern formation in *Xenopus laevis*. *The Journal of Experimental Zoology*, *310B*, 1–7.

Wilk, R., Hu, J., Blotsky, D., & Krause, H. M. (2016). Diverse and pervasive subcellular distributions for both coding and long noncoding RNAs. *Genes & Development*, *30*, 594–609.

Wilkie, G. S., & Davis, I. (2001). Drosophila wingless and pair-rule transcripts localize apically by dynein-mediated transport of RNA particles. *Cell*, *105*, 209–219.

Xiang, S., Kato, M., Wu, L. C., Lin, Y., Ding, M., Zhang, Y., et al. (2015). The LC domain of hnRNPA2 adopts similar conformations in hydrogel polymers, liquid-like droplets, and nuclei. *Cell*, *163*, 829–839.

Yisraeli, J. K., Sokol, S., & Melton, D. A. (1990). A two-step model for the localization of maternal mRNA in Xenopus oocytes: Involvement of microtubules and microfilaments in the translocation and anchoring of Vg1 mRNA. *Development*, *108*, 289–298.

Yoon, Y. J., & Mowry, K. L. (2004). Xenopus Staufen is a component of a ribonucleoprotein complex containing Vg1 RNA and kinesin. *Development*, *131*, 3035–3045.

Zaessinger, S., Busseau, I., & Simonelig, M. (2006). Oskar allows nanos mRNA translation in Drosophila embryos by preventing its deadnylation by Smaug/CCR4. *Development*, *133*, 4573–4583.

Zearfoss, N. R., Chan, A. P., Kloc, M., Allen, L. H., & Etkin, L. D. (2003). Identification of new Xlsirt family members in the Xenopus laevis oocyte. *Mechanisms of Development*, *120*, 503–509.

Zearfoss, N. R., Chan, A. P., Wu, C. F., Kloc, M., & Etkin, L. D. (2004). Hermes is a localized factor regulating cleavage of vegetal blastomeres in Xenopus laevis. *Developmental Biology*, *267*, 60–71.

Zhang, H., Elbaum-Garfinkle, S., Langdon, E. M., Taylor, N., Occhipinti, P., Bridges, A. A., et al. (2015). RNA controls polyQ protein phase transitions. *Molecular Cell*, *60*, 220–230.

Zhang, J., Houston, D. W., King, M. L., Payne, C., Wylie, C., & Heasman, J. (1998). The role of maternal VegT in establishing the primary germ layers in Xenopus embryos. *Cell*, *94*, 515–524.

Zhang, J., & King, M. L. (1996). Xenopus VegT RNA is localized to the vegetal cortex during oogenesis and encodes a novel T-box transcription factor involved in mesodermal patterning. *Development*, *122*, 4119–4129.

Zhang, H. L., Singer, R. H., & Bassell, G. J. (1999). Neurotrophin regulation of beta-actin mRNA and protein localization within growth cones. *The Journal of Cell Biology*, *147*, 59–70.

Zimyanin, V. L., Belaya, K., Pecreaux, J., Gilchrist, M. J., Clark, A., Davis, I., et al. (2008). In vivo imaging of oskar mRNA transport reveals the mechanism of posterior localization. *Cell*, *134*, 843–853.

PART II

Fertilization and germline specification

CHAPTER FIVE

Species-specific mechanisms during fertilization

Krista R. Gert, Andrea Pauli*
Research Institute of Molecular Pathology (IMP), Vienna BioCenter (VBC), Vienna, Austria
*Corresponding author: e-mail address: andrea.pauli@imp.ac.at

Contents

1. Introduction 122
2. The sperm's journey to the egg 123
   2.1 Chemotaxis and motility activation: The sperm gets moving 124
3. Sperm-egg binding 126
   3.1 Triggering the acrosome reaction 126
   3.2 Sperm-egg coat binding 129
   3.3 Sperm-egg membrane binding 131
4. Sperm-egg fusion: A lingering question mark 136
5. Conclusion 136
References 137
Further reading 143

Abstract

The perpetuation and preservation of distinct species rely on mechanisms that ensure that only interactions between gametes of the same species can give rise to viable and fertile offspring. Species-specificity can act at various stages, ranging from physical/behavioral pre-copulatory mechanisms, to pre-zygotic incompatibility during fertilization, to post-zygotic hybrid incompatibility. Herein, we focus on our current knowledge of the molecular mechanisms responsible for species-specificity during fertilization. While still poorly understood, decades of research have led to the discovery of molecules implicated in species-specific gamete interactions, starting from initial sperm-egg attraction to the binding of sperm and egg. While many of these molecules have been described as species-specific in their mode of action, relatively few have been demonstrated as such with definitive evidence. Thus, we also raise remaining questions that need to be addressed in order to characterize gamete interaction molecules as species-specific.

Current Topics in Developmental Biology, Volume 140
ISSN 0070-2153
https://doi.org/10.1016/bs.ctdb.2019.10.005

1. Introduction

Fertilization marks the beginning of life for sexually reproducing organisms. Albeit of long-standing interest to scientists, our understanding of fertilization is far from complete. One general feature of this process is that the sperm of one species fertilizes the eggs of the same species exclusively. While cross-fertilization is possible within certain genera and even among higher orders of classification, pre-zygotic isolation in the form of species-specific sperm–egg interaction is a useful strategy to prevent cross-species gamete interactions, particularly for external fertilizers. Such mechanisms exist even in mammals and other internally fertilizing organisms, ensuring that the female's costly investment in egg production or the male's sperm is not wasted by the formation of inviable hybrids. It is important to note that the term *species-specific* may not accurately describe these mechanisms in all species. In some cases, *taxon-specific* is a more apropos term for the specificity of the molecules in question.

At which steps of fertilization can species-specificity be imposed? The fertilization process can be broken down into three major events that ultimately result in the formation of a zygote. First, the sperm must find its way to the egg. Sperm motility-activating factors as well as chemoattractant molecules are important for this step, both with the potential to act species-specifically (Morita et al., 2006; Yanagimachi, Cherr, Pillai, & Baldwin, 1992; Yoshida, Hiradate, Sensui, Cosson, & Morisawa, 2013). Next, the sperm must initiate binding to the surface of the egg, first to the surrounding egg coat, and then to the egg plasma membrane (Wassarman, Jovine, & Litscher, 2001). In species whose sperm possess an acrosome, sperm–egg binding requires the induction of the acrosome reaction which exposes molecules important for sperm–egg binding (Hirohashi & Yanagimachi, 2018). Both the acrosome reaction and sperm–egg binding represent another layer of species-specificity for which molecules have been described (discussed below). Finally, the sperm and egg undergo fusion. Gamete fusion is the least understood event in fertilization, particularly in vertebrate models; thus, it remains to be seen whether species-specific mechanisms govern the final event in fertilization.

Species-specific fertilization factors have been challenging to identify and characterize given the intractability of separating essentiality from species-specificity. This begs the question, what makes a molecule species-specific? In our view, in order to definitively claim that a molecule functions species-specifically *in vivo*, it must be shown that (1) the presence of the molecule allows fertilization between conspecific, but not heterospecific, gametes (necessity) and (2) that expressing the molecule in a heterospecific gamete allows fertilization with the gametes conspecific to that molecule (sufficiency). Alternatively, use of recombinant proteins to either block conspecific fertilization or enable adhesion of the recombinant protein and conspecific gametes is another approach to demonstrate the species-specificity of a molecule in systems where genetic knockouts are not feasible. Based on these rather stringent criteria, few molecules have been shown to function species-specifically (see Fig. 1). Throughout the text, we present evidence demonstrating the species-specific nature of the discussed molecules, moving through the steps of fertilization chronologically. We also highlight what evidence is lacking in an effort to portray our understanding of species-specific fertilization across animal species.

2. The sperm's journey to the egg

The best-known examples of species-specific fertilization molecules come from broadcast spawning species including sea urchins, ascidians, and mollusks which do not actively seek out a mate to reproduce (Beekman, Nieuwenhuis, Ortiz-Barrientos, & Evans, 2016). In such species, little can be done to direct the eggs and sperm toward a conspecific gamete once they are released into the water's current—that is, unless the gametes themselves are equipped with molecules that mediate gamete recognition and/or attraction in a species-specific manner.

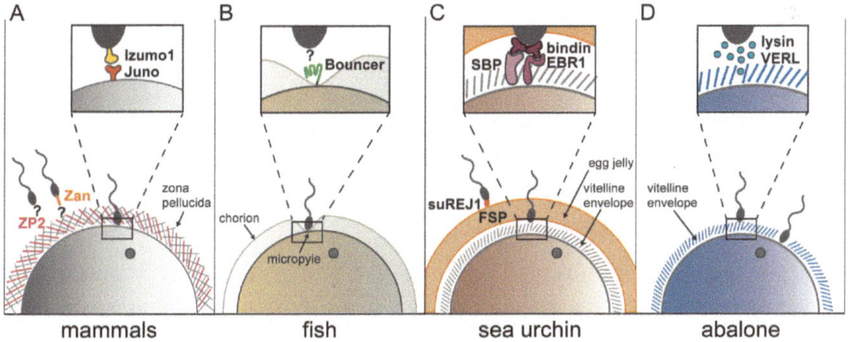

Fig. 1 Overview of species-specific fertilization molecules in representative organisms. Organism-specific egg coats are schematized for each animal group. Species-specific molecules are highlighted in color, while unknown interaction partners are indicated with a question mark. The upper boxes show magnified views of sperm-egg interaction sites.

2.1 Chemotaxis and motility activation: The sperm gets moving

The sperm, as the motile gamete, is most commonly equipped with a flagellum, allowing for directed movement toward the egg. Sperm must first be activated to gain motility, and in many species, are guided to the egg by egg-derived chemoattractants.

Such egg-secreted chemoattractants have been observed to induce sperm chemotaxis species-specifically in certain marine invertebrates. Sperm from ascidian species, such as those of the genus *Ciona*, begin vigorous, egg-directed movement when unfertilized eggs are added to the surrounding medium (Yoshida, Murata, Inaba, & Morisawa, 2002). The factor responsible for inducing both sperm activation and chemotaxis in these species is SAAF (sperm activating and attracting factor), a sulfated steroid that is released from the egg's vegetal pole (Yoshida, Inaba, Ishida, & Morisawa, 1994; Yoshida, Inaba, & Morisawa, 1993; Yoshida et al., 2002). The sulfation modifications of SAAF are essential for its function as hydrolysis of these moieties abolishes its activity (Yoshida et al., 2008). A candidate receptor for SAAF on the sperm, PMCA, has been recently reported, but its specificity has not yet been examined (Yoshida et al., 2018). SAAF is identical in two *Ciona* species (*C. intestinalis* (*Ci*) and *C. savignyi* (*Cs*)), namely, (25*S*)-3α,4β,7α,26-tetrahydroxy-5α-cholestane-3,26-disulfate) (Yoshida et al., 2002). Interestingly, SAAF from *Ascidia sydneiensis* (*As*) differs by only one double bond and the position of the hydroxyl group; these small differences are sufficient to enable genus-specific activity (Matsumori et al., 2013). Studies assessing the specificity of sperm activation and chemotaxis in heterospecific combinations of sperm and SAAFs suggest that induction of sperm chemotaxis is subject to a lower level of specificity than activation. *As* sperm exhibits the highest specificity in that it responds only to *As*-SAAF, however, *As*-SAAF can induce chemotaxis in *Cs* sperm, suggesting differential receptor-ligand affinity in *As* and *Cs* (Yoshida et al., 2013). Thus, species-specific recognition of SAAF molecules governs ascidian sperm activation and chemotaxis, but the molecular details underlying this mechanism remain to be uncovered.

In contrast to ascidians and other invertebrates as well as mammals, the eggs of fish, insects, and cephalopods are equipped with a narrow canal through the vitelline envelope that extends to the egg surface. This structure, called the micropyle, serves as the doorway into the egg which the sperm must find to enter (Yanagimachi et al., 2013). Unlike ascidians, teleost sperm activation primarily occurs in response to nonspecific external conditions including hypotonic or hypertonic media for sperm from freshwater or marine fish, respectively (Alavi & Cosson, 2006). However, while this is the case for most fish species studied so far, Pacific herring (*Clupea pallasii*) sperm are not activated upon exposure to seawater. Instead, the majority remain immotile until contacting the micropyle of conspecific eggs, as opposed to salmon or trout eggs (Yanagimachi et al., 1992).

Such observations led to the identification of egg-derived herring sperm activating proteins (HSAPs) (Oda et al., 1995) and sperm motility-initiating factor (SMIF) (Ohtake, 2003; Pillai, Shields, Yanagimachi, & Cherr, 1993), which appear to induce linear motility and circular motility, respectively (Cherr et al., 2008). HSAPs were purified from herring eggs and shown to localize in the outer layer of the chorion with monoclonal antibodies (Oda et al., 1998, 1995); they are predicted to be secreted and are homologous to Kazal-type trypsin inhibitors (Oda et al., 1998). SMIF, a 105-kDa glycoprotein, was isolated from herring chorions and localizes around the micropyle as evidenced by antibody labeling (Griffin, Vines, Pillai, Yanagimachi, & Cherr, 1996; Pillai et al., 1993). Eggs pre-incubated with anti-SMIF IgG exhibited decreased fertilization rates compared to controls, suggesting a role for SMIF in herring fertilization (Griffin et al., 1996). Consistent with this idea, depleting SMIF from herring eggs by acidic seawater greatly decreased fertilization success and could be partially restored by adding back purified SMIF, while removal of HSAPs did not affect fertility (Cherr et al., 2008). While these results suggest that SMIF is needed for herring sperm to enter the micropyle of conspecific eggs, other factors could have been removed from the egg during washing and some of the HSAPs or SMIF protein could have remained on the eggs. While genetic knockouts of these factors are needed to support the suggested roles of these proteins, activation of herring sperm by the presence of a sperm motility-initiating factor is an interesting example of a strategy to fertilize conspecific eggs selectively.

Even though the majority of teleost fish sperm do not require specific activation molecules for motility (Alavi & Cosson, 2006), egg-generated chemoattractants may play a role in sperm guidance in multiple species (Yanagimachi et al., 2013). Yanagimachi et al. (2013) identified a glycoprotein component of the chorion in flounder, herring, and steelhead eggs, termed the micropylar sperm attractant (MISA). It appears to attract conspecific sperm, as opposed to sperm from other species, to the micropyle in a species-specific manner, discriminating even between the sperm of three flounder species, starry, barfin, and black. However, MISA has not been molecularly characterized and no genetic knockout studies have been done to assess its function. Neither its necessity for fertilization nor sufficiency to attract conspecific sperm in a heterologous system have been reported to date, leaving identification of species-specific fish chemoattractant(s) as a goal for future studies.

While species-specific chemotaxis has been observed in marine organisms, mammals seem to lack a similar mechanism. Mammalian sperm do exhibit chemotactic behavior in response to factors secreted from cumulus cell-oocyte complexes (Sun et al., 2005), however, chemotaxis does not appear to be elicited in a species-specific manner (Sun et al., 2003). Because mammals employ internal fertilization in contrast to broadcast spawning invertebrates, they are provided with a robust pre-mating reproductive barrier by means of behavior and/or morphology. While mammalian hybrids can occur (Benirschke, 1967), pre-zygotic isolation mechanisms other than sperm activation and chemotaxis are in place to prevent cross-fertilization between certain species, as discussed below.

3. Sperm-egg binding

Once the sperm has gained motility and located the egg, the next step in the fertilization process entails making contact. The egg is generally surrounded by some form of extracellular matrix, ranging from egg jelly in amphibians and sea urchins, the chorion in fish and insects, to the fibrous zona pellucida (ZP) in mammals (see Fig. 1) (Shu, Suter, & Räsänen, 2015). Depending on the type of egg covering, the mode by which sperm can bind and enter also varies. However, in all cases, the first step to sperm-egg binding involves the sperm's getting past this outer egg layer to reach the plasma membrane, the site of fusion. For animals whose sperm possess an acrosome, the acrosome reaction (AR, described below) is coupled with egg coat binding and penetration.

3.1 Triggering the acrosome reaction

In many organisms, sperm are equipped with a specialized membranous structure located at the anterior end of the sperm head, the acrosome. Upon contact of the sperm with the species-appropriate trigger, the acrosomal vesicle fuses with the sperm plasma membrane to allow exocytosis of its contents (Hirohashi & Yanagimachi, 2018; Yanagimachi, 2011). This is followed by the extension of the acrosomal process (echinoderms, mollusks) or exposure of the inner acrosomal membrane (mammals, amphibians) on which essential sperm-egg binding proteins are localized

(Neill & Vacquier, 2004; Yanagimachi, 2011). In undergoing the AR, sperm is made fusion-competent and able to penetrate the outer covering of the egg. How the AR is triggered is another example of species-restricted sperm-egg communication in many species, however, mammalian AR induction is still not fully described. At least in the case of capacitated human sperm, the AR can be triggered by extracellular ATP (Foresta, Rossato, & Di Virgilio, 1992) and progesterone (Osman, Andria, Jones, & Meizel, 1989), as well as by ZP contact (Gupta, 2015; Hirohashi & Yanagimachi, 2018; Simons & Fauci, 2018).

3.1.1 Echinoderms

Sea urchin eggs are coated with egg jelly, a network of sugar-based polymers bound by glycoproteins (Bonnell, Keller, Vacquier, & Chandler, 1994). In sea urchin sperm, the AR is elicited by the concerted actions of three egg jelly molecules, sialoglycan, speract, and fucose sulfate polymer (FSP)—a primary component of the egg jelly itself and responsible for conferring species-specificity (Hirohashi & Vacquier, 2002a, 2002b; SeGall & Lennarz, 1979). Unable to induce the AR alone, sialoglycan and speract raise sperm intracellular pH, allowing for maximal AR induction (Hirohashi & Vacquier, 2002a). FSP, in contrast, is necessary for AR induction (SeGall & Lennarz, 1979) and facilitates the opening of Ca^{2+} channels by binding in a species-specific manner to its 210-kDa protein receptor, suREJ1, on the sperm plasma membrane (Hirohashi & Vacquier, 2002c; Vacquier & Moy, 1997) (Fig. 1C). The specificity of this interaction is determined by the sulfation pattern and glycosidic linkage of the sulfated fucans (Alves, Mulloy, Diniz, & Mourão, 1997; Hirohashi, Vilela-Silva, Mourão, & Vacquier, 2002; Vilela-Silva, Castro, Valente, Biermann, & Mourão, 2002), as well as the anomeric configuration of the sugar chains (Castro et al., 2009).

For example, AR induction in sperm from *Strongylocentrotus franciscanus* relies on the presence of 2-O-sulfation of FSP, while *S. purpuratus* sperm recognize 4-O-sulfated fucan (Hirohashi et al., 2002). Both species' sperm can be induced by heterospecific egg jelly endogenously containing the required sulfation modification (Hirohashi et al., 2002; Vilela-Silva, Alves, Valente, Vacquier, & Mourão, 1999). Interestingly, a sulfated fucan with identical sulfation pattern as that in *S. franciscanus* but with altered glycosidic linkage could not induce the AR in *S. franciscanus* sperm, while replacing L-fucose with L-galactose resulted in equally potent AR induction (Hirohashi et al., 2002). Comparison of the sulfated fucans in two co-occurring sea urchin species, *S. droebachiensis* and *S. pallidus*, led to the finding that differential sulfation patterns, potentially in combination with altered glycosidic linkage, contribute to the cross-fertilization barrier in these two species as well (Vilela-Silva et al., 2002). While these studies support a key role for FSP in mediating species-specific AR induction, other data is not fully consistent with this idea. Females of certain sea urchin species actually produce two types of sulfated fucan, and despite different sulfation patterns, both are equally potent in their induction of the AR in conspecific sperm (Alves, Mulloy, Moy, Vacquier, & Mourão, 1998; Vilela-Silva et al., 2002). The reason for this phenomenon remains unknown. Castro et al. (2009) proposed that the anomeric configuration of the fucan rings is a third factor in species-specific AR induction. Pre-incubation of *Glyptocidaris crenularis* sperm with its conspecific β-galactan, but not α-galactan, blocked fertilization, potentially by inducing the AR prematurely. Similarly, *Echinometra lucunter* sperm underwent the AR in response to 2-O-sulfated α-galactan and α-fucan, but not β-galactan, suggesting that the receptor in *E. lucunter* sperm specifically recognizes α anomers (Castro et al., 2009).

While induction of the AR is species-specific in sea urchins, it is less specific among another group of echinoderms, sea stars, but is also mediated by a sulfated sugar molecule. ARIS (acrosome reaction-inducing substance) is a sulfated glycoprotein that is able to induce the AR in an alkaline or high Ca^{2+} environment on its own (Nishigaki, Chiba, & Hoshi, 2000). One particular sugar fragment (Fr. 1) of ARIS contains 10–11 pentasaccharide repeats and is responsible for its activity as it still induces the AR when separated from the rest of the polysaccharide; importantly, desulfation of this fragment renders it inactive (Koyota, Swarma Wimalasiri, & Hoshi, 1997). Nakachi, Moriyama, Hoshi, and Matsumoto (2006) investigated whether the sea star AR can be induced by the egg jelly of closely related species. Within the *Asteriinae* subfamily, the AR could be induced by egg jelly of other species in a dose-dependent manner, while egg jelly from *Distolasterias nipon*, a species in a different subfamily (*Coscinasteriinae*) and from *Asterina pectinifera*, a species in a different superorder, induced the AR in only 20% of sperm or not at all, respectively (Nakachi et al., 2006). Though it remains to be experimentally tested, the same authors proposed that given the similarity of Fr. 1 glycan structure within the *Asteriinae* subfamily and because it is sufficient to induce the AR, this sugar fragment is possibly the subfamily-specific component acting in sea star AR induction. Moreover, they proposed that its specificity may be rooted in its sulfation pattern and/or glycosidic linkage (Nakachi et al., 2006).

3.1.2 Amphibians

The acrosome reaction in amphibians more closely resembles that of mammals, as fusion of the acrosomal vesicle and the sperm plasma membrane leads to exposure of the inner acrosomal membrane without formation of an acrosomal process (Ueda et al., 2007; Ueda, Yoshizaki, & Iwao, 2002). Most characterization of the AR-inducing substance in amphibians has been done in *Xenopus laevis*, belonging to Order *Anura*. Ueda et al. (2002) first identified this substance, ARISX, in extract from the pars recta (PRE) of the oviduct as well as the VE. Based on biochemical data, the AR-inducing activity of ARISX depends on a terminal α-D-GalNAc residue within its sugar chains and is not due to a protein constituent (Ueda et al., 2007; Ueda, Kubo, & Iwao, 2003). In investigating its species-specificity, Ueda et al. (2007) discovered that ARISX from *X. laevis* PRE could induce the AR in anuran sperm from species in three other genera, as well as two urodelean species, but to lower, varying degrees (2–43% vs 63% in conspecific sperm). However, despite AR induction occurring in these cases, only *Cynops pyrrhogaster* sperm was able to enter the egg cytoplasm. Thus, while there is a degree of specificity in AR induction among amphibians, the inability of most AR-reacted sperm to fertilize heterospecific eggs implies the existence of other, currently unknown mechanisms that prevent cross-fertilization in amphibians.

3.2 Sperm-egg coat binding

After the sperm has reached the egg, the next step of fertilization involves making contact. Whether the AR occurs before or after this initial contact depends upon the species and is still controversial for some. In any case, in order to penetrate and eventually fuse with the egg cell, the sperm must first initiate binding to the egg's outer covering.

The mammalian egg coat, or zona pellucida (ZP), is composed of glycoproteins that form an extracellular matrix around the egg which functions as a barrier to heterospecific sperm penetration. The human ZP consists of four such proteins, ZP1–4, while in mice, there are only three, ZP1–3 (Gupta et al., 2012). Even though ZP3 does play a role in sperm-zona binding (Bleil, Greve, & Wassarman, 1988; Bleil & Wassarman, 1980, 1983; Miller, Macek, & Shur, 1992), another ZP component, ZP2, was shown to be the primary sperm-binding factor (Rankin et al., 1998) (Fig. 1A). ZP2 exhibits taxon specificity, such that mouse sperm can bind mouse eggs transgenically expressing mouse or human ZP2, whereas human sperm exhibits greater selectivity and binds only to eggs expressing human ZP2 (Baibakov, Boggs, Yauger, Baibakov, & Dean, 2012). *Zp2* null female mice, though still able to form a ZP with transgenic expression of human ZP4, were sterile and their eggs were unable to bind mouse sperm (Avella, Baibakov, & Dean, 2014). The N-terminus of ZP2 was identified as the taxon specificity-mediating region, since an N-terminal truncation abolished sperm binding and exchanging the human for the mouse N-terminus prevented human sperm from binding to mouse eggs transgenically expressing human ZP2 (Avella et al., 2014). Furthermore, after fertilization, ZP2 loses its sperm-binding ability after the N-terminus is cleaved by ovastacin, a metalloendoprotease that is released upon fertilization (Burkart, Xiong, Baibakov, Jiménez-Movilla, & Dean, 2012; Gahlay, Gauthier, Baibakov, Epifano, & Dean, 2010; Xiong, Zhao, Beall, Sadusky, & Dean, 2017). ZP2 is therefore a critical glycoprotein for fertilization whose functions are three-fold: ZP2 is necessary for sperm-zona binding, mediates taxon specificity between mice and human gametes, and provides a block to polyspermy.

While the ZP does exhibit specificity in some cross-species fertilization examples, other instances have been demonstrated in which sperm is able to bind to the ZP of heterospecific eggs. For example, boar and horse sperm are able to undergo the AR and bind tightly to the ZP of bovine oocytes; horse sperm are even able to penetrate the ZP and enter the egg cytoplasm (Sinowatz, Wessa, Neumüller, & Palma, 2003), as are sperm from the African antelope (Roth et al., 1998). Based on the species combinations tested thus far, it appears that human sperm exhibit the most selectivity in terms of binding, with only documented cross-species binding occurring between human sperm and zona-intact gorilla and gibbon oocytes (Bedford, 1977; Lanzendorf, Holmgren, Johnson, Scobey, & Jeyendran, 1992).

In contrast to ZP2 which selectively allows binding of conspecific sperm and eggs, an alternative mechanism through which species-specificity could be imposed is by expression of a factor that selectively blocks or limits binding to heterospecific eggs. Zonadhesin was identified from pig sperm membrane as a protein that bound most strongly to pig ZP as opposed to bovine ZP or *Xenopus* egg envelopes, even when subjected to harsh washing (Hardy & Garbers, 1994) (Fig. 1A). Zonadhesin has a repetitive sequence region and domains homologous to von Willebrand factor (Hardy & Garbers, 1995). Evidence for Zonadhesin's species-specific function was obtained through experiments with *Zan*$^{-/-}$ mice (Tardif et al., 2010). While neither Zonadhesin-deficient male nor female mice exhibited decreased fertility, more *Zan*$^{-/-}$ than wild-type mouse sperm bound to heterospecific ZPs when present in equal numbers. The largest difference was observed for pig ZP, to which *Zan*$^{-/-}$ mouse sperm bound nearly five times as much as wild-type sperm. In contrast, both wild-type and *Zan*$^{-/-}$ sperm bound equally to conspecific (mouse) oocytes (Tardif et al., 2010). While these results support Zonadhesin's ability to prevent interaction of heterospecific gametes, Zonadhesin has not yet been demonstrated as sufficient to block heterospecific sperm-egg binding at this time.

3.3 Sperm-egg membrane binding

3.3.1 Mammals

The ZP seems to function as the primary barrier to cross-species fertilization in mammals, as removal of this outer layer allows sperm-egg interaction between gametes of different species (Hanada & Chang, 1972; Yanagimachi, Yanagimachi, & Rogers, 1976). However, sperm-egg binding proteins that interact after the AR appear to have some specificity in addition to the ZP. The interaction between sperm-expressed Izumo1 and the egg protein Juno was recently identified as necessary for sperm-egg interaction enabling subsequent gamete fusion, with both *Izumo1*$^{-/-}$ male and *Juno*$^{-/-}$ female mice being infertile (Bianchi, Doe, Goulding, & Wright, 2014; Inoue, Ikawa, Isotani, & Okabe, 2005) (Fig. 1A). Izumo1 is a type I membrane protein belonging to the immunoglobulin superfamily and is expressed on the inner acrosomal membrane that becomes exposed after the AR, while Juno is a folate receptor paralog that is GPI-anchored to the egg membrane (Bianchi et al., 2014; Inoue et al., 2005). Bianchi and colleagues investigated whether the Izumo1-Juno interaction imparts specificity to gamete interaction. Using a sensitive avidity-based assay able to detect transient interactions, they demonstrated that recombinant hamster Juno is able to bind human, mouse, and pig Izumo1, while recombinant human Juno does not bind mouse Izumo1, in line with previous zona-free binding assays with gametes from these species (Bianchi & Wright, 2015). While this confirms some level of specificity in the case of human fertilization molecules, broader conclusions regarding species-specific sperm-egg membrane binding in mammals will require a larger cohort of species for experimentation.

3.3.2 Teleost fish

The only sperm-egg binding protein to be characterized in teleost fish to date is the Ly6/uPAR protein Bouncer (Herberg, Gert, Schleiffer, & Pauli, 2018) (Fig. 1B). Bouncer is a GPI-anchored protein expressed in the egg and is enriched in the egg plasma membrane near the micropyle. Consistent with its egg-specific expression, female *bncr*$^{-/-}$ zebrafish lay eggs that cannot be fertilized by wild-type sperm, while mutant males exhibit normal fertility (Herberg et al., 2018). Bouncer was shown to be important for sperm-egg binding, as fewer sperm bind to *bncr*$^{-/-}$ eggs compared to wild-type. Due to its high level of sequence divergence among homologous Bouncer proteins in fish, its function in species-specific gamete recognition was investigated. Medaka and zebrafish sperm are unable to fertilize each other's eggs naturally, however, when medaka Bouncer was expressed transgenically on zebrafish eggs in the absence of zebrafish Bouncer, medaka sperm was able to enter and fertilize them (Herberg et al., 2018). Thus, Bouncer mediates species-specific sperm-egg interaction in the case of zebrafish and medaka. Bouncer is one of the few proteins whose identity as a species-specific molecule has been demonstrated through both its necessity for fertilization and its sufficiency to allow entry of heterospecific sperm into an egg. However, it remains to be seen whether and to what extent Bouncer mediates species-specific sperm entry between more closely related fish species.

3.3.3 Sea urchin

In addition to species-specific egg jelly binding and consequent AR induction in sea urchins, sperm-egg binding also occurs in a species-specific manner. With the acrosomal process exposed after the AR, an essential sperm adhesion protein comes into play. This protein, bindin, is a 24-kDa insoluble protein that coats the acrosomal process; it is associated with the membrane but is not membrane-anchored (Minor, Fromson, Britten, & Davidson, 1991; Vacquier & Moy, 1977) (Fig. 1C). Early studies found that isolated bindin particles from two different sea urchin species could species-specifically agglutinate eggs, and that this egg-agglutinating activity could be disrupted by trypsin or metaperiodate treatment, suggesting that bindin binds a carbohydrate receptor on the vitelline envelope (VE) (Glabe & Vacquier, 1977; Vacquier & Moy, 1977). Later work revealed that bindin's egg- and erythrocyte-agglutinating activity can be inhibited specifically by fucoidan (sulfated fucan) and other polysaccharides, lending support to the idea that bindin accomplishes sperm-egg adhesion through a lectin-polysaccharide interaction (Glabe, Grabel, Vacquier, & Rosen, 1982).

Bindin was found to interact species-specifically with not one, but two different receptors on the sea urchin egg. The first receptor to be identified, the 350-kDa egg receptor (or SBP, sperm-binding protein) (Fig. 1C), could be released from protease-treated eggs and bound bindin species-specifically in competition binding assays (Foltz & Lennarz, 1990; Glabe & Vacquier, 1978). Further characterization revealed this receptor to be a transmembrane glycoprotein in the vitelline layer with a preponderance of sulfated carbohydrate modifications (Hirohashi & Lennarz, 1998; Ohlendieck, Dhume, Partin, & Lennarz, 1993). When purified, the receptor was able to inhibit fertilization species-specifically (Ohlendieck et al., 1993).

The second receptor for bindin, EBR1 (Fig. 1C), was discovered by identification and cloning of species-specific cDNAs from the two co-occurring, reproductively isolated sea urchin species, *S. purpuratus* and *S. franciscanus*. Recombinant EBR1 from *S. franciscanus* bound species-specifically to bindin granules from the same species and could agglutinate conspecific eggs but was unable to bind bindin or agglutinate eggs from *S. purpuratus* (Kamei & Glabe, 2003). Investigation of EBR1's domain architecture in both species revealed that specific EBR repeats of 171 amino acids comprise the core region of the protein in both species variants, however, the species-specific domains differed. *S. purpuratus* EBR1 bears a unique HYR domain of 11 repeats, which was shown to bind *S. purpuratus* but not *S. franciscanus* sperm (Kamei & Glabe, 2003). In contrast, *S. franciscanus* EBR1 contains 10 additional EBR repeats that are highly similar to each other yet are absent in *S. purpuratus* EBR1 (Kamei & Glabe, 2003). Thus, sea urchin eggs are equipped with two receptors for bindin, potentially as a strategy to maximize the success of conspecific sperm-egg adhesion (reviewed in Vacquier, 2012).

An intriguing area of investigation with the potential to shed light on how species-specific mechanisms arose is the evolution of reproductive proteins. In contrast to other protein types, reproductive proteins are thought to evolve more quickly and are often under positive selection (Swanson & Vacquier, 2002). Indeed, bindin shows evidence for positive selection at sites outside of its conserved region, leading to a high rate of polymorphisms within species and thus maintaining a high level of genetic diversity (Biermann, 1998; Metz & Palumbi, 1996). Divergence of bindin at non-synonymous sites has also been correlated with incompatibility between gametes from different sea urchin species, specifically those in sympatric genera (Lessios, 2007; Minor et al., 1991; Zigler & Lessios, 2003; Zigler, McCartney, Levitan, & Lessios, 2005). On the other hand, positive selection has had a weaker effect on EBR1, affecting only a single amino acid position while the rest of the protein appears to have undergone purifying selection (Pujolar & Pogson, 2011). Stapper, Beerli, and Levitan (2015) found evidence for linkage disequilibrium between EBR1 and bindin driven by assortative mating in sea urchins. This provides a possible explanation as to why males and females with matching bindin loci experience higher rates of gametic compatibility and highlights the effect of the bindin-EBR1 interaction on reproductive success (Palumbi, 1999; Stapper et al., 2015).

3.3.4 Abalone

The best characterized species-specific sperm-egg gamete recognition protein pair comes from an unusual species, the abalone (genus *Haliotis*). These broadcast spawning marine gastropods employ a non-enzymatic mechanism for dissolution of the egg VE, brought about by lysin, a small sperm protein that interacts species-specifically with VERL (VE receptor for lysin), its receptor on the egg (Fig. 1D) (Kresge, Vacquier, & Sout, 2001; Swanson & Vacquier, 1997). In abalone, the AR occurs when the sperm reaches the VE, thereby releasing the chief player in sperm-egg adhesion, lysin (Kresge et al., 2001). Lysin is a secreted, 16-kDa, α-helical protein (Lewis, Talbot, & Vacquier, 1982). It contains two highly conserved tracks of basic residues and a solvent-exposed hydrophobic patch that extends the length of the protein (Kresge, Vacquier, & Stout, 2000; Shaw, Lee, Stout, & Vacquier, 1994). Lysin induces the non-enzymatic formation of a hole in the VE that is approximately three times the diameter of a sperm cell, providing passage for sperm through the VE's fibrous mesh to reach the plasma membrane (Lewis et al., 1982). Lysin's species-specific activity was evidenced in VE dissolution assays using isolated lysins and VEs (Vacquier, Carner, & Stout, 1990). When lysins from red (H. *rufescens*) and pink (*H. corrugata*) abalone were added to radioiodinated VEs from both species, pink abalone lysin dissolved both species' VEs comparably well, while red abalone lysin dissolved only conspecific VEs (Vacquier et al., 1990). Subsequent experiments with recombinant interspecies chimeras of red and pink abalone lysins revealed that substituting the N- and C-termini as well as a short internal segment of one species' lysin for the other's reduced dissolution activity on conspecific VEs, especially in the case of swapping the N-terminus (Lyon & Vacquier, 1999).

Lysin's receptor, VERL, is a 1000-kDa glycoprotein that was identified using affinity chromatography and sucrose density gradient sedimentation (Swanson & Vacquier, 1997). Isolated VERL was shown to inhibit the dissolution of VEs by lysin in a carbohydrate-independent manner, suggesting that its protein component mediates lysin binding (Raj et al., 2017; Swanson & Vacquier, 1997). Electron microscopy revealed that VERL is a long, unbranched rod with a diameter of 13nm (Swanson & Vacquier, 1997). Importantly, early characterization of the lysin-VERL interaction using fluorescence polarization assays demonstrated that it exhibits positive cooperativity, is high-affinity, and that over 100 lysin monomers bind a single VERL molecule (Swanson & Vacquier, 1997). On the protein level, VERL is comprised of 22 repeats, each 153 amino acids long, the first two of which are most different from the remaining repeats (Swanson & Vacquier, 1998). Repeat 1 is most divergent, while repeat 2 shares 75% identity with repeat 3. Repeats 3 through 22 are nearly identical, having been homogenized through concerted evolution (Galindo, Moy, Swanson, & Vacquier, 2002; Swanson & Vacquier, 1998).

As found for sea urchin reproductive proteins, evolutionary analysis of *lysin* cDNAs showed that *lysin* has diverged very rapidly at rates 2–50 times faster than rapidly evolving mammalian genes (Lee, Ota, & Vacquier, 1995; Metz, Robles-Sikisaka, & Vacquier, 1998). In addition, lysin and VERL are thought to be coevolving based on their correlated rates of amino acid substitution and evidence for linkage disequilibrium between them (Clark et al., 2009). While the majority of repeats (3 – 22) in VERL have evolved under neutral selection (Swanson & Vacquier, 1998), repeats 1 and 2 have undergone positive selection (Galindo, Vacquier, & Swanson, 2003); the latter comprises the species-specific region of the receptor as revealed by structural work on lysin-VERL repeat complexes (Raj et al., 2017). Raj et al. (2017) probed the atomic details of the lysin-VERL interaction by expressing constructs of the first three VERL repeats in mammalian cells. By co-expressing His-tagged VERL repeat (VR) 1, 2, 3, as well as VR1 + 2 and VR3 + 4 with lysin and performing a pull-down, lysin was shown to bind repeats 2 and 3, but not repeat 1. The lysin-VR3 interaction was found to be high-affinity (nM range) and non-discriminatory between red and pink abalone lysins (Raj et al., 2017). However, red lysin bound to both VR1 + 2 and VR2 more efficiently than pink lysin, demonstrating the ability of VR2 to bind preferentially to conspecific lysin. Mutational analysis of residues within the N-terminus of VR2 pinpointed critical amino acids mediating the specificity of its interaction with lysin (Raj et al., 2017).

Combining all of the data collected over the years and drawing especially from the work of Raj and colleagues, it is likely that VERL forms antiparallel homodimers that comprise the filamentous meshwork of the VE. The lysin-VERL mechanism may function in such a way that lysin dimers dissociate into monomers upon reaching the VE as a result of stronger interactions between lysin and VERL compared to the weak lysin homodimeric interface. Lysin monomers first bind to VR2 species-specifically with lower affinity, thereby beginning the dissolution of VERL fibers by repulsion of apposing basic lysin surfaces. As lysin monomers are then able to bind VR3 and the remaining repeats with higher affinity, VERL proteins would unwind completely and be repelled electrostatically, creating a hole and thus allowing the sperm to enter (Raj et al., 2017).

4. Sperm-egg fusion: A lingering question mark

Gamete fusion remains the least understood event in fertilization. To date, no vertebrate proteins have been identified as essential for fusion between sperm and eggs, though Izumo1 and Juno (Inoue, Hagihara, Wright, Takahisa, & Wada, 2015; Kato et al., 2016) and egg-expressed CD9 (Kaji et al., 2000; Le Naour, Rubinstein, Jasmin, Prenant, & Boucheix, 2000; Miyado et al., 2000) have been implicated in fusion. In invertebrates, both bindin and lysin have been purported to contain fusogenic peptides that are able to fuse lipid vesicles (Glabe, 1985; Hong & Vacquier, 1986; Ulrich, Otter, Glabe, & Hoekstra, 1998), but due to the dearth of knockout studies in these systems, this has never been confirmed *in vivo*. The only currently known gamete fusogen is HAP2/GCS1, which mediates gamete membrane fusion in protists, plants, and algae (reviewed in Hernández & Podbilewicz, 2017). Because our knowledge of the factors that mediate gamete fusion is still greatly lacking, we have little insight as to how or whether this event is regulated in a species-specific manner.

5. Conclusion

Species-specific gamete interaction has a significant influence on speciation and the maintenance of distinct species, especially for those that lack other reproductive barriers. While much is left to be characterized regarding both the general process of fertilization itself as well as how it is kept specific to a species, it appears that species-specific mechanisms are present at multiple stages of the process, enhancing interactions between conspecific sperm and eggs to bolster reproductive success and eschew the futile formation of inviable hybrids.

Mutual recognition of conspecific vs heterospecific sperm and eggs can be likened to the ability of immune cells to distinguish "self" from "nonself" or other biological self-recognition systems. For example, a small peptide (SELF-1) expressed in a group of predatory nematodes was found to mediate self-recognition via its hypervariable C-terminus, a feature reminiscent of rapidly evolving fertilization proteins (Lightfoot et al., 2019). In a similar way, the host-pathogen "arms race" leads to selection of resistance traits in the host and infection-enabling traits in the pathogen, albeit with competing rather than common interests as would occur for conspecific gametes. Indeed, coevolution of host bacteria and an infecting phage resulted in higher molecular evolution in the phage's genome, allowing it to expand its repertoire of possible hosts (Paterson et al., 2010). In the same vein, mutual optimization of reproductive proteins on both the egg and sperm enables conspecific gametes to gain a competitive edge over interactions with heterospecific gametes, ultimately contributing to the reproductive success of entire species.

References

Alavi, S. M. H., & Cosson, J. (2006). Sperm motility in fishes. (II) Effects of ions and osmolality: A review. *Cell Biology International*, *30*(1), 1–14.

Alves, A.-P., Mulloy, B., Diniz, J. A., & Mourão, P. A. S. (1997). Sulfated polysaccharides from the egg jelly layer are species-specific inducers of acrosomal reaction in sperms of sea urchins. *The Journal of Biological Chemistry*, *272*(11), 6965–6971.

Alves, A.-P., Mulloy, B., Moy, G. W., Vacquier, V. D., & Mourão, P. A. S. (1998). Females of the sea urchin *Strongylocentrotus purpuratus* differ in the structures of their egg jelly sulfated fucans. *Glycobiology*, *8*(9), 939–946.

Avella, M. A., Baibakov, B., & Dean, J. (2014). A single domain of the ZP2 zona pellucida protein mediates gamete recognition in mice and humans. *The Journal of Cell Biology*, *205*(6), 801–809.

Baibakov, B., Boggs, N. A., Yauger, B., Baibakov, G., & Dean, J. (2012). Human sperm bind to the N-terminal domain of ZP2 in humanized zonae pellucidae in transgenic mice. *The Journal of Cell Biology*, *197*(7), 897–905.

Bedford, J. M. (1977). Sperm/egg interaction: The specificity of human spermatozoa. *The Anatomical Record*, *188*, 477–488.

Beekman, M., Nieuwenhuis, B., Ortiz-Barrientos, D., & Evans, J. P. (2016). Sexual selection in hermaphrodites, sperm and broadcast spawners, plants and fungi. *Philosophical Transactions of the Royal Society B*, *371*(1706), 1–13.

Benirschke, K. (1967). Sterility and fertility of interspecific mammalian hybrids. In K. Benirschke (Ed.), *Comparative aspects of reproductive failure* (pp. 220–232). New York, NY: Springer-Verlag.

Bianchi, E., Doe, B., Goulding, D., & Wright, G. J. (2014). Juno is the egg Izumo receptor and is essential for mammalian fertilization. *Nature*, *508*(7497), 483–487.

Bianchi, E., & Wright, G. J. (2015). Cross-species fertilization: The hamster egg receptor, Juno, binds the human sperm ligand, Izumo1. *Philosophical Transactions of the Royal Society B*, *370*(1661), 1–4.

Biermann, C. H. (1998). The molecular evolution of sperm bindin in six species of sea urchins (*Echinoida: Strongylocentrotidae*). *Molecular Biology and Evolution*, *15*(12), 1761–1771.

Bleil, J. D., Greve, J. M., & Wassarman, P. M. (1988). Identification of a secondary sperm receptor in the mouse egg zona pellucida: Role in maintenance of binding of acrosome-reacted sperm to eggs. *Developmental Biology*, *128*(2), 376–385.

Bleil, J. D., & Wassarman, P. M. (1980). Mammalian sperm-egg interaction: Identification of a glycoprotein in mouse egg zonae pellucidae possessing receptor activity for sperm. *Cell*, *20*(3), 873–882.

Bleil, J. D., & Wassarman, P. M. (1983). Sperm-egg interactions in the mouse: Sequence of events and induction of the acrosome reaction by a zona pellucida glycoprotein. *Developmental Biology*, *95*(2), 317–324.

Bonnell, B. S., Keller, S. H., Vacquier, V. D., & Chandler, D. E. (1994). The sea urchin egg jelly coat consists of globular glycoproteins bound to a fibrous fucan superstructure. *Developmental Biology*, *162*(1), 313–324.

Burkart, A. D., Xiong, B., Baibakov, B., Jiménez-Movilla, M., & Dean, J. (2012). Ovastacin, a cortical granule protease, cleaves ZP2 in the zona pellucida to prevent polyspermy. *The Journal of Cell Biology*, *197*(1), 37–44.

Castro, M. O., Pomin, V. H., Santos, L. L., Vilela-Silva, A.-C. E. S., Hirohashi, N., Pol-Fachin, L., et al. (2009). A unique 2-sulfated β-galactan from the egg jelly of the sea urchin *Glyptocidaris crenularis*. *The Journal of Biological Chemistry*, *284*(28), 18790–18800.

Cherr, G. N., Morisawa, M., Vines, C. A., Yoshida, K., Smith, E. H., Matsubara, T., et al. (2008). Two egg-derived molecules in sperm motility initiation and fertilization in the Pacific herring (*Clupea pallasi*). *The International Journal of Developmental Biology*, *52*(5–6), 743–752.

Clark, N. L., Gasper, J., Sekino, M., Springer, S. A., Aquadro, C. F., & Swanson, W. J. (2009). Coevolution of interacting fertilization proteins. *PLoS Genetics*, *5*(7), e1000570-14.

Foltz, K. R., & Lennarz, W. J. (1990). Purification and characterization of an extracellular fragment of the sea urchin egg receptor for sperm. *The Journal of Cell Biology*, *111*(6), 2951–2959.

Foresta, C., Rossato, M., & Di Virgilio, F. (1992). Extracellular ATP is a trigger for the acrosome reaction in human spermatozoa. *The Journal of Biological Chemistry*, *267*(27), 19443–19447.

Gahlay, G., Gauthier, L., Baibakov, B., Epifano, O., & Dean, J. (2010). Gamete recognition in mice depends on the cleavage status of an egg's zona pellucida protein. *Science*, *329*(5988), 216–219.

Galindo, B. E., Moy, G. W., Swanson, W. J., & Vacquier, V. D. (2002). Full-length sequence of VERL, the egg vitelline envelope receptor for abalone sperm lysin. *Genes*, *288*(1–2), 111–117.

Galindo, B. E., Vacquier, V. D., & Swanson, W. J. (2003). Positive selection in the egg receptor for abalone sperm lysin. *PNAS*, *100*(8), 4639–4643.

Glabe, C. G. (1985). Interaction of the sperm adhesive protein, bindin, with phospholipid vesicles. II. Bindin induces the fusion of mixed-phase vesicles that contain phosphatidylcholine and phosphatidylserine in vitro. *The Journal of Cell Biology*, *100*(3), 800–806.

Glabe, C. G., Grabel, L. B., Vacquier, V. D., & Rosen, S. D. (1982). Carbohydrate specificity of sea urchin sperm bindin: A cell surface lectin mediating sperm-egg adhesion. *The Journal of Cell Biology*, *94*(1), 123–128.

Glabe, C. G., & Vacquier, V. D. (1977). Species specific agglutination of eggs by bindin isolated from sea urchin sperm. *Nature*, *267*, 836–838.

Glabe, C. G., & Vacquier, V. D. (1978). Egg surface glycoprotein receptor for sea urchin sperm bindin. Proceedings of the National Academy of Sciences of the United States of America, *75*(2), 881–885.

Griffin, F. J., Vines, C. A., Pillai, M. C., Yanagimachi, R., & Cherr, G. N. (1996). Sperm motility initiation factor is a minor component of the Pacific herring egg chorion. *Development, Growth & Differentiation*, *38*(2), 193–202.

Gupta, S. K. (2015). Role of zona pellucida glycoproteins during fertilization in humans. *Journal of Reproductive Immunology*, *108*, 90–97.

Gupta, S. K., Bhandari, B., Shrestha, A., Biswal, B. K., Palaniappan, C., Malhotra, S. S., et al. (2012). Mammalian zona pellucida glycoproteins: Structure and function during fertilization. *Cell and Tissue Research*, *349*(3), 665–678.

Hanada, A., & Chang, M. C. (1972). Penetration of zona-free eggs by spermatozoa of different species. *Biology of Reproduction*, *6*(2), 300–309.

Hardy, D. M., & Garbers, D. L. (1994). Species-specific binding of sperm proteins to the extracellular matrix (zona pellucida) of the egg. *The Journal of Biological Chemistry*, *269*(29), 19000–19004.

Hardy, D. M., & Garbers, D. L. (1995). A sperm membrane protein that binds in a species-specific manner to the egg extracellular matrix is homologous to von Willebrand factor. *The Journal of Biological Chemistry*, *270*(44), 26025–26028.

Herberg, S., Gert, K. R., Schleiffer, A., & Pauli, A. (2018). The Ly6/uPAR protein bouncer is necessary and sufficient for species-specific fertilization. *Science*, *361*(6406), 1029–1033.

Hernández, J. M., & Podbilewicz, B. (2017). The hallmarks of cell-cell fusion. *Development*, *144*(24), 4481–4495.

Hirohashi, N., & Lennarz, W. J. (1998). The 350-kDa sea urchin egg receptor for sperm is localized in the vitelline layer. *Developmental Biology*, *204*(1), 305–315.

Hirohashi, N., & Vacquier, V. D. (2002a). Egg fucose sulfate polymer, sialoglycan, and speract all trigger the sea urchin sperm acrosome reaction. *Biochemical and Biophysical Research Communications*, *296*(4), 833–839.

Hirohashi, N., & Vacquier, V. D. (2002b). Egg sialoglycans increase intracellular pH and potentiate the acrosome reaction of sea urchin sperm. *Journal of Biological Chemistry*, *277*(10), 8041–8047.

Hirohashi, N., & Vacquier, V. D. (2002c). High molecular mass egg fucose sulfate polymer is required for opening both Ca^{2+} channels involved in triggering the sea urchin sperm acrosome reaction. *Journal of Biological Chemistry*, *277*(2), 1182–1189.

Hirohashi, N., Vilela-Silva, A.-C. E. S., Mourão, P. A. S., & Vacquier, V. D. (2002). Structural requirements for species-specific induction of the sperm acrosome reaction by sea urchin egg sulfated fucan. *Biochemical and Biophysical Research Communications*, *298*(3), 403–407.

Hirohashi, N., & Yanagimachi, R. (2018). Sperm acrosome reaction: Its site and role in fertilization. *Biology of Reproduction*, *99*(1), 127–133.

Hong, K., & Vacquier, V. D. (1986). Fusion of liposomes induced by a cationic protein from the acrosome granule of abalone spermatozoa. *Biochemistry*, *25*(3), 543–549.

Inoue, N., Hagihara, Y., Wright, D., Takahisa, S., & Wada, I. (2015). Oocyte-triggered dimerization of sperm IZUMO1 promotes sperm-egg fusion in mice. *Nature Communications*, *6*, 1–12.

Inoue, N., Ikawa, M., Isotani, A., & Okabe, M. (2005). The immunoglobulin superfamily protein Izumo is required for sperm to fuse with eggs. *Nature*, *434*(7030), 234–238.

Kaji, K., Oda, S., Shikano, T., Ohnuki, T., Uematsu, Y., Sakagami, J., et al. (2000). The gamete fusion process is defective in eggs of CD9-deficient mice. *Nature Genetics*, *24*(3), 279–282.

Kamei, N., & Glabe, C. G. (2003). The species-specific egg receptor for sea urchin sperm adhesion is EBR1, a novel ADAMTS protein. *Genes & Development*, *17*(20), 2502–2507.

Kato, K., Satouh, Y., Nishimasu, H., Kurabayashi, A., Morita, J., Fujihara, Y., et al. (2016). Structural and functional insights into IZUMO1 recognition by JUNO in mammalian fertilization. *Nature Communications*, *7*, 1–9.

Koyota, S., Swarma Wimalasiri, K. M., & Hoshi, M. (1997). Structure of the main saccharide chain in the acrosome reaction-inducing substance of the starfish, *Asterias amurensis*. *The Journal of Biological Chemistry*, *272*(16), 10372–10376.

Kresge, N., Vacquier, V. D., & Sout, C. D. (2001). Abalone lysin: The dissolving and evolving sperm protein. *BioEssays*, *23*(1), 95–103.

Kresge, N., Vacquier, V. D., & Stout, C. D. (2000). The high resolution crystal structure of green abalone sperm lysin: Implications for species-specific binding of the egg receptor. *Journal of Molecular Biology*, *296*(5), 1225–1234.

Lanzendorf, S. E., Holmgren, W. J., Johnson, D. E., Scobey, M. J., & Jeyendran, R. S. (1992). Hemizona assay for measuring zona binding in the lowland gorilla. *Molecular Reproduction and Development*, *31*(4), 264–267.

Le Naour, F., Rubinstein, E., Jasmin, C., Prenant, M., & Boucheix, C. (2000). Severely reduced female fertility in CD9-deficient mice. *Science*, *287*(5451), 319–321.

Lee, Y.-H., Ota, T., & Vacquier, V. D. (1995). Positive selection is a general phenomenon in the evolution of abalone sperm lysin. *Molecular Biology and Evolution*, *12*(2), 231–238.

Lessios, H. A. (2007). Reproductive isolation between species of sea urchins. *Bulletin of Marine Science*, *81*(2), 191–208.

Lewis, C. A., Talbot, C. F., & Vacquier, V. D. (1982). A protein from abalone sperm dissolves the egg vitelline layer by a nonenzymatic mechanism. *Developmental Biology*, *92*(1), 227–239.

Lightfoot, J. W., Wilecki, M., Rödelsperger, C., Moreno, E., Susoy, V., Witte, H., et al. (2019). Small peptide-mediated self-recognition prevents cannibalism in predatory nematodes. *Science*, *364*(6435), 86–89.

Lyon, J. D., & Vacquier, V. D. (1999). Interspecies chimeric sperm lysins identify regions mediating species-specific recognition of the abalone egg vitelline envelope. *Developmental Biology*, *214*(1), 151–159.

Matsumori, N., Hiradate, Y., Shibata, H., Oishi, T., Shimma, S., Toyoda, M., et al. (2013). A novel sperm-activating and attracting factor from the Ascidian *Ascidia sydneiensis*. *Organic Letters*, *15*(2), 294–297.

Metz, E. C., & Palumbi, S. R. (1996). Positive selection and sequence rearrangements generate extensive polymorphism in the gamete recognition protein bindin. *Molecular Biology and Evolution*, *13*(2), 397–406.

Metz, E. C., Robles-Sikisaka, R., & Vacquier, V. D. (1998). Nonsynonymous substitution in abalone sperm fertilization genes exceeds substitution in introns and mitochondrial DNA. *Proceedings of the National Academy of Sciences of the United States of America*, *95*(18), 10676–10681.

Miller, D. J., Macek, M. B., & Shur, B. D. (1992). Complementarity between sperm surface β-1,4-galactosyl-transferase and egg-coat ZP3 mediates sperm-egg binding. *Nature*, *357*, 589–593.

Minor, J. E., Fromson, D. R., Britten, R. J., & Davidson, E. H. (1991). Comparison of the bindin proteins of *Strongylocentrotus franciscanus*, *S. purpuratus*, and *Lytechinus variegatus*: Sequences involved in the species specificity of fertilization. *Molecular Biology and Evolution*, *8*(6), 781–795.

Miyado, K., Yamada, G., Yamada, S., Hasuwa, H., Nakamura, Y., Ryu, F., et al. (2000). Requirement of CD9 on the egg plasma membrane for fertilization. *Science*, *287*(5451), 321–324.

Morita, M., Nishikawa, A., Nakajima, A., Iguchi, A., Sakai, K., Takemura, A., et al. (2006). Eggs regulate sperm flagellar motility initiation, chemotaxis and inhibition in the coral *Acropora digitifera*, *A. gemmifera* and *A. tenuis*. *The Journal of Experimental Biology*, *209*, 4574–4579.

Nakachi, M., Moriyama, H., Hoshi, M., & Matsumoto, M. (2006). Acrosome reaction is subfamily specific in sea star fertilization. *Developmental Biology*, *298*(2), 597–604.

Neill, A. T., & Vacquier, V. D. (2004). Ligands and receptors mediating signal transduction in sea urchin spermatozoa. *Reproduction*, *127*(2), 141–149.

Nishigaki, T., Chiba, K., & Hoshi, M. (2000). A 130-kDa membrane protein of sperm flagella is the receptor for asterosaps, sperm-activating peptides of starfish *Asterias amurensis*. *Developmental Biology*, *219*(1), 154–162.

Oda, S., Igarashi, Y., Manaka, K.-I., Koibuchi, N., Sakai-Sawada, M., Sakai, K., et al. (1998). Sperm-activating proteins obtained from the herring eggs are homologous to trypsin inhibitors and synthesized in follicle cells. *Developmental Biology*, *204*(1), 55–63.

Oda, S., Igarashi, Y., Ohtake, H., Sakai, K., Shimizu, N., & Morisawa, M. (1995). Sperm-activating proteins from unfertilized eggs of the Pacific herring, *Clupea pallasii*. *Development, Growth & Differentiation*, *37*(3), 257–261.

Ohlendieck, K., Dhume, S. T., Partin, J. S., & Lennarz, W. J. (1993). The sea urchin egg receptor for sperm: Isolation and characterization of the intact, biologically active receptor. *The Journal of Cell Biology*, *122*(4), 887–895.

Ohtake, H. (2003). Sperm-activating proteins obtained from the herring eggs. *Fish Physiology and Biochemistry*, *28*(1–4), 199–202.

Osman, R. A., Andria, M. L., Jones, A. D., & Meizel, S. (1989). Steroid induced exocytosis: The human sperm acrosome reaction. *Biochemical and Biophysical Research Communications*, *160*(2), 828–833.

Palumbi, S. R. (1999). All males are not created equal: Fertility differences depend on gamete recognition polymorphisms in sea urchins. *Proceedings of the National Academy of Sciences of the United States of America*, *96*(22), 12632–12637.

Paterson, S., Vogwill, T., Buckling, A., Benmayor, R., Spiers, A. J., Thomson, N. R., et al. (2010). Antagonistic coevolution accelerates molecular evolution. *Nature*, *464*, 275–278.

Pillai, M. C., Shields, T. S., Yanagimachi, R., & Cherr, G. N. (1993). Isolation and partial characterization of the sperm motility initiation factor from eggs of the Pacific herring, *Clupea pallasi*. *The Journal of Experimental Zoology*, *265*, 336–342.

Pujolar, J. M., & Pogson, G. H. (2011). Positive Darwinian selection in gamete recognition proteins of *Strongylocentrotus* sea urchins. *Molecular Ecology*, *20*(23), 4968–4982.

Raj, I., Hosseini, A., S, H., Dioguardi, E., Nishimura, K., Han, L., et al. (2017). Structural basis of egg coat-sperm recognition at fertilization. *Cell*, *169*(7), 1315–1326.e17.

Rankin, T. L., Tong, Z.-B., Castle, P. E., Lee, E., Gore-Langton, R., Nelson, L. M., et al. (1998). Human ZP3 restores fertility in Zp3 null mice without affecting order-specific sperm binding. *Development*, *125*, 2415–2424.

Roth, T. L., Weiss, R. B., Buff, J. L., Bush, L. M., Wildt, D. E., & Bush, M. (1998). Heterologous in vitro fertilization and sperm capacitation in an endangered African antelope, the scimitar-horned *Oryx* (*Oryx dammah*). *Biology of Reproduction*, *58*(2), 475–482.

SeGall, G. K., & Lennarz, W. J. (1979). Chemical characterization of the component of the jelly coat from sea urchin eggs responsible for induction of the acrosome reaction. *Developmental Biology*, *71*(1), 33–48.

Shaw, A., Lee, Y.-H., Stout, C. D., & Vacquier, V. D. (1994). The species-specificity and structure of abalone sperm lysin. *Seminars in Developmental Biology*, *5*(4), 209–215.

Shu, L., Suter, M. J.-F., & Räsänen, K. (2015). Evolution of egg coats: Linking molecular biology and ecology. *Molecular Ecology*, *24*(16), 4052–4073.

Simons, J., & Fauci, L. (2018). A model for the acrosome reaction in mammalian sperm. *Bulletin of Mathematical Biology*, *80*(9), 2481–2501.

Sinowatz, F., Wessa, E., Neumüller, C., & Palma, G. (2003). On the species specificity of sperm binding and sperm penetration of the zona pellucida. *Reproduction in Domestic Animals*, *38*(2), 141–146.

Stapper, A. P., Beerli, P., & Levitan, D. R. (2015). Assortative mating drives linkage disequilibrium between sperm and egg recognition protein loci in the sea urchin *Strongylocentrotus purpuratus*. *Molecular Biology and Evolution*, *32*(4), 859–870.

Sun, F., Bahat, A., Gakamsky, A., Girsh, E., Katz, N., Giojalas, L. C., et al. (2005). Human sperm chemotaxis: Both the oocyte and its surrounding cumulus cells secrete sperm chemoattractants. *Human Reproduction*, *20*(3), 761–767.

Sun, F., Giojalas, L. C., Rovasio, R. A., Tur-Kaspa, I., Sanchez, R., & Eisenbach, M. (2003). Lack of species-specificity in mammalian sperm chemotaxis. *Developmental Biology*, *255*(2), 423–427.

Swanson, W. J., & Vacquier, V. D. (1997). The abalone egg vitelline envelope receptor for sperm lysin is a giant multivalent molecule. *Proceedings of the National Academy of Sciences of the United States of America*, *94*(13), 6724–6729.

Swanson, W. J., & Vacquier, V. D. (1998). Concerted evolution in an egg receptor for a rapidly evolving abalone sperm protein. *Science*, *281*(5377), 710–712.

Swanson, W. J., & Vacquier, V. D. (2002). The rapid evolution of reproductive proteins. *Nature Reviews Genetics*, *3*(2), 137–144.

Tardif, S., Wilson, M. D., Wagner, R., Hunt, P., Gertsenstein, M., Nagy, A., et al. (2010). Zonadhesin is essential for species specificity of sperm adhesion to the egg zona pellucida. *Journal of Biological Chemistry*, *285*(32), 24863–24870.

Ueda, Y., Imaizumi, C., Kubo, H., Sato, K.-I., Fukami, Y., & Iwao, Y. (2007). Analysis of terminal sugar moieties and species-specificities of acrosome reaction-inducing substance in *Xenopus* (ARISX). *Development, Growth & Differentiation*, *49*(7), 591–601.

Ueda, Y., Kubo, H., & Iwao, Y. (2003). Characterization of the acrosome reaction-inducing substance in *Xenopus* (ARISX) secreted from the oviductal pars recta onto the vitelline envelope. *Developmental Biology*, *264*(1), 289–298.

Ueda, Y., Yoshizaki, N., & Iwao, Y. (2002). Acrosome reaction in sperm of the frog, *Xenopus laevis*: Its detection and induction by oviductal pars recta secretion. *Developmental Biology*, *243*(1), 55–64.

Ulrich, A. S., Otter, M., Glabe, C. G., & Hoekstra, D. (1998). Membrane fusion is induced by a distinct peptide sequence of the sea urchin fertilization protein bindin. *The Journal of Biological Chemistry*, *273*(27), 16748–16755.

Vacquier, V. D. (2012). The quest for the sea urchin egg receptor for sperm. *Biochemical and Biophysical Research Communications*, *425*(3), 583–587.

Vacquier, V. D., Carner, K. R., & Stout, C. D. (1990). Species-specific sequences of abalone lysin, the sperm protein that creates a hole in the egg envelope. *Proceedings of the National Academy of Sciences of the United States of America*, *87*(15), 5792–5796.

Vacquier, V. D., & Moy, G. W. (1977). Isolation of bindin: The protein responsible for adhesion of sperm to sea urchin eggs. *Proceedings of the National Academy of Sciences of the United States of America*, *74*(6), 2456–2460.

Vacquier, V. D., & Moy, G. W. (1997). The fucose sulfate polymer of egg jelly binds to sperm REJ and is the inducer of the sea urchin sperm acrosome reaction. *Developmental Biology*, *192*(1), 125–135.

Vilela-Silva, A.-C. E. S., Alves, A.-P., Valente, A.-P., Vacquier, V. D., & Mourão, P. A. S. (1999). Structure of the sulfated α-L-fucan from the egg jelly coat of the sea urchin *Strongylocentrotus franciscanus*: Patterns of preferential 2-O- and 4-O-sulfation determine sperm cell recognition. *Glycobiology*, *9*(9), 927–933.

Vilela-Silva, A.-C. E. S., Castro, M. O., Valente, A.-P., Biermann, C. H., & Mourão, P. A. S. (2002). Sulfated fucans from the egg jellies of the closely related sea urchins *Strongylocentrotus droebachiensis* and *Strongylocentrotus pallidus* ensure species-specific fertilization. *Journal of Biological Chemistry*, *277*(1), 379–387.

Wassarman, P. M., Jovine, L., & Litscher, E. S. (2001). A profile of fertilization in mammals. *Nature Cell Biology*, *3*(2), E59–E64.

Xiong, B., Zhao, Y., Beall, S., Sadusky, A. B., & Dean, J. (2017). A unique egg cortical granule localization motif is required for ovastacin sequestration to prevent premature ZP2 cleavage and ensure female fertility in mice. *PLoS Genetics*, *13*(1), 1–18.

Yanagimachi, R. (2011). Mammalian sperm acrosome reaction: Where does it begin before fertilization? *Biology of Reproduction*, *85*(1), 4–5.
Yanagimachi, R., Cherr, G., Matsubara, T., Andoh, T., Harumi, T., Vines, C., et al. (2013). Sperm attractant in the micropyle region of fish and insect eggs. *Biology of Reproduction*, *88*(2), 1–11.
Yanagimachi, R., Cherr, G. N., Pillai, M. C., & Baldwin, J. D. (1992). Factors controlling sperm entry into the micropyles of salmonid and herring eggs. *Development, Growth & Differentiation*, *34*(4), 447–461.
Yanagimachi, R., Yanagimachi, H., & Rogers, B. J. (1976). The use of zona-free animal ova as a test system for the assessment of the fertilizing capacity of human spermatozoa. *Biology of Reproduction*, *15*(4), 471–476.
Yoshida, M., Hiradate, Y., Sensui, N., Cosson, J., & Morisawa, M. (2013). Species-specificity of sperm motility activation and chemotaxis: A study on Ascidian species. *The Biological Bulletin*, *224*(3), 156–165.
Yoshida, M., Inaba, K., Ishida, K., & Morisawa, M. (1994). Calcium and cyclic AMP mediate sperm activation, but Ca^{2+} alone contributes sperm chemotaxis in the ascidian, *Ciona savignyi*. *Development, Growth & Differentiation*, *36*(6), 589–595.
Yoshida, M., Inaba, K., & Morisawa, M. (1993). Sperm chemotaxis during the process of fertilization in the ascidians *Ciona savignyi* and *Ciona intestinalis*. *Developmental Biology*, *157*(2), 497–506.
Yoshida, M., Murata, M., Inaba, K., & Morisawa, M. (2002). A chemoattractant for ascidian spermatozoa is a sulfated steroid. *Proceedings of the National Academy of Sciences of the United States of America*, *99*(23), 14831–14836.
Yoshida, K., Shiba, K., Sakamoto, A., Ikenaga, J., Matsunaga, S., Inaba, K., et al. (2018). Ca^{2+} efflux via plasma membrane Ca^{2+}-ATPase mediates chemotaxis in ascidian sperm. *Scientific Reports*, *8*, 1–16.
Yoshida, M., Shiba, K., Yoshida, K., Tsuchikawa, H., Ootou, O., Oishi, T., et al. (2008). Ascidian sperm activating and attracting factor: Importance of sulfate groups for the activities and implication of its putative receptor. *FEBS Letters*, *582*(23–24), 3429–3433.
Zigler, K. S., & Lessios, H. A. (2003). Evolution of bindin in the pantropical sea urchin Tripneustes: Comparisons to bindin of other genera. *Molecular Biology and Evolution*, *20*(2), 220–231.
Zigler, K. S., McCartney, M. A., Levitan, D. R., & Lessios, H. A. (2005). Sea urchin bindin divergence predicts gamete compatibility. *Evolution*, *59*(11), 2399–2404.

Further reading

Foltz, K. R., Partin, J. S., & Lennarz, W. J. (1993). Sea urchin egg receptor for sperm: Sequence similarity of binding domain and hsp70. *Science*, *259*(5100), 1421–1425.
Minor, J. E., Britten, R. J., & Davidson, E. H. (1993). Species-specific inhibition of fertilization by a peptide derived from the sperm protein bindin. *Molecular Biology of the Cell*, *4*(4), 375–387.
Moy, G. W., & Vacquier, V. D. (1979). Chapter 2 Immunoperoxidase localization of bindin during the adhesion of sperm to sea urchin eggs. In A. A. Moscona, A. Monroy, & M. Friedlander (Eds.), *Immunological approaches to embryonic development and differentiation part I*; *Vol. 13. Current topics in developmental biology* (pp. 31–44). Elsevier.
Ohlendieck, K., Partin, J. S., & Lennarz, W. J. (1994). The biologically active form of the sea urchin egg receptor for sperm is a disulfide-bonded homo-multimer. *The Journal of Cell Biology*, *125*(4), 817–824.
Vacquier, V. D. (1998). Evolution of gamete recognition proteins. *Science*, *281*(5385), 1995–1998.

Vacquier, V. D., Swanson, W. J., & Lee, Y.-H. (1997). Positive darwinian selection on two homologous fertilization proteins: What is the selective pressure driving their divergence? *Journal of Molecular Evolution*, *44*(Suppl. 1), S15–S22.
Vilela-Silva, A.-C. E. S., Hirohashi, N., & Mourão, P. A. S. (2008). The structure of sulfated polysaccharides ensures a carbohydrate-based mechanism for species recognition during sea urchin fertilization. *The International Journal of Developmental Biology*, *52*(5–6), 551–559.
Yanagimachi, R., Harumi, T., Matsubara, H., Yan, W., Yuan, S., Hirohashi, N., et al. (2017). Chemical and physical guidance of fish spermatozoa into the egg through the micropyle. *Biology of Reproduction*, *96*(4), 780–799.

CHAPTER SIX

The role of the cytoskeleton in germ plasm aggregation and compaction in the zebrafish embryo

Cara E. Moravec, Francisco Pelegri*
Laboratory of Genetics, University of Wisconsin—Madison, Madison, WI, United States
*Corresponding author: e-mail address: fjpelegri@wisc.edu

Contents

1. Introduction 146
2. Localization routes of germ plasm ribonucleoparticles during oogenesis and early embryogenesis 147
3. Composition of germ plasm masses 149
4. Stages of animal germ plasm aggregation in the early zebrafish blastodisc 151
5. Interaction of astral microtubules and dynamic F-actin in RNP pre-aggregation 156
6. Recruitment of germ plasm during furrow induction 162
7. Medial-to-distal dynamics mediating germ plasm furrow compaction 164
8. Conclusion 172
References 173

Abstract

The transmission of genetic information from one generation to another is crucial for survival of animal species. This is accomplished by the induction of primordial germ cells (PGCs) that will eventually establish the germline. In some animals the germline is induced by signals in gastrula, whereas in others it is specified by inheritance of maternal determinants, known as germ plasm. In zebrafish, aggregation and compaction of maternally derived germ plasm during the first several embryonic cell cycles is essential for generation of PGCs. These processes are controlled by cellular functions associated with the cellular division apparatus. Ribonucleoparticles containing germ plasm components are bound to both the ends of astral microtubules and a dynamic F-actin network through a mechanism integrated with that which drives the cell division program. In this chapter we discuss the role that modifications of the cell division apparatus, including the cytoskeleton and cytoskeleton-associated proteins, play in the regulation of zebrafish germ plasm assembly.

Current Topics in Developmental Biology, Volume 140
ISSN 0070-2153
https://doi.org/10.1016/bs.ctdb.2020.02.001

1. Introduction

The ability to reproduce and pass on genes to future generations is vital for the survival of animal species, and the differentiation of primordial germ cells (PGCs) from somatic cells is essential to this process. Across the animal kingdom, there are two major mechanisms for the induction of PGCs, epigenesis and preformation (Bertocchini & Chuva de Sousa Lopes, 2016; Extavour & Akam, 2003; Magnúsdóttir & Surani, 2014; Wylie, 1999). Epigenesis involves inductive cell-cell interactions, as observed in mammals. Within vertebrate species, epigenesis is also observed in several other lineages such as urodele amphibians (salamanders) (Johnson, Bachvarova, Drum, & Masi, 2001) and basal lineages within major phylogenetic groups, such as turtles within reptiles (Bachvarova et al., 2009; Bertocchini & Chuva de Sousa Lopes, 2016). On the other hand, preformation involves the inheritance of maternally derived germ plasm, a material containing ribonucleoparticles (RNPs) that are often associated with mitochondria and cytoskeletal components. Preformation is used as a mechanism for PGC specification in animal lineages across the tree of life, including *Drosophila*, *Caenorhabditis elegans*, teleost fish, anuran amphibians, and chickens (Extavour & Akam, 2003).

In species that utilize the preformation specification mechanism, transmission of germ plasm components from one generation to the next requires two steps, production of RNPs and other germ plasm components during oogenesis, and aggregation of these components into germ plasm masses. These masses segregate during the early embryonic cycles into PGCs, where they are thought to promote activation of the zygotic germ cell program and repress expression of somatic-specific genes (Strome & Updike, 2015). In addition, formation of transient localization structures involving germ line components often occurs during oogenesis. The structures include the Balbiani body, a.k.a. mitochondrial cloud, a large (ca. 5.5 μm dia. in zebrafish oocytes) subcellular membrane-less organelle conserved across vertebrates that plays a crucial role in the distribution of maternal components during oocyte development (Bontems et al., 2009; Jamieson-Lucy & Mullins, 2019; Kosaka, Kawakami, Sakamoto, & Inoue, 2007; Strome & Updike, 2015). In zebrafish, germ plasm aggregation is thought to be essential for germline specification. Interference with germ plasm inheritance through its manual removal (Hashimoto et al., 2004) or chemical treatments that interfere with germ plasm aggregation (Miranda-Rodríguez, Salas-Vidal,

Lomelí, Zurita, & Schnabel, 2017) result in embryos that lack PGCs. Conversely, overexpression of Buc protein, a component of germ plasm, in an early embryo induces ectopic germ plasm masses and an increased number of primordial germ cells (Bontems et al., 2009). Thus, zebrafish germ plasm masses are necessary and sufficient for specification of PGCs.

Even though the mechanism for creation of PGCs differs across the animal kingdom, fundamental elements are evolutionarily conserved. For example, the presence of the mRNA and/or protein for the gene *RNA helicase Vasa/DEAD box polypeptide 4* (*Vasa*) in the germline is highly conserved. Homologs for this gene are expressed at various stages of germ cell development in divergent phylogenetic lineages, from planaria to humans, and additionally are present in the form of RNA and/or protein in germ plasm of species that use the preformation specification mechanism (Gustafson & Wessel, 2010). This chapter focuses on known mechanisms involved in the progressive aggregation of germ plasm RNPs in the early vertebrate embryo, mainly through studies in the zebrafish model system and with particular emphasis on the role of the cytoskeleton in this process.

2. Localization routes of germ plasm ribonucleoparticles during oogenesis and early embryogenesis

In the zebrafish, a majority of the maternal products (RNAs and proteins) that are necessary for early development are produced in the oocyte during oogenesis and are stored until fertilization. Recent reviews have addressed processes occurring during oocyte development (Fuentes, Mullins, & Fernández, 2018; Jamieson-Lucy & Mullins, 2019), such as mechanisms for oocyte polarity cues including formation of the Balbiani body. As the oocyte develops, a distinct wedge-like region forms at the animal pole, which contains the spindle apparatus arrested in meiosis II, as well as an enrichment of ooplasm and a corresponding relative depletion of yolk particles (Fernández, Valladares, Fuentes, & Ubilla, 2006; Fuentes et al., 2018). Diametrically opposite to the animal pole, the vegetal pole contains important determinants for both somatic and germ cell lineages (Campbell, Heim, Smith, & Marlow, 2015; Ge et al., 2014; Hashimoto et al., 2004; Hino et al., 2018; Howley & Ho, 2000; Jesuthasan & Stähle, 1997; Kosaka et al., 2007; Lekven, Thorpe, Waxman, & Moon, 2001; Maegawa, Yasuda, & Inoue, 1999; Theusch, Brown, & Pelegri, 2006).

During oocyte activation, as meiosis II resumes generating the female pronucleus and a polar body, the animal pole becomes the blastodisc (Fernández et al., 2006; Fuentes & Fernández, 2010). Blastodisc formation takes advantage of the pre-existing biased distribution of cytoplasmic and yolk components of the oocyte. After egg activation, this bias is enhanced by the animal pole-oriented streaming of ooplasm, a process that results in the separation of ooplasm from the yolk and lifting of the animal cortex to form a one-cell embryo. Ooplasmic streaming occurs through both internal "axial" and cortical (meridional) paths, allowing for the transport of factors including RNAs and proteins from the yolk into the newly formed blastodisc (Fuentes & Fernández, 2010; Hashimoto et al., 2004; Howley & Ho, 2000; Theusch et al., 2006).

Three highly conserved germ plasm RNAs, *vasa/DEAD box polypeptide 4* (*vasa*) (Yoon, Kawakami, & Hopkins, 1997), *nanos3* (*nanos*) (Köprunner, Thisse, Thisse, & Raz, 2001), and *deleted in azoospermia-like* (*dazl*) (Hashimoto et al., 2004), exemplify different routes of RNA distribution and localization during oogenesis, in the mature egg, and in the early embryo (Kosaka et al., 2007). Early in oocyte development, these three RNAs are localized to the Balbiani body (Kosaka et al., 2007). However, as the oocyte undergoes development, the Balbiani body disassembles, and the localization of these germ plasm RNAs diverges into three different patterns. *nanos* RNA distribution shifts to the animal pole ooplasm, which contains the oocyte DNA content (Howley & Ho, 2000; Kosaka et al., 2007). The transport of *nanos* exemplifies a route of localization for a number of RNAs enriched in this region in the mature egg, which include additional RNAs involved in various developmental processes (Kosaka et al., 2007). Upon egg activation, mRNAs for these genes that remain in the ooplasm intermixed with the yolk are further transported through ooplasmic streaming to the forming blastodisc.

A second distribution pattern in the late oocyte, as observed with *vasa* RNA, involves mRNA localization to the cortical region surrounding the entirety of the oocyte (Howley & Ho, 2000; Kosaka et al., 2007). As the egg completes maturation, this mRNA becomes further enriched at the cytoplasmic region of the animal pole through a poorly understood process.

The third type of localization route is shown by that of *dazl*, which after its association with the Balbiani body becomes localized to the cortex at the vegetal pole of the oocyte (Howley & Ho, 2000; Kosaka et al., 2007; Maegawa et al., 1999). After egg activation, *dazl* RNA-containing particles translocate to the animal pole along cortical paths (Kosaka et al., 2007;

Maegawa et al., 1999; Theusch et al., 2006). Similar to *dazl*, RNA for the gene *CUGBP Elav-like family member 1/celf1* is localized to the vegetal pole of the oocyte (Hashimoto, Suzuki, Kageyama, Yasuda, & Inoue, 2006; Suzuki et al., 2000). However, upon egg activation, and in contrast to cortical paths for *dazl*, *celf1* RNA appears to move to the animal pole through internal (axial) ooplasmic streamers (Suzuki et al., 2000). In the case of *vasa*, *nanos*, and *dazl*, and consistent with RNA localization studies in other systems (Knaut, Steinbeisser, Schwarz, & Nüsslein-Volhard, 2002; Kosaka et al., 2007; Rangan et al., 2009), studies have shown that the 3′ UTR regions of the mRNA determines their respective distinct localization patterns in both the oocyte and the early embryo (Kosaka et al., 2007).

These different routes of localization result in the use of different paths and mechanisms of aggregation of germ plasm RNPs at the furrow. Thus, at the time of egg activation and blastodisc formation, some germ plasm RNAs, referred to as "animal germ plasm RNAs," are already present at the blastodisc in anticipation of their recruitment to the forming furrows (Theusch et al., 2006), a recruitment that occurs through the ongoing process of cell division (Eno & Pelegri, 2013, 2018; Nair et al., 2013; Theusch et al., 2006). Other RNAs, "vegetal germ plasm RNAs," are originally present at the vegetal pole of the egg and rely on additional movement pathways to reach germ plasm aggregates forming in the blastodisc (Suzuki et al., 2000; Theusch et al., 2006; Welch & Pelegri, 2014). The significance of these various paths is not yet understood. RNAs for some germ cell genes may have additional roles. For example, in *Xenopus*, vegetally localized *dead end* RNA functions not only in germ cell development (Horvay, Claussen, Katzer, Landgrebe, & Pieler, 2006; Taguchi, Watanabe, & Orii, 2014) but also in early embryonic patterning (Mei et al., 2013). Such somatic roles may result in developmental constraints that explain the existence of different germ plasm RNA localization paths. Alternatively, distinct localization routes may also serve as a mechanism for the step-wise incorporation of vegetal germ plasm RNAs into the bulk of the germ plasm aggregate (Theusch et al., 2006).

3. Composition of germ plasm masses

In zebrafish, germ plasm RNAs can be observed immediately after fertilization as roughly spherical submicron-sized RNPs at the embryonic cortex (Nair et al., 2013; Theusch et al., 2006). The majority of germ plasm RNAs are found localized to the animal pole/blastodisc, a group that includes *askopos* (*kop*) (Blaser et al., 2005), *carbonic anhydrase 15b* (*ca15b*) (Hartwig et al., 2014), *dead end* (*dnd*) (Weidinger et al., 2003), *granulito* (*gran*) (Strasser et al., 2008), *hook microtubule-tethering protein 2* (*hook2*) (Roovers et al., 2018), microRNA 202-5p (miR-202-5p) (Zhang et al., 2017), *nanos* (Köprunner et al., 2001), *piwi-like RNA-mediated gene silencing 1* (*piwil1*) (Houwing et al., 2007), *regulator of G-protein signaling 14a* (*rgs14a*) (Hartwig et al., 2014), *tudor domain protein 7* (*tdrd7a*) (Mishima et al., 2006), and *vasa* (Yoon et al., 1997). The set of known vegetal germ plasm mRNAs is smaller, including *dazl* (Hashimoto et al., 2004) and *cugbp, Elav-like family member 1/ bruno-like* (*celf1*) (Hashimoto et al., 2004). Based on a subset of tested RNAs (*ca15b, dazl, dnd, nanos, rgs14a* and *vasa/ddx4)*, RNPs appear to create clusters that are homotypic, i.e., composed of a single type of RNA, and maintain distinct boundaries from other RNPs (Eno, Hansen, & Pelegri, 2019). It was estimated that zebrafish *nanos* RNPs contain about 9 RNAs (Eno et al., 2019), an estimate in the same range as in homotypic germ plasm RNPs in other systems such as *Drosophila* (Niepielko, Eagle, & Gavis, 2018; Trcek et al., 2015).

Germ plasm also contains maternally expressed proteins, Buckyball (Buc) (Bontems et al., 2009; Campbell et al., 2015; Heim et al., 2014; Riemer, Bontems, Krishnakumar, Gömann, & Dosch, 2015), Tudor6a (Tdrd6a) (Roovers et al., 2018), and Bruno-like (Brul) (Hashimoto et al., 2006). Buc protein is found in spherical particles in the cortex before it becomes recruited with germ plasm RNPs to the furrow (Riemer et al., 2015; Roovers et al., 2018). Buc is the zebrafish homolog of Xvelo in Xenopus, with both of these proteins containing a Prion-like domain. In Xenopus, Xvelo is able to self-aggregate into membrane-less organelles that display amyloid-like features (Boke et al., 2016). Buc particles appear to be decorated with mRNA both in the cortex and the furrow (Roovers et al., 2018). In vitro, Xvelo has also been shown to sequester RNA for the Xenopus *nanos* homolog, *xcat-2*, in a manner dependent on a putative RNA-binding domain (Boke et al., 2016). Tudor6a appears to coat these spherical Buc particles in the cytoplasm as well as in the mature furrow, although in embryos lacking Tdrd6a protein Buc can still associate with germ plasm mRNAs, indicating that Tdrd6a is not required for the association between Buc and germ plasm mRNAs. Tdrd6a protein is necessary to generate the structure of the germ plasm aggregate at the furrow, since when Tdrd6a is absent the rod-like germ plasm aggregate normally present along the plane of the furrow fails to form and instead remains relatively small and fragmented (Roovers et al., 2018). Bru1 protein is also known to be present in the cytoplasm as well as in the same region of the furrow as germ plasm RNAs (Hashimoto et al., 2006), although a detailed subcellular localization pattern has yet to be reported.

Additionally, Xvelo can entrap and promote clustering of mitochondria in cytoplasmic extracts from Xenopus eggs, an activity that depends on its Prion-like domain (Boke et al., 2016) and which likely facilitates the formation of the Balbiani body during oogenesis, also enriched in mitochondria (Bontems et al., 2009; Jamieson-Lucy & Mullins, 2019; Kosaka et al., 2007; Strome & Updike, 2015). Rod-like structures containing germ plasm RNPs at the furrows in early zebrafish embryos have also been observed to be associated with mitochondria (Knaut, Pelegri, Bohmann, Schwarz, & Nüsslein-Volhard, 2000).

4. Stages of animal germ plasm aggregation in the early zebrafish blastodisc

During early zebrafish development, animal germ plasm RNPs are present as single RNPs at the cortex of the blastodisc, which subsequently undergo aggregation to generate germ plasm masses (Eno et al., 2019; Eno & Pelegri, 2013; Nair et al., 2013; Theusch et al., 2006). This process involves multimerization of RNPs into aggregates of increasingly larger numbers. These aggregates additionally appear to acquire more compact, or more tightly packed, morphology. Germ plasm RNP aggregation appears to occur gradually in a sequential set of distinct steps, which have been described as pre-aggregation, furrow recruitment, and distal compaction (see below) (Eno et al., 2019; Eno & Pelegri, 2013; Nair et al., 2013; Theusch et al., 2006). These steps all involve particle-particle interactions that lead to an increase in particle number within an aggregate. Gradually, this increase in particle number likely transitions into an increase in neighbor-neighbor interactions within an aggregate leading to its compaction (Eno et al., 2019; Eno, Gomez, Slusarski, & Pelegri, 2018; Eno & Pelegri, 2018; Nair et al., 2013; Theusch et al., 2006).

Once germ plasm RNPs are compacted into large masses in the mature furrows, they become incorporated into cells at the 16–32 cell stage. The mechanism for this process is poorly understood, but it coincides with the appearance of fully cellularized blastomeres in the early embryo (Kimmel, Ballard, Kimmel, Ullmann, & Schilling, 1995). As the zebrafish embryo continues to divide, these germ plasm aggregates maintain discrete boundaries within a cell and are inherited asymmetrically, to a single daughter cell, during cellular division (Fig. 1D). During the mid-blastula stage these germ plasm masses continue to exhibit discrete boundaries with respect

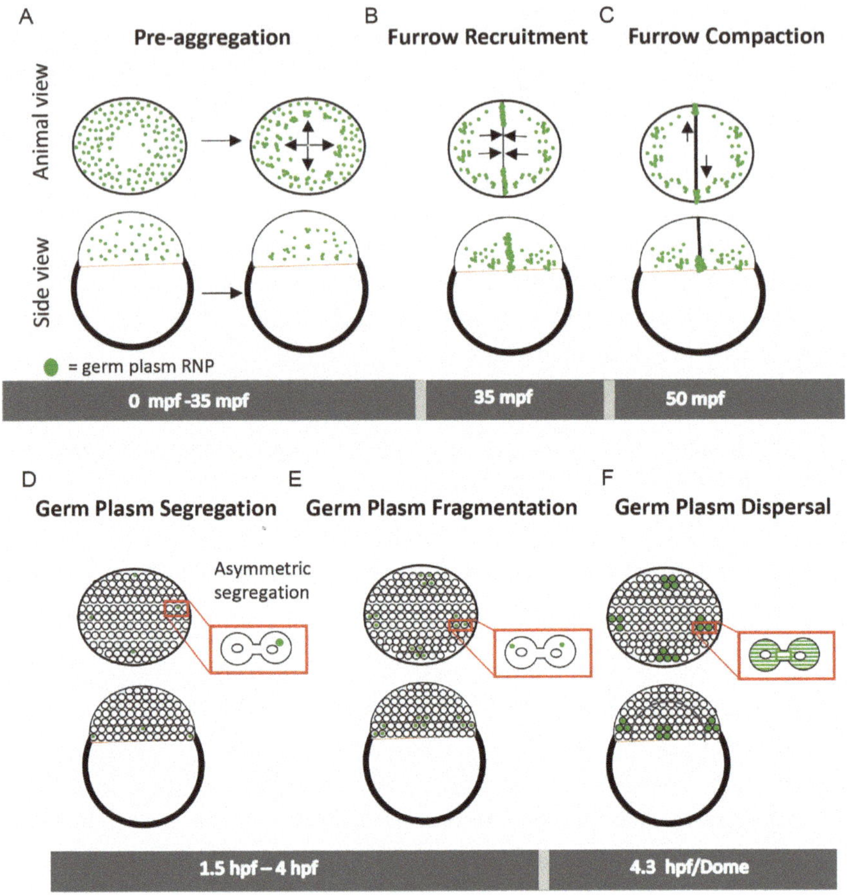

Fig. 1 Stages of germ plasm RNP movement during early development of the zebrafish. (A) Germ plasm RNPs (green) are dispersed across the blastodisc of the early one-cell embryo. As the embryo develops, these germ plasm RNPs move to the edge of the blastodisc creating an "aggregation front" and promoting the formation of larger germ plasm aggregates. (B) A subset of germ plasm aggregates is recruited to the furrow, and these RNPs extend across a fraction of the developing furrow. (C) The recruited RNPs undergo distal compaction during furrow maturation, creating a compact structure at the distal ends of the furrow. (D) The compacted germ plasm masses are incorporated into a cell around the 16–32 cell stage. These masses are inherited asymmetrically into a single daughter cell during cellular division while keeping discrete boundaries with the cytoplasm. (E) Germ plasm aggregates can undergo fragmentation but still maintain discrete boundaries. (F) During late blastula/early gastrulation, the germ plasm aggregate boundaries begin to disappear, with RNPs undergoing cytoplasmic dispersal. See text for details.

to the cytoplasm but become fragmented into multiple cells, a process of fragmentation that has been suggested to potentially help increase the pool of germ plasm-carrying cells, and therefore of PGCs (Fig. 1E) (Eno et al., 2019). During the transition from sphere stage to dome stage, the distinct boundaries of germ plasm masses begin to break apart, with individual RNPs dispersing away from the germ plasm mass into the cytoplasm (Fig. 1F). This RNP dispersal process appears to be a gradual seeping of germ plasm particles into the cytoplasm of PGCs that lasts throughout gastrulation (Eno et al., 2019). Interestingly, particles undergoing dispersal at this stage are similar in size to the RNPs initially recruited into the aggregate and, moreover, dispersing RNPs appear to have maintained their homotypic character (Eno et al., 2019). Thus, the RNPs themselves may be unchanged throughout the process of aggregation, segregation, and dispersal, and the germ plasm itself may be a mechanism to gather, inherit, and subsequently disperse distinct RNPs, constituting same-type RNA clusters, to specify the germ cell fate. This chapter focuses on the first few steps of germ plasm RNP movement, pre-aggregation, furrow recruitment, and distal compaction.

- *Pre-aggregation*: In the developing blastodisc, prior to furrow formation during the embryonic cell cycles, cellular processes generate a low-number of RNP aggregates at the cortex, typically fewer than 20 RNPs per a cluster (Nair et al., 2013). At this stage, pre-aggregates appear to form through particle-particle interaction, presumably by collisions along the plane of the cortex, followed by a persisting inter-particle association. This multimerization is coupled to movement of the germ plasm to the edge of the cortex and furrow, which is accomplished through coordination of the cytoskeletal components that mediate cell divisions (Fig. 1A). Thus, pre-aggregation is initiated before embryonic cell division, during expansion of the sperm aster (also called monoaster). This structure is derived from microtubule nucleation of sperm-derived centrioles, which have a well-known role in pronuclear congression, which culminates in pronuclear fusion (Dekens, Pelegri, Maischein, & Nüsslein-Volhard, 2003; Lindeman & Pelegri, 2012; Reinsch & Gönczy, 1998). As described below, sperm aster expansion results in the radial outward movement of RNPs, which generates an increasingly wider central RNP-free zone (Nair et al., 2013; Theusch et al., 2006). Even after initiation of the first embryonic cell divisions interphase asters continue to mediate the radially outward movement of cortical RNPs and to generate cortical RNP pre-aggregates (Nair et al., 2013; Theusch et al., 2006).

- *Furrow recruitment*: As cell division proceeds, RNP pre-aggregates progressively move outwardly within the blastodisc cortex (Nair et al., 2013; Theusch et al., 2006). At the same time, signals at the tips of astral microtubules from opposite sides of the spindle mediate furrow induction in the midpoint region between the spindle poles, where astral microtubules meet (Field, Pelletier, & Mitchison, 2019; Wühr, Tan, Parker, Detrich, & Mitchison, 2010). The opposed placement of interphase asters during the early mitoses of the zebrafish embryo additionally results in the movement of RNPs to the forming furrow, where they become recruited along the cytoplasmic side of the furrow (Eno & Pelegri, 2013; Nair et al., 2013).
- During furrow recruitment, persistent interactions between pre-aggregates appear to increase aggregate particle number and generate an elongated rod-like structure along the furrow. At furrow initiation for the first embryonic cell cycle, these rods occupy roughly two-third of each distal furrow half (Fig. 1B), with the medial region of the furrow lacking RNP aggregation. This pattern of aggregation directly reflects the RNA free-zone previously generated in the center of the embryo by the sperm aster-dependent radially outward RNP movement (Eno & Pelegri, 2013; Theusch et al., 2006). Due to the increased size of the central RNP-free zone caused by growth of the interphase asters during the earliest embryonic cell cycles, RNP recruitment in furrows becomes progressively more distally located for later cell cycles. Through this process, RNP recruitment in later cycles spans a decreasing fraction of the length of the furrow at the onset of furrow initiation. By the end of the third and fourth cell cycle, the remaining RNPs that have not been recruited to the furrows are associated with the edge of the blastodisc and are no longer available for recruitment into forming furrows.
- It has been estimated that the first furrow has on average about 1000 RNPs, with the amount of RNPs in each furrow decreasing by approximately half with each cell cycle (Eno & Pelegri, 2013). This decreasing number of RNPs in the furrows mirrors the size of cortex on both sides of a forming furrow, which due to cell division also decreases by half with each cell cycle. These observations support a model in which RNPs are recruited locally, from cortical regions immediately neighboring both sides of the furrows. Interestingly, the rod-shaped germ plasm masses composed of recruited RNPs are found in the furrows for at least the first three cell cycles, but only those germ plasm masses at the first and second

furrows are stabilized, with RNPs that are recruited to the third furrow being degraded (Eno & Pelegri, 2013). This suggests that a germ plasm mass needs to contain a certain threshold of animal RNPs to avoid degradation, so that only the most massive aggregates, those that collect at the furrows from the first and second cell cycles, are maintained (Eno & Pelegri, 2013). This degradation of smaller aggregates may ensure that only germ plasm masses large enough to influence germ cell fate effectively are inherited by prospective PGCs. Interestingly, the mechanism of pre-aggregate formation and recruitment at the furrows, which are mediated by the radially symmetric, outward growth of astral microtubules, result in only a fraction of initial RNPs becoming collected into the furrows cleavage planes, with the remaining fraction being driven to the periphery of the blastodisc where they do not integrate into germ plasm masses and eventually are degraded. Thus, the cell division apparatus is co-opted for the systematic gathering of germ plasm RNPs at the furrows for consecutive cell cycles, a systematic process that also involves loss of a fraction of maternally-inherited germ plasm RNPs.

- *Distal compaction*: As the cell cycle progresses each elongated, rod-like RNP aggregate that is present along the furrow becomes shortened in length and increases in diameter, acquiring a more globular shape (Eno & Pelegri, 2013; Pelegri, Knaut, Maischein, Schulte-Merker, & Nüsslein-Volhard, 1999; Theusch et al., 2006). This reorganization occurs as the aggregates accumulate at the furrow distal ends; hence, this process has been referred to as distal compaction. At this stage, the number of particles forming the aggregate does not appear to increase, yet particle-particle interactions within the aggregate likely do so, resulting in the more compact overall structure of the aggregates (Fig. 1C). The compaction of the animal pole germ plasm in the distal end of the first furrow coincides with the arrival and incorporation of the vegetal RNPs into the blastodisc and the furrow, with *dazl* incorporation into the germ plasm occurring at the distal-most end of the furrow (Theusch et al., 2006; Welch & Pelegri, 2014). The resulting aggregate now contains both animal and vegetal RNPs and appears to be the mature form of the germ plasm mass. These masses subsequently ingress into PGCs, undergo asymmetric segregation during the cleavage stages, and later disassemble to specify the germline (Fig. 1 and below) (Braat, Zandbergen, van de Water, Goos, & Zivkovic, 1999; Eno et al., 2019; Knaut et al., 2000; Theusch et al., 2006; Yoon et al., 1997).

Improper RNP localization is observed when one-cell embryos are treated with drugs inhibiting cytoskeleton activity (Eno & Pelegri, 2018; Nair et al., 2013; Pelegri et al., 1999; Theusch et al., 2006), showing that the cytoskeleton is necessary for aggregation, movement and compaction of germ plasm: inhibition of the F-actin network results in loss of RNP association to the blastodisc cortex (Eno & Pelegri, 2018; Theusch et al., 2006) while inhibition of microtubules causes defective RNP aggregation (Nair et al., 2013). Several studies using maternal mutants either isolated from ENU screens (Eno et al., 2018; Eno & Pelegri, 2018; Nair et al., 2013; Pelegri et al., 1999) or recently made with CRISPR-Cas9 (Campbell et al., 2015), have allowed a more in-depth study of the role of the cytoskeleton, cytoskeletal regulators and other key factors in germ plasm RNP segregation. In the following sections, we provide a more detailed account of cellular events that facilitate germ plasm aggregation and compaction in these early embryonic stages.

5. Interaction of astral microtubules and dynamic F-actin in RNP pre-aggregation

As mentioned above, during the step of pre-aggregation, single RNPs that are anchored to the cortex of the blastodisc become clustered together in small aggregates (Nair et al., 2013). This process is accomplished by an outward clearance of single RNPs that are associated with the tips of growing sperm asters and interphase spindle astral microtubules (Nair et al., 2013; Theusch et al., 2006). As this process unfolds, RNPs are gradually cleared from the center of the embryo while at the same time, an RNP multimerization "wave-front" forms at the boundary between the central RNP-free zone and a band-like RNP-containing region. During this process, and in addition to their association with the tips of microtubules, germ plasm RNPs are associated with short F-actin fragments arranged into bundles that form circumferentially arranged peripheral rings (Nair et al., 2013; Theusch et al., 2006). As the early embryonic cell cycles proceed, these F-actin rings appear to migrate toward the edge of the blastodisc (Fig. 2A). The products of two maternal genes, *motley* and *aura*, have key regulatory roles in this process, and their molecular and cellular analysis has provided insights into regulation of germ plasm aggregation at these stages.

The *motley* mutation is located in *birc5b*, a duplicated copy of the gene coding for Survivin that has a dedicated maternal-only expression. Survivin is a member of the Baculoviral inhibitor of Apoptosis Repeat containing

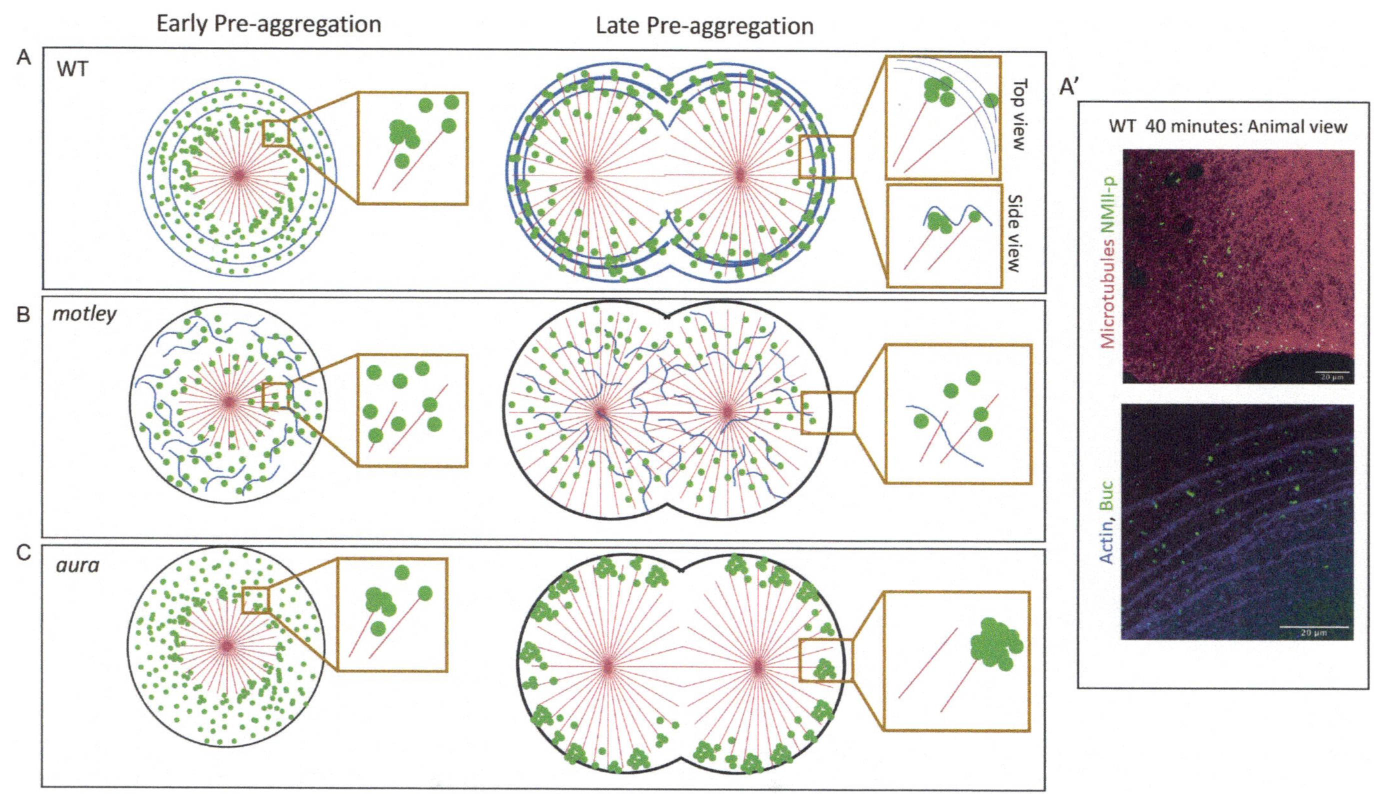

Fig. 2 See figure legend on next page

protein family and contains a single conserved domain characteristic of this family, the BIR domain (Ambrosini, Adida, & Altieri, 1997). The *motley* mutant allele contains a transition mutation of a conserved splice donor base pair, causing alternative splicing that results in loss of part of the BIR domain and the remaining C-terminal portion of the protein (Nair et al., 2013). Survivin is a well-known component of the Chromosomal Passenger Complex (CPC), which additionally involves the regulatory kinase Aurora B and the scaffold and regulatory factors INCENP and Borealin (Carmena, Wheelock, Funabiki, & Earnshaw, 2012). The CPC is known to have multiple rules during cell division including but not limited to chromosome bi-orientation during mitosis as well as cytokinesis progression (Carmena et al., 2012; Kitagawa & Lee, 2015; Wheatley & Altieri, 2019). Currently, it is unknown if the role of Birc5b in germ plasm segregation occurs within the context of the CPC or if it acts independent of this complex, as has been reported with regards to other cellular functions (Wheatley & Altieri, 2019).

The most readily observable phenotype in the *motley* mutation is a failure to develop a furrow, indicating a defect in initiation of cytokinesis (Nair et al., 2013). Immunostaining for microtubules shows a defect in the generation of a microtubule-free zone at the site of furrowing, an early step in furrow formation consistent with the early cytokinesis defect (Jesuthasan, 1998; Wühr et al., 2010). Labeling for actin also showed that a contractile actin ring fails to form along the furrow, also consistent with the lack of

Fig. 2 The role of astral microtubules and F-actin in aggregation of germ plasm RNPs in the blastodisc of the one-cell zebrafish embryo. (A) Radial astral microtubule growth from both the sperm aster and the bipolar spindle results in a radially outward movement of RNPs and is associated with multimerization of germ plasm RNPs at the tips of the microtubules. F-actin rings attenuate the microtubule-dependent movement of RNPs, presumably through direct interaction between the actin rings and germ plasm RNPs. (A′) Animal pole images of a wild-type embryo at 40 min post-fertilization showing the interaction between microtubules (purple) and germ plasm RNPs aggregates (non-muscle myosin (NMII-p)) (green) or F-actin (blue) and germ plasm RNPs aggregates (Buc) (green). Scale bar = 20 μM. (B) The germ plasm RNPs fail to interact with the tips of the microtubules in the *motley/birc5b* mutants, leading to failed aggregation in these embryos. *Motley* mutant embryos also display disorganized F-actin, suggesting that Birc5b is not only the linker between microtubules but also promotes reorganization of the cytoskeleton required for RNP aggregation. (C) In *aura* mutants, germ plasm RNP aggregates are found at the edge of the blastodisc in large ectopic clusters. In these mutants, germ plasm RNPs may be outwardly directed by growing microtubules but this movement is not regulated by the actin rings, causing the aggregates to accumulate at the periphery of the embryo. See text for details.

furrows. Unexpectedly, *motley* mutants showed cytoskeletal abnormalities even before furrow formation, namely in the formation of circumferentially arranged F-actin rings associated with germ plasm RNPs. In *motley* mutants, these F-actin rings fail to form; instead, F-actin is present as short aggregates throughout the cortex (Nair et al., 2013).

The failure of cytokinesis in *motley* embryos is accompanied by a significant decrease in germ plasm RNP multimerization during the pre-aggregation step (Fig. 2B) (Nair et al., 2013). While wild-type embryos form aggregates that contain roughly 20 germ plasm RNPs, the largest clusters observed in *motley* mutants are significantly smaller (7 RNPs), and a greater fraction of RNPs are present as single particles in these mutants. This RNP multimerization defect is comparable to that observed in embryos treated with nocodazole, which prevents the polymerization of astral microtubules (Nair et al., 2013), hinting at a link between microtubule growth, Birc5b function and germ plasm RNP multimerization. Indeed, immunolabeling of Birc5b in the early embryo shows co-localization of Birc5b protein with germ plasm RNPs at the tips of astral microtubules. Moreover, in *motley/birc5b* mutants, germ plasm RNPs cannot interact with these astral microtubule tips (Nair et al., 2013). Together, these observations indicate that Birc5b acts as an essential linker between microtubules and germ plasm RNPs. Since germ plasm RNPs are associated with both Birc5b and F-actin rings, and *motley/birc5b* mutants exhibit an F-actin reorganization defect (Fig. 2B), it seems plausible that Birc5b also has a key role in the reorganization of the cytoskeleton that accompanies RNP aggregation. This likely reflects a link between F-actin and microtubule network dynamics in the early embryo, a relationship that is increasingly being recognized as a general phenomenon in cell biological processes (Henty-Ridilla, Juanes, & Goode, 2017; Henty-Ridilla, Rankova, Eskin, Kenny, & Goode, 2016).

Despite a general reduction in the RNP aggregation process, it is intriguing that some degree of germ plasm RNP aggregation is observed under all conditions that interfere with either microtubules or F-actin dynamics, e.g., in embryos treated with nocodazole or in *motley/birc5b* mutants (Nair et al., 2013). It is possible that under these conditions, RNPs exhibit a low level of aggregation from random inter-particle collisions, caused by particle movement along the plane of the cortex, a movement that may be partially inhibited by association to the cortex itself. Thus, the role of the cytoskeleton- and Birc5b-based events may be to increase the efficiency of inter-particle collisions to a level that allows pre-aggregate formation at a rate sufficient for incorporation of a large fraction of RNPs into pre-aggregates and subsequently the forming furrow (see also below).

While the study of *motley* mutants has increased our understanding of the role of microtubules and linker factors in the pre-aggregation of germ plasm RNPs, the study of *aura* mutants has revealed a role for F-actin dynamics in germ plasm RNP pre-aggregation (Eno & Pelegri, 2018). The maternal-effect *aura* mutation disrupts the *mid1-interacting protein 1 like* (*mid1ip1l*) gene, where a single base pair mutation changes a lysine to a stop codon before a conserved C-terminal region, generating a complete loss of function mutation (Eno, Solanki, & Pelegri, 2016). Similar to *motley/birc5b*, the *mid1ip1l* gene is expressed only during oogenesis and has a maternal function dedicated to processes in the early zebrafish zygote (Eno et al., 2016). *aura/mid1ip1l* mutant embryos show a wide variety of phenotypic defects, including cortical granule release, cortical integrity/wound repair, and failure to complete cytokinesis. At the sub-cellular level, *aura* mutant embryos show defects in both F-actin dynamics at the cortex and forming furrow and in the reorganization of the microtubule cytoskeleton during furrow maturation (see below; Eno & Pelegri, 2018; Eno et al., 2016).

The F-actin defects in *aura* mutants become apparent soon after fertilization. In the cortex of the early wild-type embryo, F-actin initially appears as punctate structures that form transient aggregates. These aggregates grow until approximately 30 min after fertilization when the network reorganizes into the above-mentioned peripheral rings composed of parallel short F-actin fragments (Fig. 2A). As expected, treatments with cytochalasin D, an inhibitor of actin polymerization, abolishes these rings (Eno & Pelegri, 2018). In *aura* mutants, there is no observable difference in the early punctate F-actin. However, instead of reorganizing into well-developed ring structures, F-actin forms a fine meshwork that develops instead into a honeycomb cell-like pattern, with any partially formed arches appearing thinner and more disorganized than wild type (Eno & Pelegri, 2018). When *aura* embryos are treated with phalloidin, an actin stabilizer, F-actin ring formation appears closer to wild type in both the number of rings and their thickness, suggesting a partial rescue of the phenotype. Live imaging with the LifeAct transgene, which labels F-actin, shows a dynamic wave-like pattern in the cortex and cycles of F-actin polymerization in wild-type embryos (Eno & Pelegri, 2018). On the other hand, *aura* mutants appear to lack such F-actin polymerization waves, instead showing the gradual appearance of the repeated cell-like pattern observed in fixed embryos. The partial rescue by phalloidin treatment and the differences observed in actin dynamics in live embryos suggest that Mid1ip1lL mediates F-actin reorganization in the early embryo by promoting F-actin polymerization and/or stabilization (Eno & Pelegri, 2018).

During the developmental period corresponding to the first several embryonic cell cycles, *aura* embryos fail to recruit germ plasm RNPs to the furrow (Eno & Pelegri, 2018). Instead, these RNPs accumulate into ectopic aggregates at the edge of the blastodisc, in excess relative to wild type (Fig. 2C). A time course analysis suggests that *aura/mid1ip1l* mutants exhibit defects leading to germ plasm RNP mislocalization even before furrow initiation, specifically the germ plasm RNP distribution reflects an apparent increase in the rate of outward RNP clearance in *aura/mid1ip1l* mutants (Eno & Pelegri, 2018). An increase in clearance rate and ectopic accumulation of larger aggregates at the edge of the blastodiscs was also observed in wild-type embryos treated with cytochalasin D, similar to *aura/mid1ip1l* mutants (Eno & Pelegri, 2018). When wild type and *aura* embryos were treated with nocodazole to inhibit microtubule polymerization, a slower rate of outward movement of germ plasm RNPs was observed, together with a reduction of ectopic germ plasm aggregate formation at the edge of the *aura/mid1ip1l* mutant blastodiscs (Eno & Pelegri, 2018). Overall, the studies suggest that cortical F-actin attenuates the rate of microtubule-dependent outward movement of germ plasm RNPs. This attenuation appears to be essential to prevent germ plasm RNPs from moving to the edge of the blastodisc before furrow formation, which would otherwise preclude their recruitment along these structures.

Super-resolution imaging and 3-D reconstruction of the F-actin rings in early one-cell wild-type embryos sheds additional light on events associated with the formation of these structures (Eno & Pelegri, 2018). These rings appear to consist of cortical contractions, or "trenches," with RNP aggregates associated with the F-actin cortex in between these trenches. These trenches appear to be dynamic, and as development proceeds, they become more peripherally located in the blastodisc, mirroring the outward movement of the germ plasm RNP aggregation front. In *aura/mid1ip1l* mutant embryos, germ plasm RNPs appear to associate with the cortical F-actin cortex which, consistent with the reduction in F-actin rings, lacks cortical trenches (Eno & Pelegri, 2018). As in the case of the F-actin rings in the cortex, treatment of *aura* embryos with phalloidin rescues formation of F-actin cortical trenches (Eno & Pelegri, 2018). These observations suggest that F-actin cortical trenches may help maintain an appropriate rate of net outward movement of germ plasm RNPs by counteracting outward

RNP clearance caused by astral microtubule growth. This modulated outward flow retains germ plasm RNPs in the cortex, thus allowing RNP recruitment to occur at the forming furrows.

Altogether, the pre-aggregation stage appears to be dependent on coordination of two antagonistic forces, each dependent on dynamic changes in the cytoskeleton. Early on, astral microtubules are responsible for moving germ plasm RNPs to the edge of the blastodisc, accomplished by an outward radial movement of RNP aggregates associated with the tips of growing astral microtubules. As astral microtubules grow outwardly, RNP aggregates become larger due to the increased RNP number within the aggregate, presumably through RNP-RNP collision followed by inter-particle adhesion. As these processes occur, cortical F-actin rings, corresponding to cortical trenches, act as a counteracting force by attenuating RNP movement to the edge of the blastodisc, thus maintaining germ plasm RNP aggregates at the cortex and allowing their incorporation into the forming furrows (Fig. 2A).

6. Recruitment of germ plasm during furrow induction

As RNPs aggregate in the cortex, they also become localized to the forming furrows (Fig. 3A). Observations and aggregate quantification suggest that aggregates are brought to the furrow through the action of astral microtubules, which gather at the furrow from flanking cortical regions (Eno & Pelegri, 2013; Nair et al., 2013). Phenotypes in maternal genes currently known to cause germ plasm aggregation defects also cause abnormalities in the furrow-associated cytoskeleton, a phenotype that interferes with a clear interpretation of their role as regulators of the cytoskeleton during germ plasm reorganization in the furrow. A recent report characterized a CRISPR-Cas9 generated mutant in *kif5ba*, a maternally expressed homolog of Kinesin 1, a major anterograde motor for transport along microtubules (Campbell et al., 2015). A loss-of-function mutation in *kif5ba* produce embryos that undergo cell division during the cleavage phase of development, displaying normal formation of furrow-associated cytoskeletal structures such as the furrow microtubule array (see below) and F-actin contractile band. In spite of these normal furrow structures, this maternal mutant exhibits defects in the localization of germ plasm to the furrow (Fig. 3C), thus allowing to dissociate a role of germ plasm movement from that of the underlying cytoskeletal structure. In addition, *kif5ba* mutant embryos show an embryonic axis induction phenotype caused by defects in the

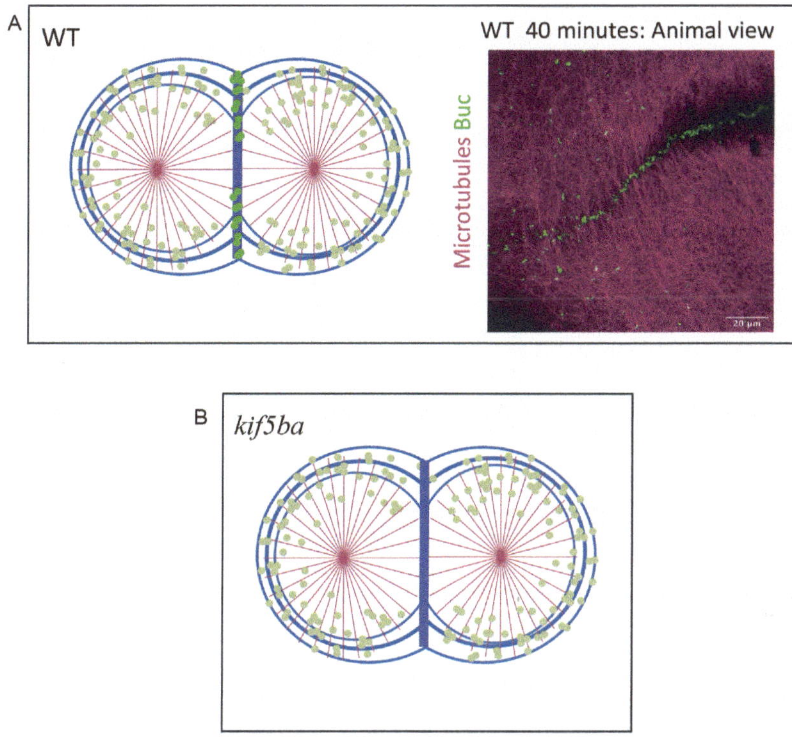

Fig. 3 A maternally expressed kinesin is necessary for recruitment of germ plasm during furrow induction. (A) Left: Germ plasm RNP aggregates (green) are collected in the furrow through the action of astral microtubules (purple), resulting in their accumulation along the approximately distalmost two-third of the furrow. Right: an animal pole view of a 40-min old wild-type embryo shows dispersed RNPs (Buc) (green) in an immature furrow. Furrow initiation is associated with cleaving of microtubules (purple) to generate a microtubule exclusion zone. Scale bar = 20 μM. (B) Depletion of a maternally expressed kinesin, Kif5ba, results in failure to recruit germ plasm aggregates to the furrow in spite of the lack of apparent cytoskeletal abnormalities. See text for details.

formation of aligned microtubule tracks at the vegetal cortex (Campbell et al., 2015), which normally occurs in the first embryonic cell cycle and is required for the re-localization of the dorsal determinates and axis induction (Ge et al., 2014; Jesuthasan & Stähle, 1997; Tran et al., 2012). Thus, this motor acts in the movement of ribonucleoparticles along an organized microtubule network, as in the case of germ plasm RNPs, as well as in changes in the microtubule-based cytoskeleton itself, as in the case of aligned vegetal cortex microtubules.

The failure of *kif5ba* mutants to recruit germ plasm RNAs to the furrow during the first few cell cycle correlates with the absence of primordial germ cells at 1 day post-fertilization in these embryos, indicating that *kif5ba* is essential for germ plasm RNA recruitment to the cleavage furrows and germ cell specification in the embryo (Campbell et al., 2015). Kif5ba was initially identified as a binding partner of Buc via a co-immunoprecipitation pull down analysis, which suggests a role for this motor protein in the localization of Buc in the early embryo (Campbell et al., 2015). Indeed, *kif5ba* mutants also failed to accumulate ectopically expressed Buc-GFP to the furrows and did not exhibit the increase in PGC cell numbers that occurs upon Buc overexpression in wild type (Campbell et al., 2015). These findings indicate that maternal Kif5Ba enriches Buc at the cleavage furrows. In addition, the small percentage of *kif5ba* mutant embryos where Buc protein is recruited to the furrow shows a defect in the enrichment of this protein at furrow distal ends. The precise role of Kif5Ba in germ plasm RNP recruitment to the furrow remains to be determined. It is possible that Kif5Ba begins to transport germ plasm RNPs along the underlying network of furrow-associated microtubules at the furrow. Alternatively, Kif5Ba may act early in the process of microtubule reorganization at the developing furrow in a step coupled to RNP anchoring (see below). The distal compaction defect observed in the small portion of the embryos that exhibit some aggregate recruitment at the furrow suggests an additional role for this motor protein in FMA reorganization during furrow maturation.

7. Medial-to-distal dynamics mediating germ plasm furrow compaction

Once germ plasm RNPs are recruited to the furrow, they undergo compaction at the distal ends of the furrow (Eno & Pelegri, 2013; Eno et al., 2019, 2018; Pelegri et al., 1999; Theusch et al., 2006). One of the main cytoskeletal changes associated with this compaction is the creation and reorganization of the furrow microtubule array (FMA) (Danilchik, Funk, Brown, & Larkin, 1998; Jesuthasan, 1998). Initially, when the furrow is established, bundled microtubule remnants of the anaphase asters appear along the furrow, parallel to each other and perpendicular to the furrow, which are thought to have a fundamental role in transport of vesicles to the membrane during cellular division (Danilchik, Bedrick, Brown, & Ray, 2003; Danilchik et al., 1998; Feng, Schwarz, & Jesuthasan, 2002;

Jesuthasan, 1998). As furrow maturation proceeds, the ends of these microtubule bundles appear to translocate toward the distal ends of the furrow, as they create a "V"-like shape pointing distally (Jesuthasan, 1998; Pelegri et al., 1999) (Fig. 4A). Either disruption or stabilization of the FMA prevents enrichment of germ plasm RNPs to the distal ends of furrows, suggesting that the reorganization and enrichment of FMA tubules at the distal ends of the furrow is required for the process of distal compaction (Fig. 4).

One of the proteins that regulates reorganization of the FMA is non-muscle Myosin-2, a motor protein known to mediate the contractile ring at the division furrow (Glotzer, 2005; Reichl et al., 2008), which in zebrafish is also localized to the furrow of the cleaving embryo (Urven, Yabe, & Pelegri, 2006). Inhibition of Myosin II in zebrafish embryos with myosin-2-inhibitors such as blebbistatin and ML7 does not appear to affect furrow ingression (Urven et al., 2006), a phenomenon that remains unexplained. However, Myosin-2-inhibited embryos fail to create a mature furrow, with defects in F-actin band formation, recruitment of β-Catenin to the furrow, and FMA reorganization (Urven et al., 2006).

The effects on FMA reorganization are particularly revealing. In Myosin-2-inhibited embryos, establishment of the FMA is normal, displaying microtubules that are perpendicular to the furrow (Urven et al., 2006). The defects occur during furrow maturation, with Myosin-2-inhibited embryos exhibiting FMA tubules that lack their characteristic "V"-shaped rearrangement and enrichment at the distal tips of the mature furrows, persisting instead abnormally in their initial arrangement, perpendicular to the plane of the furrow. Concomitant with defects in FMA reorganization, Myosin-2-inhibited embryos exhibit normal germ plasm RNP recruitment to the furrow but RNP compaction fails to occur (Fig. 4B) (Urven et al., 2006). This lack of distal compaction of germ plasm supports the notion that reorganization of the FMA is essential for germ plasm compaction into the distal regions of the embryo and that non-muscle Myosin-2 is required for this process.

A key regulator of cytokinesis that participates in germ plasm compaction is the RhoA/Rock pathway. RhoA is a member of the Rho-GTPase that helps regulate cytoskeletal dynamics acting as a molecular switch by cycling between a GDP-bound inactive state, RhoA-GDP, and a GTP-bound active state, RhoA-GTP. During the RhoA active state, it can interact with downstream effectors, such as the kinases Rock and Citron kinase, and the

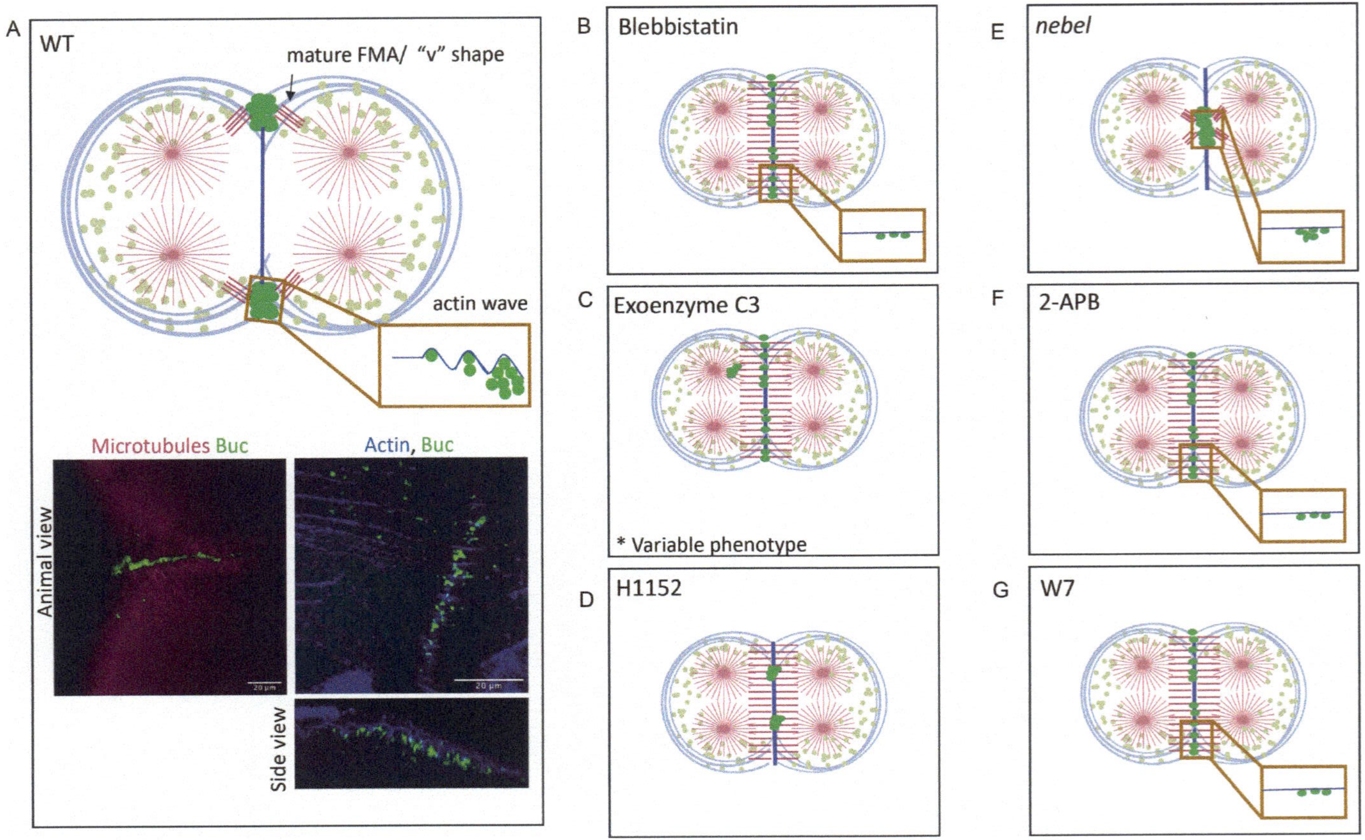

Fig. 4 See figure legend on opposite page

formin mDia, allowing for regulation of the contractile ring (Jaffe & Hall, 2005). Besides playing a role in controlling cytokinesis, the RhoA/Rock/myosin II pathway has been shown to regulate targeting of mRNA to particular cellular domains (Stuart et al., 2008) and to participate in formation of stress granules (Tsai & Wei, 2010), which share similarities to germ plasm such as the presence of disordered domain-containing proteins (Protter & Parker, 2016), suggesting a widespread use of this pathway in RNA localization and sequestration mechanisms. In zebrafish, RhoA and Rock are both localized to the furrow during the first cellular cleavage events, with RhoA being enriched across the cleavage furrow but not in the distally compacted germ plasm RNPs (Miranda-Rodríguez et al., 2017). Zebrafish contain three Rock genes, *rock1*, *rock2a*, and *rock2b*, and all are expressed throughout development (White et al., 2017). Rock protein is uniformly distributed in the blastomeres but is enriched at the distal tips of the cleavage furrows, in a similar pattern to germ plasm RNPs, and subsequently localizes primarily with germ plasm masses in the distal furrow region (Miranda-Rodríguez et al., 2017). The localization pattern of RhoA along the furrow could reflect a broader role of RhoA in furrow development, while the more localized pattern for Rock is consistent with direct interaction of this factor with germ plasm RNPs to promote their distally-oriented compaction.

Fig. 4 Distal compaction of germ plasm RNPs is dependent on the rearrangement of microtubules and a dynamic F-actin wave. (A) Once germ plasm RNPs are located in the furrow, they become compacted to the distal ends of the furrow. This distal compaction is dependent on the reorganization of the FMA in the maturing furrow, with microtubules arranged as a "V"-shape with their tips pointing distally, as well as on a dynamic F-actin wave and SCWs. Animal pole images of wild-type embryos show the interaction between the mature FMA (purple) and distally compact germ plasm aggregates (Buc) (green) or between F-actin and compacting germ plasm. Note in the side view germ plasm RNPs are located between F-actin indentations. Scale bar = 20 μM. (B–D) Pharmacological disruption of the activity of non-muscle myosin-II (B) or Rho-A (C)/Rock (D) during furrow formation prevents distal compaction of germ plasm, with the furrow failing to develop a mature FMA and actin indentations. (E–G) Embryos lacking SCWs exhibit defects in distal compaction of germ plasm RNPs. In *nebel* mutants, the germ plasm undergoes compaction to the medial furrow region together with bundling FMA tubules, and no actin indentations are observed. Drug treatments that prevent SCWs (F) or block the function of Calmodulin (G) allow recruitment of germ plasm RNPs to the furrow, but they fail to undergo distal compaction, and no FMA rearrangement occurs. See text for details.

Pharmacological inhibition of RhoA and Rock in zebrafish embryos results in mislocalization of germ plasm RNPs along the furrow (Eno et al., 2018; Miranda-Rodríguez et al., 2017). Embryos treated with either Exoenzyme C3 transferase or Rhosin, drugs that inactivate RhoA, exhibit a lack of RhoA accumulation at the furrows as well as a failure to localize germ plasm mRNAs to the distal furrow regions (Eno et al., 2018; Miranda-Rodríguez et al., 2017).

Embryos treated with these agents showed a variable germ plasm localization phenotype, such as aggregates localized outside the cleavage furrow, reduced aggregate accumulation at the furrow, lack of compaction in the cleavage furrow, and localization of compacted germ plasm mRNAs in the center of the embryo (Fig. 4C). Although embryos treated with Exoenzyme C3 do not survive beyond early development, embryos treated with Rhosin survive and display decreased numbers of primordial germ cells in the prospective gonad, as well as a range of germ cell distributions, consistent with the localization phenotypes observed in 4 cell embryos (Miranda-Rodríguez et al., 2017). Embryos treated with a specific Rock inhibitor, h1152, expected to affect the function of all Rock genes, undergo normal development but exhibit defects in the localization of germ plasm mRNA aggregates at the furrow (Miranda-Rodríguez et al., 2017). These fail to undergo distal compaction, with RNP compaction occurring instead in the medial region of the furrow (Fig. 4D). At 1 day post-fertilization, Rock inhibited embryos also exhibit a reduced number of primordial germ cells (Miranda-Rodríguez et al., 2017). Overall the evidence suggests that the RhoA/Rock pathway might participate in the processes of recruitment and distal compaction of germ plasm RNPs at the furrow.

In addition to F-actin regulation at the contractile ring, the RhoA/Rock pathway also has cellular roles including regulating microtubule dynamics (Amano, Nakayama, & Kaibuchi, 2010; Etienne-Manneville & Hall, 2002; Jaffe & Hall, 2005). Actin and microtubules were visualized in RhoA and Rock inhibited zebrafish embryos in order to understand the influence that Rho/Rock had on cytoskeleton dynamics, specifically in the context of germ plasm compaction. In both sets of treated embryos, the F-actin cytoskeleton appeared to be similar to wild-type embryos, while microtubules showed defects in FMA reorganization (Miranda-Rodríguez et al., 2017). The treated embryos displayed a phenotype similar to that observed in myosin-2-inhibited embryos (Urven et al., 2006): FMA tubules fail to undergo the characteristic enrichment at the distal furrow ends and reorganization into a distally-pointing "V"-shape, maintaining instead their initial arrangement perpendicular to the furrow (Fig. 4B–D) (Miranda-Rodríguez et al., 2017). These experiments indicate that both Rho-A and Rock are necessary for reorganization of the FMA and for germ plasm RNP compaction to the furrow distal ends. It will be essential in the future to identify cellular targets of this signaling pathway and their role in dynamic regulation of the furrow and germ plasm-associated cytoskeleton, which drives RNP movement and aggregation

Defects in germ plasm segregation are also observed in embryos mutant for the maternal gene *nebel* (Eno et al., 2018; Pelegri et al., 1999), a gene whose molecular identity remains unknown. These embryos undergo normal nuclear division and furrow initiation but fail to develop a mature furrow, indicative of a defect in completion of cytokinesis. These mutant embryos are unable to develop a fully adhesive membrane between dividing cells, causing the furrow to undergo regression such that the resulting embryos become syncytial (Pelegri et al., 1999). Visualization of microtubules in *nebel* mutants indicates defects in FMA reorganization, with FMA tubules in *nebel* mutants becoming enriched in the medial region rather than at the distal furrow regions (Fig. 4E) (Eno et al., 2018; Pelegri et al., 1999). Medial enrichment of FMA microtubules in *nebel* mutants is associated with germ plasm RNPs creating an aberrantly placed compact mass in this same medial region (Fig. 4E) (Eno et al., 2018; Pelegri et al., 1999), reflecting a coupling between FMA and germ plasm RNP aggregate reorganization during furrow maturation.

During furrow development, cytoskeletal rearrangements such as those of the FMA and F-actin are accompanied by waves of intercellular calcium that are necessary for cytokinesis during the first few cell cycles (Chang & Meng, 1995; Créton, Speksnijder, & Jaffe, 1998; Fluck, 1991; Webb, Lee, Karplus, & Miller, 1997). These waves have a relatively low speed of propagation (0.1–1 μm/s), leading to their designation as slow calcium waves (SCWs) (Jaffe, 1993). Embryos undergoing cellular cleavage contain two sets of calcium waves during the first several cell cycles. The first wave occurs before furrow initiation; this wave is a superficial "initiation wave" and is necessary for furrow formation. The second wave is a notably larger "maturation wave" that infiltrates deeper into the cytoplasm, is correlated with furrow maturation, and is thought to be required for processes such as exocytosis of internal membrane vesicles at the furrow (Lee, Webb, & Miller, 2003; Li, Webb, Chan, & Miller, 2008).

The *nebel* mutant phenotype appears to occur because the SCW is necessary for the coupled process of FMA reorganization and germ plasm compaction (Eno et al., 2018). Inhibition of intracellular calcium with the calcium chelator BAPTA results in reduced furrow formation, which is expectably accompanied by defects in germ plasm RNP recruitment. When embryos are treated with specific inhibitors of calcium release via inositol phosphate 3 (IP3), such as 2-APB (Bilmen & Michelangeli, 2002), or U72122, furrows do initiate but FMA microtubules appear to be arrested in their initial parallel conformation and do not undergo reorganization or distal enrichment (Fig. 4F). Under these conditions, germ plasm RNPs that become recruited to the furrow remain distributed along its length, without undergoing distal compaction. These phenotypes were also observed in embryos that were exposed to an inhibitor of the mitochondrial respiratory complex I, 3-NP, likely because of the role of mitochondria, which are enriched along the furrow in early zebrafish embryos (Eno et al., 2018), in modulating intracellular calcium levels (Jouaville, Ichas, Holmuhamedov, Camacho, & Lechleiter, 1995). The calcium-dependent protein Calmodulin appears to mediate downstream events triggered by SCWs in the early embryo. Calmodulin is localized to the furrow in wild-type embryos but not *nebel* mutants or embryos with reduced SCWs, and embryos treated with Calmodulin inhibitors, W7 and W5, also fail to undergo FMA reorganization as well as the associated distal enrichment of germ plasm RNPs (Fig. 4G) (Eno et al., 2018). Altogether, these observations indicate that SCWs, mediated by Calmodulin activity, provide polarity cues for the medial-to-distal reorganization of the furrow-associated cytoskeleton and distal enrichment of germ plasm aggregates.

Distal compaction of germ plasm aggregates relies on both microtubules and F-actin, similar to the pre-aggregation of germ plasm RNPs in the cortex. High magnification 3D reconstructions of actin in the furrow revealed unexpected F-actin contractions, or furrow indentations, which are similar and structurally continuous with cortical actin rings (Fig. 4A) (Eno & Pelegri, 2018). Treatment of wild-type embryos with cytochalasin D or phalloidin decreases or increases, respectively, the number of waves along the furrow. Interestingly, neither treatment allows for germ plasm compaction, suggesting that both the integrity and dynamic nature of F-actin at the furrow are essential for this process (Eno & Pelegri, 2018). In the case of phalloidin-treated embryos, it is possible that F-actin has been stabilized, increasing the numbers of contractions in the furrow but not allowing movement. Similar to the case of the trenches in the cortex prior to furrow formation, germ plasm RNPs are located in between F-actin indentations in the furrow (Eno et al., 2018; Eno & Pelegri, 2018).

Moreover, as these furrows undergo maturation, the indentations become condensed to the furrow distal ends (Eno & Pelegri, 2018). This data led to an "encroachment" model which posits that as the furrow develops a gradual increase in the frequency of F-actin indentations generates an increasing density of associated germ plasm RNPs in the distal furrow region, thereby promoting fusion of RNP aggregates and leading to compaction.

Similar to its effect at the cortex, *aura* embryos lack furrow-associated F-actin indentations, as well as the reorganization of FMA tubules associated with distal compaction of germ plasm aggregates. As in the case of F-actin dynamics at the cortex, treatment of *aura* embryos with phalloidin stabilizes F-actin in the furrow, allowing for partial formation of these indentations (Eno & Pelegri, 2018). These observations suggest that Mid1ip1lL is necessary to promote actin waves and germ plasm compaction not only in the cortex but also at the furrow (Eno & Pelegri, 2018). In addition to Mid1ip1lL function, SCWs and myosin II activity appear to be essential for dynamic reorganization of furrow-associated cytoskeletal networks. *nebel* mutant embryos as well as wild-type embryos treated with 2-APB, W7, or blebbistatin, which inhibit SCWs or Myosin-2 activity, respectively, do not display cortical indentations in the furrow, suggesting a lack of F-actin dynamics under these conditions (Eno et al., 2018) (Fig. 4B, E–G). Thus, F-actin indentations found at the furrow appear to be necessary for germ plasm RNP compaction and are dependent on Mid1ip1lL, SCWs, Calmodulin, and Myosin-2 activity.

During furrow maturation and concurrent with germ plasm RNP aggregate compaction, F-actin indentations in the furrow appear to be associated with FMA tubule tips (Eno et al., 2018). Moreover, these two types of structures appear to translocate coordinately along the furrow during FMA reorganization and simultaneously with germ plasm distal compaction. It is currently unknown if the F-actin and the FMA both provide forces to promote germ plasm aggregation to the distal end of the furrow, or if the force is primarily provided by one cytoskeletal system alone, with the other system following passively. Interestingly, in *nebel* mutants, an FMA can develop that reorganizes into microtubules creating a "v" shape, but in this case, microtubules become enriched in the medial region of the furrow and microtubule tips point inwardly. This likely reflects inward-directed microtubule bundling, reminiscent of the bundling of midzone microtubule remnants during formation of the midbody in smaller cell types (Otegui, Verbrugghe, & Skop, 2005), as opposed to distal bundling microtubules that normally form in the larger cells of early wild-type zebrafish embryos. These observations provide further support for the hypothesis that SCWs are able to re-direct microtubule organization to facilitate distally-oriented microtubule bundling and germ plasm compaction in the large early blastomeres of wild-type zebrafish. Considering the effect of inhibitors of the F-actin network, it is likely that a malleable, dynamic F-actin network is also essential for these coupled processes.

8. Conclusion

The cytoskeleton in the early embryo has multiple roles, such as establishing and implementing cellular division, biorientation and chromosome segregation, and general transport of proteins and RNAs; a wide range of activities whose specificity is likely facilitated by various cellular cofactors. In this chapter, we provided an overview of the role of the cytoskeleton in formation of germ plasm RNP masses, specifically during the early stages of germ plasm segregation: pre-aggregation, furrow recruitment, and distal compaction. These initial but crucial steps are necessary for specification of germ cells at later stages of embryonic development. These early processes appear to occur via coordination of the microtubule and F-actin networks present at both the cortex and the furrow. These networks appear to both antagonize each other and to act coordinately at various stages, with any imbalances in cytoskeletal components or cofactors disrupting germ plasm aggregation and compaction. Interestingly, different steps of germ plasm segregation appear to rely on similar underlying processes, such as the presence of a dynamic F-actin cortical network and inter-RNP aggregation, processes that are modified to confer different properties and directionality to the network. For example, the F-actin network is regulated to translocate radially outward prior to furrow formation, likely driven by the outward growth of astral microtubules, whereas in the furrow this network is modified through medial-to-distal directional cues imparted by SCWs and their effectors.

During the cleavage stages microtubule and F-actin networks likely continue to facilitate later stages of germ plasm aggregation. Thus, an anchor for the germ plasm mass to a spindle pole could coordinate asymmetric inheritance of the germ plasm mass, a mechanism that has been observed in other organisms (Lambert & Nagy, 2002; Liu et al., 2011; Shlyakhtina, Moran, & Portal, 2019). Further studies will require in-depth imaging and functional studies of microtubules, F-actin, germ plasm, and interacting factors through all stages of germ plasm segregation, from their initial higher-order aggregation during furrow maturation to their asymmetric inheritance and dispersal into the cytoplasm at later stages.

It is likely that our current knowledge only scratches the surface of a complex set of interactions leading to germ plasm mass formation in the early embryo. As targeted genomic engineering is becoming more accessible, specifically with the innovation of CRISPR-Cas9, and as we develop more precise tools for temporal and spatial regulation of gene products, it will be possible to identify and understand the function of additional factors that participate in this integrated process. These studies will allow for a deeper knowledge of the role of cytoskeleton dynamics in the segregation of germ line determinants and will more generally provide insights into the role of interacting cytoskeletal networks in early embryonic development and other cellular systems.

References

Amano, M., Nakayama, M., & Kaibuchi, K. (2010). Rho-kinase/ROCK: A key regulator of the cytoskeleton and cell polarity. *Cytoskeleton (Hoboken, N.J.)*, *67*, 545–554. https://doi.org/10.1002/cm.20472.

Ambrosini, G., Adida, C., & Altieri, D. C. (1997). A novel anti-apoptosis gene, survivin, expressed in cancer and lymphoma. *Nature Medicine*, *3*, 917–921. https://doi.org/10.1038/nm0897-917.

Bachvarova, R. F., Crother, B. I., Manova, K., Chatfield, J., Shoemaker, C. M., Crews, D. P., et al. (2009). Expression of Dazl and Vasa in turtle embryos and ovaries: Evidence for inductive specification of germ cells. *Evolution & Development*, *11*, 525–534. https://doi.org/10.1111/j.1525-142X.2009.00360.x.

Bertocchini, F., & Chuva de Sousa Lopes, S. M. (2016). Germline development in amniotes: A paradigm shift in primordial germ cell specification. *BioEssays*, *38*, 791–800. https://doi.org/10.1002/bies.201600025.

Bilmen, J. G., & Michelangeli, F. (2002). Inhibition of the type 1 inositol 1,4,5-trisphosphate receptor by 2-aminoethoxydiphenylborate. *Cellular Signalling*, *14*, 955–960. https://doi.org/10.1016/s0898-6568(02)00042-6.

Blaser, H., Eisenbeiss, S., Neumann, M., Reichman-Fried, M., Thisse, B., Thisse, C., et al. (2005). Transition from non-motile behaviour to directed migration during early PGC development in zebrafish. *Journal of Cell Science*, *118*, 4027–4038. https://doi.org/10.1242/jcs.02522.

Boke, E., Ruer, M., Wühr, M., Coughlin, M., Lemaitre, R., Gygi, S. P., et al. (2016). Amyloid-like self-assembly of a cellular compartment. *Cell*, *166*, 637–650. https://doi.org/10.1016/j.cell.2016.06.051.

Bontems, F., Stein, A., Marlow, F., Lyautey, J., Gupta, T., Mullins, M. C., et al. (2009). Bucky ball organizes germ plasm assembly in zebrafish. *Current Biology*, *19*, 414–422. https://doi.org/10.1016/j.cub.2009.01.038.

Braat, A. K., Zandbergen, T., van de Water, S., Goos, H. J., & Zivkovic, D. (1999). Characterization of zebrafish primordial germ cells: Morphology and early distribution of vasa RNA. *Developmental Dynamics*, *216*, 153–167. https://doi.org/10.1002/(SICI)1097-0177(199910)216:2<153::AID-DVDY6>3.0.CO;2-1.

Campbell, P. D., Heim, A. E., Smith, M. Z., & Marlow, F. L. (2015). Kinesin-1 interacts with Bucky ball to form germ cells and is required to pattern the zebrafish body axis. *Development*, *142*, 2996–3008. https://doi.org/10.1242/dev.124586.

Carmena, M., Wheelock, M., Funabiki, H., & Earnshaw, W. C. (2012). The chromosomal passenger complex (CPC): From easy rider to the godfather of mitosis. *Nature Reviews. Molecular Cell Biology*, *13*, 789–803. https://doi.org/10.1038/nrm3474.

Chang, D. C., & Meng, C. (1995). A localized elevation of cytosolic free calcium is associated with cytokinesis in the zebrafish embryo. *The Journal of Cell Biology*, *131*, 1539–1545. https://doi.org/10.1083/jcb.131.6.1539.

Créton, R., Speksnijder, J. E., & Jaffe, L. F. (1998). Patterns of free calcium in zebrafish embryos. *Journal of Cell Science*, *111*(Pt. 12), 1613–1622.

Danilchik, M. V., Bedrick, S. D., Brown, E. E., & Ray, K. (2003). Furrow microtubules and localized exocytosis in cleaving *Xenopus laevis* embryos. *Journal of Cell Science*, *116*, 273–283. https://doi.org/10.1242/jcs.00217.

Danilchik, M. V., Funk, W. C., Brown, E. E., & Larkin, K. (1998). Requirement for microtubules in new membrane formation during cytokinesis of Xenopus embryos. *Developmental Biology*, *194*, 47–60. https://doi.org/10.1006/dbio.1997.8815.

Dekens, M. P. S., Pelegri, F. J., Maischein, H.-M., & Nüsslein-Volhard, C. (2003). The maternal-effect gene futile cycle is essential for pronuclear congression and mitotic spindle assembly in the zebrafish zygote. *Development*, *130*, 3907–3916. https://doi.org/10.1242/dev.00606.

Eno, C., Gomez, T., Slusarski, D. C., & Pelegri, F. (2018). Slow calcium waves mediate furrow microtubule reorganization and germ plasm compaction in the early zebrafish embryo. *Development 145*, dev156604. https://doi.org/10.1242/dev.156604.

Eno, C., Hansen, C. L., & Pelegri, F. (2019). Aggregation, segregation, and dispersal of homotypic germ plasm RNPs in the early zebrafish embryo. *Developmental Dynamics*, *248*, 306–318. https://doi.org/10.1002/dvdy.18.

Eno, C., & Pelegri, F. (2013). Gradual recruitment and selective clearing generate germ plasm aggregates in the zebrafish embryo. *BioArchitecture*, *3*, 125–132. https://doi.org/10.4161/bioa.26538.

Eno, C., & Pelegri, F. (2018). Modulation of F-actin dynamics by maternal Mid1ip1L controls germ plasm aggregation and furrow recruitment in the zebrafish embryo. *Development145.* , dev156596https://doi.org/10.1242/dev.156596.

Eno, C., Solanki, B., & Pelegri, F. (2016). Aura (mid1ip1l) regulates the cytoskeleton at the zebrafish egg-to-embryo transition. *Development*, *143*, 1585–1599. https://doi.org/10.1242/dev.130591.

Etienne-Manneville, S., & Hall, A. (2002). Rho GTPases in cell biology. *Nature*, *420*, 629–635. https://doi.org/10.1038/nature01148.

Extavour, C. G., & Akam, M. (2003). Mechanisms of germ cell specification across the metazoans: Epigenesis and preformation. *Development*, *130*, 5869–5884. https://doi.org/10.1242/dev.00804.

Feng, B., Schwarz, H., & Jesuthasan, S. (2002). Furrow-specific endocytosis during cytokinesis of zebrafish blastomeres. *Experimental Cell Research*, *279*, 14–20. https://doi.org/10.1006/excr.2002.5579.

Fernández, J., Valladares, M., Fuentes, R., & Ubilla, A. (2006). Reorganization of cytoplasm in the zebrafish oocyte and egg during early steps of ooplasmic segregation. *Developmental Dynamics*, *235*, 656–671. https://doi.org/10.1002/dvdy.20682.

Field, C. M., Pelletier, J. F., & Mitchison, T. J. (2019). Disassembly of actin and keratin networks by aurora B kinase at the midplane of cleaving *Xenopus laevis* eggs. *Current Biology*, *29*, 1999–2008.e4. https://doi.org/10.1016/j.cub.2019.05.016.

Fluck, R. A. (1991). Slow calcium waves accompany cytokinesis in medaka fish eggs. *The Journal of Cell Biology*, *115*, 1259–1265. https://doi.org/10.1083/jcb.115.5.1259.

Fuentes, R., & Fernández, J. (2010). Ooplasmic segregation in the zebrafish zygote and early embryo: Pattern of ooplasmic movements and transport pathways. *Developmental Dynamics*, *239*, 2172–2189. https://doi.org/10.1002/dvdy.22349.

Fuentes, R., Mullins, M. C., & Fernández, J. (2018). Formation and dynamics of cytoplasmic domains and their genetic regulation during the zebrafish oocyte-to-embryo transition. *Mechanisms of Development*, *154*, 259–269. https://doi.org/10.1016/j.mod.2018.08.001.

Ge, X., Grotjahn, D., Welch, E., Lyman-Gingerich, J., Holguin, C., Dimitrova, E., et al. (2014). Hecate/Grip 2a acts to reorganize the cytoskeleton in the symmetry-breaking event of embryonic axis induction. *PLoS Genetics*, *10*, e1004422. https://doi.org/10.1371/journal.pgen.1004422.

Glotzer, M. (2005). The molecular requirements for cytokinesis. *Science*, *307*, 1735–1739. https://doi.org/10.1126/science.1096896.

Gustafson, E. A., & Wessel, G. M. (2010). Vasa genes: Emerging roles in the germ line and in multipotent cells. *BioEssays*, *32*, 626–637. https://doi.org/10.1002/bies.201000001.

Hartwig, J., Tarbashevich, K., Seggewiß, J., Stehling, M., Bandemer, J., Grimaldi, C., et al. (2014). Temporal control over the initiation of cell motility by a regulator of G-protein signaling. *Proceedings of the National Academy of Sciences of the United States of America*, *111*, 11389–11394. https://doi.org/10.1073/pnas.1400043111.

Hashimoto, Y., Maegawa, S., Nagai, T., Yamaha, E., Suzuki, H., Yasuda, K., et al. (2004). Localized maternal factors are required for zebrafish germ cell formation. *Developmental Biology*, *268*, 152–161. https://doi.org/10.1016/j.ydbio.2003.12.013.

Hashimoto, Y., Suzuki, H., Kageyama, Y., Yasuda, K., & Inoue, K. (2006). Bruno-like protein is localized to zebrafish germ plasm during the early cleavage stages. *Gene Expression Patterns*, *6*, 201–205. https://doi.org/10.1016/j.modgep.2005.06.006.

Heim, A. E., Hartung, O., Rothhämel, S., Ferreira, E., Jenny, A., & Marlow, F. L. (2014). Oocyte polarity requires a Bucky ball-dependent feedback amplification loop. *Development*, *141*, 842–854. https://doi.org/10.1242/dev.090449.

Henty-Ridilla, J. L., Juanes, M. A., & Goode, B. L. (2017). Profilin directly promotes microtubule growth through residues mutated in amyotrophic lateral sclerosis. *Current Biology*, *27*, 3535–3543.e4. https://doi.org/10.1016/j.cub.2017.10.002.

Henty-Ridilla, J. L., Rankova, A., Eskin, J. A., Kenny, K., & Goode, B. L. (2016). Accelerated actin filament polymerization from microtubule plus ends. *Science*, *352*, 1004–1009. https://doi.org/10.1126/science.aaf1709.

Hino, H., Nakanishi, A., Seki, R., Aoki, T., Yamaha, E., Kawahara, A., et al. (2018). Roles of maternal wnt8a transcripts in axis formation in zebrafish. *Developmental Biology*, *434*, 96–107. https://doi.org/10.1016/j.ydbio.2017.11.016.

Horvay, K., Claussen, M., Katzer, M., Landgrebe, J., & Pieler, T. (2006). Xenopus dead end mRNA is a localized maternal determinant that serves a conserved function in germ cell development. *Developmental Biology*, *291*, 1–11. https://doi.org/10.1016/j.ydbio.2005.06.013.

Houwing, S., Kamminga, L. M., Berezikov, E., Cronembold, D., Girard, A., van den Elst, H., et al. (2007). A role for Piwi and piRNAs in germ cell maintenance and transposon silencing in Zebrafish. *Cell*, *129*, 69–82. https://doi.org/10.1016/j.cell.2007.03.026.

Howley, C., & Ho, R. K. (2000). mRNA localization patterns in zebrafish oocytes. *Mechanisms of Development*, *92*, 305–309. https://doi.org/10.1016/s0925-4773(00)00247-1.

Jaffe, L. F. (1993). Classes and mechanisms of calcium waves. *Cell Calcium*, *14*, 736–745. https://doi.org/10.1016/0143-4160(93)90099-r.

Jaffe, A. B., & Hall, A. (2005). Rho GTPases: Biochemistry and biology. *Annual Review of Cell and Developmental Biology*, *21*, 247–269. https://doi.org/10.1146/annurev.cellbio.21.020604.150721.

Jamieson-Lucy, A., & Mullins, M. C. (2019). The vertebrate Balbiani body, germ plasm, and oocyte polarity. *Current Topics in Developmental Biology*, *135*, 1–34. https://doi.org/10.1016/bs.ctdb.2019.04.003.

Jesuthasan, S. (1998). Furrow-associated microtubule arrays are required for the cohesion of zebrafish blastomeres following cytokinesis. *Journal of Cell Science*, *111*(Pt. 24), 3695–3703.

Jesuthasan, S., & Stähle, U. (1997). Dynamic microtubules and specification of the zebrafish embryonic axis. *Current Biology*, *7*, 31–42. https://doi.org/10.1016/s0960-9822(06)00025-x.

Johnson, A. D., Bachvarova, R. F., Drum, M., & Masi, T. (2001). Expression of axolotl DAZL RNA, a marker of germ plasm: Widespread maternal RNA and onset of expression in germ cells approaching the gonad. *Developmental Biology*, *234*, 402–415. https://doi.org/10.1006/dbio.2001.0264.

Jouaville, L. S., Ichas, F., Holmuhamedov, E. L., Camacho, P., & Lechleiter, J. D. (1995). Synchronization of calcium waves by mitochondrial substrates in *Xenopus laevis* oocytes. *Nature*, *377*, 438–441. https://doi.org/10.1038/377438a0.

Kimmel, C. B., Ballard, W. W., Kimmel, S. R., Ullmann, B., & Schilling, T. F. (1995). Stages of embryonic development of the zebrafish. *Developmental Dynamics*, *203*, 253–310. https://doi.org/10.1002/aja.1002030302.

Kitagawa, M., & Lee, S. H. (2015). The chromosomal passenger complex (CPC) as a key orchestrator of orderly mitotic exit and cytokinesis. *Frontiers in Cell and Development Biology*, *3*, 14. https://doi.org/10.3389/fcell.2015.00014.

Knaut, H., Pelegri, F., Bohmann, K., Schwarz, H., & Nüsslein-Volhard, C. (2000). Zebrafish vasa RNA but not its protein is a component of the germ plasm and segregates asymmetrically before germline specification. *The Journal of Cell Biology*, *149*, 875–888. https://doi.org/10.1083/jcb.149.4.875.

Knaut, H., Steinbeisser, H., Schwarz, H., & Nüsslein-Volhard, C. (2002). An evolutionary conserved region in the vasa 3′UTR targets RNA translation to the germ cells in the zebrafish. *Current Biology*, *12*, 454–466. https://doi.org/10.1016/s0960-9822(02)00723-6.

Köprunner, M., Thisse, C., Thisse, B., & Raz, E. (2001). A zebrafish nanos-related gene is essential for the development of primordial germ cells. *Genes & Development*, *15*, 2877–2885. https://doi.org/10.1101/gad.212401.

Kosaka, K., Kawakami, K., Sakamoto, H., & Inoue, K. (2007). Spatiotemporal localization of germ plasm RNAs during zebrafish oogenesis. *Mechanisms of Development*, *124*, 279–289. https://doi.org/10.1016/j.mod.2007.01.003.

Lambert, J. D., & Nagy, L. M. (2002). Asymmetric inheritance of centrosomally localized mRNAs during embryonic cleavages. *Nature*, *420*, 682–686. https://doi.org/10.1038/nature01241.

Lee, K. W., Webb, S. E., & Miller, A. L. (2003). Ca^{2+} released via IP3 receptors is required for furrow deepening during cytokinesis in zebrafish embryos. *The International Journal of Developmental Biology*, *47*, 411–421.

Lekven, A. C., Thorpe, C. J., Waxman, J. S., & Moon, R. T. (2001). Zebrafish wnt8 encodes two wnt8 proteins on a bicistronic transcript and is required for mesoderm and neurectoderm patterning. *Developmental Cell*, *1*, 103–114. https://doi.org/10.1016/s1534-5807(01)00007-7.

Li, W. M., Webb, S. E., Chan, C. M., & Miller, A. L. (2008). Multiple roles of the furrow deepening Ca^{2+} transient during cytokinesis in zebrafish embryos. *Developmental Biology*, *316*, 228–248. https://doi.org/10.1016/j.ydbio.2008.01.027.

Lindeman, R. E., & Pelegri, F. (2012). Localized products of futile cycle/lrmp promote centrosome-nucleus attachment in the zebrafish zygote. *Current Biology*, *22*, 843–851. https://doi.org/10.1016/j.cub.2012.03.058.

Liu, B., Larsson, L., Franssens, V., Hao, X., Hill, S. M., Andersson, V., et al. (2011). Segregation of protein aggregates involves actin and the polarity machinery. *Cell*, *147*, 959–961. https://doi.org/10.1016/j.cell.2011.11.018.

Maegawa, S., Yasuda, K., & Inoue, K. (1999). Maternal mRNA localization of zebrafish DAZ-like gene. *Mechanisms of Development*, *81*, 223–226. https://doi.org/10.1016/s0925-4773(98)00242-1.

Magnúsdóttir, E., & Surani, M. A. (2014). How to make a primordial germ cell. *Development*, *141*, 245–252. https://doi.org/10.1242/dev.098269.

Mei, W., Jin, Z., Lai, F., Schwend, T., Houston, D. W., King, M. L., et al. (2013). Maternal Dead-End1 is required for vegetal cortical microtubule assembly during Xenopus axis specification. *Development*, *140*, 2334–2344. https://doi.org/10.1242/dev.094748.

Miranda-Rodríguez, J. R., Salas-Vidal, E., Lomelí, H., Zurita, M., & Schnabel, D. (2017). RhoA/ROCK pathway activity is essential for the correct localization of the germ plasm mRNAs in zebrafish embryos. *Developmental Biology*, *421*, 27–42. https://doi.org/10.1016/j.ydbio.2016.11.002.

Mishima, Y., Giraldez, A. J., Takeda, Y., Fujiwara, T., Sakamoto, H., Schier, A. F., et al. (2006). Differential regulation of germline mRNAs in soma and germ cells by zebrafish miR-430. *Current Biology*, *16*, 2135–2142. https://doi.org/10.1016/j.cub.2006.08.086.

Nair, S., Marlow, F., Abrams, E., Kapp, L., Mullins, M. C., & Pelegri, F. (2013). The chromosomal passenger protein Birc5b organizes microfilaments and germ plasm in the Zebrafish embryo. *PLoS Genetics*, *9*, e1003448. https://doi.org/10.1371/journal.pgen.1003448.

Niepielko, M. G., Eagle, W. V. I., & Gavis, E. R. (2018). Stochastic seeding coupled with mRNA self-recruitment generates heterogeneous Drosophila germ granules. *Current Biology*, *28*, 1872–1881.e3. https://doi.org/10.1016/j.cub.2018.04.037.

Otegui, M. S., Verbrugghe, K. J., & Skop, A. R. (2005). Midbodies and phragmoplasts: Analogous structures involved in cytokinesis. *Trends in Cell Biology*, *15*, 404–413. https://doi.org/10.1016/j.tcb.2005.06.003.

Pelegri, F., Knaut, H., Maischein, H. M., Schulte-Merker, S., & Nüsslein-Volhard, C. (1999). A mutation in the zebrafish maternal-effect gene nebel affects furrow formation and vasa RNA localization. *Current Biology*, *9*, 1431–1440. https://doi.org/10.1016/s0960-9822(00)80112-8.

Protter, D. S. W., & Parker, R. (2016). Principles and properties of stress granules. *Trends in Cell Biology*, *26*, 668–679. https://doi.org/10.1016/j.tcb.2016.05.004.

Rangan, P., DeGennaro, M., Jaime-Bustamante, K., Coux, R.-X., Martinho, R. G., & Lehmann, R. (2009). Temporal and spatial control of germ-plasm RNAs. *Current Biology*, *19*, 72–77. https://doi.org/10.1016/j.cub.2008.11.066.

Reichl, E. M., Ren, Y., Morphew, M. K., Delannoy, M., Effler, J. C., Girard, K. D., et al. (2008). Interactions between myosin and actin crosslinkers control cytokinesis contractility dynamics and mechanics. *Current Biology*, *18*, 471–480. https://doi.org/10.1016/j.cub.2008.02.056.

Reinsch, S., & Gönczy, P. (1998). Mechanisms of nuclear positioning. *Journal of Cell Science*, *111*(Pt. 16), 2283–2295.

Riemer, S., Bontems, F., Krishnakumar, P., Gömann, J., & Dosch, R. (2015). A functional Bucky ball-GFP transgene visualizes germ plasm in living zebrafish. *Gene Expression Patterns*, *18*, 44–52. https://doi.org/10.1016/j.gep.2015.05.003.

Roovers, E. F., Kaaij, L. J. T., Redl, S., Bronkhorst, A. W., Wiebrands, K., de Jesus Domingues, A. M., et al. (2018). Tdrd6a regulates the aggregation of Buc into functional subcellular compartments that drive germ cell specification. *Developmental Cell*, *46*, 285–301.e9. https://doi.org/10.1016/j.devcel.2018.07.009.

Shlyakhtina, Y., Moran, K. L., & Portal, M. M. (2019). Asymmetric inheritance of cell fate determinants: Focus on RNA. *Non-coding RNA*, *5*, E38. https://doi.org/10.3390/ncrna5020038.

Strasser, M. J., Mackenzie, N. C., Dumstrei, K., Nakkrasae, L.-I., Stebler, J., & Raz, E. (2008). Control over the morphology and segregation of Zebrafish germ cell granules during embryonic development. *BMC Developmental Biology*, *8*, 58. https://doi.org/10.1186/1471-213X-8-58.

Strome, S., & Updike, D. (2015). Specifying and protecting germ cell fate. *Nature Reviews. Molecular Cell Biology*, *16*, 406–416. https://doi.org/10.1038/nrm4009.

Stuart, H. C., Jia, Z., Messenberg, A., Joshi, B., Underhill, T. M., Moukhles, H., et al. (2008). Localized Rho GTPase activation regulates RNA dynamics and compartmentalization in tumor cell protrusions. *The Journal of Biological Chemistry*, *283*, 34785–34795. https://doi.org/10.1074/jbc.M804014200.

Suzuki, H., Maegawa, S., Nishibu, T., Sugiyama, T., Yasuda, K., & Inoue, K. (2000). Vegetal localization of the maternal mRNA encoding an EDEN-BP/Bruno-like protein in zebrafish. *Mechanisms of Development*, *93*, 205–209. https://doi.org/10.1016/s0925-4773(00)00270-7.

Taguchi, A., Watanabe, K., & Orii, H. (2014). Intracellular localizations of the Dead End protein in Xenopus primordial germ cells. *The International Journal of Developmental Biology*, *58*, 793–798. https://doi.org/10.1387/ijdb.140308ho.

Theusch, E. V., Brown, K. J., & Pelegri, F. (2006). Separate pathways of RNA recruitment lead to the compartmentalization of the zebrafish germ plasm. *Developmental Biology*, *292*, 129–141. https://doi.org/10.1016/j.ydbio.2005.12.045.

Tran, L. D., Hino, H., Quach, H., Lim, S., Shindo, A., Mimori-Kiyosue, Y., et al. (2012). Dynamic microtubules at the vegetal cortex predict the embryonic axis in zebrafish. *Development*, *139*, 3644–3652. https://doi.org/10.1242/dev.082362.

Trcek, T., Grosch, M., York, A., Shroff, H., Lionnet, T., & Lehmann, R. (2015). Drosophila germ granules are structured and contain homotypic mRNA clusters. *Nature Communications*, *6*, 7962. https://doi.org/10.1038/ncomms8962.

Tsai, N.-P., & Wei, L.-N. (2010). RhoA/ROCK1 signaling regulates stress granule formation and apoptosis. *Cellular Signalling*, *22*, 668–675. https://doi.org/10.1016/j.cellsig.2009.12.001.

Urven, L. E., Yabe, T., & Pelegri, F. (2006). A role for non-muscle myosin II function in furrow maturation in the early zebrafish embryo. *Journal of Cell Science*, *119*, 4342–4352. https://doi.org/10.1242/jcs.03197.

Webb, S. E., Lee, K. W., Karplus, E., & Miller, A. L. (1997). Localized calcium transients accompany furrow positioning, propagation, and deepening during the early cleavage period of zebrafish embryos. *Developmental Biology*, *192*, 78–92. https://doi.org/10.1006/dbio.1997.8724.

Weidinger, G., Stebler, J., Slanchev, K., Dumstrei, K., Wise, C., Lovell-Badge, R., et al. (2003). Dead end, a novel vertebrate germ plasm component, is required for zebrafish primordial germ cell migration and survival. *Current Biology*, *13*, 1429–1434. https://doi.org/10.1016/s0960-9822(03)00537-2.

Welch, E., & Pelegri, F. (2014). Cortical depth and differential transport of vegetally localized dorsal and germ line determinants in the zebrafish embryo. *BioArchitecture*, *5*, 13–26. https://doi.org/10.1080/19490992.2015.1080891.

Wheatley, S. P., & Altieri, D. C. (2019). Survivin at a glance. *Journal of Cell Science*, *132*, jcs223826. https://doi.org/10.1242/jcs.223826.

White, R. J., Collins, J. E., Sealy, I. M., Wali, N., Dooley, C. M., Digby, Z., et al. (2017). A high-resolution mRNA expression time course of embryonic development in zebrafish. *eLife*, *6*, e30860. https://doi.org/10.7554/eLife.30860.

Wühr, M., Tan, E. S., Parker, S. K., Detrich, H. W., & Mitchison, T. J. (2010). A model for cleavage plane determination in early amphibian and fish embryos. *Current Biology, 20*, 2040–2045. https://doi.org/10.1016/j.cub.2010.10.024.
Wylie, C. (1999). Germ cells. *Cell, 96*, 165–174. https://doi.org/10.1016/s0092-8674(00)80557-7.
Yoon, C., Kawakami, K., & Hopkins, N. (1997). Zebrafish vasa homologue RNA is localized to the cleavage planes of 2- and 4-cell-stage embryos and is expressed in the primordial germ cells. *Development, 124*, 3157–3165.
Zhang, J., Liu, W., Jin, Y., Jia, P., Jia, K., & Yi, M. (2017). MiR-202-5p is a novel germ plasm-specific microRNA in zebrafish. *Scientific Reports, 7*, 7055. https://doi.org/10.1038/s41598-017-07675-x.

CHAPTER SEVEN

Dead end and Detour: The function of the RNA-binding protein Dnd in posttranscriptional regulation in the germline

Theresa Gross-Thebing[a], Erez Raz[b,*]
[a]Institute of Anatomy and Vascular Biology, University of Münster, Münster, Germany
[b]Institute of Cell Biology, University of Münster, Münster, Germany
*Corresponding author: e-mail address: erez.raz@uni-muenster.de

Contents

1. Introduction 182
2. Dead end: A conserved vertebrate-specific germ cell protein 182
3. Dead end is important for development and function of the germline 187
4. The role of Dead end in germ cell fate 189
5. Tumor formation associated with loss of Dead end 192
6. Regulation of RNA expression, stability and translation by Dead end 193
7. Dead end acts as a molecular hub for diverse posttranscriptional regulation events 199
Acknowledgments 201
References 201

Abstract

Posttranscriptional regulation is a key part of controlling gene expression in different cell types, in particular in the context of specification, maintenance and differentiation of germline cells. A central regulator of these processes is the vertebrate protein Dead end (Dnd). This RNA-binding protein is important for the survival and preservation of the fate of primordial germ cells (PGCs) and for subsequent development of the male germline. In this chapter, we review the biological and molecular functions of the protein and suggest a model that takes into account the diverse roles described for Dnd in the germline. According to this model, Dnd functions as a scaffold that can bind a wide range of RNA molecules and, at the same time, provides a platform for a variety of proteins that affect posttranscriptional processes such as RNA stability and translation. This scenario offers a mechanistic basis for the control of diverse molecular processes in different contexts in germline development by the Dnd protein.

Current Topics in Developmental Biology, Volume 140
ISSN 0070-2153
https://doi.org/10.1016/bs.ctdb.2019.12.003

1. Introduction

One of the earliest developmental processes during embryonic development in sexually reproducing organisms is the specification of primordial germ cells (PGCs), the precursors of the germline. PGC specification either relies on the inheritance of localized, maternally provided RNAs and proteins, collectively termed the germ plasm, or is regulated by inductive signals provided by the environment (Aguero, Kassmer, Alberio, Johnson, & King, 2017; Lehmann, 2012; Sybirna, Wong, & Surani, 2019; Tan & Tee, 2019; Trcek & Lehmann, 2019). Despite the different mechanisms controlling PGC specification, subsequent development of the germline is, to a large extent, conserved among different species, as reflected by gene expression, cell behavior and, importantly, by cellular differentiation into gametes, sperm and egg.

Irrespective of the mode of specification, germ cells have to suppress somatic differentiation programs and, thereby, protect their fate (Robert, Garvis, & Palladino, 2015; Strome & Updike, 2015). This is especially important before the cells reach the gonad and become committed to produce only gametes (Nicholls et al., 2019) as they interact with somatic cells along their migration path (Barton, LeBlanc, & Lehmann, 2016). A central mechanism for maintaining the germline fate involves cytoplasmic determinants that help control RNA stability and function in germ cells (Robert et al., 2015; Strome & Updike, 2015). In this context, a key player is the vertebrate RNA-binding protein Dead end (Dnd). Here, we review the current knowledge regarding the function of Dead end in controlling germline fate and survival and concerning the involvement of the protein in the development of germline tumors.

2. Dead end: A conserved vertebrate-specific germ cell protein

Dead end was identified in a large-scale screen for mRNAs expressed in zebrafish embryos (Thisse et al., 2001; Weidinger et al., 2003). Subsequently, *dnd* was found to be expressed in germ cells of a wide range of other vertebrates including humans (Table 1), but direct homologs have not been identified in invertebrates. In zebrafish and *Xenopus*, *dnd* RNA is maternally provided within the germ plasm (Horvay et al., 2006; Slanchev et al., 2009; Weidinger et al., 2003). In addition, the *Xenopus* Dnd protein

Expression and function of Dead end in different vertebrates.

Organism	Name	Expression	Biological function	Molecular function	References
Xenopus laevis	xDnd1, XDE	Maternal, germ plasm, PGCs, ovary	Body axis specification, survival of PGCs	Control of RNA stability and translation, counteracting miRNAs, ATPase activity	Aguero, Jin, et al. (2017), Aguero et al. (2018), Horvay, Claußen, Katzer, Landgrebe, and Pieler (2006), Koebernick, Loeber, Arthur, Tarbashevich, and Pieler (2010), and Mei et al. (2013)
Danio rerio	Dnd	Maternal, germ plasm, PGCs, ovary	Survival and motility of PGCs, inhibition of somatic differentiation of PGCs	Control of RNA stability and translation. Counteracting miRNAs, ATPase activity	Goudarzi et al. (2012), Gross-Thebing et al. (2017), Kedde et al. (2007), Liu and Collodi (2010), Mickoleit, Banisch, and Raz (2011), Slanchev et al. (2009), and Weidinger et al. (2003)
Oryzias latipes	Dnd	Germ plasm, PGCs, ovary and testis	PGC survival and specification	Protecting germ plasm RNAs from degradation	Hong et al. (2016) and Liu et al. (2009)
Gallus gallus	CDH	PGCs, ovary and testis	–	–	Aramaki, Kubota, Soh, Yamauchi, and Hattori (2009) and Aramaki et al. (2007)
Rattus norvegicus	Dnd1	PGCs	Survival of PGCs, putative tumor suppressing activity	–	Northrup et al. (2011, 2012)
Mus musculus	Dnd1	PGCs, male embryonic germ cells and adult testis, uniform expression of one isoform in the early embryo	Survival of PGCs, differentiation and cell cycle arrest of male germ cells, putative tumor suppressing activity, embryonic viability	Promoting RNA decay by polyadenylation, counteracting miRNA function to stabilize RNAs	Bhattacharya et al. (2007), Cook, Coveney, Batchvarov, Nadeau, and Capel (2009), Cook, Munger, Nadeau, and Capel (2011), Gu et al. (2018), Niimi et al. (2019), Ruthig et al. (2019), Sakurai, Iguchi, Moriwaki, and Noguchi (1995), Suzuki et al. (2016), Yabuta, Kurimoto, Ohinata, Seki, and Saitou (2006), Yamaji et al. (2017), Youngren et al. (2005), Zechel et al. (2013), and Zhu, Bhattacharya, and Matin (2007)

Continued

Table 1 Expression and function of Dead end in different vertebrates.—Cont'd

Organism	Name	Expression	Biological function	Molecular function	References
Homo sapiens	DND1	PGCs, cancer tissues	–	Counteracting miRNA function to stabilize RNAs	Cheng, Pan, Lu, Zhu, and Chen (2017), Kedde et al. (2007), and Tang et al. (2015)
Acipenser ruthenus	ArDnd1	Ovary and testis	PGC survival	–	Baloch et al. (2019) and Linhartová et al. (2015)
Acipenser sinensis	AsDnd	PGCs, ovary and testis	–	–	Yang, Yue, Ye, Li, and Wei (2015)
Carassius gibelio	CgDnd	PGCs, ovary and testis	–	–	Li, Liu, et al. (2016)
Colossoma macropomum	tDnd	Ovary and testis	–	–	Vasconcelos et al. (2019)
Cyprinus carpio	Dnd	PGCs	PGC survival	–	Su et al. (2014)
Epinephelus coioides	EcDnd	PGCs, mature ovary and testis	PGC survival	–	Sun et al. (2017)
Gadus morhua	Dnd	PGCs	PGC survival	–	Škugor, Slanchev, Torgersen, Tveiten, and Andersen (2014) and Škugor, Tveiten, Krasnov, and Andersen (2014)
Gobiocypris rarus	Dnd	PGCs, ovary and testis	–	–	Duan et al. (2015)
Ictalurus punctatus	Dnd	PGCs	PGC survival	–	Su et al. (2015)
Monopterus albus	Dnd	PGCs	–	–	Xiao et al. (2019)

Oreochromis niloticus	Dnd	Maternal, ovary and testis	–	–	Jin, Davie, and Migaud (2019)
Oryzias celebensis	OcDnd	PGCs, ovary and testis	PGC survival	–	Zhu, Gui, Zhu, Li, and Li (2018)
Paralichthys olivaceus	PoDnd	PGCs, ovary and testis	–	–	Wang et al. (2015)
Scophthalmus maximus	SmDnd	PGCs, ovary and testis	–	–	Lin et al. (2013)
Salmo Salar	Dnd	Maternal, PGCs, ovary and testis	PGC survival	–	Nagasawa, Fernandes, Yoshizaki, Miwa, and Babiak (2013) and Wargelius et al. (2016)
Thunnus orientalis	BFTDnd	PGCs, type A spermatogonia	–	–	Yazawa, Takeuchi, Morita, Ishida, and Yoshizaki (2013)

itself is expressed in oocytes and fertilized eggs and is thus likely to be present in early PGCs (Aguero, Jin, et al., 2017). In mice and humans, and by extension in other mammals, rather than being provided maternally, *dnd* RNA is zygotically transcribed in early germ cells soon after their specification (in mice at embryonic stage E7.25) (Tang et al., 2015; Yabuta et al., 2006). *dnd* continues to be expressed in migrating PGCs and in germ cells that reach the developing gonad (Weidinger et al., 2003; Youngren et al., 2005). The expression of Dnd during more advanced stages of germline development was most extensively described in the mouse: In adult mice, Dnd1 was found to be expressed in male embryonic germ cells (Suzuki et al., 2016) and in all subpopulations of undifferentiated spermatogonia, but it is subsequently downregulated in more differentiated spermatogonia (Niimi et al., 2019).

The mouse *dnd1* RNA has two alternatively spliced isoforms that encode for proteins harboring the conserved RNA-binding domain and differ by 12 amino acids at the amino terminus (a 340 and a 352 amino acid version, Fig. 1). The larger isoform is ubiquitously expressed within the early embryo, and later is expressed in embryonic germline stages and early postnatal testes, whereas the other isoform is expressed only within adult testes (Bhattacharya et al., 2007). Interestingly, three alternatively spliced RNA isoforms that are differentially expressed in germ cells were also identified in tambaqui, a freshwater fish (Vasconcelos et al., 2019).

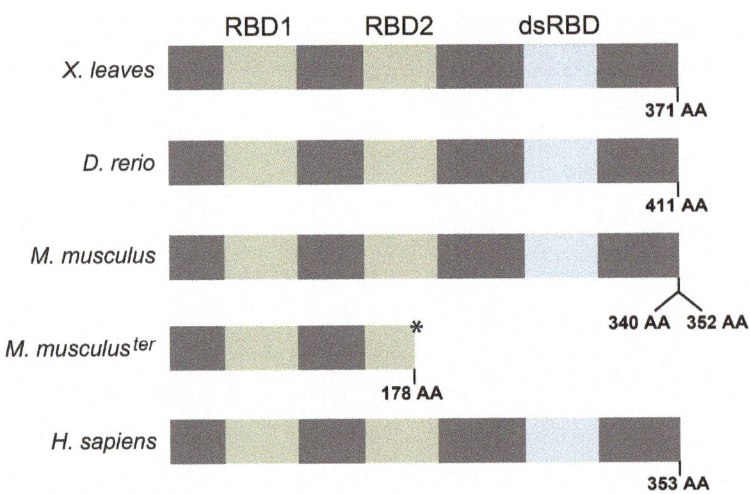

Fig. 1 Conserved modular structure of the RNA-binding protein Dnd in vertebrates. The protein sequence of Dnd harbors two RNA-binding domains (RBD, green) as well as a double-strand RBD (dsRBD, blue). The *ter* mutation in the mouse causes a premature stop (asterisk) in the second RNA-binding domain of Dnd1. AA = amino acids.

Targeted expression of a fluorescent Dnd fusion protein in zebrafish PGCs revealed localization of the protein to perinuclear germ granules (nonmembrane bound organelles unique to the germline that contain RNA and proteins important for germ cell development) (Slanchev et al., 2009; Weidinger et al., 2003) that also contain the universal germ cell marker Vasa (Raz, 2000) and the Nanos protein (Köprunner, Thisse, Thisse, & Raz, 2001). The localization of Dnd to germ cell granules was found to depend on the integrity of the RNA recognition motif (RRM) of the protein (Fig. 1). Interestingly, studies in zebrafish revealed that in addition to its localization to germ cell granules, Dnd protein is also found in the nucleus, suggesting that Dnd is involved in controlling shuttling and subcellular localization of RNAs within the cell (Slanchev et al., 2009). Specifically, mutating amino acids within Dnd's RRM shifted the subcellular localization of the protein toward the nucleus, suggesting that upon binding RNA Dnd can be exported from the nucleus to the cytoplasm (Slanchev et al., 2009).

The finding that mouse Dnd1 fused to a fluorescent protein can localize to germinal granules when expressed in zebrafish germ cells (Slanchev et al., 2009) is in line with the idea that Dnd fulfills similar functions in the two species. Consistently, mouse Dnd1 protein is found in the nucleus and in the cytoplasm of adult testes, and co-localizes with the conserved germ cell protein Nanos in p-bodies, which are organelles involved in RNA processing (Niimi et al., 2019). Taken together, the conserved expression and subcellular localization of Dead end in the vertebrate germline suggest that the protein plays important roles during PGC development.

3. Dead end is important for development and function of the germline

Loss of Dead end severely affects germline development at different stages, with phenotypes ranging from loss of cell motility to cell death and improper differentiation during embryonic stages, to infertility and tumor formation in the mature animal. The function of Dead end during early developmental stages was first analyzed in zebrafish. In this organism, knock-down of the protein led to failure of the germ cells to acquire motility, loss of molecular germline markers and death of the PGCs before they reached the region where the gonad develops (Weidinger et al., 2003), as well as defects in fate maintenance (Gross-Thebing et al., 2017, see below). Germ cell loss as a result of Dead end knockdown was also observed in other vertebrates such as frogs (Horvay et al., 2006) and rodents (Youngren et al., 2005).

This served as a tool in basic research and species conservation by allowing the generation of animals deficient of germline cells that are used as hosts for transplantation of germline cells (Baloch et al., 2019; Ciruna et al., 2002; Li, Hong, Xu, Song, & Hong, 2016; Linhartová et al., 2015; Nakagawa, Kobayashi, & Ueno, 2002; Yoshizaki et al., 2016).

Loss of PGCs in Dead end-deficient mice during migratory stages and upon entry into the gonadal ridges was presumed to be caused by programmed cell death (apoptosis), since the degree of cell loss was correlated with the function of the apoptotic factor Bax (Cook et al., 2009; Noguchi & Noguchi, 1985). Loss of PGCs led to sterility in male mice, while in female mice a few oocytes emerged from some PGCs that survived and reached the gonad (Noguchi & Noguchi, 1985). Upon analyzing Dead end function at more advanced stages of germline development, studies have shown that Dnd is important for survival and differentiation of spermatogonia (Niimi et al., 2019; Yamaji et al., 2017). Accordingly, conditional knockout of Dnd1 in juvenile mouse testis results in a gradual loss of spermatogonia, a decrease in testis size, and sterility (Niimi et al., 2019). Immunostaining of spermatogonia lacking Dnd function revealed strong upregulation of active Caspase3, consistent with the idea that the cells undergo apoptosis (Niimi et al., 2019).

Interestingly, while Dnd is essential for development of the germline at different stages, there is currently no evidence that the protein is required for the actual specification of the germline (Gross-Thebing et al., 2017; Weidinger et al., 2003). Knockdown of Dnd in zebrafish does not affect the formation of PGCs that contain the characteristic germinal granules, nor does it influence the initial expression of PGC-specific RNA molecules (Gross-Thebing et al., 2017; Weidinger et al., 2003). These results are consistent with the notion that Dnd is not required for PGC specification. Indeed, the number of PGCs in mouse embryos homozygous for a mutation in the *dnd1* gene was initially similar to that in wild-type siblings (Sakurai et al., 1995), as was also seen during induction of PGC-like cells (PGCLC) from Dnd1-deficient mouse embryonic stem cells (ESC) (Yamaji et al., 2017).

Nevertheless, in the case of zebrafish and *Xenopus* embryos, it is possible that maternally provided Dnd plays a role in PGC specification. This uncertainty persists because functional analyses of Dnd in these two model organisms during germline development relied on inhibiting RNA translation in fertilized eggs, a procedure that does not affect any maternally provided Dnd protein present in the egg and early embryo. Indeed, maternally provided Dnd protein has been shown to be important for body axis specification in *Xenopus* (Mei et al., 2013), suggesting that in different organisms Dnd protein is involved in additional processes rather than acting strictly in germ cell development. Together, the information presented above indicates that Dnd function is required for maintaining germline cells and inhibiting their apoptosis.

4. The role of Dead end in germ cell fate

In addition to the studies demonstrating the importance of Dnd for germ cell survival, studies in zebrafish have shown that Dnd controls maintenance of germline fate before the germ cells arrive at the region where the gonad develops (Gross-Thebing et al., 2017). Intriguingly, treating zebrafish embryos with Morpholinos directed against *dnd* RNA resulted in morphological aberrations of PGCs, which is consistent with the idea that Dnd function is important for maintaining germline fate (Wong & Zohar, 2015). Indeed, more detailed analysis of this phenomenon revealed that Dnd prevents the differentiation of PGCs into somatic cells, thereby maintaining the PGC fate following specification of germ cells (Gross-Thebing et al., 2017). Specifically, upon loss of Dnd the expression of somatic RNAs increases in PGCs that differentiate into somatic cells according to the identity of the specific tissues within which they were located. Indeed, directed differentiation of germ cells into other cell types could be achieved by expression of specific somatic fate inducing determinants in Dnd-depleted PGCs (Gross-Thebing et al., 2017). Thus, in this context Dnd helps suppress somatic differentiation pathways during early germline development, which is particularly crucial during stages of germ cell migration through developing somatic tissues (Gross-Thebing et al., 2017). Consistent with a function of Dnd in controlling PGC fate and protecting the early germline from somatic differentiation, Dnd localizes to perinuclear granules that are considered to contain germline determinants important for preserving germ cell fate (Ouyang et al., 2019; Strome & Updike, 2015).

Thus, based on the findings in zebrafish embryos, a significant number of germ cells that appear to be lost in Dnd-deficient embryos may have actually undergone somatic differentiation (Gross-Thebing et al., 2017). Similarly, loss of Dnd-deficient germ cells during migration in mouse embryos could also be due to differentiation into somatic cells (Youngren et al., 2005). Thus, loss of Dnd function reveals the PGCs' "latent pluripotency," which is their inherent potential to differentiate into derivatives of all germ layers. Dnd function thus maintains PGCs in a functionally unipotent state such that they solely contribute to germline fates.

Following their arrival at the developing male gonad, PGCs downregulate the expression of proteins that are associated with a pluripotent state (Pesce, Wang, Wolgemuth, & Schöler, 1998; Yamaguchi, Kimura, Tada, Nakatsuji, & Tada, 2005). Eventually, PGCs differentiate into gametes in response to signals from their surrounding gonadal niche (Kocer, Reichmann, Best, & Adams, 2009). Based on this sequence of events, it could be expected that upon arrival at the gonad, PGCs downregulate Dnd, so they can successfully differentiate into gametes. However, Dnd continues to be expressed in the germline (Niimi et al., 2019; Yamaji et al., 2017) and its function is essential for subsequent differentiation into gametes (Cook et al., 2011; Ruthig et al., 2019). Indeed, transcriptome analysis of male germ cells isolated from gonads of *dnd1* mutant mice revealed that the germ cells were unable to downregulate pluripotency factors, leading to defects in differentiation of male germ cells (Cook et al., 2011; Ruthig et al., 2019). Another study, in which Dnd1 was specifically eliminated at the time of germ cell differentiation within the embryonic gonad of mice, shows similar results: In this case, eliminating Dnd1 also led to defects in differentiation of germline in males (Suzuki et al., 2016). The phenotype observed upon loss of Dnd at these particular stages thus suggests that Dnd helps control entry into the germline differentiation program in the gonad.

The phenotypes observed upon loss of Dnd at the different stages of germline development appear to be contradictory. Loss of Dnd during the early migratory stages of germline development results in differentiation in response to cues in the environment. In contrast, eliminating Dnd function in germ cells following their arrival at the gonad interferes with differentiation of the male germline. These seemingly contradicting results show that Dnd performs multiple functions in the germline as it inhibits the somatic differentiation programs prior to arrival at the gonad, whereas it plays a positive role in promoting differentiation of germ cells into male gametes. Accordingly, lack of Dnd function in germ cells that reached the gonad interferes with the progression of germline differentiation, thereby maintaining a state of "latent pluripotency" in PGCs.

Nevertheless, a common denominator for the roles of Dnd revealed in loss-of-function studies is the finding that Dnd appears to control the fate of germ cells by regulating genes related to pluripotency and fate plasticity. Interestingly, the apparent change in Dnd function during germline development suggests that Dnd plays a different role as the germline transits from the embryonic program to the fetal program. Before the germline initiates its differentiation program within the gonad, it has to maintain its pluripotent potential along its route of migration through the developing embryo, while simultaneously suppressing this potential (with the help of Dnd) to avoid differentiating into somatic cells. Once in the gonad, the inhibitory effect of Dnd on differentiation is released, allowing the germ cells to differentiate into gametes in response to differentiation cues provided by gonadal cells. At this stage, Dnd presumably interacts with other proteins and assumes a different role, in which it helps cells progress in their differentiation into gametes.

In support of the notion that PGCs maintain their "latent pluripotency" prior to arrival in the embryonic gonad, germline stem cells can be derived from germ cells before they colonize the embryonic gonad. These cells can, in turn, be reprogrammed to pluripotent stem cells by cytokines (Matsui, Zsebo, & Hogan, 1992; Resnick, Bixler, Cheng, & Donovan, 1992), but the efficiency of this procedure declines after the cells arrive at the gonad (Kimura et al., 2008; Matsui et al., 2014).

The precise mechanisms that allow Dnd to be involved in the different processes of maintaining germline fate and controlling germline plasticity are currently unknown. Understanding this issue is further complicated by the fact that this RNA-binding protein interacts with many different RNAs and proteins (see below). Nonetheless, in very general terms, during the migratory stages of germline development Dnd may either interfere with germ cells' ability to sense extracellular signaling cues or shut down their intracellular response to such cues. Conversely, at later stages of germline development, Dnd may interact with other proteins to promote differentiation and encourage cells to exit the pluripotent state.

An additional phenotype observed in Dnd-compromised PGCs is apoptosis (Gross-Thebing et al., 2017; Weidinger et al., 2003). A possible explanation for this phenotype is that while loss of Dnd results in defects in cell differentiation, a partial loss of function might lead to the development of cells with an intermediate constitution of fate, which is resolved by apoptosis. Such a scenario could thus be more pronounced in the zebrafish model, in which knockdown (in contrast with complete knockout) experiments using morpholinos were conducted, compared to experiments with complete loss of protein function. Similarly, the mutant Dnd1 protein in mice represents a truncation rather than a complete loss of the protein (see below), which might lead to a scenario, in which most of the abnormal mutant cells are eliminated by apoptosis (Cook et al., 2009). Along these lines, specific quality control mechanisms were found to operate during early embryogenesis to remove abnormally developing cells in the embryo (Fuchs & Steller, 2011; Yamaguchi & Miura, 2015). For example, blastomeres in the inner cell mass of the early blastocyst mouse embryo were observed to undergo apoptosis as a consequence of failing to adopt the correct position and thus jeopardizing embryo integrity during development (Bedzhov & Zernicka-Goetz, 2015; Plusa, Piliszek, Frankenberg, Artus, & Hadjantonakis, 2008). Likewise, mitotic cerebral cortical cells during embryonic neurogenesis in mice are removed by programmed cell death upon acquisition of a defective karyotype (Peterson et al., 2012).

5. Tumor formation associated with loss of Dead end

Relevant to Dnd's role in controlling germline fate, studies using a specific genetic background in combination with a mutation in the mouse *dnd1* linked the function of Dnd to transformation of germ cells and teratoma formation (Youngren et al., 2005; Zhu et al., 2007). Teratomas are germline tumors that are most commonly found in the gonad and are composed of cell types of more than one germ layer (Bustamante-Marín, Garness, & Capel, 2013; Oosterhuis & Looijenga, 2005; Schneider et al., 2001; Solter, 2006). The diversity of cell types found in teratomas is believed to reflect the latent pluripotency of germ cells (Solter, 2006) and is reminiscent of the phenotype identified for zebrafish germ cells knocked down for Dnd (Gross-Thebing et al., 2017).

Intriguingly, combining a specific teratoma-susceptible mice strain (129SvJ family of inbred strains) that is normally associated with a low rate of teratoma development (Stevens & Hummel, 1957; Stevens & Little, 1954), with a premature stop mutation in the *dnd1* locus (Fig. 1), named *ter* (referring to *ter*atoma), dramatically increased the tumor frequency in the fetal testis, such that one-third of the offspring was affected (Asada, Varnum, Frankel, & Nadeau, 1994; Stevens, 1973; Youngren et al., 2005; reviewed in Zhu et al., 2007). While the Dnd-deficient germ cells that reached the gonad were prone to initiate teratoma, the majority of the PGCs were lost by apoptosis shortly following their specification (Cook et al., 2009; Sakurai et al., 1995).

Considering the role of Dnd in the processes mentioned above, it is likely that misregulation of germ cell fate acquisition and maintenance in *ter* mutants (e.g., failure to downregulate pluripotent markers; Ruthig et al., 2019) in combination with defects in regulating apoptosis and mitotic activity underlie germ cell tumor formation, as has been shown for other types of tumors (Hanahan & Weinberg, 2011). Relevant for such defects in mitotic activity is the finding that Dnd directly binds and controls expression of regulators of the cell cycle (Cook et al., 2011; Kedde et al., 2007; Ruthig et al., 2019; Zhu, Iacovino, Mahen, Kyba, & Matin, 2011). Specifically, Dnd stabilizes the expression of p27 and/or p21, cyclin-dependent kinase inhibitors that regulate the cell cycle, in human tumor cells or germ cells in the mouse testis (Cook et al., 2011; Kedde et al., 2007). Such abnormal expression of cell cycle regulators and chromatin-modifying proteins due to direct and indirect (Gu et al., 2018) involvement of Dnd could lead to genomic instability (e.g., Berton et al., 2017), which could, in turn, promote tumor formation.

It is noteworthy that the truncated protein produced by the mutant *ter* allele possesses activity that might increase the risk of teratoma formation, while the complete loss-of-function allele has no effect on the formation of testicular teratomas (Zechel et al., 2013). Therefore, residual activity of the mutant *ter* allele that contains an intact RRM domain (Fig. 1) could keep the mutant cells in a PGC-like state, thus inhibiting further differentiation. Overall, studies of the *ter* mutant suggest that loss of Dnd contributes to teratoma formation in multiple ways, such as via abnormal control of apoptosis, mitosis, and histone modifications, as well as via defects in regulating the expression of cell fate-determining factors, which promote the transition to a cancerous cell state (Berdasco & Esteller, 2010; Hanahan & Weinberg, 2011) and the formation of tumors within the germline.

6. Regulation of RNA expression, stability and translation by Dead end

Dead end has been shown to control the levels and function of mRNAs. Specifically, the protein regulates RNA stability by protecting it from degradation (e.g., by counteracting miRNA function; Kedde et al., 2007), or, conversely, by promoting RNA decay (e.g., by inhibiting polyadenylation; Yamaji et al., 2017) (Fig. 2). At the same time, by blocking miRNA function, Dnd can enhance the translation of mRNAs it interacts with (Aguero, Jin, et al., 2017). As a secondary effect, Dnd can affect RNA transcription by regulating the activity of RNAs involved in controlling chromatin structure (Gu et al., 2018) (Fig. 2). Many different RNAs bind to and are regulated by Dnd (Ruthig et al., 2019; Yamaji et al., 2017), and the proteins encoded by these RNAs are involved in diverse processes. As a result, we still lack a comprehensive model for the precise role of Dnd

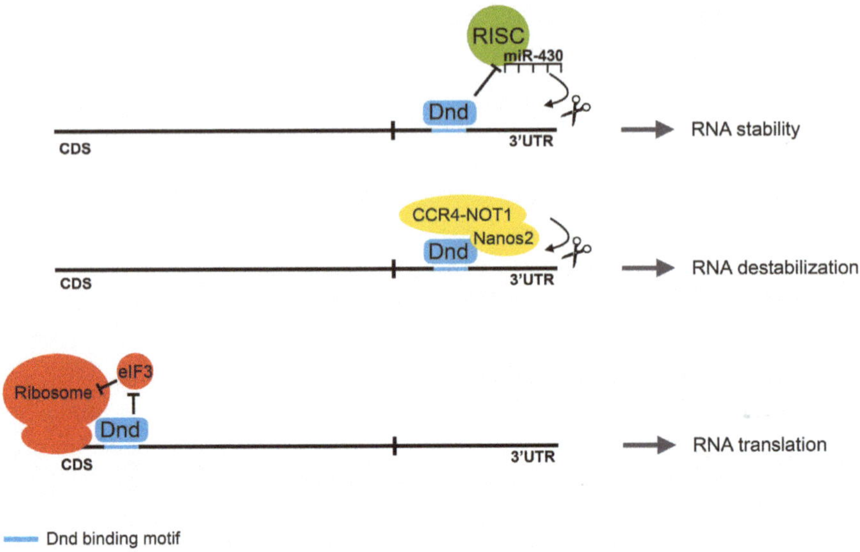

Fig. 2 Molecular effects of the RNA-binding protein Dnd on RNA stability, decay and translation. Dnd promotes either RNA stability, decay or translation depending on the proteins it interacts with, in combination with the position of the Dnd binding motif (blue) in the RNA sequence. Upper scheme—Dnd can interfere with microRNA (miR)-430 mediated degradation of its target RNAs to promote RNA stability. During early embryonic development, expression of miRNA-430 promotes degradation of maternal RNAs via the RNA-induced silencing complex (RISC). Binding of Dnd to the untranslated region (UTR) inhibits RNA decay. Middle scheme—Association of Dnd with Nanos2 and the CCR4-NOT1 complex induces deadenylation and decay of Dnd target RNAs. Lower scheme—Following binding of Dnd to binding sites within the coding sequence (CDS) of its target RNAs, it releases the inhibitory effect of eIF3 on the ribosome and promotes translation of target RNAs.

in controlling germ cell development that considers the functions of the protein in different contexts. Formulating such a model is further complicated by the fact that effects observed in Dnd-compromised cells can represent direct roles of the protein or secondary consequences of primary defects in regulating the function of specific RNAs (Gu et al., 2018; Kedde et al., 2007; Ruthig et al., 2019; Yamaji et al., 2017; Zhu et al., 2011). This issue is particularly critical when analyzing late stages of germline development under conditions in which the germ cells had already lacked Dnd at earlier stages.

Below we review findings relevant for the molecular basis for the function of Dnd as a positive regulator of RNA function, followed by presentation of cases where Dnd was shown to act in the opposite manner.

The RNA-stabilizing activity of Dnd was first observed in zebrafish PGCs, where the protein was shown to protect germline-specific RNAs from miR-430 mediated degradation (Kedde et al., 2007). While miR-430 promotes global deadenylation and clearance of maternal mRNAs in the early zebrafish embryo (Giraldez et al., 2006; Mishima et al., 2006), certain germline mRNAs such as *nanos, tdrd7* and *hub* were maintained in the PGCs (Kedde et al., 2007; Mickoleit et al., 2011), even though they contain miR-430 binding sites at their 3′UTR. This phenomenon was shown to depend on the RNA-binding function of Dnd. Specifically, Dnd binds those RNAs and prevents miR-430 binding, thereby relieving inhibition of their translation and increasing their stability within PGCs (Kedde et al., 2007). In addition to stabilizing the RNAs that encode proteins with functions relevant for PGC fate, Dnd was also found to positively regulate RNAs encoding proteins that confer cell motility (RNAs encoding for the myosin light chain kinase (MLCK) and the Zinc Finger E-Box Binding Homeobox 1 (Zeb1)) (Goudarzi et al., 2012). In early *Xenopus* embryos, *in vivo* studies provided further support for the notion that Dnd functions in RNA stabilization, as its overexpression together with the RNA-binding protein ElrB1 in the soma prevented clearance of its own RNA by miR-18 (Koebernick et al., 2010). Dnd's ability to counteract the function of miRNAs was also suggested to be relevant at more advanced stages of germline development. For example, Dnd1 was found to bind the RNA encoding for the methylation promoting factor Ezh2 in mouse testis (Gu et al., 2018). As Dnd binds to *ezh2* 3′UTR and counteracts the negative effect of miR-26a on the function of this RNA *in vitro*, it is likely that Dnd plays a role in facilitating Ezh2 expression during gametogenesis (Gu et al., 2018). Additional observations in support of the notion that Dnd functions by counteracting miRNA function were obtained in the context of human tumor formation. Here, Dnd was shown to stabilize RNAs such as those encoding for the tumor suppressors LArge Tumor Suppressor2 (LATS2) and p27(Kip1) (Kedde et al., 2007), as well as the RNA encoding for the apoptosis-promoting factor Bim (Cheng et al., 2017). Interestingly, LATS2 was recently shown to suppress breast cancer progression by maintaining cell identity (Furth et al., 2018).

Several mechanisms might account for the ability of Dnd to counteract miRNA function. Based on the close proximity of miRNA-binding sequences and the predicted mRNA sequences that Dnd binds (Kedde et al., 2007), it was suggested that Dnd sterically obstructs the assembly of the RNA-induced silencing complex (RISC) that would otherwise promote RNA degradation. Considering the localization of Dnd to specific cellular compartments, zebrafish germinal granules (Slanchev et al., 2009; Weidinger et al., 2003) or p-bodies in mice (Suzuki et al., 2016), another possibility might be that Dnd plays a role in localizing RNAs to these structures where they are sequestered and thus protected from the action of the miRNA machinery. Last, as presented below, the function of Dnd could involve direct or indirect interaction between Dnd and the RISC.

Thus far, a direct role for Dnd in promoting RNA stability by inhibiting the action of miRNAs was shown for specific candidate RNA molecules rather than at the level of an unbiased whole transcriptome wide screen. Comparing the transcriptome of germ cells from gonads of the 129SvJ/ter mouse strain with wild-type cells did not reveal enrichment for RNAs harboring a signature for miRNA-regulation, although the previously described target RNAs were re-identified (Ruthig et al., 2019). Intriguingly, the results showed that a subset of RNAs whose level is regulated by Dnd encode for regulators of chromatin structure (Ruthig et al., 2019), further highlighting the possibility that Dnd might indirectly affect the transcriptome of germ cells. Defining the precise molecular mechanisms of Dnd function would thus require discrimination between direct and indirect targets.

Related to this issue, the RNA-binding domains of Dnd themselves do not appear to be involved in recognition of unique nucleotide sequence motifs; instead, Dnd was initially shown to interact with uridine-rich motifs in the RNA (Kedde et al., 2007). Subsequently, sequencing of RNAs that were co-immunoprecipitated with Dnd1 expressed in human HEK293T revealed a UU[A/U] triplet (Yamaji et al., 2017) or [A/G/U]AU[C/G/U]A[A/U] (Ruthig et al., 2019) as the motifs enriched within RNAs that Dnd1 binds, with the number of motifs positively correlated with the degree of regulation by Dnd (Yamaji et al., 2017).

Early studies as well as more recent ones have aimed at providing mechanistic and structural insights into the molecular mechanism of Dnd function in the context of its RNA-binding activity (Duszczyk et al., 2019; Li et al., 2018; Slanchev et al., 2009). Analysis of the human DND1 structure revealed a combined function of two consecutive RNA-binding domains (RBDs) with a double-stranded RBD (dsRBD) (Duszczyk et al., 2019) (Fig. 1). Based on structural and functional analyses, the authors of this study suggested that the first RBD binds the target at low affinity, thereby allowing the protein to scan the RNA until the [C/U]A[C/U]U motif is recognized, subsequently engaging both RBDs in a more stable interaction.

Following the establishment of a stable interaction with Dnd, target-specific effects such as CCR4-NOT1 mediated RNA degradation are thought to involve the dsRBD (Duszczyk et al., 2019). Consistently, mutating critical amino acids in the RBDs and the dsRBD interferes with Dnd function *in vitro* and *in vivo* (Duszczyk et al., 2019; Li et al., 2018; Slanchev et al., 2009). Significantly, while Dnd binding regulates mRNA fate and function, the localization of Dnd itself could be regulated by its target RNAs, since Dnd protein versions that contain only the N-terminal RBD lose their subcellular localization to germinal granules in zebrafish PGCs (Slanchev et al., 2009), pointing to RNA-dependent subcellular localization of Dnd in germ cells.

Interestingly, while Dnd binding sites are primarily located within the 3′UTR, weaker binding sites were also found within the coding region of target RNAs (Aguero, Jin, et al., 2017; Ruthig et al., 2019). It might thus be speculated that Dnd mediates target-specific effects as a result of the binding position within the RNA and interaction with different RNA regulatory machineries.

In addition to Dnd's role in promoting translation via protecting certain mRNAs from miRNA-related degradation, *Xenopus* Dnd has also been shown to interact with and to relieve the inhibitory function of the eIF3f protein to enhance *nanos1* mRNA translation (Aguero, Jin, et al., 2017). Dnd was shown to bind the mRNA next to a translational control element (TCE) (Aguero et al., 2018), a secondary structure motif located downstream of the *nanos1* start codon that physically blocks its translation (Luo, Nerlick, An, & King, 2011). Here it was suggested that Dnd acts as an ATP-dependent helicase that promotes structural remodeling of the TCE, thereby relieving the translational inhibition of *nanos1* (Aguero et al., 2018). Likewise, zebrafish Dnd has been shown to possess ATPase activity located at the C-terminus of the protein that is important for early PGC development (Liu & Collodi, 2010).

The effect of Dnd on chromatin structure was described in the context of teratoma formation in mice carrying the *ter* mutation (Gu et al., 2018). Here, Dnd1 was found to positively regulate the level of the RNA encoding for the histone methyltransferase enhancer of zeste homolog 2 (Ezh2) by counteracting the function of miRNAs (Gu et al., 2018). In mammals, PGCs undergo a wave of DNA demethylation following their specification, an activity that erases parental imprints on the genome (Hajkova et al., 2002; Reik, 2001). Of note, this is not the case in zebrafish (Skvortsova et al., 2019). In mice, suppressive H3K27 (H3 lysine 27 (H3K27) trimethylation (me3)) histone modifications are reintroduced during PGC migration between E8.0 and E12.5 (Hajkova et al., 2008). Dnd positively controls this process by promoting translation of Ezh2 that attaches suppressive methyl groups to histones of somatic and meiotic genes in germ cells (Gu et al., 2018). In the *ter* mutant, abnormal changes in histone modifications might thus contribute to formation of teratoma.

In contrast to the role of Dnd in protecting mRNAs from degradation and facilitating their translation through its effect on miRNA function, Dnd1 has also been found to act as a negative element in this context. Specifically, it was recently shown that Dnd interacts with Nanos2, which together with the CCR4/NOT1 complex promotes deadenylation in male gonadal extracts or adult testis and spermatogonia of mice (Niimi et al., 2019; Suzuki et al., 2016; Yamaji et al., 2017). The interaction of Dnd with the deadenylation machinery could confer specificity to the complex, allowing differential processing of certain RNAs and not others. In doing so, Dnd recruits the CCR4/NOT1 deadenylation complex to p-bodies, cytoplasmic centers of RNA degradation (Decker & Parker, 2012), in male gonocytes as well as *in vitro* (Niimi et al., 2019; Suzuki et al., 2016). This function of Dnd could involve the interaction with different isoforms of Nanos protein that could modify the function of the complex, since Nanos2 is only expressed in germ cells of the male gonad, whereas Nanos3 is primarily expressed during early stages of germline development (Suzuki, Niimi, & Saga, 2014; Tsuda et al., 2003).

Analysis of Dnd1 target RNAs after overexpression of Dnd1 in HEK293T and mouse fibroblast cells resulted in reduced levels of Dnd1 target RNAs. The reverse effect was observed upon knockdown of Dnd1 in cultured germline stem cells and in PGCLCs from Dnd1-deficient mESC *in vitro* (Yamaji et al., 2017). Similarly, transcriptome analysis of male gonadal germ cells in the Svj129/*ter* mouse strain revealed that Dnd is required for downregulation of pluripotency marker genes that play a role in progression of the male-type specific differentiation program (Ruthig et al., 2019).

An issue that should be considered when analyzing the molecular functions of different Dnd mutants is the discrimination between phenotypes resulting from a complete loss of Dnd protein and those that result from the expression of mutant versions of the protein. For example, in contrast to the complete loss of Dnd protein, a truncated version of Dnd is expressed in the *ter* mouse strain. The Dnd *ter* version harbors an RNA-binding domain (RBD) and could thus bind RNAs (Aguero et al., 2018;

Ruthig et al., 2019), yet it lacks other domains important for Dnd1 function (Duszczyk et al., 2019). Expression of truncated protein versions in germline cells could thus give rise to phenotypes that differ from those observed in cells that completely lack the protein (Zechel et al., 2013).

Taken together, Dnd regulates mRNA stability and translation in early germline development and male-type differentiation of germ cells in the gonad. Dnd can affect the level and function of mRNAs both positively and negatively by employing different interaction partners and protein domains. It is plausible that target specificity and the specific effect on the target are achieved via different interactions of Dnd with other proteins that are expressed at distinct stages of germline development. Dnd can regulate a large number of RNAs directly and indirectly, and some of these targets can be grouped as RNAs affecting specific processes such as pluripotency, chromatin modification or cell migration. At the level of individual RNAs, the number and position of Dnd binding sites can dictate the actual effect of Dnd binding. As such, Dnd can be considered a key node for posttranscriptional control of RNA function that contributes to the plasticity of germline development at different stages.

7. Dead end acts as a molecular hub for diverse posttranscriptional regulation events

Dnd contains several RNA-binding domains that facilitate interaction of the protein with RNA (Fig. 1); here, the particular combination and number of RBDs can increase the affinity and specificity of Dnd binding to its target RNAs, as suggested for other RNA-binding proteins with a modular structure (Lunde, Moore, & Varani, 2007). Interestingly, apart from its ATPase activity, Dnd does not contain additional domains with potential enzymatic activities, suggesting that its specific function at the molecular level is dictated by the variety of proteins it interacts with. Indeed, a similar function of recruiting different regulatory protein complexes to RNAs was also reported for other RNA-binding proteins such as the PUF family proteins in yeast and worm (Haupt et al., 2020; Kershaw et al., 2015).

Accordingly, as an RNA-binding protein that is expressed specifically within the germline of vertebrates, Dnd could control the differential expression of RNAs between somatic and germ cells (Kedde et al., 2007; Köprunner et al., 2001; Mishima et al., 2006). In this context, Dnd might provide a platform for binding specific proteins that harbor activities

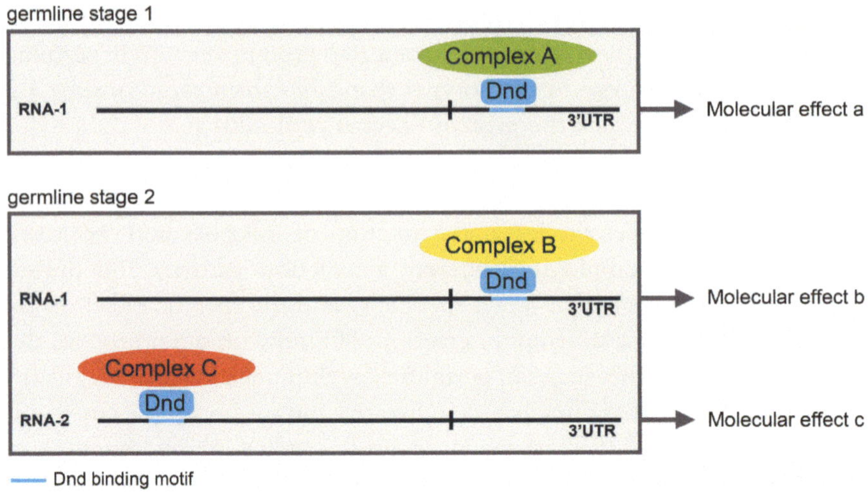

Fig. 3 Dnd acts as a hub for posttranscriptional regulation in the germline. Depending on its association with different stage-specific RNA regulatory proteins (complex A, B or C) and different RNAs (RNA-1 or RNA-2), Dnd exerts different effects on the function of its target RNAs (molecular effect a and b on the same RNA molecule, or c on a different molecule).

regulating RNA stability and translation (Figs. 2 and 3). In the soma, where Dnd is not expressed, the proteins interacting with Dnd in the germline would not affect the properties of the same RNAs. It is possible that at different stages of germline development, Dnd recruits different RNA regulatory machineries present at that particular stage, thereby facilitating different processes relevant for RNA function (Fig. 3). In the same direction, Dnd could interact with its RNA targets based on their temporal expression pattern in the differentiating germline. Together, Dnd could function as a node that connects different posttranscriptional regulation activities with different RNA molecules that are expressed at specific stages of germline development. By controlling groups of functionally related RNAs in different ways, Dnd could act as a major coordinator of specific cellular activities relevant for germ cell development (e.g., avoiding somatic differentiation (Gross-Thebing et al., 2017)), controlling behavior (e.g., acquiring motility (Goudarzi et al., 2012) or regulating pluripotency and reprogramming (Ruthig et al., 2019)). This function was suggested for other RNA-binding proteins, for example, in the context of biological clocks and response of yeast cells to iron depletion (reviewed in Keene, 2007). Last, the precise location of the Dnd binding motif within the RNA (e.g., in the 3′UTR, 5′UTR or in the coding sequence of the RNA) could determine the actual effect Dnd and the proteins it recruits have on the RNA (Fig. 3). Together, we suggest that Dnd can be considered as a key hub for posttranscriptional regulation of RNAs specifically in the germline. Within germ cells, Dnd mediates a range of posttranscriptional effects by linking RNA-modifying proteins that affect RNA function to specific RNA molecules. In this way, Dnd serves as a central element controlling the expression of proteins with essential functions for different stages of germline development.

Acknowledgments

We thank Celeste Brennecka for editing of the manuscript and Nina Knubel for graphic design. This work was supported by funds from the German Research Foundation (DFG, Clinical Research Unit 326, Male germ cells).

References

Aguero, T., Jin, Z., Chorghade, S., Kalsotra, A., King, M. L., & Yang, J. (2017). Maternal dead-end 1 promotes translation of nanos1 through binding the eIF3 complex. *Development*, *144*, 3755–3765. https://doi.org/10.1242/dev.152611.

Aguero, T., Jin, Z., Owens, D., Malhotra, A., Newman, K., Yang, J., et al. (2018). Combined functions of two RRMs in Dead-end1 mimic helicase activity to promote *nanos1* translation in the germline. *Molecular Reproduction and Development*, *85*, 896–908. https://doi.org/10.1002/mrd.23062.

Aguero, T., Kassmer, S., Alberio, R., Johnson, A., & King, M. L. (2017). *Mechanisms of vertebrate germ cell determination. In F. Pelegri, M. Danilchik, & A. Sutherland (Eds.), Vertebrate development* (pp. 383–440). Cham: Springer International Publishing. https://doi.org/10.1007/978-3-319-46095-6_8.

Aramaki, S., Kubota, K., Soh, T., Yamauchi, N., & Hattori, M. (2009). Chicken dead end homologue protein is a nucleoprotein of germ cells including primordial germ cells. *The Journal of Reproduction and Development*, *55*, 214–218. https://doi.org/10.1262/jrd.20154.

Aramaki, S., Sato, F., Kato, T., Soh, T., Kato, Y., & Hattori, M. (2007). Molecular cloning and expression of dead end homologue in chicken primordial germ cells. *Cell and Tissue Research*, *330*, 45–52. https://doi.org/10.1007/s00441-007-0435-1.

Asada, Y., Varnum, D. S., Frankel, W. N., & Nadeau, J. H. (1994). A mutation in the Ter gene causing increased susceptibility to testicular teratomas maps to mouse chromosome 18. *Nature Genetics*, *6*, 363–368. https://doi.org/10.1038/ng0494-363.

Baloch, A. R., Franěk, R., Tichopád, T., Fučíková, M., Rodina, M., & Pšenička, M. (2019). Dnd1 knockout in sturgeons by CRISPR/Cas9 generates germ cell free host for surrogate production. *Animals*, *9*, 174. https://doi.org/10.3390/ani9040174.

Barton, L. J., LeBlanc, M. G., & Lehmann, R. (2016). Finding their way: themes in germ cell migration. *Current Opinion in Cell Biology*, *42*, 128–137. https://doi.org/10.1016/j.ceb.2016.07.007.

Bedzhov, I., & Zernicka-Goetz, M. (2015). Cell death and morphogenesis during early mouse development: Are they interconnected?: Insights & Perspectives. *BioEssays*, *37*, 372–378. https://doi.org/10.1002/bies.201400147.

Berdasco, M., & Esteller, M. (2010). Aberrant epigenetic landscape in cancer: How cellular identity goes awry. *Developmental Cell*, *19*, 698–711. https://doi.org/10.1016/j.devcel.2010.10.005.

Berton, S., Cusan, M., Segatto, I., Citron, F., D'Andrea, S., Benevol, S., et al. (2017). Loss of p27kip1 increases genomic instability and induces radio-resistance in luminal breast cancer cells. *Scientific Reports*, 7, 595. https://doi.org/10.1038/s41598-017-00734-3.

Bhattacharya, C., Aggarwal, S., Zhu, R., Kumar, M., Zhao, M., Meistrich, M. L., et al. (2007). The mouse dead-end gene isoform α is necessary for germ cell and embryonic viability. *Biochemical and Biophysical Research Communications*, *355*, 194–199. https://doi.org/10.1016/j.bbrc.2007.01.138.

Bustamante-Marín, X., Garness, J. A., & Capel, B. (2013). Testicular teratomas: An intersection of pluripotency, differentiation and cancer biology. *The International Journal of Developmental Biology*, *57*, 201–210. https://doi.org/10.1387/ijdb.130136bc.

Cheng, F., Pan, Y., Lu, Y.-M., Zhu, L., & Chen, S. (2017). RNA-binding protein Dnd1 promotes breast cancer apoptosis by stabilizing the Bim mRNA in a miR-221 binding site. *BioMed Research International*, *2017*, 1–10. https://doi.org/10.1155/2017/9596152.

Ciruna, B., Weidinger, G., Knaut, H., Thisse, B., Thisse, C., Raz, E., et al. (2002). Production of maternal-zygotic mutant zebrafish by germ-line replacement. *Proceedings of the National Academy of Sciences of the United States of America*, *99*, 14919–14924. https://doi.org/10.1073/pnas.222459999.

Cook, M. S., Coveney, D., Batchvarov, I., Nadeau, J. H., & Capel, B. (2009). BAX-mediated cell death affects early germ cell loss and incidence of testicular teratomas in Dnd1Ter/Ter mice. *Developmental Biology*, *328*, 377–383. https://doi.org/10.1016/j.ydbio.2009.01.041.

Cook, M. S., Munger, S. C., Nadeau, J. H., & Capel, B. (2011). Regulation of male germ cell cycle arrest and differentiation by DND1 is modulated by genetic background. *Development*, *138*, 23–32. https://doi.org/10.1242/dev.057000.

Decker, C. J., & Parker, R. (2012). P-bodies and stress granules: Possible roles in the control of translation and mRNA degradation. *Cold Spring Harbor Perspectives in Biology*, *4*, a012286. https://doi.org/10.1101/cshperspect.a012286.

Duan, J., Feng, G., Chang, P., Zhang, X., Zhou, Q., Zhong, X., et al. (2015). Germ cell-specific expression of dead end (dnd) in rare minnow (Gobiocypris rarus). *Fish Physiology and Biochemistry*, *41*, 561–571. https://doi.org/10.1007/s10695-015-0029-x.

Duszczyk, M. M., Wischnewski, H., Kazeeva, T., Loughlin, F. E., von Schroetter, C., Pradère, U., Hall, J., Ciaudo, C., & Allain, F. H.-T. (2019). The solution structure of Dead End bound to AU-rich RNA reveals an unprecedented mode of tandem RRM-RNA recognition required for mRNA regulation. *BioRxiv*. https://doi.org/10.1101/572156.

Fuchs, Y., & Steller, H. (2011). Programmed cell death in animal development and disease. *Cell*, *147*, 742–758. https://doi.org/10.1016/j.cell.2011.10.033.

Furth, N., Pateras, I. S., Rotkopf, R., Vlachou, V., Rivkin, I., Schmitt, I., et al. (2018). LATS1 and LATS2 suppress breast cancer progression by maintaining cell identity and metabolic state. *Life Science Alliance*, *1*, e201800171. https://doi.org/10.26508/lsa.201800171.

Giraldez, A. J., Mishima, Y., Rihel, J., Grocock, R. J., Van Dongen, S., Inoue, K., et al. (2006). Zebrafish MiR-430 promotes deadenylation and clearance of maternal mRNAs. *Science*, *312*, 75–79. https://doi.org/10.1126/science.1122689.

Goudarzi, M., Banisch, T. U., Mobin, M. B., Maghelli, N., Tarbashevich, K., Strate, I., et al. (2012). Identification and regulation of a molecular module for bleb-based cell motility. *Developmental Cell*, *23*, 210–218. https://doi.org/10.1016/j.devcel.2012.05.007.

Gross-Thebing, T., Yigit, S., Pfeiffer, J., Reichman-Fried, M., Bandemer, J., Ruckert, C., et al. (2017). The vertebrate protein dead end maintains primordial germ cell fate by inhibiting somatic differentiation. *Developmental Cell*, *43*, 704-715.e5. https://doi.org/10.1016/j.devcel.2017.11.019.

Gu, W., Mochizuki, K., Otsuka, K., Hamada, R., Takehara, A., & Matsui, Y. (2018). Dnd1-mediated epigenetic control of teratoma formation in mouse. *Biology Open*, *7*, bio032318. https://doi.org/10.1242/bio.032318.

Hajkova, P., Ancelin, K., Waldmann, T., Lacoste, N., Lange, U. C., Cesari, F., et al. (2008). Chromatin dynamics during epigenetic reprogramming in the mouse germ line. *Nature*, *452*, 877–881. https://doi.org/10.1038/nature06714.

Hajkova, P., Erhardt, S., Lane, N., Haaf, T., El-Maarri, O., Reik, W., et al. (2002). Epigenetic reprogramming in mouse primordial germ cells. *Mechanisms of Development*, *117*, 15–23. https://doi.org/10.1016/S0925-4773(02)00181-8.

Hanahan, D., & Weinberg, R. A. (2011). Hallmarks of cancer: The next generation. *Cell*, *144*, 646–674. https://doi.org/10.1016/j.cell.2011.02.013.

Haupt, K. A., Law, K. T., Enright, A. L., Kanzler, C. R., Shin, H., Wickens, M., et al. (2020). A PUF Hub drives self-renewal in *Caenorhabditis elegans* germline stem cells. *Genetics*, *21*, 147–161. genetics.302772.2019. https://doi.org/10.1534/genetics.119.302772.

Hong, N., Li, M., Yuan, Y., Wang, T., Yi, M., Xu, H., et al. (2016). Dnd is a critical specifier of primordial germ cells in the medaka fish. *Stem Cell Reports*, *6*, 411–421. https://doi.org/10.1016/j.stemcr.2016.01.002.

Horvay, K., Claußen, M., Katzer, M., Landgrebe, J., & Pieler, T. (2006). Xenopus Dead end mRNA is a localized maternal determinant that serves a conserved function in germ cell development. *Developmental Biology*, *291*, 1–11. https://doi.org/10.1016/j.ydbio.2005.06.013.

Jin, Y. H., Davie, A., & Migaud, H. (2019). Expression pattern of nanos, piwil, dnd, vasa and pum genes during ontogenic development in Nile tilapia Oreochromis niloticus. *Gene*, *688*, 62–70. https://doi.org/10.1016/j.gene.2018.11.078.

Kedde, M., Strasser, M. J., Boldajipour, B., Vrielink, J. A. F. O., Slanchev, K., le Sage, C., et al. (2007). RNA-binding protein Dnd1 inhibits microRNA access to target mRNA. *Cell*, *131*, 1273–1286. https://doi.org/10.1016/j.cell.2007.11.034.

Keene, J. D. (2007). RNA regulons: Coordination of post-transcriptional events. *Nature Reviews. Genetics*, *8*, 533–543. https://doi.org/10.1038/nrg2111.

Kershaw, C. J., Costello, J. L., Talavera, D., Rowe, W., Castelli, L. M., Sims, P. F. G., et al. (2015). Integrated multi-omics analyses reveal the pleiotropic nature of the control of gene expression by Puf3p. *Scientific Reports*, *5*, 15518. https://doi.org/10.1038/srep15518.

Kimura, T., Tomooka, M., Yamano, N., Murayama, K., Matoba, S., Umehara, H., et al. (2008). AKT signaling promotes derivation of embryonic germ cells from primordial germ cells. *Development*, *135*, 869–879. https://doi.org/10.1242/dev.013474.

Kocer, A., Reichmann, J., Best, D., & Adams, I. R. (2009). Germ cell sex determination in mammals. *Molecular Human Reproduction*, *15*, 205–213. https://doi.org/10.1093/molehr/gap008.

Koebernick, K., Loeber, J., Arthur, P. K., Tarbashevich, K., & Pieler, T. (2010). Elr-type proteins protect Xenopus Dead end mRNA from miR-18-mediated clearance in the soma. *Proceedings of the National Academy of Sciences of the United States of America*, *107*, 16148–16153. https://doi.org/10.1073/pnas.1004401107.

Köprunner, M., Thisse, C., Thisse, B., & Raz, E. (2001). A zebrafish nanos-related gene is essential for the development of primordial germ cells. *Genes & Development*, *15*, 2877–2885. https://doi.org/10.1101/gad.212401.

Lehmann, R. (2012). Germline stem cells: Origin and destiny. *Cell Stem Cell*, *10*, 729–739. https://doi.org/10.1016/j.stem.2012.05.016.

Li, M., Hong, N., Xu, H., Song, J., & Hong, Y. (2016). Germline replacement by blastula cell transplantation in the fish medaka. *Scientific Reports*, *6*, 29658. https://doi.org/10.1038/srep29658.

Li, Q., Li, Y., Yang, S., Huang, S., Yan, M., Ding, Y., et al. (2018). CRISPR–Cas9-mediated base-editing screening in mice identifies DND1 amino acids that are critical for primordial germ cell development. *Nature Cell Biology*, *20*, 1315–1325. https://doi.org/10.1038/s41556-018-0202-4.

Li, S.-Z., Liu, W., Li, Z., Wang, Y., Zhou, L., Yi, M.-S., et al. (2016). Molecular characterization and expression pattern of a germ cell marker gene dnd in gibel carp (Carassius gibelio). *Gene*, *591*, 183–190. https://doi.org/10.1016/j.gene.2016.07.027.

Lin, F., Zhao, C. Y., Xu, S. H., Ma, D. Y., Xiao, Z. Z., Xiao, Y. S., et al. (2013). Germline-specific and sexually dimorphic expression of a dead end gene homologue in turbot (Scophthalmus maximus). *Theriogenology*, *80*, 665–672. https://doi.org/10.1016/j.theriogenology.2013.06.016.

Linhartová, Z., Saito, T., Kašpar, V., Rodina, M., Prášková, E., Hagihara, S., et al. (2015). Sterilization of sterlet Acipenser ruthenus by using knockdown agent, antisense morpholino oligonucleotide, against dead end gene. *Theriogenology*, *84*, 1246–1255.e1. https://doi.org/10.1016/j.theriogenology.2015.07.003.

Liu, W., & Collodi, P. (2010). Zebrafish dead end possesses ATPase activity that is required for primordial germ cell development. *The FASEB Journal*, *24*, 2641–2650. https://doi.org/10.1096/fj.09-148403.

Liu, L., Hong, N., Xu, H., Li, M., Yan, Y., Purwanti, Y., et al. (2009). Medaka dead end encodes a cytoplasmic protein and identifies embryonic and adult germ cells. *Gene Expression Patterns*, *9*, 541–548. https://doi.org/10.1016/j.gep.2009.06.008.

Lunde, B. M., Moore, C., & Varani, G. (2007). RNA-binding proteins: Modular design for efficient function. *Nature Reviews. Molecular Cell Biology*, *8*, 479–490. https://doi.org/10.1038/nrm2178.

Luo, X., Nerlick, S., An, W., & King, M. L. (2011). Xenopus germline nanos1 is translationally repressed by a novel structure-based mechanism. *Development*, *138*, 589–598. https://doi.org/10.1242/dev.056705.

Matsui, Y., Takehara, A., Tokitake, Y., Ikeda, M., Obara, Y., Morita-Fujimura, Y., et al. (2014). The majority of early primordial germ cells acquire pluripotency by AKT activation. *Development*, *141*, 4457–4467. https://doi.org/10.1242/dev.113779.

Matsui, Y., Zsebo, K., & Hogan, B. L. (1992). Derivation of pluripotential embryonic stem cells from murine primordial germ cells in culture. *Cell*, *70*, 841–847.

Mei, W., Jin, Z., Lai, F., Schwend, T., Houston, D. W., King, M. L., et al. (2013). Maternal Dead-End1 is required for vegetal cortical microtubule assembly during Xenopus axis specification. *Development*, *140*, 2334–2344. https://doi.org/10.1242/dev.094748.

Mickoleit, M., Banisch, T. U., & Raz, E. (2011). Regulation of hub mRNA stability and translation by miR430 and the dead end protein promotes preferential expression in zebrafish primordial germ cells. *Developmental Dynamics*, *240*, 695–703. https://doi.org/10.1002/dvdy.22571.

Mishima, Y., Giraldez, A. J., Takeda, Y., Fujiwara, T., Sakamoto, H., Schier, A. F., et al. (2006). Differential regulation of germline mRNAs in soma and germ cells by zebrafish miR-430. *Current Biology*, *16*, 2135–2142. https://doi.org/10.1016/j.cub.2006.08.086.

Nagasawa, K., Fernandes, J. M. O., Yoshizaki, G., Miwa, M., & Babiak, I. (2013). Identification and migration of primordial germ cells in Atlantic salmon, *Salmo salar*: Characterization of *Vasa*, *Dead End*, and *Lymphocyte antigen 75* genes. *Molecular Reproduction and Development*, *80*, 118–131. https://doi.org/10.1002/mrd.22142.

Nakagawa, M., Kobayashi, T., & Ueno, K. (2002). Production of germline chimera in loach (Misgurnus anguillicaudatus) and proposal of new method for preservation of endangered fish species. *The Journal of Experimental Zoology*, *293*, 624–631. https://doi.org/10.1002/jez.10184.

Nicholls, P. K., Schorle, H., Naqvi, S., Hu, Y.-C., Fan, Y., Carmell, M. A., et al. (2019). Mammalian germ cells are determined after PGC colonization of the nascent gonad. *Proceedings of the National Academy of Sciences of the United States of America, 116*, 25677–25687. https://doi.org/10.1073/pnas.1910733116.

Niimi, Y., Imai, A., Nishimura, H., Yui, K., Kikuchi, A., Koike, H., et al. (2019). Essential role of mouse Dead end1 in the maintenance of spermatogonia. *Developmental Biology, 445*, 103–112. https://doi.org/10.1016/j.ydbio.2018.11.003.

Noguchi, T., & Noguchi, M. (1985). A recessive mutation (ter) causing germ cell deficiency and a high incidence of congenital testicular teratomas in 129/Sv-ter mice. *Journal of the National Cancer Institute, 75*, 385–392.

Northrup, E., Eisenblätter, R., Glage, S., Rudolph, C., Dorsch, M., Schlegelberger, B., et al. (2011). Loss of Dnd1 facilitates the cultivation of genital ridge-derived rat embryonic germ cells. *Experimental Cell Research, 317*, 1885–1894. https://doi.org/10.1016/j.yexcr.2011.04.013.

Northrup, E., Zschemisch, N.-H., Eisenblätter, R., Glage, S., Wedekind, D., Cuppen, E., et al. (2012). The ter mutation in the Rat Dnd1 gene initiates gonadal teratomas and infertility in both genders. *PLoS One, 7*, e38001. https://doi.org/10.1371/journal.pone.0038001.

Oosterhuis, J. W., & Looijenga, L. H. J. (2005). Testicular germ-cell tumours in a broader perspective. *Nature Reviews Cancer, 5*, 210–222. https://doi.org/10.1038/nrc1568.

Ouyang, J. P. T., Folkmann, A., Bernard, L., Lee, C.-Y., Seroussi, U., Charlesworth, A. G., et al. (2019). P Granules protect RNA interference genes from silencing by piRNAs. *Developmental Cell, 50,* 716-728.e6. https://doi.org/10.1016/j.devcel.2019.07.026.

Pesce, M., Wang, X., Wolgemuth, D. J., & Schöler, H. R. (1998). Differential expression of the Oct-4 transcription factor during mouse germ cell differentiation. *Mechanisms of Development, 71*, 89–98. https://doi.org/10.1016/S0925-4773(98)00002-1.

Peterson, S. E., Yang, A. H., Bushman, D. M., Westra, J. W., Yung, Y. C., Barral, S., et al. (2012). Aneuploid cells are differentially susceptible to caspase-mediated death during embryonic cerebral cortical development. *The Journal of Neuroscience, 32*, 16213–16222. https://doi.org/10.1523/JNEUROSCI.3706-12.2012.

Plusa, B., Piliszek, A., Frankenberg, S., Artus, J., & Hadjantonakis, A.-K. (2008). Distinct sequential cell behaviours direct primitive endoderm formation in the mouse blastocyst. *Development, 135*, 3081–3091. https://doi.org/10.1242/dev.021519.

Raz, E. (2000). The function and regulation of vasa-like genes in germ-cell development. *Genome Biol, 1*, 1–6. REVIEWS1017. https://doi.org/10.1186/gb-2000-1-3-reviews1017.

Reik, W. (2001). Epigenetic reprogramming in mammalian development. *Science, 293*, 1089–1093. https://doi.org/10.1126/science.1063443.

Resnick, J. L., Bixler, L. S., Cheng, L., & Donovan, P. J. (1992). Long-term proliferation of mouse primordial germ cells in culture. *Nature, 359*, 550–551. https://doi.org/10.1038/359550a0.

Robert, V. J., Garvis, S., & Palladino, F. (2015). Repression of somatic cell fate in the germline. *Cellular and Molecular Life Sciences, 72*, 3599–3620. https://doi.org/10.1007/s00018-015-1942-y.

Ruthig, V. A., Friedersdorf, M. B., Garness, J. A., Munger, S. C., Bunce, C., Keene, J. D., et al. (2019). The RNA-binding protein DND1 acts sequentially as a negative regulator of pluripotency and a positive regulator of epigenetic modifiers required for germ cell reprogramming. *Development, 146,* dev.175950. https://doi.org/10.1242/dev.175950.

Sakurai, T., Iguchi, T., Moriwaki, K., & Noguchi, M. (1995). The ter mutation first causes primordial germ cell deficiency in ter/ter mouse embryos at 8 days of gestation. *Development, Growth & Differentiation, 37*, 293–302. https://doi.org/10.1046/j.1440-169X.1995.t01-2-00007.x.

Schneider, D. T., Schuster, A. E., Fritsch, M. K., Hu, J., Olson, T., Lauer, S., et al. (2001). Multipoint imprinting analysis indicates a common precursor cell for gonadal and nongonadal pediatric germ cell tumors. *Cancer Research, 61,* 7268–7276.

Škugor, A., Slanchev, K., Torgersen, J. S., Tveiten, H., & Andersen, Ø. (2014). Conserved mechanisms for germ cell-specific localization of nanos3 Transcripts in teleost species with aquaculture significance. *Marine Biotechnology, 16,* 256–264. https://doi.org/10.1007/s10126-013-9543-y.

Škugor, A., Tveiten, H., Krasnov, A., & Andersen, Ø. (2014). Knockdown of the germ cell factor Dead end induces multiple transcriptional changes in Atlantic cod (Gadus morhua) hatchlings. *Animal Reproduction Science, 144,* 129–137. https://doi.org/10.1016/j.anireprosci.2013.12.010.

Skvortsova, K., Tarbashevich, K., Stehling, M., Lister, R., Irimia, M., Raz, E., et al. (2019). Retention of paternal DNA methylome in the developing zebrafish germline. *Nature Communications, 10,* 3054. https://doi.org/10.1038/s41467-019-10895-6.

Slanchev, K., Stebler, J., Goudarzi, M., Cojocaru, V., Weidinger, G., & Raz, E. (2009). Control of Dead end localization and activity—Implications for the function of the protein in antagonizing miRNA function. *Mechanisms of Development, 126,* 270–277. https://doi.org/10.1016/j.mod.2008.10.006.

Solter, D. (2006). From teratocarcinomas to embryonic stem cells and beyond: A history of embryonic stem cell research. *Nature Reviews. Genetics,* 7, 319–327. https://doi.org/10.1038/nrg1827.

Stevens, L. C. (1973). A new inbred subline of mice (129/terSv) with a high incidence of spontaneous congenital testicular teratomas2. *Journal of the National Cancer Institute, 50,* 235–242. https://doi.org/10.1093/jnci/50.1.235.

Stevens, L. C., & Hummel, K. P. (1957). A description of spontaneous congenital testicular teratomas in strain 129 mice. *Journal of the National Cancer Institute, 18,* 719–747.

Stevens, L. C., & Little, C. C. (1954). Spontaneous testicular teratomas in an inbred strain of mice. *Proceedings of the National Academy of Sciences of the United States of America, 40,* 1080–1087. https://doi.org/10.1073/pnas.40.11.1080.

Strome, S., & Updike, D. (2015). Specifying and protecting germ cell fate. *Nature Reviews. Molecular Cell Biology, 16,* 406–416. https://doi.org/10.1038/nrm4009.

Su, B., Peatman, E., Shang, M., Thresher, R., Grewe, P., Patil, J., et al. (2014). Expression and knockdown of primordial germ cell genes, vasa, nanos and dead end in common carp (Cyprinus carpio) embryos for transgenic sterilization and reduced sexual maturity. *Aquaculture, 420–421,* S72–S84. https://doi.org/10.1016/j.aquaculture.2013.07.008.

Su, B., Shang, M., Grewe, P. M., Patil, J. G., Peatman, E., Perera, D. A., et al. (2015). Suppression and restoration of primordial germ cell marker gene expression in channel catfish, Ictalurus punctatus, using knockdown constructs regulated by copper transport protein gene promoters: Potential for reversible transgenic sterilization. *Theriogenology, 84,* 1499–1512. https://doi.org/10.1016/j.theriogenology.2015.07.037.

Sun, Z.-H., Zhou, L., Li, Z., Liu, X.-C., Li, S.-S., Wang, Y., et al. (2017). Sexual dimorphic expression of dnd in germ cells during sex reversal and its requirement for primordial germ cell survival in protogynous hermaphroditic grouper. *Comparative Biochemistry and Physiology. Part B, Biochemistry & Molecular Biology, 208–209,* 47–57. https://doi.org/10.1016/j.cbpb.2017.04.003.

Suzuki, A., Niimi, Y., & Saga, Y. (2014). Interaction of NANOS2 and NANOS3 with different components of the CNOT complex may contribute to the functional differences in mouse male germ cells. *Biology Open, 3,* 1207–1216. https://doi.org/10.1242/bio.20149308.

Suzuki, A., Niimi, Y., Shinmyozu, K., Zhou, Z., Kiso, M., & Saga, Y. (2016). Dead end1 is an essential partner of NANOS2 for selective binding of target RNAs in male germ cell development. *EMBO Reports, 17,* 37–46. https://doi.org/10.15252/embr.201540828.

Sybirna, A., Wong, F. C. K., & Surani, M. A. (2019). Genetic basis for primordial germ cells specification in mouse and human: Conserved and divergent roles of PRDM and SOX transcription factors. In R. Lehmann (Ed.), *Current topics in developmental biology* (pp. 35–89). Elsevier. https://doi.org/10.1016/bs.ctdb.2019.04.004.
Tan, H., & Tee, W. (2019). Committing the primordial germ cell: An updated molecular perspective. *Wiley Interdisciplinary Reviews. Systems Biology and Medicine, 11*, e1436. https://doi.org/10.1002/wsbm.1436.
Tang, W. W. C., Dietmann, S., Irie, N., Leitch, H. G., Floros, V. I., Bradshaw, C. R., et al. (2015). A unique gene regulatory network resets the human germline epigenome for development. *Cell, 161*, 1453–1467. https://doi.org/10.1016/j.cell.2015.04.053.
Thisse, B., Pflumio, S., Fürthauer, M., Loppin, B., Heyer, V., Degrave, A., et al. (2001). *Expression of the zebrafish genome during embryogenesis (NIH R01 RR15402)*. ZFIN Direct Data Submission. http://zfin.org.
Trcek, T., & Lehmann, R. (2019). Germ granules in *Drosophila*. *Traffic, 20*, 650–660. https://doi.org/10.1111/tra.12674.
Tsuda, M., Sasaoka, Y., Kiso, M., Abe, K., Haraguchi, S., Kobayashi, S., et al. (2003). Conserved role of nanos proteins in germ cell development. *Science, 301*, 1239–1241. https://doi.org/10.1126/science.1085222.
Vasconcelos, A. C. N., Streit, D. P., Octavera, A., Miwa, M., Kabeya, N., & Yoshizaki, G. (2019). The germ cell marker dead end reveals alternatively spliced transcripts with dissimilar expression. *Scientific Reports, 9*, 2407. https://doi.org/10.1038/s41598-019-39101-9.
Wang, X., Liu, Q., Xiao, Y., Yang, Y., Wang, Y., Song, Z., et al. (2015). The dnd RNA identifies germ cell origin and migration in olive flounder (Paralichthys olivaceus). *BioMed Research International, 2015*, 1–9. https://doi.org/10.1155/2015/428591.
Wargelius, A., Leininger, S., Skaftnesmo, K. O., Kleppe, L., Andersson, E., Taranger, G. L., et al. (2016). Dnd knockout ablates germ cells and demonstrates germ cell independent sex differentiation in Atlantic salmon. *Scientific Reports, 6*, 21284. https://doi.org/10.1038/srep21284.
Weidinger, G., Stebler, J., Slanchev, K., Dumstrei, K., Wise, C., Lovell-Badge, R., et al. (2003). dead end, a novel vertebrate germ plasm component, is required for zebrafish primordial germ cell migration and survival. *Current Biology, 13*, 1429–1434. https://doi.org/10.1016/S0960-9822(03)00537-2.
Wong, T.-T., & Zohar, Y. (2015). Production of reproductively sterile fish by a non-transgenic gene silencing technology. *Scientific Reports, 5*, 15822. https://doi.org/10.1038/srep15822.
Xiao, Q., Sun, Y., Liang, X., Zhang, L., Onxayvieng, K., Li, Z., et al. (2019). Visualizing primordial germ cell migration in embryos of rice field eel (Monopterus albus) using fluorescent protein tagged 3′ untranslated regions of nanos3, dead end and vasa. *Comparative Biochemistry and Physiology. Part B, Biochemistry & Molecular Biology, 235*, 62–69. https://doi.org/10.1016/j.cbpb.2019.06.002.
Yabuta, Y., Kurimoto, K., Ohinata, Y., Seki, Y., & Saitou, M. (2006). Gene expression dynamics during germline specification in mice identified by quantitative single-cell gene expression profiling. *Biology of Reproduction, 75*, 705–716. https://doi.org/10.1095/biolreprod.106.053686.
Yamaguchi, S., Kimura, H., Tada, M., Nakatsuji, N., & Tada, T. (2005). Nanog expression in mouse germ cell development. *Gene Expression Patterns, 5*, 639–646. https://doi.org/10.1016/j.modgep.2005.03.001.
Yamaguchi, Y., & Miura, M. (2015). Programmed cell death in neurodevelopment. *Developmental Cell, 32*, 478–490. https://doi.org/10.1016/j.devcel.2015.01.019.
Yamaji, M., Jishage, M., Meyer, C., Suryawanshi, H., Der, E., Yamaji, M., et al. (2017). DND1 maintains germline stem cells via recruitment of the CCR4–NOT complex to target mRNAs. *Nature, 543*, 568–572. https://doi.org/10.1038/nature21690.

Yang, X., Yue, H., Ye, H., Li, C., & Wei, Q. (2015). Identification of a germ cell marker gene, the dead end homologue, in Chinese sturgeon Acipenser sinensis. *Gene*, *558*, 118–125. https://doi.org/10.1016/j.gene.2014.12.059.

Yazawa, R., Takeuchi, Y., Morita, T., Ishida, M., & Yoshizaki, G. (2013). The Pacific bluefin tuna (Thunnus orientalis) dead end gene is suitable as a specific molecular marker of type A spermatogonia. *Molecular Reproduction and Development*, *80*, 871–880. https://doi.org/10.1002/mrd.22224.

Yoshizaki, G., Takashiba, K., Shimamori, S., Fujinuma, K., Shikina, S., Okutsu, T., et al. (2016). Production of germ cell-deficient salmonids by dead end gene knockdown, and their use as recipients for germ cell transplantation. *Molecular Reproduction and Development*, *83*, 298–311. https://doi.org/10.1002/mrd.22625.

Youngren, K. K., Coveney, D., Peng, X., Bhattacharya, C., Schmidt, L. S., Nickerson, M. L., et al. (2005). The Ter mutation in the dead end gene causes germ cell loss and testicular germ cell tumours. *Nature*, *435*, 360–364. https://doi.org/10.1038/nature03595.

Zechel, J. L., Doerner, S. K., Lager, A., Tesar, P. J., Heaney, J. D., & Nadeau, J. H. (2013). Contrasting effects of Deadend1 (Dnd1) gain and loss of function mutations on allelic inheritance, testicular cancer, and intestinal polyposis. *BMC Genetics*, *14*, 54. https://doi.org/10.1186/1471-2156-14-54.

Zhu, R., Bhattacharya, C., & Matin, A. (2007). The role of dead-end in germ-cell tumor development. *Annals of the New York Academy of Sciences*, *1120*, 181–186. https://doi.org/10.1196/annals.1411.006.

Zhu, T., Gui, L., Zhu, Y., Li, Y., & Li, M. (2018). Dnd is required for primordial germ cell specification in Oryzias celebensis. *Gene*, *679*, 36–43. https://doi.org/10.1016/j.gene.2018.08.068.

Zhu, R., Iacovino, M., Mahen, E., Kyba, M., & Matin, A. (2011). Transcripts that associate with the RNA binding protein, DEAD-END (DND1), in embryonic stem (ES) cells. *BMC Molecular Biology*, *12*, 37. https://doi.org/10.1186/1471-2199-12-37.

CHAPTER EIGHT

From mother to embryo: A molecular perspective on zygotic genome activation

Edlyn Wu, Nadine L. Vastenhouw*
Max Planck Institute of Molecular Cell Biology and Genetics, Dresden, Germany
*Corresponding author: e-mail address: vastenhouw@mpi-cbg.de

Contents

1. Introduction 210
2. The many-layers of transcriptional regulation in ZGA 211
   2.1 Transcription basics 211
   2.2 Transcription in the context of early embryo development 214
3. The molecular landscape of ZGA 217
   3.1 RNA polymerase II 217
   3.2 Transcription factors in ZGA 218
   3.3 Chromatin 229
4. Compartmentalizing transcription 238
5. Concluding remarks and future perspectives 240
Acknowledgments 242
References 242

Abstract

In animals, the early embryo is mostly transcriptionally silent and development is fueled by maternally supplied mRNAs and proteins. These maternal products are important not only for survival, but also to gear up the zygote's genome for activation. Over the last three decades, research with different model organisms and experimental approaches has identified molecular factors and proposed mechanisms for how the embryo transitions from being transcriptionally silent to transcriptionally competent. In this chapter, we discuss the molecular players that shape the molecular landscape of ZGA and provide insights into their mode of action in activating the transcription program in the developing embryo.

Abbreviations

ESC embryonic stem cell
MZT maternal-to-zygotic transition

Current Topics in Developmental Biology, Volume 140
ISSN 0070-2153
https://doi.org/10.1016/bs.ctdb.2020.02.002

Pol II RNA polymerase II
PTM post-translational modification
TF transcription factor
ZGA zygotic genome activation

1. Introduction

A fundamental transition during embryogenesis is the transfer of developmental control from the mother to the zygote. During this transition, known as the maternal-to-zygotic transition (MZT), developmental control is transferred via two tightly coordinated processes: the clearance of maternally loaded products and the gradual onset of transcription during zygotic genome activation (ZGA) (Vastenhouw, Cao, & Lipshitz, 2019). The molecules and molecular events that shape maternal clearance are better defined. Specifically, maternal mRNAs are destabilized through the activity of RNA-binding proteins, small RNAs, and conserved degradation machinery that trigger deadenylation or terminal uridylation (Vastenhouw et al., 2019). Modifications such as N6-methyladenosine (m^6A) and 5-methylcytosine (m^5C) can also assist in regulating the stability of maternal mRNAs (Yang et al., 2019; Zhao et al., 2017). In contrast, such mechanistic details have not yet been mapped out for the onset of zygotic transcription. Notwithstanding this, over the years, the use of different model organisms and experimental approaches have significantly contributed toward our understanding of ZGA and transcriptional control during early embryonic development.

This review focuses on the molecular features of ZGA. We begin with a brief overview of transcription and the different layers of its regulation. We then place transcriptional control in the context of early embryonic development. We highlight the major regulators identified to have a critical role in ZGA. Several featured players, such as OCT4, SOX2, and NANOG, are key proteins that also function in pluripotency, reprogramming, and cellular differentiation. Therefore, some of the characterization and mechanistic studies from stem cell research are brought into consideration, as they lend insight into potentially similar behavior and modes of activity in the context of the developing embryo. Next, we discuss relevant proteins that make up chromatin architecture (histones, histone modifications, and cohesin), as well as chromatin remodelers that are implicated in transcriptional activation in the early embryo. We also describe recent insights into compartmentalization as a means to organize transcription events during the onset of ZGA. Finally, we highlight the technical challenges faced and molecular strategies that can be used toward a better mechanistic understanding of ZGA.

2. The many-layers of transcriptional regulation in ZGA

Understanding transcriptional control first requires basic knowledge of transcription and chromatin organization. Here, we briefly revisit the molecular events that give rise to the initiation of transcription (Fig. 1) and then discuss transcriptional control in the context of the developing embryo.

2.1 Transcription basics

During transcription, the multi-subunit enzyme RNA polymerase II (Pol II) catalyzes the synthesis of messenger RNA (mRNA) from a DNA template. To initiate transcription, Pol II cannot act on its own to recognize and bind the promoters of target genes. It must rely on DNA-binding proteins that recognize DNA elements within the promoter and enhancer. The general transcription factors (TFs), also known as transcription initiation factors, TFIIA-H, are collectively part of the core transcriptional machinery that guide Pol II to the promoter. The binding of Pol II also relies on specific TFs. These TFs can recognize and bind to specific DNA motifs within the enhancer region, allowing them to regulate transcription by selectively activating the expression of target genes in distinct cells. Together, the initiation factors and sequence-specific TFs guide Pol II to the promoter. Along with the TATA-binding protein (TBP), the TFs and initiation factors assemble with Pol II near the transcription start site (TSS) to form the pre-initiation complex (PIC) (Cramer, 2019). Mediator is a multi-protein complex that helps bridge PIC to sequence-specific TFs bound to upstream regulatory elements, thereby stabilizing the formed transcription complexes at the promoters of target genes (Soutourina, 2018).

The RNA polymerase II subunit B1 (RPB1) is the largest subunit of Pol II. The C-terminal domain (CTD) of RPB1 is unique, highly conserved, and flexible. It contains heptapeptide repeats ($Tyr^1Ser^2Pro^3Thr^4Ser^5Pro^6Ser^7$ or $Y^1S^2P^3T^4S^5P^6S^7$) which serve as a binding platform for factors involved in transcription, mRNA processing, and chromatin remodeling (Harlen & Churchman, 2017). Residues in this repeat are subjected to post-translational modifications (PTMs), such as phosphorylation, which reflects the activity state of Pol II in the transcription cycle. Once Pol II

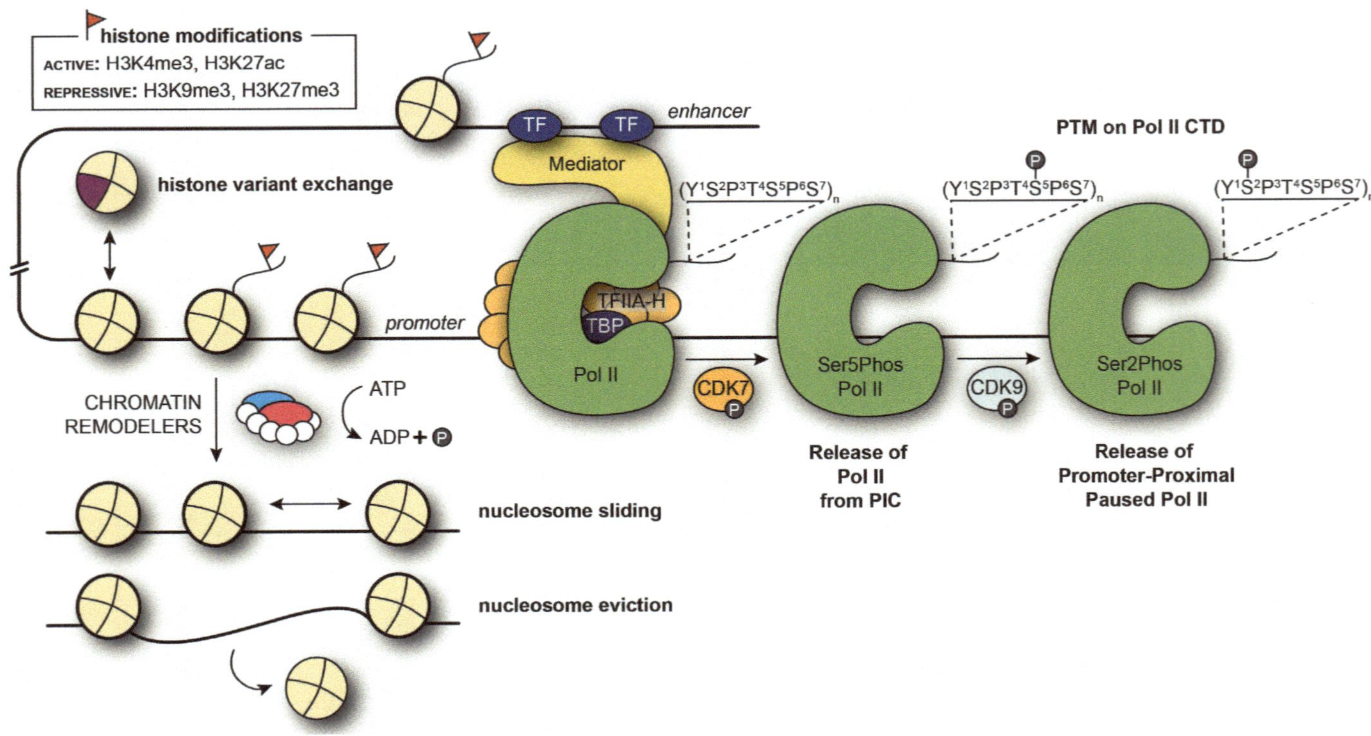

Fig. 1 The regulation of transcription initiation. The different layers of regulation in transcription initiation are indicated, with details discussed in the text. The number of heptapeptide repeats present at the CTD of Pol II varies between organisms and is denoted by n.

is assembled into the PIC, the Ser5 residue within the heptapeptide repeat is phosphorylated by the cyclin-dependent kinase 7 (CDK7), a component of TFIIH. Pol II is then released from the PIC at the promoter (Wong, Jin, & Struhl, 2014). After reaching ~30–50 base pairs (bp) downstream of the TSS, Pol II pauses and the CTD is yet again phosphorylated, this time on Ser2, by CDK9. Pol II then enters the elongation stage to efficiently proceed with transcription (Harlen & Churchman, 2017).

The process of transcription happens in the context of chromatin. In eukaryotes, the DNA template for transcription is packaged into chromatin within the nucleus. The nucleosome is the basic repeating unit of chromatin and is composed of ~147 bp of DNA usually wrapped around an octamer of the four core histones: a hetero-tetramer of H3/H4 and two H2A/H2B heterodimers. The organization of chromatin can impact DNA accessibility, and in turn, the network of molecular and physical interactions needed for transcription. Nucleosome composition, histone modifications, chromatin remodeling, and DNA methylation can directly influence the chromatin state and organization (Becker & Workman, 2013; Hyun, Jeon, Park, & Kim, 2017; Klemm, Shipony, & Greenleaf, 2019; Talbert & Henikoff, 2017). For example, nucleosomes can be comprised of histone variants that can impact the local chromatin structure. Unlike the canonical H2A/H2B and H3/H4 histones, histone variants are typically incorporated into nucleosomes throughout the cell cycle in a replication-independent manner. Similar to the CTD of Pol II, histones bear a flexible tail that can be post-translationally modified by enzymes known as writers. Specific histone modification marks have been associated with active and repressed transcription states. For instance, trimethylation of Histone H3 Lysine 4 (H3K4me3) and acetylation of Histone 3 Lysine 27 (H3K27ac) are considered to mark active chromatin, whereas H3K9me3 and H3K27me3 are associated with repressed chromatin (Skvortsova, Iovino, & Bogdanovic, 2018). Thus, histone variants and histone modifications can affect local chromatin structure directly, by affecting histone-DNA contacts (in *cis*), or indirectly, through reader proteins that recognize these PTMs (in *trans*).

Readers often recruit (or are themselves) chromatin remodelers. Chromatin remodeling complexes are key components in regulating chromatin accessibility. They regulate accessibility by modifying the nucleosome position or local chromatin structure through sliding or eviction of bound nucleosomes by ATP hydrolysis (Clapier, Iwasa, Cairns, & Peterson, 2017). The displacement of nucleosomes can render promoters more accessible for additional regulatory factors and transcription machinery to bind.

For example, the SWI/SNF complexes are among the most well-defined remodelers that act on chromatin by repositioning or evicting nucleosomes (Clapier et al., 2017). SWI/SNF subfamilies share core subunits and also contain subunits specific to each complex. In mammals, about 30 genes encode the ATPase subunits of chromatin remodeling complexes, many of which are genetically non-redundant (Ho & Crabtree, 2010). The evolution of so many subunits seems to point to combinatorial diversity accompanied by specialized roles to meet the demands of context-dependent functions, whether it is tissue specificity, developmental stage-specific gene expression, or different TF partners (Ho & Crabtree, 2010).

2.2 Transcription in the context of early embryo development

Transcriptional activation in embryos has long been thought to occur in two waves: a minor wave in which a subset of genes is transcribed during the early cleavage divisions, and a major wave that constitutes widespread transcriptional activation of zygotic genes coinciding with lengthening of the cell cycle. However, genome-wide analysis of transcriptome dynamics across early developmental stages in *Drosophila*, zebrafish, and mouse has revealed that the genome is gradually and continuously activated, reflecting ZGA as a continuum of genes that are transcriptionally activated rather than as distinct waves (Vastenhouw et al., 2019). Nevertheless, as the embryo gears up for transcription, it coordinates this activity with rapid cell divisions, as well as with other biological processes such as DNA replication, and the dynamics of maternally loaded RNA and protein (Fig. 2). Here we discuss the effects of these processes on the transcription state of the developing embryo.

Rapid cellular division cycles are a shared characteristic of early embryonic development in many species. This is followed by the slowing down of the cell cycle that roughly coincides with the onset of bulk activation of zygotic genes. The early embryo cycles between DNA synthesis (S) phase and mitosis (M) phase, with no gap phases (Foe & Alberts, 1983). The ensuing number of division cycles and the timeline for developmental control to be passed from mother to zygote differs across animals (Fig. 2) (Vastenhouw et al., 2019). For example, the *Drosophila* embryo initially undergoes 13 rapid nuclear division cycles. Three hours after fertilization, the embryo enters gastrulation. In contrast, the mouse embryo takes 1.5 days to go through a single cell division cycle and gastrulation occurs only 6 days after fertilization. The cell cycle length itself

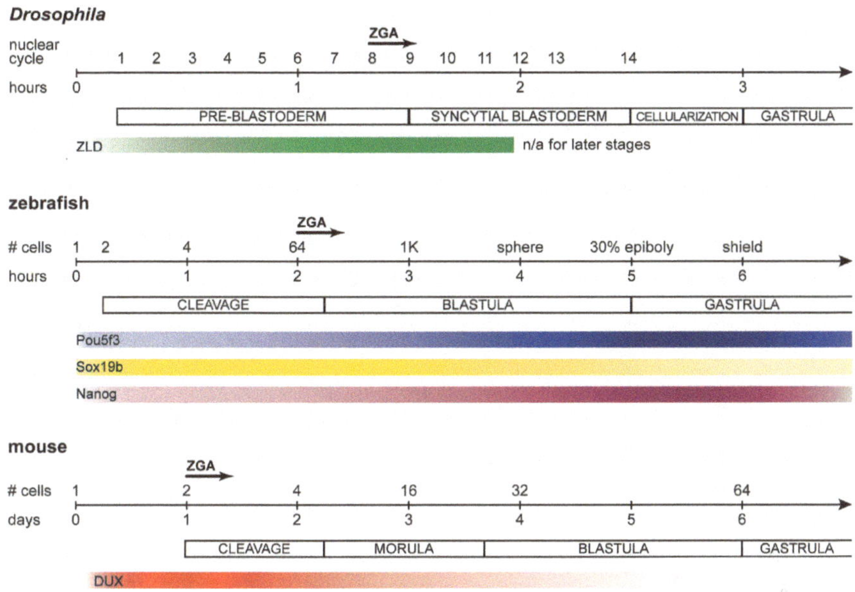

Fig. 2 Transcription in the context of embryo development. Timelines of early embryo development in *Drosophila*, zebrafish, and mouse. "→ ZGA" denotes the onset of the first reported transcription events, with more widespread transcription activity occurring at later stages. Expression profiles of key regulators are indicated, and are based on analyses by Western blot for ZLD (Nien et al., 2011) and Pou5f3 (Lippok, Song, & Driever, 2014), *in situ* hybridization and/or real time-PCR for Sox19b (Okuda et al., 2006) and Nanog (Schuff et al., 2012; Xu et al., 2012), and RNAseq for *Dux* (Hendrickson et al., 2017).

has been suggested to be too short to accommodate for transcription (Edgar & Schubiger, 1986; Kimelman, Kirschner, & Scherson, 1987). In *Xenopus*, it was indeed shown that lengthening the cell cycle or DNA replication time results in premature activation of zygotic genes (Collart, Allen, Bradshaw, Smith, & Zegerman, 2013; Kimelman et al., 1987). However, evidence from zebrafish and *Drosophila* indicate that altering the cell cycle does not affect the onset of transcription in these species (McCleland & O'Farrell, 2008; Zhang, Kothari, Mullins, & Lampson, 2014). In *Drosophila*, cell cycle lengthening actually depends on the onset of transcription, as sites of stalled DNA replication and components of the replication machinery have been shown to overlap with Pol II-occupied genomic regions at nuclear cycle 13 (Blythe & Wieschaus, 2015). Moreover, in both zebrafish and *Drosophila*, the cell cycle length might influence transcription output, as early transcribed genes tend to be short and lack introns (Heyn et al., 2014; Kwasnieski, Orr-Weaver, & Bartel, 2019). Furthermore, aborted transcripts and retained introns are frequently detected in the early *Drosophila* embryo, suggesting that abortion of nascent transcripts may serve as a means to regulate the expression of large genes (Kwasnieski et al., 2019; Rothe, Pehl, Taubert, & Jackle, 1992; Shermoen & O'Farrell, 1991). Thus, cell cycle length seems to influence both transcript quality and quantity, while its effect on the timing of zygotic transcription is less clear.

As many proteins are maternally loaded, they may be expected to play a role in maintaining the genome in a transcriptionally repressed state. Indeed, both sequence-specific repressors as well as more general transcriptional repressors have been identified (Almouzni & Wolffe, 1995; Amodeo, Jukam, Straight, & Skotheim, 2015; Brown & Wu, 1993; Chari, Wilky, Govindan, & Amodeo, 2019; Dunican, Ruzov, Hackett, & Meehan, 2008; Joseph et al., 2017; Pritchard & Schubiger, 1996; Ruzov et al., 2004, 2009). For example, in *Drosophila*, the maternally loaded DNA-binding protein, Tramtrack (TTK) regulates the transcription of the segmentation gene *fushi tarazu* (*ftz*) in a dose-response manner (Brown & Wu, 1993; Pritchard & Schubiger, 1996). Increasing the level of TTK protein in the embryo affects the transcription initiation of *ftz*, whereas reducing the amount of TTK leads to premature *ftz* transcription (Pritchard & Schubiger, 1996). For histones, the fertilized embryo contains massive histone stores in the form of maternal mRNAs and proteins (Adamson & Woodland, 1974; Woodland & Adamson, 1977). Histones can bind DNA with high affinity and can hinder both transcription and the binding of other factors (Campos & Reinberg, 2009). Together, these features would make histones, both free and DNA-bound, general repressors of transcription. Indeed, work in *Xenopus*, zebrafish and *Drosophila* has shown that histones repress transcription in the early embryo (Almouzni, Mechali, & Wolffe, 1990, 1991; Almouzni & Wolffe, 1995; Amodeo et al., 2015; Joseph et al., 2017; Prioleau, Huet, Sentenac, & Mechali, 1994).

Since the early embryo is mostly transcriptionally silent, maternal genes also need to provide the zygote with the mRNAs and proteins important for transcription activation. As a consequence, the components needed for transcription, such as the transcription machinery or a transcription activator, may initially not be present in sufficient amounts to support transcription. Indeed, in *Xenopus*, TBP translation (of maternally loaded RNA) is strongly up-regulated just before the onset of widespread genome activation, and precocious expression of TBP has been shown to facilitate transcription of reporter genes (Veenstra, Destree, & Wolffe, 1999).

Thus, in terms of transcription regulation, transcription can only be activated when enough transcription activator has accumulated over time. Similarly, the TFs required for ZGA in *Drosophila* and zebrafish are maternally loaded as mRNA and are highly translated when embryos approach the onset of zygotic transcription (Lee et al., 2013; Liang, Nien, et al., 2008; Okuda et al., 2006; Takeda, Matsuzaki, Oki, Miyagawa, & Amanuma, 1994; Xu et al., 2012).

Taking into consideration the repressors and activators, it is reasonable to view them together as two competing mechanisms in regulating transcription: first, the presence of maternally loaded repressors favors the absence of transcription. As transcription activators begin to be translated from maternal mRNAs, transcription can be initiated once sufficient levels are reached. Work in *Xenopus* and zebrafish provide supporting evidence for a dynamic competition for DNA binding between transcription repressors and activators. In the context of ZGA, the timing for onset of transcription is reached when the concentration of non-DNA-bound histones in the nucleus has dropped, and the transcription machinery is sufficiently expressed to outcompete histones for DNA binding and subsequent assembly of transcription complexes (Almouzni et al., 1990; Almouzni & Wolffe, 1995; Amodeo et al., 2015; Joseph et al., 2017; Palfy, Joseph, & Vastenhouw, 2017; Prioleau et al., 1994; Veenstra et al., 1999).

3. The molecular landscape of ZGA

Here, we highlight some of the key molecular players in transcription, many of which are encoded by maternal genes. We describe their molecular features, their impact on the biology of the developing embryo, followed by insights into their mode of action during ZGA.

3.1 RNA polymerase II

Subunits of the Pol II complex, along with many basal transcription machinery, such as TBP and general TFs, are maternally deposited as mRNA or protein into *Drosophila*, *Xenopus*, zebrafish, and mammalian embryos (Bogolyubova & Bogolyubov, 2014; Hart, Raha, Lawson, & Green, 2007; Owens et al., 2016; Tomancak et al., 2002, 2007; Veenstra et al., 1999). At least in zebrafish, Pol II is present as early as the 8-cell stage, and distinct phosphorylated forms of Pol II (Ser2Phos and Ser5Phos) can be detected at a time coinciding with the onset of ZGA (Vastenhouw et al., 2010). A similar coordinated pattern is observed in *Caenorhabditis elegans* and *Drosophila* embryos, with the presence of Ser2Phos-Pol II being restricted to somatic nuclei at a time coinciding with transcriptional activity of somatic nuclei (Edgar & Schubiger, 1986; Seydoux & Dunn, 1997; Seydoux & Fire, 1994; Seydoux et al., 1996). In *C. elegans*, the initial absence of transcription in the zygote and soma is thought to be due to the unavailability of the general transcription factor TAF-4, which is sequestered to the cytoplasm by the maternally-inherited OMA-1/2 zinc-finger proteins (Guven-Ozkan, Nishi, Robertson, & Lin, 2008). In the early germline blastomeres, where transcription starts later than in somatic cells, Ser2Phos-Pol II is absent in a manner that depends on the presence and activity of the germline-specific transcription repressor, PIE-1, suggesting that the lack of transcription in the early embryonic germ lineage is due to the differential distribution of phosphorylated Pol II and thus a restriction in Pol II activity (Seydoux & Dunn, 1997; Tenenhaus, Schubert, & Seydoux, 1998). In the early embryo of zebrafish, *Drosophila*, and *C. elegans*, no transcription takes place in spite of the fact that Pol II is present. In zebrafish, genome-wide profiling of Pol II on pre- and post-MZT embryos reveals that Pol II is not associated with the genome before ZGA, suggesting that Pol II becomes associated with genomic loci only during the onset of ZGA (Vastenhouw et al., 2010). In *Drosophila*, comprehensive analysis of Pol II occupancy during the MZT point to Pol II already bound to DNA but remains in a paused state (Chen et al., 2013; Zeitlinger et al., 2007). The pausing of Pol II is a widespread feature of metazoan transcriptional control, whereby the polymerase pauses at the promoter-proximal site 25–50 bp downstream of the TSS at the end of transcription initiation (Gaertner & Zeitlinger, 2014). Specifically, in *Drosophila* embryos at the mid-blastula transition (nuclear cycle 13 and 14), Pol II is paused at the promoters of thousands of genes near the TSS (Chen et al., 2013; Zeitlinger et al., 2007). This spatiotemporal binding dynamics of Pol II can be seen in even earlier embryos that are transitioning from nuclear cycle 12 to 13. During this window, a large number of previously unbound genes now become bound by Pol II, suggesting a massive recruitment of Pol II for subsequent establishment of transcriptional poising (Blythe & Wieschaus, 2015). These findings indicate that Pol II pausing could serve as an underlying mechanism for preparing the activation of later zygotic genes while preventing Pol II from going into active elongation.

3.2 Transcription factors in ZGA

In the last decade, work from many groups has identified key TFs that are responsible for ZGA. Several lines of evidence support their role as major

TF	Organism	Binding motif	DNA-binding domain
ZLD	*Drosophila*	CAGGTAG	
Pou5f3		ATGC[A/T]AAT	
Sox19b	*zebrafish*	CATTGTT	
Nanog		CAATTA[A/G]CA	
DUX	*mouse*	[A/T]GATTCAATC[A/T]	
DUX4	*human*	TAA[T/C][T/C][T/C]AATCA	

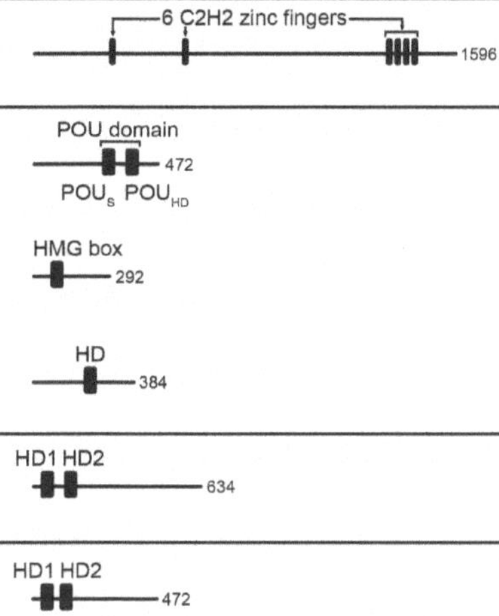

Fig. 3 The molecular features of the key TFs involved in ZGA in *Drosophila*, zebrafish, mouse, and human. The schematic representation of the DNA-binding domains of ZLD, Pou5f3, Sox19b, Nanog, and DUX are shown, along with their binding motifs found to be enriched in zygotic genes activated at ZGA. Note that there are related variants in most cases.

ZGA regulators. First, specific motifs corresponding to these TFs are strongly enriched in the promoters and enhancers of zygotic genes activated at ZGA (Fig. 3). Second, they are highly expressed prior to ZGA, suggesting their accumulation is crucial for setting the stage for the transcription engagement of zygotic genes (Fig. 2). And third, the depletion of these TFs in embryos results in early developmental defects and misregulation of zygotic genes (De Iaco et al., 2017; Lee et al., 2013; Liang, Nien, et al., 2008). Although these TFs are not homologs of one another and contain different DNA-binding domain types (Fig. 3), they share an analogous function in initiating transcriptional competency of the genome in the early embryo. Here, we discuss key activators of ZGA, followed by additional TFs that contribute to regulating gene expression in early embryos.

3.2.1 *Zelda in* Drosophila

The first key activator of ZGA was identified in *Drosophila*. The first clue that there might be a key TF dedicated for ZGA was the observation that many early transcribed genes are enriched for a heptamer motif (CAGGTAG) and related sequences (referred to as TAGteam elements) in their enhancers (ten Bosch, Benavides, & Cline, 2006). Using yeast one-hybrid, the Zinc-finger early *Drosophila* activator (Zelda, ZLD) was identified as the TF that binds to these TAGteam elements (Liang, Nien, et al., 2008). ZLD is a 1596-amino acid protein which contains six C2H2 (Cys-Cys-His-His motif) zinc fingers (Fig. 3). Four of these are clustered at the C-terminus and have been shown to be required for DNA-binding activity to TAGteam elements (Hamm, Bondra, & Harrison, 2015; Liang, Nien, et al., 2008). In addition, it contains a transcriptional activation domain that is characterized by a low complexity region of highly repetitive amino acid sequences (Hamm et al., 2015). *Zld* is maternally deposited in the embryo as mRNA (Liang, Nien, et al., 2008). ZLD is detectable in 1-h-old embryos and increases in the next hour of development (nuclear cycle 8 onward), coinciding with the expression of early genes (Nien et al., 2011) (Fig. 2). The protein needs to be continuously present to drive the expression of early and late zygotic genes (McDaniel et al., 2019). *Drosophila* embryos lacking *zld* show defects in cellular blastoderm formation and fail to activate genes essential for cellularization, sex determination, and patterning (Liang, Nien, et al., 2008; Nien et al., 2011).

What is ZLD's mode of action for potentiating the transcription of hundreds of early zygotic genes? One proposed mechanism is that ZLD may function as a so-called "pioneer" TF (Harrison, Li, Kaplan, Botchan, & Eisen, 2011). Pioneer TFs were originally defined as a unique class of TFs that can bind nucleosomal DNA, modify the local chromatin to an "open" conformation, and allow for the subsequent recruitment and binding of additional TFs (Zaret & Carroll, 2011). In agreement with this definition, ZLD was recently shown to exhibit sequence-specific binding to nucleosomal DNA (McDaniel et al., 2019). Furthermore, using Chromatin ImmunoPrecipitation coupled to high-throughput sequencing (ChIP-seq), it was shown that ZLD is already bound to thousands of CAGGTAG sites upstream of genes that are not yet actively transcribed at nuclear cycle 8. In addition, there is strong correlation between early ZLD binding, highly accessible chromatin, and transcriptional activity of these genes by nuclear cycle 14 (Harrison et al., 2011). Finally, ZLD has been reported to facilitate the binding of TF such as Bicoid and Dorsal to their regulatory sequences (Foo et al., 2014; Kanodia et al., 2012; Sun et al., 2015; Xu et al., 2014; Yanez-Cuna, Dinh, Kvon, Shlyueva, & Stark, 2012). For Dorsal, regions normally bound by the protein become highly occupied with nucleosomes in the absence of *zld*, suggesting that ZLD can overcome high nucleosome barrier at its target sites and assists the binding of other TFs (Schulz et al., 2015;

Sun et al., 2015). Together, this supports the notion that ZLD, as a pioneer factor, can shape the chromatin landscape during MZT and prime genes for future activation: the binding of ZLD establishes and maintains chromatin in an accessible state, likely by lowering the high nucleosome barrier, which in turn facilitates the binding of additional TFs to their motifs (Schulz et al., 2015; Sun et al., 2015).

One important question in the context of embryonic development is how ZLD deals with the establishment and maintenance of chromatin accessibility during the rapid nuclear divisions? During mitosis, ZLD is not detected on chromosomes. Instead, ZLD is distributed in the cytoplasm until the end of mitosis when it rapidly re-enters the nucleus and associates with chromatin (Dufourt et al., 2018). A fine-grained time series of the chromatin landscape over multiple cell cycles revealed ZLD's essential role, along with the TF GAGA factor (GAF, see Section 3.2.4), in establishing chromatin accessibility (Blythe & Wieschaus, 2016; Foo et al., 2014; Sun et al., 2015). These findings suggest a mechanism whereby ZLD, with its pioneering nucleosome-binding capacity, can efficiently re-establish and maintain chromatin accessible regions following DNA replication by binding to its target sites distributed across the genome (Blythe & Wieschaus, 2016). Together with other factors, such as GAF and the chromatin remodelers that it recruits, ZLD can exhibit such dynamic behavior and drive genomic reprogramming in the early embryo.

3.2.2 Pou5f3, Sox19b, Nanog in zebrafish

The discovery of ZLD as the master regulator of ZGA in *Drosophila* sparked the search for a TF with an analogous function in vertebrate ZGA. However, there is no ZLD homolog found outside of the insect clade (Ribeiro et al., 2017). A combination of transcriptomics, ribosome profiling data, and loss-of-function analyses identified Pou5f3, Sox19b, and Nanog, the three homologs of the core TFs in the maintenance of stem cell pluripotency (OCT4, SOX2, and NANOG, respectively) as the zebrafish ZGA master regulators (Lee et al., 2013; Leichsenring, Maes, Mossner, Driever, & Onichtchouk, 2013; Onichtchouk et al., 2010). These three TFs are maternally deposited as mRNA (Okuda et al., 2006; Schuff et al., 2012; Takeda et al., 1994; Xu et al., 2012) and are the most highly translated sequence-specific TFs by the 64-cell stage (Lee et al., 2013). Here, we first describe the three factors individually, then discuss their function in ZGA.

POU class 5 homeobox 3 (*Pou5f3*) is part of the POU family of TFs, which are marked by a bipartite POU domain consisting of an N-terminal

POU-specific region and a C-terminal POU-type homeodomain joined by a flexible linker (Herr et al., 1988). The N-terminal POU-specific (POU_S) region binds the 5′ half of the "octamer (oct) motif" ATGCAAAT and related variants, while the C-terminal POU homeodomain (POU_{HD}) binds the 3′ half (Klemm, Rould, Aurora, Herr, & Pabo, 1994) (Fig. 3). Work on the mammalian proteins (OCT TFs) has revealed that depending on the motif sequence, POU proteins exhibit different DNA-binding or dimerization properties and can assemble into higher-order structures with distinct interacting partners, such as SOX2 and chromatin remodeler components like Brahma-related gene 1 (BRG1), also known as SWI/SNF-related, matrix associated, actin-dependent regulator of chromatin, subfamily a, member 4 (SMARCA4), and chromodomain helicase DNA-binding protein 4 (CHD4) (Nieto et al., 2007; Remenyi et al., 2001; Tantin, 2013). Such molecular flexibility for different binding partners could also be explained by the linker (Tantin, 2013). The linker has been reported to differ in residues and in lengths among OCT TFs and among various species, suggesting that OCT TFs may have structurally evolved to cooperate with a variety of factors through its linker (Esch et al., 2013; Tantin, 2013). In zebrafish, the Pou5f3 linker only shares several conserved residues with that found in mammals, and a chimeric mouse OCT4 protein with the zebrafish Pou5f3 linker fails to show reprogramming activity when expressed in mouse embryonic fibroblasts (Esch et al., 2013). Despite the structural differences with its mammalian counterpart, Pou5f3 plays a critical function in zebrafish. Embryos lacking maternal and zygotic Pou5f3 are developmentally delayed, exhibit extreme patterning defects, and eventually arrest during gastrulation (Lunde, Belting, & Driever, 2004; Reim & Brand, 2006).

SRY-box transcription factor 19b (Sox19b) belongs to the SoxB1 family of TFs (Okuda et al., 2006; Vriz & Lovell-Badge, 1995). All four SoxB1 family members are expressed during zebrafish embryo development with either localized or ubiquitous expression, with *sox19b* being the only member that is maternally loaded as mRNA (Okuda et al., 2006). Sox19b contains a High Mobility Group (HMG) domain that mediates DNA binding to the sequence CATTGTT or related motifs (Fig. 3), and a transcriptional activation domain at the C-terminus (Kamachi & Kondoh, 2013; Leichsenring et al., 2013). Structural studies based on the mouse SOX2 and other HMG domain-containing proteins show that binding of the HMG domain to DNA induces widening of the minor groove, leading to subsequent bending toward the major groove (Remenyi et al., 2003). SOX proteins have a higher affinity for DNA binding when present with other DNA-binding partners (Kamachi, Uchikawa, & Kondoh, 2000). For example, in the presence of POU domain proteins SOX2 seems to fit tightly into the minor groove (Remenyi et al., 2003) while in the absence of an interacting partner, the C-terminus of the SOX2-HMG domain is unstructured and has a lower affinity for DNA. These findings indicate that the HMG box domain likely gives SOX proteins the versatility to bind and bend DNA and to facilitate recruitment of various interacting protein partners (Remenyi et al., 2003). Although zebrafish SoxB1 TFs have not been biochemically characterized, loss-of-function analysis has indicated their role in early embryogenesis: embryos depleted of *sox19b* are viable and appear normal, likely because the other SoxB1 genes can compensate for *sox19b*, while quadruple SoxB1 gene knockdown results in severe developmental defects, including gastrulation and dorsoventral patterning abnormalities, as well as early embryonic lethality in some instances (Okuda, Ogura, Kondoh, & Kamachi, 2010).

Nanog is a homeodomain-containing TF that recognizes and binds the CATTAACA motif and related sequences through its homeodomain (Fig. 3) (Chambers et al., 2003; Mitsui et al., 2003; Xu et al., 2012). In addition to the homeodomain, mouse NANOG also contains a dimerization domain that consists of a tryptophan repeat (WR) at every fifth residue (Mullin et al., 2017). Homodimers of NANOG make up a major fraction of NANOG protein complexes in mouse ESCs, and the dimerization domain is required to convey the self-renewal properties in stem cells (Wang, Levasseur, & Orkin, 2008). In zebrafish, no WR sequence can be found and fusion of the mouse dimerization domain to zebrafish Nanog does not confer self-renewal properties to the zebrafish protein (Xu et al., 2012). Although the zebrafish ortholog seems to lack the self-renewal function, the protein retains an essential role during embryonic development. This role is evident from the defects in the extraembryonic yolk syncytial layer (YSL), increased apoptosis at the end of gastrulation, and arrested development of *nanog* mutants (Gagnon, Obbad, & Schier, 2018; Schuff et al., 2012; Veil et al., 2018; Xu et al., 2012).

As described above, individual mutants of *pou5f3*, *sox19b* (in the absence of all of its family members), and *nanog* display severe developmental defects. A combined loss of maternal Pou5f3, Sox19b, and Nanog leads to arrested development before gastrulation, as well as failure to activate their target genes (Lee et al., 2013; Leichsenring et al., 2013). These findings suggest potential cooperation between the three TFs in regulating zygotic genes important for further development (Lee et al., 2013; Leichsenring et al., 2013). While the molecular basis for potential cooperativity in gene regulation is currently unknown, structural studies of SOX proteins indicate that SOX can partner with POU domain-containing proteins, thereby imparting higher affinity DNA binding (Kamachi et al., 2000). For Nanog, recent studies indicate that the protein is not strictly needed in embryonic cells but rather is primarily required for proper development of the YSL (Gagnon et al., 2018). The reduced expression of Nanog-dependent zygotic genes and subsequent embryonic defects could be due to inactivation or failure to activate the YSL gene expression program (Gagnon et al., 2018; Veil et al., 2018). Here, Nanog's observed effect on transcription is hypothesized to be due to its cooperativity with other regulators, such as Pou5f3 and Sox19b (Lee et al., 2013; Perez-Camps et al., 2016).

Mechanistically, how do Pou5f3, Sox19b, and Nanog activate transcription of zygotic genes in the early embryo? Similar to ZLD in *Drosophila* and to its mammalian homologs in pluripotency control (Soufi et al., 2015; Takahashi & Yamanaka, 2006; Yu et al., 2007), Pou5f3, Sox19b, and Nanog may function as pioneer factors. The zebrafish genome harbors thousands of Pou5f3, Sox19b, and Nanog sites within putative regulatory elements, with reduced expression of many associated genes when these TFs are lost (Gagnon et al., 2018; Lee et al., 2013; Leichsenring et al., 2013; Xu et al., 2012). A closer look into these TF-bound genomic regions revealed that they change their state from being nucleosome-occupied before ZGA to nucleosome-depleted after ZGA, a change that depends on TF binding (Palfy, Schulze, Valen, & Vastenhouw, 2020; Veil, Yampolsky, Gruning, & Onichtchouk, 2019). Nucleosome-depletion during ZGA is most noticeable in sequences that are bound by Pou5f3, Sox19b, and Nanog, whereas single or dual occupancy sites were less affected (Palfy et al., 2020; Veil et al., 2019). Co-occupancy by multiple TFs has been reported in the context of ZGA, for example in the case of ZLD (Harrison et al., 2011; Nien et al., 2011). This is also a recurring feature of pluripotency programming, in which composite OCT-SOX binding motifs are commonly found within promoters and enhancers of many OCT4- and SOX2-activated genes in ESCs (Boyer et al., 2005). Importantly, the establishment of accessible regulatory regions by Pou5f3, Sox19b, and Nanog primes genes for transcription activation (Palfy et al., 2020; Veil et al., 2019). The relationship between chromatin accessibility and forthcoming transcription echoes closely the changing chromatin landscape and transcriptional shift observed in other organisms. In *Drosophila*, *Xenopus*, mouse and human embryos, chromatin accessibility profiling during MZT also revealed the progressive establishment of a permissive chromatin state at regulatory elements accompanying ZGA (Blythe & Wieschaus, 2016; Gentsch, Owens, & Smith, 2019; Lu et al., 2016; Wu et al., 2018).

If Pou5f3, Sox19b, and Nanog bind their targets and poise genes for transcription activation, chromatin remodelers could be recruited to facilitate further opening of regulatory regions, which in turn would promote recruitment of additional TFs for widespread activation of zygotic genes (Liu, Wang, Hu, Wang, & Zhang, 2018; Schulz et al., 2015; Sun et al., 2015; Veil et al., 2019). Evidence from mouse ESCs indicate an interdependence between OCT4 and the SWI/SNF component, BRG1. BRG1 is enriched at sites where OCT4 facilitates opening of chromatin (King & Klose, 2017, further discussed in Section 3.3.3). Furthermore, recent proteomic studies on these pluripotency factors in mouse ESCs identified a network of interacting proteins that include chromatin-modifying complexes (SWI/SNF and BRG1 subunit), histone chaperones and modifiers (FACT, LSD1, MLL complex), transcription machinery, transcription repression complexes (Polycomb, NuRD), together with cell cycle and DNA replication machinery (Ding, Xu, Faiola, Ma'ayan, & Wang, 2012; Gagliardi et al., 2013; Liang, Wan, et al., 2008; Pardo et al., 2010; van den Berg et al., 2010). These proteomic studies provide a framework for the range of interacting partners involved in pluripotency programming. It remains to be determined whether similar networks and interactions are in place in transcriptional programming of the embryo. We are only starting to biochemically characterize the TFs with a major role in ZGA. Far more work is needed on Pou5f3, Nanog, and Sox19b to mechanistically understand how they bind to nucleosomes and interact with downstream transcription regulators in controlling ZGA, and to determine how stable regulatory complexes are maintained to ensure precise timing of activation of different zygotic genes during the MZT.

3.2.3 DUX in mammals

The double homeobox (*Dux*) genes were recently discovered to be capable of widespread transcriptional activation in mouse and human. *Dux* (in mouse) and *DUX4* (in human) genes encode a double homeodomain TF specific to placental mammals (Clapp et al., 2007; Leidenroth et al., 2012). Unlike the previously described TFs, DUX/DUX4 is not maternally loaded and is instead expressed in the early embryo before widespread ZGA (De Iaco et al., 2017; Eckersley-Maslin et al., 2019; Hendrickson et al., 2017). The *Dux* genes are found as multi-copy loci in telomeric and pericentromeric regions, which are characteristically transcriptionally silent (Clapp et al., 2007; Leidenroth et al., 2012). How then are the *Dux* genes activated from a highly repressed region of chromatin at the 1-cell stage? In a recent screen for positive regulators of transcription of ZGA genes in mouse ESCs, developmental pluripotency-associated 2 and 4 (DPPA2 and DPPA4, both maternally loaded) were identified as activators of ZGA by directly binding to the *Dux* promoter and gene body to drive mouse *Dux* expression (De Iaco, Coudray, Duc, & Trono, 2019; Eckersley-Maslin et al., 2019). In turn, the DUX TF activates hundreds of genes and retroviral elements that define the cleavage-specific transcriptional program in humans and in mice (Fig. 3) (De Iaco et al., 2017; Hendrickson et al., 2017). These target genes are enriched for the DUX-binding motif: [A/T]GATTCAATC[A/T] for DUX in mouse and TAA[T/C][T/C][T/C]AATCA for DUX4 in humans (Geng et al., 2012; Hendrickson et al., 2017). In loss-of-function analysis, observed phenotypes include developmental defects in both pre- and post-implantation embryos and failure of embryos to reach the morula stage (De Iaco et al., 2017; De Iaco, Verp, Offner, Grun, & Trono, 2020). These findings indicate that *Dux* is critical for embryogenesis, although not absolutely required, as prolonged monitoring of *Dux* homozygotes revealed some viable pups that escape these developmental and fatal defects (De Iaco, Verp, et al., 2020).

The mechanism by which DUX activates the transcription of hundreds of genes in the early embryo remains unknown. In a study conducted with human myoblast cells, P300/CBP was identified as a DUX4-interacting partner (Tang et al., 2013). Furthermore, enrichment of active H3K27ac marks were detected flanking the regions bound by DUX4 (Tang et al., 2013). These findings suggest a mechanism wherein DUX4 recruits P300/CBP to remodel and open up chromatin. Whether this interaction is established in the early mammalian embryo for ZGA remains to be investigated. Although the role of DUX as a major ZGA activator has only been recently discovered, the stage is set for further investigation of DUX function. Does DUX exhibit pioneer TF features and function in a manner analogous to its *Drosophila* and zebrafish counterparts? What is the relationship between chromatin organization and transcriptional activity in mammals? What are the interacting partners of DUX?

3.2.4 *Other notable TFs: GAF, NF-YA, and STELLA/DPPA3*

In flies, zebrafish, mouse, and human, ZLD, Pou5f3, Sox19b, Nanog, and DUX have been identified as the master regulators of ZGA. Recent studies have uncovered other factors that can also influence ZGA, thereby supporting the idea that gradual ZGA involves the combined action of multiple transcriptional regulators. We highlight several notable maternally loaded transcription factors that could partake in the activation of zygotic genes in the same or a distinct pathway as the master regulators.

3.2.4.1 Transcription intermediary factor 1-alpha (TIF1α)

TIF1α was among the earliest factors identified in transcriptional regulation of early zygotic genes in the mouse embryo (Torres-Padilla & Zernicka-Goetz, 2006). TIF1α, also referred to as TRIM24, is maternally loaded and contains a tripartite motif (TRIM) made up of a coiled-coiled, a RING, and a B-box domain, as well as a bromodomain at the C-terminus. At the onset of early zygotic gene transcription, TIF1α translocates from the cytoplasm to the pronuclei. In the pronuclei, TIF1α colocalizes to discrete foci with initiating Ser5Phos-Pol II and the catalytic subunits of two chromatin remodeling complexes: BRG1 (SMARCA4) of the SWI/SNF complex and SNF2H (SMARCA5) of the ISWI complex (Torres-Padilla & Zernicka-Goetz, 2006). When TIF1α is inhibited, the localization of Pol II, BRG1, and SNF2H in foci is impaired, embryos arrest between the 2- and 4-cell stage, and TIF1α target genes are misregulated. Some of these genes also depend on SNF2H for expression (Torres-Padilla & Zernicka-Goetz, 2006). These findings suggest that the maternal factor TIF1α is an essential regulator of ZGA in the early mouse embryo, and that activation of its target genes depends on the function of the ISWI chromatin remodeler subunit, SNF2H. Whether TIF1α and DUX use similar or distinct mechanisms and interacting partners to initiate transcription of zygotic genes remains to be elucidated.

3.2.4.2 GAGA factor

Although ZLD is essential for activating the transcription of hundreds of genes during the *Drosophila* MZT, chromatin accessibility of many of these genes does not entirely depend on ZLD. This observation has led to suggestions that other TFs may also play a role in ZGA (Raff, Kellum, & Alberts, 1994; Schulz et al., 2015). Motif analysis in ZLD target genes revealed enriched GAGA binding sites occupied by GAGA factor/ Trithorax-like (GAF/TRL) (Moshe & Kaplan, 2017; Schulz et al., 2015;

Sun et al., 2015). GAF is thought to partner with the nucleosome remodeler multi-protein complex, NURF, to activate transcription (Tsukiyama, Becker, & Wu, 1994; Tsukiyama & Wu, 1995). GAF has also been shown to be associated with paused Pol II at promoter-proximal regions in *Drosophila* S2 cells (Lee et al., 2008). Taken together, these findings suggest GAF may also have a pioneering role in early *Drosophila* development, by controlling chromatin accessibility, nucleosome organization, and transcriptional pausing of Pol II upstream of zygotic genes. Further characterization of GAF will tease out how it participates in these events, as well as its cooperativity with ZLD.

3.2.4.3 Nuclear transcription factor Y subunit alpha (NF-YA)

Recently, NF-YA has been identified as another TF with a role in ZGA. NF-YA is the sequence-specific DNA-binding subunit of the conserved heterotrimeric NF-Y complex that also contains NF-YB and NF-YC (Maity & De Crombrugghe, 1996). The binding motif of NF-YA, CCATT, is present in ~30% of eukaryotic promoters (Dolfini, Gatta, & Mantovani, 2012). Structurally, NF-YA forms a histone-like structure when binding to DNA, allowing it to bend DNA, thereby creating an open configuration that likely facilitates binding of other TFs (Nardini et al., 2013; Oldfield et al., 2014). In mammals and zebrafish, NF-YA is maternally loaded (Ladam et al., 2018; Lu et al., 2016). Knockdown of NF-YA in mouse results in embryos arresting at the morula stage and down-regulation of ~15% of ZGA genes (Lu et al., 2016). The mechanism by which NF-YA activates zygotic genes remains unclear. Genome-wide analysis in mouse and zebrafish revealed that the CCATT binding motif is highly enriched in accessible chromatin regions (Ladam et al., 2018; Lu et al., 2016; Palfy et al., 2020). In mouse ESCs, nucleosome-depleted regions at active enhancers are not only occupied by NF-YA, but also by OCT4, SOX2, and NANOG (Oldfield et al., 2014). These findings provide additional insight into co-occupancy and cooperativity between multiple TFs in potentiating transcriptional activation: NF-YA could act as a pioneer TF and cooperate with DUX (in mouse) and Pou5f3, Sox19b, and Nanog (in zebrafish) in priming genes for transcription in ZGA (Lu et al., 2016; Palfy et al., 2020; Veil et al., 2019). Additionally, NF-YA could function in maintaining the integrity of nucleosome-depleted regions and ensure correct positioning of PIC at promoters, as was reported in mouse ESCs (Oldfield et al., 2019). Further studies are needed to dissect the coordinated function of NF-YA and other TFs in promoting and maintaining an accessible chromatin structure and regulating transcription initiation.

3.2.4.4 STELLA/DPPA3

In mouse, STELLA/DPPA3 has been identified as a maternal factor important for regulating a subset of transposable elements during MZT (Huang et al., 2017). *Stella*-deficient mice exhibit developmental defects after the 2-cell stage, with only a few embryos reaching the blastocyst stage (Bortvin, Goodheart, Liao, & Page, 2004). At the molecular level, the long terminal repeat-retrotransposons that are normally activated in mouse ZGA, such as the mouse endogenous retrovirus with leucine tRNA primer (MERVL), are significantly impaired in the absence of *Stella*. STELLA contains a SAP-like domain and a splicing factor motif-like structure, and can bind both DNA and RNA *in vitro*. These features highlight STELLA as a candidate co-activator that functions at the DNA-RNA interface, in chromatin reorganization, and RNA processing (Nakamura et al., 2007; Payer et al., 2003). Although the mechanism for endogenous retrovirus regulation by STELLA remains unknown, these findings emphasize the contribution of a maternally-encoded protein in the regulation of transposable elements during embryogenesis.

3.3 Chromatin

In addition to Pol II and TFs, the global and local organization of chromatin also influences transcription as it impacts the accessibility of DNA. In the context of ZGA, genome-wide analyses of the chromatin landscape have revealed correlations between gene expression and changes in chromatin organization, including nucleosome positioning, histone variants and histone modifications, DNA methylation, chromatin remodelers, and architectural proteins (Becker & Workman, 2013; Hyun et al., 2017; Klemm et al., 2019; Talbert & Henikoff, 2017). Thus, there is reason to believe that histones, epigenetic regulators, and architectural proteins play a role in ZGA. Here, we focus on those that have been more closely studied in the context of ZGA. We discuss the role of core histone levels, several histone variants and modifications, as well as chromatin remodelers, and the architectural proteins CTCF and cohesin. A recent review provides further details into the epigenetic landscape during ZGA (Eckersley-Maslin, Alda-Catalinas, & Reik, 2018).

3.3.1 Histones

DNA is usually wrapped around an octamer of the four core histones (two each of H2A, H2B, H3, and H4) that make up a nucleosome. DNA that enters and exits the nucleosome interacts with linker H1 histones. Core histones contain a basic unstructured N-terminal tail domain and a histone-fold C-terminal domain (Wolffe & Pruss, 1996), while linker histones contain a globular domain instead of a histone-fold domain and both N- and C-terminal tails are unstructured (Khochbin, 2001). The core histones and H1 are encoded by histone genes present in multiple copies that lack introns. Instead of a poly(A) tail, transcripts synthesized from histone genes bear a 3′ stem loop that regulates mRNA processing and promotes translation efficiency (Dominski & Marzluff, 1999). Unlike core histones, histone variants are encoded by single-copy genes that contain introns, and their mRNAs are polyadenylated (Talbert & Henikoff, 2017).

3.3.1.1 Histone levels

Generally, histones are produced and deposited in a replication-coupled manner so as to meet the demand of proper packaging of the newly replicated DNA (Marzluff, 2005). In contrast, massive histone stores are supplied to the fertilized embryo in the form of maternal mRNAs and proteins (Adamson & Woodland, 1974; Woodland & Adamson, 1977). As histones can bind DNA with high affinity and are present in excess in the embryo, histones have been long thought to be repressors of transcription. In 1982, Newport and Kirschner showed that in *Xenopus* embryos, transcriptional repression of an injected reporter gene could be alleviated by increasing the amount of DNA. They hypothesized that transcriptional activation was due to titration of a DNA-binding transcriptional repressor, which would be depleted through binding to the added DNA (Newport & Kirschner, 1982). Histones would fit this transcriptional repressor profile. Work over the years has shown indeed that excessive histone levels can regulate the transcription of reporter genes in *Xenopus*, through a dynamic competition between chromatin assembly and transcription complexes (Almouzni et al., 1990, 1991; Almouzni & Wolffe, 1995; Amodeo et al., 2015; Prioleau et al., 1994). Extending on this competition model, recent work in zebrafish and *Drosophila* showed that addition of more histones in the embryo delays the onset of transcription of zygotic genes, while depleting histone levels or inducing earlier expression of key TFs leads to premature expression of zygotic genes (Chari et al., 2019; Joseph et al., 2017). Furthermore, analysis of endogenous Pou5f3 target genes and heterologous reporters revealed that TF binding is sensitive to histone levels present in the embryo (Joseph et al., 2017). Taken together, these findings support the competition model in which histones and transcription activators compete for DNA binding to regulate transcriptional activation of zygotic genes.

3.3.1.2 Linker histones

Linker histones promote DNA folding and maintain higher-order chromatin structure (Harshman, Young, Parthun, & Freitas, 2013). A defining feature of linker histone H1 is that its subtypes are expressed in a stage and cell-type specific manner. In *Drosophila*, *Xenopus*, and zebrafish, embryonic H1 is replaced by somatic H1 subtypes over the course of embryo development (Khochbin, 2001). Specifically, in *Drosophila*, embryo-specific BigH1 is abundantly expressed up until cellularization and is then replaced by H1, coinciding with widespread ZGA (Perez-Montero, Carbonell, Moran, Vaquero, & Azorin, 2013). Loss-of-function mutation in BigH1 results in embryonic lethality, along with signs of premature ZGA in both soma and primordial germ cells, including increased levels of active Pol II and elevated zygotic transcripts. Although it is unclear how BigH1 and H1 affect transcription, studies in *Xenopus* might provide some molecular insight into this exchange.

In *Xenopus*, the embryonic linker histone (H1M/B4) is maternally loaded while somatic H1 subtypes (H1A-C) only appear at the beginning of gastrulation, replacing the embryonic form (Dimitrov, Almouzni, Dasso, & Wolffe, 1993; Dworkin-Rastl, Kandolf, & Smith, 1994; Hock, Moorman, Fischer, & Scheer, 1993). It was suggested that as H1M is intrinsically less basic compared to somatic H1 (based on amino acid composition), the presence of H1M in the early embryo could maintain loose H1-DNA interactions, thereby allowing chromatin to exist in a less condensed state (Dimitrov et al., 1993). *In vitro* assays have revealed that the embryonic subtype indeed renders chromatin more accessible than the somatic subtype, further supporting the notion that embryonic H1 allows for permissive changes to the chromatin template (Saeki et al., 2005). Maintaining chromatin in a fluid state could be achieved together with the help of architectural High Mobility Group (HMG) proteins, which have been reported to be enriched with the embryonic H1 subtype on chromosomes in both *Drosophila* and *Xenopus* (Ner & Travers, 1994; Ura, Nightingale, & Wolffe, 1996). In the context of ZGA, this would suggest that the embryonic linker histone does not play a role in repressing transcription, but rather renders it permissive for remodeling.

3.3.1.3 Histone variants

Histone variants are encoded by single-copy genes that contain introns, and their mRNAs are polyadenylated (Talbert & Henikoff, 2017). Unlike core histones, histone variants are typically expressed and incorporated into nucleosomes throughout the cell cycle in a replication-independent manner (Talbert & Henikoff, 2017). In *Drosophila*, incorporation of H2A variant H2Av into chromatin coincides with the onset of ZGA, although its deposition is not limited to sites of active transcription (Clarkson & Saint, 1999; Leach et al., 2000). Additionally, the amount of H3 bound to chromatin is reduced with each cell cycle in the early *Drosophila* embryo and this decrease is correlated with a gradual increase of its variant, H3.3, on chromatin (Shindo & Amodeo, 2019). Taken together, these correlative studies implicate histone variants in ZGA, yet functional analysis is needed to understand their relevance for ZGA. Specifically, how are histone-DNA contacts affected and how does this influence the accessibility for TF binding? Do histone variants, for example, call upon specific chromatin remodeling complexes and subunits, such as specialized SWI/SNF complexes, which have been shown to have developmentally distinct functions (Ho & Crabtree, 2010)?

3.3.2 Histone modifications

Histones are subjected to a range of PTMs on their flexible tails. These modifications can either affect chromatin accessibility directly, or indirectly through the effect of readers. Acetylation, for example, alters the charge of histones, which, in turn, modifies histone-DNA contacts. This can result in changes in local chromatin structure and hence chromatin accessibility for TF binding (Klemm et al., 2019). In contrast, methylation is recognized by reader proteins that change chromatin structure. Specific modifications have been linked with active or repressed transcription states. For example, H3K4me3 and H3K27ac are associated with active chromatin, while H3K9me3 and H3K27me3 are typically associated with repressed chromatin (Skvortsova et al., 2018). In the context of ZGA, ChIP-seq data have revealed both distinct histone modification patterns and changes in their levels that contribute to shaping global and local chromatin in embryos transitioning from pre-ZGA to ZGA (Akkers et al., 2009; Li, Harrison, Villalta, Kaplan, & Eisen, 2014; Lindeman et al., 2011; Liu et al., 2016;

Vastenhouw et al., 2010). Below, we discuss our current molecular understanding of the epigenetic marks H3K27ac, H3K4me3, and H3K27me3 in ZGA.

3.3.2.1 H3K27ac

In *Drosophila*, zebrafish, and mouse embryos, H3K27ac has been associated with genes that are activated during ZGA. For *Drosophila*, histone acetylation marks are enriched at the TSS of hundreds of gene loci before ZGA (Li et al., 2014). Many of these are also enriched for ZLD's TAGteam elements suggesting that ZLD binding is strongly coupled to early histone acetylation and thus to shaping the early chromatin landscape (Li et al., 2014). In zebrafish embryos, the pre-ZGA chromatin landscape also displays widespread deposition of H3K27ac at promoters, which is strongly correlated with transcription of the corresponding zygotic genes at ZGA (Zhang et al., 2018). Finally, the transition from oocytes to 2-cell mouse embryo is also accompanied by an increase in H3K27ac (Dahl et al., 2016). Thus, the establishment of H3K27ac at zygotic genes prior to their activation is a common feature across different organisms.

What is the function of the H3K27ac and how does it relate to transcription activation? In a recent study conducted in zebrafish, the acetyl-lysine writer (Brd4) and reader (p300/Cbp) were identified as regulators of genome activation. When Brd4 or p300/Cbp are precociously expressed, hundreds of zygotic genes are prematurely transcribed and up-regulated, whereas chemical inhibition of Brd4 or p300/Cbp results in a reduction of zygotic transcription and arrested gastrulation (Chan et al., 2019). In agreement with this, another study showed that maternal depletion of the most abundantly expressed histone acetyltransferases, including p300, significantly impaired both transcriptional activation of zygotic genes and maternal mRNA clearance (Zhang et al., 2018). In zebrafish, both Brd4 and p300/Cbp are maternally loaded and abundantly expressed in the early embryo before widespread zygotic transcription (Toyama, Rebbert, Dey, Ozato, & Dawid, 2008). These findings suggest that H3K27ac deposition and recognition are required and precede transcriptional activation. One possibility is that H3K27ac could serve to nucleate transcription complex assembly. In support of this, it was reported that mouse ESCs treated with JQ1, a small molecule that competes with BRD4 acetyl-lysine recognition, abolishes the assembly of Pol II and BRD4 on enhancers of normally active genes, suggesting that Pol II depends on H3K27ac recognition by BRD4 (Filippakopoulos et al., 2010; Li et al., 2019). However, SOX2 remained detectable, indicating that SOX2 can still assemble on genes destined for activation (Li et al., 2019). Further studies are needed to dissect the sequence of molecular events relating to histone acetylation and recognition, TF binding, and assembly of transcription machinery during ZGA.

3.3.2.2 H3K4me3

Besides acetylation, histone methylation is also thought to play a prominent role in ZGA. In *Xenopus*, H3K4me3 emerges on genes prior to genome activation, increases when embryos approach ZGA, and becomes significantly enriched by gastrulation stage (Akkers et al., 2009; Hontelez et al., 2015). Similarly, in zebrafish, H3K4me3 appears before genome activation and is present on more than 80% of genes during ZGA (Lindeman et al., 2011; Vastenhouw et al., 2010). This includes both active and inactive genes, suggesting that H3K4me3 on promoters might poise embryonic genes for activation, especially those involved in homeostatic and developmental regulation (Lindeman et al., 2011; Vastenhouw et al., 2010). Many inactive genes are marked by the repressive mark H3K27me3 in addition to the activating mark H3K4me3, forming bivalent domains (Vastenhouw et al., 2010). Bivalent chromatin domains were identified in ESCs and are thought to poise genes for activation while keeping them repressed (Bernstein et al., 2006; Vastenhouw & Schier, 2012). The landscape of H3K4me3 in mouse revealed that H3K4me3 marks are already prevalent in the oocyte as broad domains that span more than 10 kb (Dahl et al., 2016). The levels of H3K4me3 are drastically reduced in 2-cell stage embryos and appear to be restricted to TSS of genes. Although the dynamics of H3K4me3 differs from *Xenopus* and zebrafish, H3K4me3 could serve to first pre-mark genes, then share a similar role in specifying promoters and poising genes for activation (Dahl et al., 2016). In contrast to what has been observed in *Xenopus*, zebrafish and mouse, H3K4me3 appears largely absent in the *Drosophila* genome prior to ZGA and emerges across the genome by cycle 14, when widespread transcription begins (Chen et al., 2013; Li et al., 2014).

What is the role of H3K4me3 in the onset of zygotic transcription? At least in *Drosophila*, since H3K4me3 marks do not emerge until widespread transcription has already started, zygotic transcription does not seem to require H3K4me3 (Chen et al., 2013; Li et al., 2014). In contrast, in mouse, zebrafish, and *Xenopus*, the regulation of H3K4me3 and the transcription landscape of the zygotic genome seem to be tightly coupled. In the early mouse embryo, the enzymes responsible for the methylation and demethylation of H3K4 are mixed lineage leukemia 2 (MLL2), also known as lysine

(K)-specific methyltransferase 2B (KMT2B), and K-specific demethylase 5A/B (KDM5A/B), respectively (Andreu-Vieyra et al., 2010; Dahl et al., 2016). MLL2 is present in the oocyte and its expression increases approaching ovulation, suggesting an important role in early embryogenesis (Andreu-Vieyra et al., 2010). Indeed, depletion of MLL2 results in significant loss of H3K4me3 and impairs activation of zygotic genes (Andreu-Vieyra et al., 2010). Furthermore, embryos arrest between the 1- and 4-cell stages. For KDM5A/B, loss-of-function analysis has shown that embryos retain high levels of H3K4me3 on thousands of ZGA-related genes, indicating that KDM5A/B are responsible for the narrowing of H3K4me3 domains (Dahl et al., 2016). Many of these ZGA-related genes were less abundant and observed phenotypes included developmental defects in both pre- and post-implantation embryos and failure of embryos to reach the blastocyst stage. In mouse, *Xenopus*, and zebrafish, H3K4me3 marks on promoters seem to correlate with a lack of DNA methylation at the TSS, indicating that H3K4me3 marks may be established on hypomethylated TSS of zygotic genes prior to their activation (Bogdanovic et al., 2011; Dahl et al., 2016; Hontelez et al., 2015; Jiang et al., 2013; Liu et al., 2018; Potok, Nix, Parnell, & Cairns, 2013). Such coordinated methylation activity may contribute to recruitment of remodelers to maintain chromatin accessibility and binding of key TFs for ZGA (Dahl et al., 2016; Liu et al., 2018).

3.3.3 Chromatin remodelers

Chromatin remodeling complexes are important modulators of chromatin accessibility. Chromatin remodelers use energy from ATP to displace nucleosomes, thereby altering the local chromatin structure and influencing chromatin accessibility for TFs and transcription machinery (Clapier et al., 2017). In different organisms, mutations in genes encoding the ATPase subunits of chromatin remodelers often lead to lethality, severe developmental defects, or maternal-effect phenotypes (Bultman et al., 2006; Burns et al., 2003; Cheung et al., 2018; Loppin, Docquier, Bonneton, & Couble, 2000; Philipps et al., 2008; Torres-Padilla & Zernicka-Goetz, 2006). However, very few genes have been further characterized to decipher their functional contribution to transcriptional activation of specific genes. The ISWI chromatin remodeler subunit SNF2H (SMARCA5) was discussed earlier: together with TIF1α, it is involved in the transcriptional activation of a subset of zygotic genes in the early mouse embryo

(Torres-Padilla & Zernicka-Goetz, 2006). Here, we highlight a second chromatin remodeler with a critical role in ZGA.

Brahma-related gene 1 (*Brg1*) was the first gene identified with an essential role in ZGA in mammals (Bultman et al., 2006). It encodes the catalytic ATPase subunit of the SWI/SNF complex (Bultman, Gebuhr, & Magnuson, 2005). In mouse, BRG1 is maternally provided and the protein is essential for development. Null homozygotes were first reported to die at the blastocyst stage (Bultman et al., 2005). It was further shown that embryos conceived from *Brg1*-depleted eggs arrest at the two-cell stage and that transcription of 30% of expressed genes was reduced (Bultman et al., 2006). These findings suggested that maternal BRG1 could be a critical transcriptional regulator during ZGA. When examining the effect of maternally-depleted BRG1 on histone modifications, no effect was detected on the global levels of histone acetylation, but the levels of the active H3K4me2 mark were decreased to 60% (Bultman et al., 2006). Precisely how BRG1 and SWI/SNF are recruited to early genes and how they affect their transcriptional activation remains to be elucidated. In mouse ESCs, BRG1 has been reported to be required for reinforcing OCT4 binding to regulatory elements (King & Klose, 2017). This interaction facilitates recruitment of SOX2 and stabilizes SOX2 on pluripotency-associated OCT4 target genes (King & Klose, 2017). Although direct interaction between OCT4 and BRG1 remains to be validated, these results illustrate the cooperativity between OCT4 and a chromatin remodeler. They also suggest a mechanism whereby BRG1 serves to support a functionally mature and accessible regulatory region bound by OCT4 and assists in engaging genes for transcriptional activation during cellular reprogramming at developmental transitions (King & Klose, 2017). In addition, a recent study in zebrafish showed that in embryos depleted of *smarca4a* (*brg1* homolog in zebrafish), there was a significant decrease in chromatin accessibility at Pou5f3 and Nanog binding sites, further supporting the notion that Brg1 could be recruited by Pou5f3 to further establish chromatin accessibility (Liu et al., 2018). Whether this interaction between Brg1 and Pou5f3 is retained on zygotic genes in zebrafish ZGA will need to be addressed.

3.3.4 Chromatin organization

Within the nucleus, the genome is spatially organized into compartments (active and inactive chromatin) and topologically-associated domains (TADs) (Dekker & Mirny, 2016). Such organization can facilitate long range chromatin contacts, so that specific enhancers and promoters are brought into proximity and can engage in transcription (van Steensel & Furlong, 2019). In the case of TADs, domains can span between 50 kb and 2 Mb depending on the species (Hug & Vaquerizas, 2018). Key components in 3D genome organization are the CCCTC-binding factor (CTCF) and its interacting partner, cohesin (Vietri Rudan & Hadjur, 2015). CTCF is a DNA-binding protein that recognizes the CCCTC motif widely distributed along the genome (Lee & Iyer, 2012). Together with cohesin, it serves as the major chromatin organizer and is enriched at TAD borders. Several recent reviews provide detailed description of the interplay between transcription and the genome-wide 3D architecture of chromatin (Hug & Vaquerizas, 2018; van Steensel & Furlong, 2019). Here, we give a short overview of TAD formation during development and highlight two molecular factors implicated in chromatin organization and ZGA.

Recent chromosome conformation capture studies have revealed progressive changes in the architecture of chromatin during the MZT. In *Drosophila* and mammals, organization and the emergence of TADs coincide with the onset of ZGA (Du et al., 2017; Flyamer et al., 2017; Gao et al., 2018; Ke et al., 2017; Wu et al., 2018). Interestingly, establishment of TADs occurs even in the presence of transcriptional inhibitors, although the insulation of TADs was affected (Du et al., 2017; Hug, Grimaldi, Kruse, & Vaquerizas, 2017; Ke et al., 2017). These findings suggest that while not necessary for TAD formation, transcription does have a role in maintaining chromatin organization. In contrast, in zebrafish, the genome is highly structured and organized into TADs even before ZGA, loses structural features during ZGA, and regains organization as the embryo progresses through development (Kaaij, van der Weide, Ketting, & de Wit, 2018). Thus, establishment and dynamics of TADs with respect to ZGA does not seem to be conserved across different organisms.

The molecular aspects underlying 3D chromatin remodeling during ZGA remain largely unknown as there have been only a few studies into the roles of CTCF and cohesin in ZGA. In mouse, zygotes depleted of the cohesin subunit SCC1 are largely devoid of TADs and chromatin loops, indicating that cohesin is already essential to organize the zygotic genome into chromatin loops and TADs in the 1-cell embryo (Gassler et al., 2017). The size of loops and domains depends on the cohesin-associated protein, WAPL, implicating an additional regulator in 3D genome organization in mouse embryos (Gassler et al., 2017). In zebrafish, ChIP studies on Rad21 (the zebrafish homolog of SCC1) reveal a shift in its binding distribution during zebrafish MZT (Meier et al., 2018). It has been shown that CTCF and cohesin binding are not limited to TAD boundaries and can also associate with regions within TADs and insulate specific genes or regulatory regions of the genome (Lupianez et al., 2015). In zebrafish, prior to ZGA, cohesin is preferentially bound to pericentromeric satellite, repetitive, and non-coding regions. From ZGA to post-ZGA, Rad21 shifts to being near TSS and gene-rich regions. A small subset of Rad21 binding sites within gene loci coincides with the co-occupation of Pou5f3 and Sox19b, as well as active histone marks (Meier et al., 2018). In addition, Rad21 knockdown leads to a delay in activation of zygotic genes, as well as disruption in the formation of nucleoli and Pol II foci (Meier et al., 2018). This suggests that in zebrafish, cohesin marks genes co-bound and co-regulated by the major ZGA TFs, so as to keep these regions open and transcription-competent potentially in collaboration with histone modifiers.

4. Compartmentalizing transcription

Membrane-less compartments are a type of intracellular structures that can provide spatial organization within the cell. In the nucleus, examples of these compartments include nucleoli, Cajal bodies, splicing speckles, histone locus bodies, and paraspeckles, which have been characterized based on their formation under specific conditions, high concentrations of specific markers, or their function (Mao, Zhang, & Spector, 2011). In a sense, compartmentalization provides a way to contain biochemical reactions within cells and to buffer fluctuations (Banani, Lee, Hyman, & Rosen, 2017; Klosin et al., 2020). Here, we focus on the recent insights into compartmentalization as a potential means to organize transcription events for ZGA.

Thanks to advances in microscopy and imaging tools, sites of transcription activity have been detected as clusters within the nucleus and can be visualized by following transcription machinery, TFs, transcription regulators, and other mRNA-associated factors (Buckley & Lis, 2014; Liu & Tjian, 2018). In the context of the developing embryo, the earliest transcription events can be detected as two distinct foci of Pol II in both *Drosophila* and zebrafish (Blythe & Wieschaus, 2016; Chan et al., 2019; Chen et al., 2013; Hadzhiev et al., 2019; Hilbert et al., 2018). In addition, ZLD was shown to form clusters within the nucleus (Dufourt et al., 2018; Mir et al., 2017). In these clusters, ZLD binds to DNA transiently, with a residence time ranging from hundreds of milliseconds (corresponding to non-specific binding) to seconds (specific binding). ZLD signal does not always colocalize with elongating Pol II, suggesting that it does not form long-lasting stable contacts during transcription (Dufourt et al., 2018). In *Drosophila*, clusters of the TF Ultrabithorax (UBX) were also detected in nuclei of the developing embryo (Tsai et al., 2017). Regions containing high local concentrations of UBX were found to colocalize with actively transcribed UBX targets, enhancers bearing low-affinity binding sites for UBX, as well as its cofactor Homothorax (Farley, Olson, Zhang, Rokhsar, & Levine, 2016; Tsai et al., 2017).

What can the function of such Pol II and TF clusters be? It has been suggested that a crowded local environment of transcription machinery and activators can increase the probability of protein-DNA binding, promote cooperative interactions, and allow for rapid multi-molecular assembly of a transcription compartment (Hnisz, Shrinivas, Young, Chakraborty, & Sharp, 2017; Mir, Bickmore, Furlong, & Narlikar, 2019). As such, ZLD accumulation could serve to potentiate recruitment of other TFs, chromatin remodelers, and transcription machinery leading up to transcriptional activation. Such hierarchy of molecular interactions for transcription foci formation has been suggested as another property of pioneer factors (Chen et al., 2014; Dufourt et al., 2018; Liu et al., 2014; Mir et al., 2017; Swinstead et al., 2016). Similarly, high local UBX concentrations may modulate transcription by overcoming low-affinity protein-DNA interactions, as well as promoting more frequent and stable binding and cooperative events.

How transcription foci are assembled during embryogenesis is unclear, but studies in other systems have provided some insight. From *in vitro* and cell culture studies, transcription activators have been suggested to form liquid-like condensates, or phase-separated droplets, via interactions between low complexity regions that make up the transcriptional activation domain (Chong et al., 2018). Such high density of homo- and heterotypic interactions has been suggested to lead to the assembly of higher-order structures within non-membrane bound subcellular compartments where transcription takes place (Hnisz et al., 2017). This model is supported by recent studies in which master TFs (such as OCT4 and SOX2), Pol II, Mediator, and BRD4 have been found to form foci in nuclei (Boehning et al., 2018; Boija et al., 2018; Cho et al., 2016; Chong et al., 2018; Ghamari et al., 2013; Li et al., 2019; Liu et al., 2014; Sabari et al., 2018). These foci often localize at clusters of enhancers (referred to as super-enhancers), suggesting that high numbers of TF binding sites are important for cluster formation (Sabari et al., 2018; Whyte et al., 2013). Since ZLD is detected as discrete clusters within the nucleus and contains an activation domain composed of low complexity regions, it could also nucleate the assembly of higher-order structures (Dufourt et al., 2018; Mir et al., 2018). The binding of ZLD can stabilize interactions at low-affinity sites, increase the probability of protein–DNA interactions, facilitate the binding of other TFs to accessible chromatin and the subsequent assembly of a transcription compartment (Schulz et al., 2015; Sun et al., 2015; Yamada et al., 2019). As such, it may modulate the timing and output of transcriptional activation of target genes within transcription compartments (Yamada et al., 2019). In zebrafish, the two distinct foci of Pol II that precede all other transcription events colocalize with the *miR-430* locus and its transcripts (Chan et al., 2019; Hilbert et al., 2018; Sato et al., 2019). As *miR-430* transcription is co-regulated by Pou5f3, Sox19b, and Nanog, and the genomic locus is highly repetitive (Lee et al., 2013), it may exhibit super-enhancer features. This may render the locus prone to assemble into a multi-molecular transcription compartment to initiate the first zygotic transcription events.

Taken together, compartmentalization appears to play an important role in ZGA, but many questions remain. Perhaps the most intriguing one is how compartmentalization might affect the stereotyped activation of genes during embryogenesis. The observation that co-expressed genes are in close proximity suggests they might "share a neighborhood" to economically and efficiently utilize transcription resources (Rieder et al., 2014) and the exciting possibility that the first foci in the embryo could nucleate subsequent transcription events.

5. Concluding remarks and future perspectives

Much progress has been made in our understanding of how the embryo transitions from being transcriptionally silent to transcriptionally competent and active. A clear picture of the molecular mechanism underlying ZGA, however, will require single molecule imaging approaches, cell-free systems, and proteomics.

Given the nucleosome-association features of pioneer TFs, *in vitro* and single molecule-based approaches will be useful in studying the binding properties of the major TFs and their impact on histone-DNA contacts. Such high-resolution experiments will shed light on how ZLD, Pou5f3, Sox19b, Nanog, and DUX overcome the chromatin barrier in the context of ZGA. Recent studies conducted with yeast TFs have exploited fluorescence resonance energy transfer (FRET)-based approaches to study the "pioneering" functions of these specific TFs. It was reported that TF binding reduced the degree of chromatin compaction and induced partial unwrapping of nucleosomes, rather than disrupting nucleosome structure or evicting histones (Donovan, Chen, Jipa, Bai, & Poirier, 2019; Mivelaz et al., 2020). Furthermore, increasing TF concentration induced local opening of chromatin (Donovan et al., 2019; Mivelaz et al., 2020). These findings provide a mechanistic view of the impact of TF binding on nucleosomes and the architecture of chromatin in facilitating the binding of additional factors. A similar effect could also be exerted by the major ZGA TFs on their bound regions of targeted zygotic genes, and thus allow for chromatin remodelers and additional gene-specific TFs to bind.

Cell-free systems will allow complex activities to be recapitulated so that the contribution of individual components and complexes can be examined. Such tools have been instrumental in uncovering the molecular details of other cellular processes related to mRNA regulation, as well as the proteins and small RNA machineries participating in post-transcriptional gene regulation during *Drosophila* and *C. elegans* development (Jeske, Meyer, Temme, Freudenreich, & Wahle, 2006; Wu et al., 2017). Some of the characterization studies of general and specific TFs and their assembly into an initiation complex with Pol II came from nuclear extracts derived from *Drosophila* embryos (Kadonaga, 2004). Although a technical challenge with cell-free systems is the acquisition of enough material from early embryo stages, such a tool will help tease out the molecular mechanism underlying ZGA, and more importantly, provide tight experimental control over the molecular players in question. Recently, a nucleoplasmic extract derived from *Xenopus* eggs was shown to support transcription and used to analyze the role of histone levels on the recruitment of transcription machinery to promoters (Barrows & Long, 2019). With this and similar systems, further questions regarding the relationship between the levels of different molecular players and transcription output, co-occupancy and cooperativity between TFs, and the sequences of events leading up to hierarchical organization of a transcription hub can be addressed.

Finally, proteomics will be required to understand the changing landscape of proteins and protein interactions in the early embryo. In mouse, *Xenopus*, and *Drosophila*, proteomics analyses across different embryonic stages has started to provide a comprehensive landscape of the protein expression changes during early development (Gao et al., 2017; Israel et al., 2019; Peshkin et al., 2015; Zhang, Ahmed-Braimah, Goldberg, & Wolfner, 2019). A recent study performed in mouse ESCs combined genomics and proteomics to dissect the subcomplexes of the polycomb repressive complex 2 (PRC2) and their differential recruitment in the epigenetic program underlying cell fate specification (Perino, van Mierlo, Wardle, Marks, & Veenstra, 2019). In the context of ZGA, such studies will shed light on the recruitment mechanisms of ZGA regulators, identify new cofactors and effectors, as well as bridge our understanding of converging biological processes of transcription activation, DNA replication, and cell cycle.

Acknowledgments

We thank Drs. Máté Palfy, Thomas Quail, and Johannes Stratmann for critical reading of the manuscript, as well as members of the Vastenhouw lab for discussions. E.W. is supported by the Fonds de recherche du Québec—Santé (FRQS). Research in the Vastenhouw lab is supported by Max Planck Society core funding, Human Frontiers Science Program, Deutsche Forschungsgemeinschaft (DFG), and the Volkswagen Foundation.

References

Adamson, E. D., & Woodland, H. R. (1974). Histone synthesis in early amphibian development: Histone and DNA syntheses are not co-ordinated. *Journal of Molecular Biology*, *88*, 263–285.

Akkers, R. C., van Heeringen, S. J., Jacobi, U. G., Janssen-Megens, E. M., Francoijs, K. J., Stunnenberg, H. G., et al. (2009). A hierarchy of H3K4me3 and H3K27me3 acquisition in spatial gene regulation in Xenopus embryos. *Developmental Cell*, *17*, 425–434.

Almouzni, G., Mechali, M., & Wolffe, A. P. (1990). Competition between transcription complex assembly and chromatin assembly on replicating DNA. *The EMBO Journal*, *9*, 573–582.

Almouzni, G., Mechali, M., & Wolffe, A. P. (1991). Transcription complex disruption caused by a transition in chromatin structure. *Molecular and Cellular Biology*, *11*, 655–665.

Almouzni, G., & Wolffe, A. P. (1995). Constraints on transcriptional activator function contribute to transcriptional quiescence during early Xenopus embryogenesis. *The EMBO Journal*, *14*, 1752–1765.

Amodeo, A. A., Jukam, D., Straight, A. F., & Skotheim, J. M. (2015). Histone titration against the genome sets the DNA-to-cytoplasm threshold for the Xenopus midblastula transition. *Proceedings of the National Academy of Sciences of the United States of America*, *112*, E1086–E1095.

Andreu-Vieyra, C. V., Chen, R., Agno, J. E., Glaser, S., Anastassiadis, K., Stewart, A. F., et al. (2010). MLL2 is required in oocytes for bulk histone 3 lysine 4 trimethylation and transcriptional silencing. *PLoS Biology*, *8*, e1000453.

Banani, S. F., Lee, H. O., Hyman, A. A., & Rosen, M. K. (2017). Biomolecular condensates: Organizers of cellular biochemistry. *Nature Reviews. Molecular Cell Biology*, *18*, 285–298.

Barrows, J. K., & Long, D. T. (2019). Cell-free transcription in Xenopus egg extract. *The Journal of Biological Chemistry*, *294*, 19645–19654.

Becker, P. B., & Workman, J. L. (2013). Nucleosome remodeling and epigenetics. *Cold Spring Harbor Perspectives in Biology*, *5*, a017905.

Bernstein, B. E., Mikkelsen, T. S., Xie, X., Kamal, M., Huebert, D. J., Cuff, J., et al. (2006). A bivalent chromatin structure marks key developmental genes in embryonic stem cells. *Cell*, *125*, 315–326.

Blythe, S. A., & Wieschaus, E. F. (2015). Zygotic genome activation triggers the DNA replication checkpoint at the midblastula transition. *Cell*, *160*, 1169–1181.
Blythe, S. A., & Wieschaus, E. F. (2016). Establishment and maintenance of heritable chromatin structure during early Drosophila embryogenesis. *eLife*, *5*, e20148.
Boehning, M., Dugast-Darzacq, C., Rankovic, M., Hansen, A. S., Yu, T., Marie-Nelly, H., et al. (2018). RNA polymerase II clustering through carboxy-terminal domain phase separation. *Nature Structural & Molecular Biology*, *25*, 833–840.
Bogdanovic, O., Long, S. W., van Heeringen, S. J., Brinkman, A. B., Gomez-Skarmeta, J. L., Stunnenberg, H. G., et al. (2011). Temporal uncoupling of the DNA methylome and transcriptional repression during embryogenesis. *Genome Research*, *21*, 1313–1327.
Bogolyubova, I. O., & Bogolyubov, D. S. (2014). Nuclear distribution of RNA polymerase II and mRNA processing machinery in early mammalian embryos. *BioMed Research International*, *2014*, 681596.
Boija, A., Klein, I. A., Sabari, B. R., Dall'Agnese, A., Coffey, E. L., Zamudio, A. V., et al. (2018). Transcription factors activate genes through the phase-separation capacity of their activation domains. *Cell*, *175*, 1842–1855.e1816.
Bortvin, A., Goodheart, M., Liao, M., & Page, D. C. (2004). Dppa3/Pgc7/stella is a maternal factor and is not required for germ cell specification in mice. *BMC Developmental Biology*, *4*, 2.
Boyer, L. A., Lee, T. I., Cole, M. F., Johnstone, S. E., Levine, S. S., Zucker, J. P., et al. (2005). Core transcriptional regulatory circuitry in human embryonic stem cells. *Cell*, *122*, 947–956.
Brown, J. L., & Wu, C. (1993). Repression of Drosophila pair-rule segmentation genes by ectopic expression of tramtrack. *Development*, *117*, 45–58.
Buckley, M. S., & Lis, J. T. (2014). Imaging RNA polymerase II transcription sites in living cells. *Current Opinion in Genetics & Development*, *25*, 126–130.
Bultman, S. J., Gebuhr, T. C., & Magnuson, T. (2005). A Brg1 mutation that uncouples ATPase activity from chromatin remodeling reveals an essential role for SWI/SNF-related complexes in beta-globin expression and erythroid development. *Genes & Development*, *19*, 2849–2861.
Bultman, S. J., Gebuhr, T. C., Pan, H., Svoboda, P., Schultz, R. M., & Magnuson, T. (2006). Maternal BRG1 regulates zygotic genome activation in the mouse. *Genes & Development*, *20*, 1744–1754.
Burns, K. H., Viveiros, M. M., Ren, Y., Wang, P., DeMayo, F. J., Frail, D. E., et al. (2003). Roles of NPM2 in chromatin and nucleolar organization in oocytes and embryos. *Science*, *300*, 633–636.
Campos, E. I., & Reinberg, D. (2009). Histones: Annotating chromatin. *Annual Review of Genetics*, *43*, 559–599.
Chambers, I., Colby, D., Robertson, M., Nichols, J., Lee, S., Tweedie, S., et al. (2003). Functional expression cloning of Nanog, a pluripotency sustaining factor in embryonic stem cells. *Cell*, *113*, 643–655.
Chan, S. H., Tang, Y., Miao, L., Darwich-Codore, H., Vejnar, C. E., Beaudoin, J. D., et al. (2019). Brd4 and P300 confer transcriptional competency during zygotic genome activation. *Developmental Cell*, *49*, 867–881, e868.
Chari, S., Wilky, H., Govindan, J., & Amodeo, A. A. (2019). Histone concentration regulates the cell cycle and transcription in early development. *Development*, *146*, dev177402.
Chen, K., Johnston, J., Shao, W., Meier, S., Staber, C., & Zeitlinger, J. (2013). A global change in RNA polymerase II pausing during the Drosophila midblastula transition. *eLife*, *2*, e00861.
Chen, J., Zhang, Z., Li, L., Chen, B. C., Revyakin, A., Hajj, B., et al. (2014). Single-molecule dynamics of enhanceosome assembly in embryonic stem cells. *Cell*, *156*, 1274–1285.

Cheung, C. T., Pasquier, J., Bouleau, A., Nguyen, T., Chesnel, F., Guiguen, Y., et al. (2018). Double maternal-effect: Duplicated nucleoplasmin 2 genes, npm2a and npm2b, with essential but distinct functions are shared by fish and tetrapods. *BMC Evolutionary Biology*, *18*, 167.

Cho, W. K., Jayanth, N., English, B. P., Inoue, T., Andrews, J. O., Conway, W., et al. (2016). RNA Polymerase II cluster dynamics predict mRNA output in living cells. *eLife*, *5*, e13617.

Chong, S., Dugast-Darzacq, C., Liu, Z., Dong, P., Dailey, G. M., Cattoglio, C., et al. (2018). Imaging dynamic and selective low-complexity domain interactions that control gene transcription. *Science*, *361*, eaar2555.

Clapier, C. R., Iwasa, J., Cairns, B. R., & Peterson, C. L. (2017). Mechanisms of action and regulation of ATP-dependent chromatin-remodelling complexes. *Nature Reviews. Molecular Cell Biology*, *18*, 407–422.

Clapp, J., Mitchell, L. M., Bolland, D. J., Fantes, J., Corcoran, A. E., Scotting, P. J., et al. (2007). Evolutionary conservation of a coding function for D4Z4, the tandem DNA repeat mutated in facioscapulohumeral muscular dystrophy. *American Journal of Human Genetics*, *81*, 264–279.

Clarkson, M., & Saint, R. (1999). A His2AvDGFP fusion gene complements a lethal His2AvD mutant allele and provides an in vivo marker for Drosophila chromosome behavior. *DNA and Cell Biology*, *18*, 457–462.

Collart, C., Allen, G. E., Bradshaw, C. R., Smith, J. C., & Zegerman, P. (2013). Titration of four replication factors is essential for the Xenopus laevis midblastula transition. *Science*, *341*, 893–896.

Cramer, P. (2019). Organization and regulation of gene transcription. *Nature*, *573*, 45–54.

Dahl, J. A., Jung, I., Aanes, H., Greggains, G. D., Manaf, A., Lerdrup, M., et al. (2016). Broad histone H3K4me3 domains in mouse oocytes modulate maternal-to-zygotic transition. *Nature*, *537*, 548–552.

De Iaco, A., Coudray, A., Duc, J., & Trono, D. (2019). DPPA2 and DPPA4 are necessary to establish a 2C-like state in mouse embryonic stem cells. *EMBO Reports*, *20*, e47382.

De Iaco, A., Planet, E., Coluccio, A., Verp, S., Duc, J., & Trono, D. (2017). DUX-family transcription factors regulate zygotic genome activation in placental mammals. *Nature Genetics*, *49*, 941–945.

De Iaco, A., Verp, S., Offner, S., Grun, D., & Trono, D. (2020). DUX is a non-essential synchronizer of zygotic genome activation. *Development*, *147*, dev177725.

Dekker, J., & Mirny, L. (2016). The 3D genome as moderator of chromosomal communication. *Cell*, *164*, 1110–1121.

Dimitrov, S., Almouzni, G., Dasso, M., & Wolffe, A. P. (1993). Chromatin transitions during early Xenopus embryogenesis: Changes in histone H4 acetylation and in linker histone type. *Developmental Biology*, *160*, 214–227.

Ding, J., Xu, H., Faiola, F., Ma'ayan, A., & Wang, J. (2012). Oct4 links multiple epigenetic pathways to the pluripotency network. *Cell Research*, *22*, 155–167.

Dolfini, D., Gatta, R., & Mantovani, R. (2012). NF-Y and the transcriptional activation of CCAAT promoters. *Critical Reviews in Biochemistry and Molecular Biology*, *47*, 29–49.

Dominski, Z., & Marzluff, W. F. (1999). Formation of the 3' end of histone mRNA. *Gene*, *239*, 1–14.

Donovan, B. T., Chen, H., Jipa, C., Bai, L., & Poirier, M. G. (2019). Dissociation rate compensation mechanism for budding yeast pioneer transcription factors. *eLife*, *8*, e43008.

Du, Z., Zheng, H., Huang, B., Ma, R., Wu, J., Zhang, X., et al. (2017). Allelic reprogramming of 3D chromatin architecture during early mammalian development. *Nature*, *547*, 232–235.

Dufourt, J., Trullo, A., Hunter, J., Fernandez, C., Lazaro, J., Dejean, M., et al. (2018). Temporal control of gene expression by the pioneer factor Zelda through transient interactions in hubs. *Nature Communications*, *9*, 5194.

Dunican, D. S., Ruzov, A., Hackett, J. A., & Meehan, R. R. (2008). xDnmt1 regulates transcriptional silencing in pre-MBT Xenopus embryos independently of its catalytic function. *Development*, *135*, 1295–1302.

Dworkin-Rastl, E., Kandolf, H., & Smith, R. C. (1994). The maternal histone H1 variant, H1M (B4 protein), is the predominant H1 histone in Xenopus pregastrula embryos. *Developmental Biology*, *161*, 425–439.

Eckersley-Maslin, M., Alda-Catalinas, C., Blotenburg, M., Kreibich, E., Krueger, C., & Reik, W. (2019). Dppa2 and Dppa4 directly regulate the Dux-driven zygotic transcriptional program. *Genes & Development*, *33*, 194–208.

Eckersley-Maslin, M. A., Alda-Catalinas, C., & Reik, W. (2018). Dynamics of the epigenetic landscape during the maternal-to-zygotic transition. *Nature Reviews. Molecular Cell Biology*, *19*, 436–450.

Edgar, B. A., & Schubiger, G. (1986). Parameters controlling transcriptional activation during early Drosophila development. *Cell*, *44*, 871–877.

Esch, D., Vahokoski, J., Groves, M. R., Pogenberg, V., Cojocaru, V., Vom Bruch, H., et al. (2013). A unique Oct4 interface is crucial for reprogramming to pluripotency. *Nature Cell Biology*, *15*, 295–301.

Farley, E. K., Olson, K. M., Zhang, W., Rokhsar, D. S., & Levine, M. S. (2016). Syntax compensates for poor binding sites to encode tissue specificity of developmental enhancers. *Proceedings of the National Academy of Sciences of the United States of America*, *113*, 6508–6513.

Filippakopoulos, P., Qi, J., Picaud, S., Shen, Y., Smith, W. B., Fedorov, O., et al. (2010). Selective inhibition of BET bromodomains. *Nature*, *468*, 1067–1073.

Flyamer, I. M., Gassler, J., Imakaev, M., Brandao, H. B., Ulianov, S. V., Abdennur, N., et al. (2017). Single-nucleus Hi-C reveals unique chromatin reorganization at oocyte-to-zygote transition. *Nature*, *544*, 110–114.

Foe, V. E., & Alberts, B. M. (1983). Studies of nuclear and cytoplasmic behaviour during the five mitotic cycles that precede gastrulation in Drosophila embryogenesis. *Journal of Cell Science*, *61*, 31–70.

Foo, S. M., Sun, Y., Lim, B., Ziukaite, R., O'Brien, K., Nien, C. Y., et al. (2014). Zelda potentiates morphogen activity by increasing chromatin accessibility. *Current Biology*, *24*, 1341–1346.

Gaertner, B., & Zeitlinger, J. (2014). RNA polymerase II pausing during development. *Development*, *141*, 1179–1183.

Gagliardi, A., Mullin, N. P., Ying Tan, Z., Colby, D., Kousa, A. I., Halbritter, F., et al. (2013). A direct physical interaction between Nanog and Sox2 regulates embryonic stem cell self-renewal. *The EMBO Journal*, *32*, 2231–2247.

Gagnon, J. A., Obbad, K., & Schier, A. F. (2018). The primary role of zebrafish nanog is in extra-embryonic tissue. *Development*, *145*, dev147793.

Gao, Y., Liu, X., Tang, B., Li, C., Kou, Z., Li, L., et al. (2017). Protein expression landscape of mouse embryos during pre-implantation development. *Cell Reports*, *21*, 3957–3969.

Gao, L., Wu, K., Liu, Z., Yao, X., Yuan, S., Tao, W., et al. (2018). Chromatin accessibility landscape in human early embryos and its association with evolution. *Cell*, *173*, 248–259. e215.

Gassler, J., Brandao, H. B., Imakaev, M., Flyamer, I. M., Ladstatter, S., Bickmore, W. A., et al. (2017). A mechanism of cohesin-dependent loop extrusion organizes zygotic genome architecture. *The EMBO Journal*, *36*, 3600–3618.

Geng, L. N., Yao, Z., Snider, L., Fong, A. P., Cech, J. N., Young, J. M., et al. (2012). DUX4 activates germline genes, retroelements, and immune mediators: Implications for facioscapulohumeral dystrophy. *Developmental Cell*, *22*, 38–51.

Gentsch, G. E., Owens, N. D. L., & Smith, J. C. (2019). The spatiotemporal control of zygotic genome activation. *iScience*, *16*, 485–498.

Ghamari, A., van de Corput, M. P., Thongjuea, S., van Cappellen, W. A., van Ijcken, W., van Haren, J., et al. (2013). In vivo live imaging of RNA polymerase II transcription factories in primary cells. *Genes & Development*, *27*, 767–777.

Guven-Ozkan, T., Nishi, Y., Robertson, S. M., & Lin, R. (2008). Global transcriptional repression in C. elegans germline precursors by regulated sequestration of TAF-4. *Cell*, *135*, 149–160.

Hadzhiev, Y., Qureshi, H. K., Wheatley, L., Cooper, L., Jasiulewicz, A., Van Nguyen, H., et al. (2019). A cell cycle-coordinated polymerase II transcription compartment encompasses gene expression before global genome activation. *Nature Communications*, *10*, 691.

Hamm, D. C., Bondra, E. R., & Harrison, M. M. (2015). Transcriptional activation is a conserved feature of the early embryonic factor Zelda that requires a cluster of four zinc fingers for DNA binding and a low-complexity activation domain. *The Journal of Biological Chemistry*, *290*, 3508–3518.

Harlen, K. M., & Churchman, L. S. (2017). The code and beyond: Transcription regulation by the RNA polymerase II carboxy-terminal domain. *Nature Reviews. Molecular Cell Biology*, *18*, 263–273.

Harrison, M. M., Li, X. Y., Kaplan, T., Botchan, M. R., & Eisen, M. B. (2011). Zelda binding in the early Drosophila melanogaster embryo marks regions subsequently activated at the maternal-to-zygotic transition. *PLoS Genetics*, *7*, e1002266.

Harshman, S. W., Young, N. L., Parthun, M. R., & Freitas, M. A. (2013). H1 histones: Current perspectives and challenges. *Nucleic Acids Research*, *41*, 9593–9609.

Hart, D. O., Raha, T., Lawson, N. D., & Green, M. R. (2007). Initiation of zebrafish haematopoiesis by the TATA-box-binding protein-related factor Trf3. *Nature*, *450*, 1082–1085.

Hendrickson, P. G., Dorais, J. A., Grow, E. J., Whiddon, J. L., Lim, J. W., Wike, C. L., et al. (2017). Conserved roles of mouse DUX and human DUX4 in activating cleavage-stage genes and MERVL/HERVL retrotransposons. *Nature Genetics*, *49*, 925–934.

Herr, W., Sturm, R. A., Clerc, R. G., Corcoran, L. M., Baltimore, D., Sharp, P. A., et al. (1988). The POU domain: A large conserved region in the mammalian pit-1, oct-1, oct-2, and Caenorhabditis elegans unc-86 gene products. *Genes & Development*, *2*, 1513–1516.

Heyn, P., Kircher, M., Dahl, A., Kelso, J., Tomancak, P., Kalinka, A. T., et al. (2014). The earliest transcribed zygotic genes are short, newly evolved, and different across species. *Cell Reports*, *6*, 285–292.

Hilbert, L., Sato, Y., Kimura, H., Jülicher, F., Honigmann, A., Zaburdaev, V., et al. (2018). Transcription organizes euchromatin similar to an active microemulsion. *bioRxiv*, 234112, [preprint].

Hnisz, D., Shrinivas, K., Young, R. A., Chakraborty, A. K., & Sharp, P. A. (2017). A phase separation model for transcriptional control. *Cell*, *169*, 13–23.

Ho, L., & Crabtree, G. R. (2010). Chromatin remodelling during development. *Nature*, *463*, 474–484.

Hock, R., Moorman, A., Fischer, D., & Scheer, U. (1993). Absence of somatic histone H1 in oocytes and preblastula embryos of Xenopus laevis. *Developmental Biology*, *158*, 510–522.

Hontelez, S., van Kruijsbergen, I., Georgiou, G., van Heeringen, S. J., Bogdanovic, O., Lister, R., et al. (2015). Embryonic transcription is controlled by maternally defined chromatin state. *Nature Communications*, *6*, 10148.

Huang, Y., Kim, J. K., Do, D. V., Lee, C., Penfold, C. A., Zylicz, J. J., et al. (2017). Stella modulates transcriptional and endogenous retrovirus programs during maternal-to-zygotic transition. *eLife*, *6*, e22345.

Hug, C. B., Grimaldi, A. G., Kruse, K., & Vaquerizas, J. M. (2017). Chromatin architecture emerges during zygotic genome activation independent of transcription. *Cell*, *169*, 216–228.e219.

Hug, C. B., & Vaquerizas, J. M. (2018). The birth of the 3D genome during early embryonic development. *Trends in Genetics*, *34*, 903–914.

Hyun, K., Jeon, J., Park, K., & Kim, J. (2017). Writing, erasing and reading histone lysine methylations. *Experimental & Molecular Medicine*, *49*, e324.

Israel, S., Ernst, M., Psathaki, O. E., Drexler, H. C. A., Casser, E., Suzuki, Y., et al. (2019). An integrated genome-wide multi-omics analysis of gene expression dynamics in the preimplantation mouse embryo. *Scientific Reports*, *9*, 13356.

Jeske, M., Meyer, S., Temme, C., Freudenreich, D., & Wahle, E. (2006). Rapid ATP-dependent deadenylation of nanos mRNA in a cell-free system from Drosophila embryos. *The Journal of Biological Chemistry*, *281*, 25124–25133.

Jiang, L., Zhang, J., Wang, J. J., Wang, L., Zhang, L., Li, G., et al. (2013). Sperm, but not oocyte, DNA methylome is inherited by zebrafish early embryos. *Cell*, *153*, 773–784.

Joseph, S. R., Palfy, M., Hilbert, L., Kumar, M., Karschau, J., Zaburdaev, V., et al. (2017). Competition between histone and transcription factor binding regulates the onset of transcription in zebrafish embryos. *eLife*, *6*, e23326.

Kaaij, L. J. T., van der Weide, R. H., Ketting, R. F., & de Wit, E. (2018). Systemic loss and gain of chromatin architecture throughout zebrafish development. *Cell Reports*, *24*, 1–10.e14.

Kadonaga, J. T. (2004). Regulation of RNA polymerase II transcription by sequence-specific DNA binding factors. *Cell*, *116*, 247–257.

Kamachi, Y., & Kondoh, H. (2013). Sox proteins: Regulators of cell fate specification and differentiation. *Development*, *140*, 4129–4144.

Kamachi, Y., Uchikawa, M., & Kondoh, H. (2000). Pairing SOX off: With partners in the regulation of embryonic development. *Trends in Genetics*, *16*, 182–187.

Kanodia, J. S., Liang, H. L., Kim, Y., Lim, B., Zhan, M., Lu, H., et al. (2012). Pattern formation by graded and uniform signals in the early Drosophila embryo. *Biophysical Journal*, *102*, 427–433.

Ke, Y., Xu, Y., Chen, X., Feng, S., Liu, Z., Sun, Y., et al. (2017). 3D chromatin structures of mature gametes and structural reprogramming during mammalian embryogenesis. *Cell*, *170*, 367–381.e320.

Khochbin, S. (2001). Histone H1 diversity: Bridging regulatory signals to linker histone function. *Gene*, *271*, 1–12.

Kimelman, D., Kirschner, M., & Scherson, T. (1987). The events of the midblastula transition in Xenopus are regulated by changes in the cell cycle. *Cell*, *48*, 399–407.

King, H. W., & Klose, R. J. (2017). The pioneer factor OCT4 requires the chromatin remodeller BRG1 to support gene regulatory element function in mouse embryonic stem cells. *eLife*, *6*, e22631.

Klemm, J. D., Rould, M. A., Aurora, R., Herr, W., & Pabo, C. O. (1994). Crystal structure of the Oct-1 POU domain bound to an octamer site: DNA recognition with tethered DNA-binding modules. *Cell*, *77*, 21–32.

Klemm, S. L., Shipony, Z., & Greenleaf, W. J. (2019). Chromatin accessibility and the regulatory epigenome. *Nature Reviews. Genetics*, *20*, 207–220.

Klosin, A., Oltsch, F., Harmon, T., Honigmann, A., Jülicher, F., Hyman, A. A., & Zechner, C. (2020). Phase separation provides a mechanism to reduce noise in cells. *Science*, *367*, 464–468.

Kwasnieski, J. C., Orr-Weaver, T. L., & Bartel, D. P. (2019). Early genome activation in Drosophila is extensive with an initial tendency for aborted transcripts and retained introns. *Genome Research*, *29*, 1188–1197.

Ladam, F., Stanney, W., Donaldson, I. J., Yildiz, O., Bobola, N., & Sagerstrom, C. G. (2018). TALE factors use two distinct functional modes to control an essential zebrafish gene expression program. *eLife*, *7*, e36144.

Leach, T. J., Mazzeo, M., Chotkowski, H. L., Madigan, J. P., Wotring, M. G., & Glaser, R. L. (2000). Histone H2A.Z is widely but nonrandomly distributed in chromosomes of Drosophila melanogaster. *The Journal of Biological Chemistry*, *275*, 23267–23272.

Lee, M. T., Bonneau, A. R., Takacs, C. M., Bazzini, A. A., DiVito, K. R., Fleming, E. S., et al. (2013). Nanog, Pou5f1 and SoxB1 activate zygotic gene expression during the maternal-to-zygotic transition. *Nature*, *503*, 360–364.

Lee, B. K., & Iyer, V. R. (2012). Genome-wide studies of CCCTC-binding factor (CTCF) and cohesin provide insight into chromatin structure and regulation. *The Journal of Biological Chemistry*, *287*, 30906–30913.

Lee, C., Li, X., Hechmer, A., Eisen, M., Biggin, M. D., Venters, B. J., et al. (2008). NELF and GAGA factor are linked to promoter-proximal pausing at many genes in Drosophila. *Molecular and Cellular Biology*, *28*, 3290–3300.

Leichsenring, M., Maes, J., Mossner, R., Driever, W., & Onichtchouk, D. (2013). Pou5f1 transcription factor controls zygotic gene activation in vertebrates. *Science*, *341*, 1005–1009.

Leidenroth, A., Clapp, J., Mitchell, L. M., Coneyworth, D., Dearden, F. L., Iannuzzi, L., et al. (2012). Evolution of DUX gene macrosatellites in placental mammals. *Chromosoma*, *121*, 489–497.

Li, J., Dong, A., Saydaminova, K., Chang, H., Wang, G., Ochiai, H., et al. (2019). Single-molecule nanoscopy elucidates RNA polymerase II transcription at single genes in live cells. *Cell*, *178*, 491–506.e428.

Li, X. Y., Harrison, M. M., Villalta, J. E., Kaplan, T., & Eisen, M. B. (2014). Establishment of regions of genomic activity during the Drosophila maternal to zygotic transition. *eLife*, *3*, e03737.

Liang, H. L., Nien, C. Y., Liu, H. Y., Metzstein, M. M., Kirov, N., & Rushlow, C. (2008). The zinc-finger protein Zelda is a key activator of the early zygotic genome in Drosophila. *Nature*, *456*, 400–403.

Liang, J., Wan, M., Zhang, Y., Gu, P., Xin, H., Jung, S. Y., et al. (2008). Nanog and Oct4 associate with unique transcriptional repression complexes in embryonic stem cells. *Nature Cell Biology*, *10*, 731–739.

Lindeman, L. C., Andersen, I. S., Reiner, A. H., Li, N., Aanes, H., Ostrup, O., et al. (2011). Prepatterning of developmental gene expression by modified histones before zygotic genome activation. *Developmental Cell*, *21*, 993–1004.

Lippok, B., Song, S., & Driever, W. (2014). Pou5f1 protein expression and posttranslational modification during early zebrafish development. *Developmental Dynamics*, *243*, 468–477.

Liu, Z., Legant, W. R., Chen, B. C., Li, L., Grimm, J. B., Lavis, L. D., et al. (2014). 3D imaging of Sox2 enhancer clusters in embryonic stem cells. *eLife*, *3* e04236.

Liu, Z., & Tjian, R. (2018). Visualizing transcription factor dynamics in living cells. *The Journal of Cell Biology*, *217*, 1181–1191.

Liu, G., Wang, W., Hu, S., Wang, X., & Zhang, Y. (2018). Inherited DNA methylation primes the establishment of accessible chromatin during genome activation. *Genome Research*, *28*, 998–1007.

Liu, X., Wang, C., Liu, W., Li, J., Li, C., Kou, X., et al. (2016). Distinct features of H3K4me3 and H3K27me3 chromatin domains in pre-implantation embryos. *Nature*, *537*, 558–562.

Loppin, B., Docquier, M., Bonneton, F., & Couble, P. (2000). The maternal effect mutation sesame affects the formation of the male pronucleus in Drosophila melanogaster. *Developmental Biology*, *222*, 392–404.

Lu, F., Liu, Y., Inoue, A., Suzuki, T., Zhao, K., & Zhang, Y. (2016). Establishing chromatin regulatory landscape during mouse preimplantation development. *Cell*, *165*, 1375–1388.

Lunde, K., Belting, H. G., & Driever, W. (2004). Zebrafish pou5f1/pou2, homolog of mammalian Oct4, functions in the endoderm specification cascade. *Current Biology*, *14*, 48–55.

Lupianez, D. G., Kraft, K., Heinrich, V., Krawitz, P., Brancati, F., Klopocki, E., et al. (2015). Disruptions of topological chromatin domains cause pathogenic rewiring of gene-enhancer interactions. *Cell*, *161*, 1012–1025.

Maity, S. N., & De Crombrugghe, B. (1996). Purification, characterization, and role of CCAAT-binding factor in transcription. *Methods in Enzymology*, *273*, 217–232.

Mao, Y. S., Zhang, B., & Spector, D. L. (2011). Biogenesis and function of nuclear bodies. *Trends in Genetics*, *27*, 295–306.

Marzluff, W. F. (2005). Metazoan replication-dependent histone mRNAs: A distinct set of RNA polymerase II transcripts. *Current Opinion in Cell Biology*, *17*, 274–280.

McCleland, M. L., & O'Farrell, P. H. (2008). RNAi of mitotic cyclins in Drosophila uncouples the nuclear and centrosome cycle. *Current Biology*, *18*, 245–254.

McDaniel, S. L., Gibson, T. J., Schulz, K. N., Fernandez Garcia, M., Nevil, M., Jain, S. U., et al. (2019). Continued activity of the pioneer factor Zelda is required to drive zygotic genome activation. *Molecular Cell*, *74*, 185–195.e184.

Meier, M., Grant, J., Dowdle, A., Thomas, A., Gerton, J., Collas, P., et al. (2018). Cohesin facilitates zygotic genome activation in zebrafish. *Development*, *145*, dev156521.

Mir, M., Bickmore, W., Furlong, E. E. M., & Narlikar, G. (2019). Chromatin topology, condensates and gene regulation: Shifting paradigms or just a phase? *Development*, *146*, dev182766.

Mir, M., Reimer, A., Haines, J. E., Li, X. Y., Stadler, M., Garcia, H., et al. (2017). Dense Bicoid hubs accentuate binding along the morphogen gradient. *Genes & Development*, *31*, 1784–1794.

Mir, M., Stadler, M. R., Ortiz, S. A., Hannon, C. E., Harrison, M. M., Darzacq, X., et al. (2018). Dynamic multifactor hubs interact transiently with sites of active transcription in Drosophila embryos. *eLife*, *7*, e40497.

Mitsui, K., Tokuzawa, Y., Itoh, H., Segawa, K., Murakami, M., Takahashi, K., et al. (2003). The homeoprotein Nanog is required for maintenance of pluripotency in mouse epiblast and ES cells. *Cell*, *113*, 631–642.

Mivelaz, M., Cao, A. M., Kubik, S., Zencir, S., Hovius, R., Boichenko, I., et al. (2020). Chromatin fiber invasion and nucleosome displacement by the Rap1 transcription factor. *Molecular Cell*, *77*, 488–500.e9.

Moshe, A., & Kaplan, T. (2017). Genome-wide search for Zelda-like chromatin signatures identifies GAF as a pioneer factor in early fly development. *Epigenetics & Chromatin*, *10*, 33.

Mullin, N. P., Gagliardi, A., Khoa, L. T. P., Colby, D., Hall-Ponsele, E., Rowe, A. J., et al. (2017). Distinct contributions of tryptophan residues within the dimerization domain to Nanog function. *Journal of Molecular Biology*, *429*, 1544–1553.

Nakamura, T., Arai, Y., Umehara, H., Masuhara, M., Kimura, T., Taniguchi, H., et al. (2007). PGC7/Stella protects against DNA demethylation in early embryogenesis. *Nature Cell Biology*, *9*, 64–71.

Nardini, M., Gnesutta, N., Donati, G., Gatta, R., Forni, C., Fossati, A., et al. (2013). Sequence-specific transcription factor NF-Y displays histone-like DNA binding and H2B-like ubiquitination. *Cell*, *152*, 132–143.

Ner, S. S., & Travers, A. A. (1994). HMG-D, the Drosophila melanogaster homologue of HMG 1 protein, is associated with early embryonic chromatin in the absence of histone H1. *The EMBO Journal*, *13*, 1817–1822.

Newport, J., & Kirschner, M. (1982). A major developmental transition in early Xenopus embryos: II. Control of the onset of transcription. *Cell*, *30*, 687–696.

Nien, C. Y., Liang, H. L., Butcher, S., Sun, Y., Fu, S., Gocha, T., et al. (2011). Temporal coordination of gene networks by Zelda in the early Drosophila embryo. *PLoS Genetics*, *7*, e1002339.

Nieto, L., Joseph, G., Stella, A., Henri, P., Burlet-Schiltz, O., Monsarrat, B., et al. (2007). Differential effects of phosphorylation on DNA binding properties of N Oct-3 are dictated by protein/DNA complex structures. *Journal of Molecular Biology*, *370*, 687–700.

Okuda, Y., Ogura, E., Kondoh, H., & Kamachi, Y. (2010). B1 SOX coordinate cell specification with patterning and morphogenesis in the early zebrafish embryo. *PLoS Genetics*, *6*, e1000936.

Okuda, Y., Yoda, H., Uchikawa, M., Furutani-Seiki, M., Takeda, H., Kondoh, H., et al. (2006). Comparative genomic and expression analysis of group B1 sox genes in zebrafish indicates their diversification during vertebrate evolution. *Developmental Dynamics*, *235*, 811–825.

Oldfield, A. J., Henriques, T., Kumar, D., Burkholder, A. B., Cinghu, S., Paulet, D., et al. (2019). NF-Y controls fidelity of transcription initiation at gene promoters through maintenance of the nucleosome-depleted region. *Nature Communications*, *10*, 3072.

Oldfield, A. J., Yang, P., Conway, A. E., Cinghu, S., Freudenberg, J. M., Yellaboina, S., et al. (2014). Histone-fold domain protein NF-Y promotes chromatin accessibility for cell type-specific master transcription factors. *Molecular Cell*, *55*, 708–722.

Onichtchouk, D., Geier, F., Polok, B., Messerschmidt, D. M., Mossner, R., Wendik, B., et al. (2010). Zebrafish Pou5f1-dependent transcriptional networks in temporal control of early development. *Molecular Systems Biology*, *6*, 354.

Owens, N. D. L., Blitz, I. L., Lane, M. A., Patrushev, I., Overton, J. D., Gilchrist, M. J., et al. (2016). Measuring absolute RNA copy numbers at high temporal resolution reveals transcriptome kinetics in development. *Cell Reports*, *14*, 632–647.

Palfy, M., Joseph, S. R., & Vastenhouw, N. L. (2017). The timing of zygotic genome activation. *Current Opinion in Genetics & Development*, *43*, 53–60.

Palfy, M., Schulze, G., Valen, E., & Vastenhouw, N. L. (2020). Chromatin accessibility established by Pou5f3, Sox19b and Nanog primes genes for activity during zebrafish genome activation. *PLoS Genetics*, *16*, e1008546.

Pardo, M., Lang, B., Yu, L., Prosser, H., Bradley, A., Babu, M. M., et al. (2010). An expanded Oct4 interaction network: Implications for stem cell biology, development, and disease. *Cell Stem Cell*, *6*, 382–395.

Payer, B., Saitou, M., Barton, S. C., Thresher, R., Dixon, J. P., Zahn, D., et al. (2003). Stella is a maternal effect gene required for normal early development in mice. *Current Biology*, *13*, 2110–2117.

Perez-Camps, M., Tian, J., Chng, S. C., Sem, K. P., Sudhaharan, T., Teh, C., et al. (2016). Quantitative imaging reveals real-time Pou5f3-Nanog complexes driving dorsoventral mesendoderm patterning in zebrafish. *eLife*, *5*, e11475.

Perez-Montero, S., Carbonell, A., Moran, T., Vaquero, A., & Azorin, F. (2013). The embryonic linker histone H1 variant of Drosophila, dBigH1, regulates zygotic genome activation. *Developmental Cell*, *26*, 578–590.

Perino, M., van Mierlo, G., Wardle, S. M. T., Marks, H., & Veenstra, G. J. (2019). Two distinct functional axes of positive feedback-enforced PRC2 recruitment in mouse embryonic stem cells. *bioRxiv*, 669960, [preprint].

Peshkin, L., Wuhr, M., Pearl, E., Haas, W., Freeman, R. M., Jr., Gerhart, J. C., et al. (2015). On the relationship of protein and mRNA dynamics in vertebrate embryonic development. *Developmental Cell*, *35*, 383–394.

Philipps, D. L., Wigglesworth, K., Hartford, S. A., Sun, F., Pattabiraman, S., Schimenti, K., et al. (2008). The dual bromodomain and WD repeat-containing mouse protein BRWD1 is required for normal spermiogenesis and the oocyte-embryo transition. *Developmental Biology*, *317*, 72–82.

Potok, M. E., Nix, D. A., Parnell, T. J., & Cairns, B. R. (2013). Reprogramming the maternal zebrafish genome after fertilization to match the paternal methylation pattern. *Cell*, *153*, 759–772.

Prioleau, M. N., Huet, J., Sentenac, A., & Mechali, M. (1994). Competition between chromatin and transcription complex assembly regulates gene expression during early development. *Cell*, *77*, 439–449.

Pritchard, D. K., & Schubiger, G. (1996). Activation of transcription in Drosophila embryos is a gradual process mediated by the nucleocytoplasmic ratio. *Genes & Development*, *10*, 1131–1142.

Raff, J. W., Kellum, R., & Alberts, B. (1994). The Drosophila GAGA transcription factor is associated with specific regions of heterochromatin throughout the cell cycle. *The EMBO Journal*, *13*, 5977–5983.

Reim, G., & Brand, M. (2006). Maternal control of vertebrate dorsoventral axis formation and epiboly by the POU domain protein Spg/Pou2/Oct4. *Development*, *133*, 2757–2770.

Remenyi, A., Lins, K., Nissen, L. J., Reinbold, R., Scholer, H. R., & Wilmanns, M. (2003). Crystal structure of a POU/HMG/DNA ternary complex suggests differential assembly of Oct4 and Sox2 on two enhancers. *Genes & Development*, *17*, 2048–2059.

Remenyi, A., Tomilin, A., Pohl, E., Lins, K., Philippsen, A., Reinbold, R., et al. (2001). Differential dimer activities of the transcription factor Oct-1 by DNA-induced interface swapping. *Molecular Cell*, *8*, 569–580.

Ribeiro, L., Tobias-Santos, V., Santos, D., Antunes, F., Feltran, G., de Souza Menezes, J., et al. (2017). Evolution and multiple roles of the Pancrustacea specific transcription factor zelda in insects. *PLoS Genetics*, *13*, e1006868.

Rieder, D., Ploner, C., Krogsdam, A. M., Stocker, G., Fischer, M., Scheideler, M., et al. (2014). Co-expressed genes prepositioned in spatial neighborhoods stochastically associate with SC35 speckles and RNA polymerase II factories. *Cellular and Molecular Life Sciences*, *71*, 1741–1759.

Rothe, M., Pehl, M., Taubert, H., & Jackle, H. (1992). Loss of gene function through rapid mitotic cycles in the Drosophila embryo. *Nature*, *359*, 156–159.

Ruzov, A., Dunican, D. S., Prokhortchouk, A., Pennings, S., Stancheva, I., Prokhortchouk, E., et al. (2004). Kaiso is a genome-wide repressor of transcription that is essential for amphibian development. *Development*, *131*, 6185–6194.

Ruzov, A., Savitskaya, E., Hackett, J. A., Reddington, J. P., Prokhortchouk, A., Madej, M. J., et al. (2009). The non-methylated DNA-binding function of Kaiso is not required in early Xenopus laevis development. *Development*, *136*, 729–738.

Sabari, B. R., Dall'Agnese, A., Boija, A., Klein, I. A., Coffey, E. L., Shrinivas, K., et al. (2018). Coactivator condensation at super-enhancers links phase separation and gene control. *Science*, *361*, eaar3958.

Saeki, H., Ohsumi, K., Aihara, H., Ito, T., Hirose, S., Ura, K., et al. (2005). Linker histone variants control chromatin dynamics during early embryogenesis. *Proceedings of the National Academy of Sciences of the United States of America*, *102*, 5697–5702.

Sato, Y., Hilbert, L., Oda, H., Wan, Y., Heddleston, J. M., Chew, T. L., et al. (2019). Histone H3K27 acetylation precedes active transcription during zebrafish zygotic genome activation as revealed by live-cell analysis. *Development*, *146*, dev179127.

Schuff, M., Siegel, D., Philipp, M., Bundschu, K., Heymann, N., Donow, C., et al. (2012). Characterization of Danio rerio Nanog and functional comparison to Xenopus Vents. *Stem Cells and Development*, *21*, 1225–1238.

Schulz, K. N., Bondra, E. R., Moshe, A., Villalta, J. E., Lieb, J. D., Kaplan, T., et al. (2015). Zelda is differentially required for chromatin accessibility, transcription factor binding, and gene expression in the early Drosophila embryo. *Genome Research*, *25*, 1715–1726.

Seydoux, G., & Dunn, M. A. (1997). Transcriptionally repressed germ cells lack a subpopulation of phosphorylated RNA polymerase II in early embryos of Caenorhabditis elegans and Drosophila melanogaster. *Development*, *124*, 2191–2201.

Seydoux, G., & Fire, A. (1994). Soma-germline asymmetry in the distributions of embryonic RNAs in Caenorhabditis elegans. *Development*, *120*, 2823–2834.

Seydoux, G., Mello, C. C., Pettitt, J., Wood, W. B., Priess, J. R., & Fire, A. (1996). Repression of gene expression in the embryonic germ lineage of C. elegans. *Nature*, *382*, 713–716.

Shermoen, A. W., & O'Farrell, P. H. (1991). Progression of the cell cycle through mitosis leads to abortion of nascent transcripts. *Cell*, *67*, 303–310.

Shindo, Y., & Amodeo, A. A. (2019). Dynamics of free and chromatin-bound histone H3 during early embryogenesis. *Current Biology*, *29*, 359–366.e354.

Skvortsova, K., Iovino, N., & Bogdanovic, O. (2018). Functions and mechanisms of epigenetic inheritance in animals. *Nature Reviews. Molecular Cell Biology*, *19*, 774–790.

Soufi, A., Garcia, M. F., Jaroszewicz, A., Osman, N., Pellegrini, M., & Zaret, K. S. (2015). Pioneer transcription factors target partial DNA motifs on nucleosomes to initiate reprogramming. *Cell*, *161*, 555–568.

Soutourina, J. (2018). Transcription regulation by the mediator complex. *Nature Reviews. Molecular Cell Biology*, *19*, 262–274.

Sun, Y., Nien, C. Y., Chen, K., Liu, H. Y., Johnston, J., Zeitlinger, J., et al. (2015). Zelda overcomes the high intrinsic nucleosome barrier at enhancers during Drosophila zygotic genome activation. *Genome Research*, *25*, 1703–1714.

Swinstead, E. E., Miranda, T. B., Paakinaho, V., Baek, S., Goldstein, I., Hawkins, M., et al. (2016). Steroid receptors reprogram FoxA1 occupancy through dynamic chromatin transitions. *Cell*, *165*, 593–605.

Takahashi, K., & Yamanaka, S. (2006). Induction of pluripotent stem cells from mouse embryonic and adult fibroblast cultures by defined factors. *Cell*, *126*, 663–676.

Takeda, H., Matsuzaki, T., Oki, T., Miyagawa, T., & Amanuma, H. (1994). A novel POU domain gene, zebrafish pou2: Expression and roles of two alternatively spliced twin products in early development. *Genes & Development*, *8*, 45–59.

Talbert, P. B., & Henikoff, S. (2017). Histone variants on the move: Substrates for chromatin dynamics. *Nature Reviews. Molecular Cell Biology*, *18*, 115–126.

Tang, Z., Chen, W. Y., Shimada, M., Nguyen, U. T., Kim, J., Sun, X. J., et al. (2013). SET1 and p300 act synergistically, through coupled histone modifications, in transcriptional activation by p53. *Cell*, *154*, 297–310.

Tantin, D. (2013). Oct transcription factors in development and stem cells: Insights and mechanisms. *Development*, *140*, 2857–2866.

ten Bosch, J. R., Benavides, J. A., & Cline, T. W. (2006). The TAGteam DNA motif controls the timing of Drosophila pre-blastoderm transcription. *Development*, *133*, 1967–1977.

Tenenhaus, C., Schubert, C., & Seydoux, G. (1998). Genetic requirements for PIE-1 localization and inhibition of gene expression in the embryonic germ lineage of Caenorhabditis elegans. *Developmental Biology*, *200*, 212–224.

Tomancak, P., Beaton, A., Weiszmann, R., Kwan, E., Shu, S., Lewis, S. E., et al. (2002). Systematic determination of patterns of gene expression during Drosophila embryogenesis. *Genome Biology*, *3*, RESEARCH0088.

Tomancak, P., Berman, B. P., Beaton, A., Weiszmann, R., Kwan, E., Hartenstein, V., et al. (2007). Global analysis of patterns of gene expression during Drosophila embryogenesis. *Genome Biology*, *8*, R145.

Torres-Padilla, M. E., & Zernicka-Goetz, M. (2006). Role of TIF1alpha as a modulator of embryonic transcription in the mouse zygote. *The Journal of Cell Biology*, *174*, 329–338.

Toyama, R., Rebbert, M. L., Dey, A., Ozato, K., & Dawid, I. B. (2008). Brd4 associates with mitotic chromosomes throughout early zebrafish embryogenesis. *Developmental Dynamics*, *237*, 1636–1644.

Tsai, A., Muthusamy, A. K., Alves, M. R., Lavis, L. D., Singer, R. H., Stern, D. L., et al. (2017). Nuclear microenvironments modulate transcription from low-affinity enhancers. *eLife*, *6*, e28975.

Tsukiyama, T., Becker, P. B., & Wu, C. (1994). ATP-dependent nucleosome disruption at a heat-shock promoter mediated by binding of GAGA transcription factor. *Nature*, *367*, 525–532.

Tsukiyama, T., & Wu, C. (1995). Purification and properties of an ATP-dependent nucleosome remodeling factor. *Cell*, *83*, 1011–1020.

Ura, K., Nightingale, K., & Wolffe, A. P. (1996). Differential association of HMG1 and linker histones B4 and H1 with dinucleosomal DNA: Structural transitions and transcriptional repression. *The EMBO Journal*, *15*, 4959–4969.

van den Berg, D. L., Snoek, T., Mullin, N. P., Yates, A., Bezstarosti, K., Demmers, J., et al. (2010). An Oct4-centered protein interaction network in embryonic stem cells. *Cell Stem Cell*, *6*, 369–381.

van Steensel, B., & Furlong, E. E. M. (2019). The role of transcription in shaping the spatial organization of the genome. *Nature Reviews. Molecular Cell Biology*, *20*, 327–337.

Vastenhouw, N. L., Cao, W. X., & Lipshitz, H. D. (2019). The maternal-to-zygotic transition revisited. *Development*, *146*, dev161471.

Vastenhouw, N. L., & Schier, A. F. (2012). Bivalent histone modifications in early embryogenesis. *Current Opinion in Cell Biology*, *24*, 374–386.

Vastenhouw, N. L., Zhang, Y., Woods, I. G., Imam, F., Regev, A., Liu, X. S., et al. (2010). Chromatin signature of embryonic pluripotency is established during genome activation. *Nature*, *464*, 922–926.

Veenstra, G. J., Destree, O. H., & Wolffe, A. P. (1999). Translation of maternal TATA-binding protein mRNA potentiates basal but not activated transcription in Xenopus embryos at the midblastula transition. *Molecular and Cellular Biology*, *19*, 7972–7982.

Veil, M., Schaechtle, M. A., Gao, M., Kirner, V., Buryanova, L., Grethen, R., et al. (2018). Maternal Nanog is required for zebrafish embryo architecture and for cell viability during gastrulation. *Development*, *145*, dev155366.

Veil, M., Yampolsky, L. Y., Gruning, B., & Onichtchouk, D. (2019). Pou5f3, SoxB1, and Nanog remodel chromatin on high nucleosome affinity regions at zygotic genome activation. *Genome Research*, *29*, 383–395.

Vietri Rudan, M., & Hadjur, S. (2015). Genetic tailors: CTCF and cohesin shape the genome during evolution. *Trends in Genetics*, *31*, 651–660.

Vriz, S., & Lovell-Badge, R. (1995). The zebrafish Zf-Sox 19 protein: A novel member of the Sox family which reveals highly conserved motifs outside of the DNA-binding domain. *Gene*, *153*, 275–276.

Wang, J., Levasseur, D. N., & Orkin, S. H. (2008). Requirement of Nanog dimerization for stem cell self-renewal and pluripotency. *Proceedings of the National Academy of Sciences of the United States of America*, *105*, 6326–6331.

Whyte, W. A., Orlando, D. A., Hnisz, D., Abraham, B. J., Lin, C. Y., Kagey, M. H., et al. (2013). Master transcription factors and mediator establish super-enhancers at key cell identity genes. *Cell*, *153*, 307–319.

Wolffe, A. P., & Pruss, D. (1996). Hanging on to histones. Chromatin. *Current Biology*, *6*, 234–237.

Wong, K. H., Jin, Y., & Struhl, K. (2014). TFIIH phosphorylation of the Pol II CTD stimulates mediator dissociation from the preinitiation complex and promoter escape. *Molecular Cell, 54*, 601–612.

Woodland, H. R., & Adamson, E. D. (1977). The synthesis and storage of histones during the oogenesis of Xenopus laevis. *Developmental Biology, 57*, 118–135.

Wu, E., Vashisht, A. A., Chapat, C., Flamand, M. N., Cohen, E., Sarov, M., et al. (2017). A continuum of mRNP complexes in embryonic microRNA-mediated silencing. *Nucleic Acids Research, 45*, 2081–2098.

Wu, J., Xu, J., Liu, B., Yao, G., Wang, P., Lin, Z., et al. (2018). Chromatin analysis in human early development reveals epigenetic transition during ZGA. *Nature, 557*, 256–260.

Xu, Z., Chen, H., Ling, J., Yu, D., Struffi, P., & Small, S. (2014). Impacts of the ubiquitous factor Zelda on Bicoid-dependent DNA binding and transcription in Drosophila. *Genes & Development, 28*, 608–621.

Xu, C., Fan, Z. P., Muller, P., Fogley, R., DiBiase, A., Trompouki, E., et al. (2012). Nanog-like regulates endoderm formation through the Mxtx2-Nodal pathway. *Developmental Cell, 22*, 625–638.

Yamada, S., Whitney, P. H., Huang, S. K., Eck, E. C., Garcia, H. G., & Rushlow, C. A. (2019). The Drosophila pioneer factor Zelda modulates the nuclear microenvironment of a dorsal target enhancer to potentiate transcriptional output. *Current Biology, 29*, 1387–1393.e1385.

Yanez-Cuna, J. O., Dinh, H. Q., Kvon, E. Z., Shlyueva, D., & Stark, A. (2012). Uncovering cis-regulatory sequence requirements for context-specific transcription factor binding. *Genome Research, 22*, 2018–2030.

Yang, Y., Wang, L., Han, X., Yang, W. L., Zhang, M., Ma, H. L., et al. (2019). RNA 5-methylcytosine facilitates the maternal-to-zygotic transition by preventing maternal mRNA decay. *Molecular Cell, 75*, 1188–1202.e11.

Yu, J., Vodyanik, M. A., Smuga-Otto, K., Antosiewicz-Bourget, J., Frane, J. L., Tian, S., et al. (2007). Induced pluripotent stem cell lines derived from human somatic cells. *Science, 318*, 1917–1920.

Zaret, K. S., & Carroll, J. S. (2011). Pioneer transcription factors: Establishing competence for gene expression. *Genes & Development, 25*, 2227–2241.

Zeitlinger, J., Stark, A., Kellis, M., Hong, J. W., Nechaev, S., Adelman, K., et al. (2007). RNA polymerase stalling at developmental control genes in the Drosophila melanogaster embryo. *Nature Genetics, 39*, 1512–1516.

Zhang, Z., Ahmed-Braimah, Y. H., Goldberg, M. L., & Wolfner, M. F. (2019). Calcineurin-dependent protein phosphorylation changes during egg activation in Drosophila melanogaster. *Molecular & Cellular Proteomics, 18*, S145–S158.

Zhang, M., Kothari, P., Mullins, M., & Lampson, M. A. (2014). Regulation of zygotic genome activation and DNA damage checkpoint acquisition at the mid-blastula transition. *Cell Cycle, 13*, 3828–3838.

Zhang, B., Wu, X., Zhang, W., Shen, W., Sun, Q., Liu, K., et al. (2018). Widespread enhancer dememorization and promoter priming during parental-to-zygotic transition. *Molecular Cell, 72*, 673–686.e676.

Zhao, B. S., Wang, X., Beadell, A. V., Lu, Z., Shi, H., Kuuspalu, A., et al. (2017). m(6)A-dependent maternal mRNA clearance facilitates zebrafish maternal-to-zygotic transition. *Nature, 542*, 475–478.

PART III

Maternal regulation of embryonic patterning and morphogenesis

CHAPTER NINE

Maternal regulation of seed growth and patterning in flowering plants

Allison R. Phillips[a], Matthew M.S. Evans[b,*]
[a]Biology Department, Wisconsin Lutheran College, Milwaukee, WI, United States
[b]Department of Plant Biology, Carnegie Institution for Science, Stanford, CA, United States
*Corresponding author: e-mail address: mevans@carnegiescience.edu

Contents

1. Introduction 258
2. Female gametophyte development 259
3. Seed development 260
4. Gametophytic maternal effects 262
   4.1 Maternal effects caused by abnormal embryo sac morphology 263
   4.2 Perdurance of maternal products 265
   4.3 Imprinting effects 266
   4.4 Maternal effects of undetermined mechanism 268
5. Sporophytic maternal effects on seed development 269
   5.1 Control of seed size and composition by maternal transfer tissues 270
   5.2 Sporophytic effects on embryo patterning 272
   5.3 Effect of DNA methylation and small RNAs on seed development 273
6. Conclusion 274
Acknowledgments 274
References 274

Abstract

The plant haploid generation is specified late in higher plant development, and post-meiotic haploid plant cells divide mitotically to produce a haploid gametophyte, in which a subset of cells differentiates into the gametes. The immediate mother of the angiosperm seed is the female gametophyte, also called the embryo sac. In most flowering plants the embryo sac is comprised of two kinds of gametes (egg and central cell) and two kinds of subsidiary cells (antipodals and synergids) all of which descend from a single haploid spore produced by meiosis. The embryo sac develops within a specialized organ of the flower called the ovule, which supports and controls many steps in the development of both the embryo sac and the seed. Double fertilization of the central cell and egg cell by the two sperm cells of a pollen grain produce the endosperm and embryo of the seed, respectively. The endosperm and embryo develop under the influence of their precursor gametes and the surrounding tissues of the ovule

Current Topics in Developmental Biology, Volume 140
ISSN 0070-2153
https://doi.org/10.1016/bs.ctdb.2019.10.008

and the gametophyte. The final size and pattern of the angiosperm seed then is the result of complex interactions across multiple tissues of three different generations (maternal sporophyte, maternal gametophyte, and the fertilization products) and three different ploidies (haploid gametophyte, diploid parental sporophyte and embryo, and triploid endosperm).

1. Introduction

The life cycle of angiosperms alternates between diploid sporophyte and haploid gametophyte generations (Walbot & Evans, 2003). In animals, meiosis is followed immediately by maturation of the gametes (egg or sperm) without subsequent divisions. Conversely, after meiosis, plant spores undergo a period of growth to produce a multicellular gametophyte, which may be independent of (e.g., ferns) or dependent on (e.g., flowering plants) the diploid mother sporophyte. Flowering plant gametophytes are sexually dimorphic. The female gametophyte, or embryo sac, contains two gametes, the egg cell and the central cell, and subsidiary antipodal and synergid cells, whereas the male gametophyte, or pollen grain, contains three cells: two sperm cells inside the vegetative cell. Tip growth of the vegetative cell produces a pollen tube that carries the sperm cells through floral tissues to the embryo sac. Double fertilization results in the formation of the seed and consists of two events: one sperm cell fuses with the egg cell to form the embryo and the second sperm cell with the diploid central cell to form the endosperm. Because genes are actively expressed in the haploid generation to control development and function of the gametes and subsidiary cells, mutations in genes essential for these processes have characteristic effects: mutations have reduced transmission efficiency compared to the wild-type allele (in crosses with heterozygous plants) and female gametophyte mutants have reduced fertility (but not males because there is an excess of pollen in most crosses) (Evans & Grossniklaus, 2009; Hackenberg & Twell, 2019; Serbes, Palovaara, & Gross-Hardt, 2019; Walbot & Evans, 2003). Many mutant screens have taken advantage of the reduced transmission and/or fertility to identify gametophyte mutants. Most commonly, female mutants exhibit early embryo sac arrest suggesting the gametophyte is largely under the control of the female haploid genome (Serbes et al., 2019). Because of this active gene expression and the fact that the female gametophyte and the seed develop imbedded within the diploid floral organs, the diploid sporophyte has maternal effects on the daughter embryo sac and (grand)daughter seeds and the haploid embryo sac has maternal effects on the daughter seed (Fig. 1) (Chettoor, Phillips, Coker, Dilkes, & Evans, 2016; Evans & Kermicle, 2001; Golden et al., 2002; Hulskamp, Schneitz, & Pruitt, 1995). This review will focus primarily on maternal effects on the seed from the maternal sporophyte and gametophyte with an emphasis on maize and Arabidopsis as model systems.

2. Female gametophyte development

The embryo sac of both *Arabidopsis* and maize follows the *Polygonum* type arrangement, which is the most common developmental pattern in angiosperms (Drews & Yadegari, 2002; Evans & Grossniklaus, 2009). The embryo sac develops within the ovule surrounded by the nucellus, which is in turn surrounded by the integuments. One cell of the nucellus undergoes meiosis to produce four megaspores. One megaspore survives

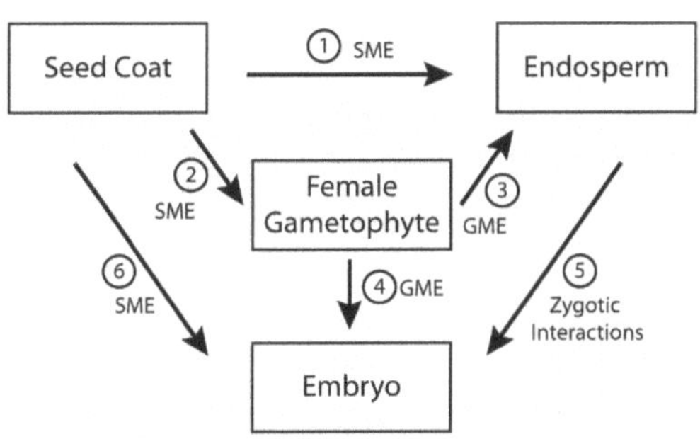

Category (*mutant example*)
1. Maternal transfer tissues (*ttg2*)
2. Maternal transfer tissues (*amp1*)
3. Imprinting (*mea*); Embryo sac morphology (*bsl1*); Perdurance (*PRL*)
4. Embryo sac morphology (*ssc1*); Perdurance (*ESF1*);
5. Maternal endosperm defects (*MEE)*
6. Maternal transfer tissues (*mkk4/5*)

Fig. 1 Types of maternal effects between different seed and floral tissues. The female gametophyte contains the two maternal gametes the egg cell and the central cell. The seed coat, endosperm and embryo are tissues that comprise the mature seed. The seed coat is derived from tissues of the maternal diploid flower, while the endosperm and embryo are zygotic tissues. SME, sporophytic maternal effect; GME, gametophytic maternal effect.

and undergoes three rounds of free nuclear division followed by cellularization to produce an embryo sac consisting of 8 nuclei in 7 cells (Evans & Grossniklaus, 2009; Serbes et al., 2019). At the micropylar end, two peripheral cells develop into synergids (which attract the pollen tube for sperm delivery), and the third cell differentiates into the egg cell. In the center, the large homodiploid central cell (containing two haploid polar nuclei) is fertilized to produce the endosperm. At the chalazal end, opposite to the micropyle, the cluster of three cells differentiates into the antipodal cells. Polarity is already evident during the syncytial stages of embryo sac development with asymmetric distribution of plastids, mitochondria, and microtubules (Huang & Sheridan, 1994; Russell, 1979). Positioning of the nuclei along the micropylar-chalazal (m-c) axis defines cellular differentiation within the mature embryo sac (Chettoor & Evans, 2015; Ebel, Mariconti, & Gruissem, 2004; Guo, Huang, Han, & Zee, 2004; Sundaresan & Alandete-Saez, 2010; Tekleyohans, Nakel, & Gross-Hardt, 2017). In maize, the antipodal cells proliferate forming a cluster of 20–100 cells at maturity, and while they do not divide in Arabidopsis, they persist after fertilization in both species (Randolph, 1936; Song, Yuan, & Sundaresan, 2014). In maize, the antipodal cells have papillate cell walls suggesting they can act as a transfer tissue for fluid and nutrient movement (Diboll & Larson, 1966), and in both species there is the possibility that they provide positional information post-fertilization for seed development. Auxin signaling in the micropylar nucellus appears to be involved in embryo sac patterning, and auxin signaling is also associated with antipodal cell proliferation in maize (Ceccato et al., 2013; Chettoor & Evans, 2015; Lituiev et al., 2013).

3. Seed development

The first asymmetric division of the zygote establishes the apical-basal axis of the embryo. The larger basal cell (at the micropylar pole of the embryo sac) will divide to form part of the root meristem and a column of cells that will differentiate into an extra embryonic transfer tissue, the suspensor. The smaller apical cell will divide in all planes to produce most of the tissue types of the embryo, including the root meristem, the embryonic shoot or hypocotyl, the cotyledon(s) (called the scutellum in maize), and the shoot meristem (Nardmann & Werr, 2009). The endosperm—the sister of the embryo—is the nutritive tissue that supports the embryo during germination and/or embryogenesis. In maize and Arabidopsis (and many other angiosperms) the endosperm is triploid with two genome copies from the mother and only one from the father—deviations from this ratio typically lead to abnormal endosperm development (Charlton et al., 1995; Haig & Westoby, 1991; Scott, Spielman, Bailey, & Dickinson, 1998). In both species, the endosperm begins development as a syncytium and then cellularizes centripetally. The mature endosperm of maize consists of multiple tissue types (Fig. 2): the embryo surrounding region (ESR) and basal endosperm transfer layer (BETL) are adjacent to the maternal sporophyte tissues at the base of the endosperm; apical to that is the basal intermediate zone (BIZ) followed by the conducting zone (CZ); the outermost two layers are the

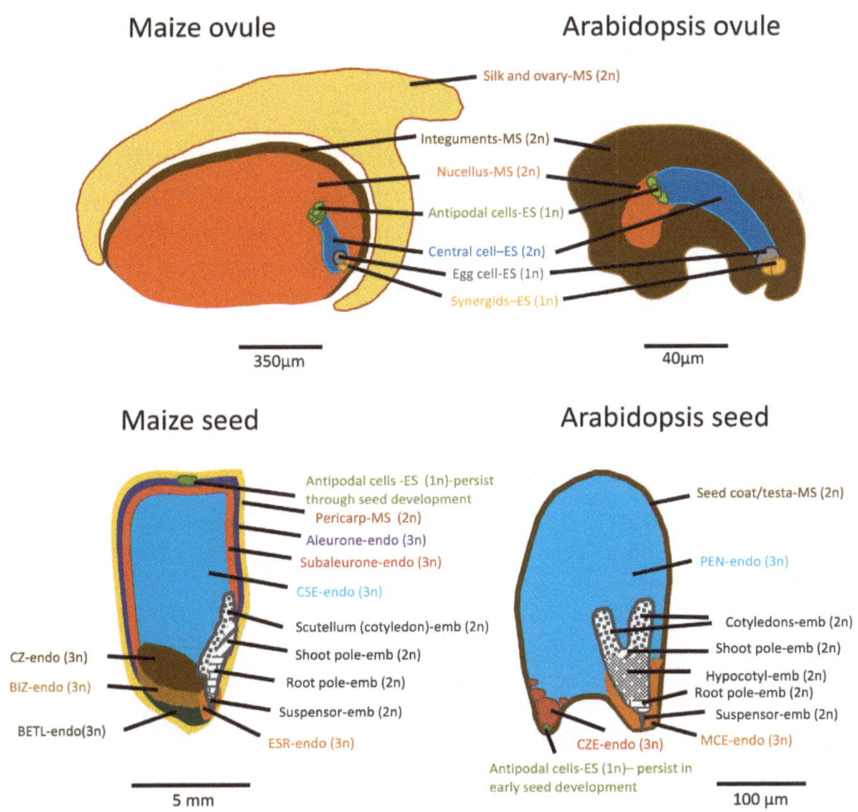

Fig. 2 Tissue types in maize and Arabidopsis ovules, embryo sacs, and seeds. The ploidy of each tissue is indicated. Whether each cell type is part of the maternal sporophyte (MS), embryo sac (ES), endosperm (endo), or embryo (emb) is indicated. BETL, basal endosperm transfer layer; BIZ, basal intermediate zone; CSE, central starchy endosperm; CZ, conducting zone; CZE, chalazal endosperm (cyst and nodes); ESR, embryo surrounding region; MCE, micropylar endosperm; PEN, peripheral endosperm.

aluerone (AL) and subaleurone (SAL); and the central starchy endosperm (CSE) makes up the bulk of the endosperm. These regions are defined by distinct morphology and/or transcriptional domains (Leroux et al., 2014; Li et al., 2014; Olsen, 2004; Olsen, Linnestad, & Nichols, 1999). In Arabidopsis, the embryo develops similarly but with two cotyledons and a more distinct hypocotyl. The endosperm has fewer defined domains than in maize with a chalazal cyst and nodules (CZE) next to the antipodal cells, the micropylar endosperm (MCE) next to the embryo, and the peripheral endosperm (PEN) making up the bulk of the endosperm (Li, 2017; Sørensen, Chaudhury, Robert, Bancharel, & Berger, 2001). The seed coat, or testa, is derived from the maternal integuments of the ovule.

Seed morphology and seed size are coordinated by interactions between maternal gametophytic tissues (embryo sac), maternal sporophytic tissues (integuments and nucellus), and zygotic tissues (endosperm and embryo). Gametophytic and sporophytic maternal effects can be distinguished by their mode of transmission (Chettoor et al., 2016; Evans & Kermicle, 2001; Grossniklaus & Schneitz, 1998). For a gametophytic maternal effect, abnormal seeds always carry the mutation by inheritance through the maternal embryo sac and up to half of the seeds from heterozygotes can be defective. For recessive sporophytic effects, progeny only display the phenotype if the mother plant is homozygous, unlike gametophytic maternal effects. For a dominant sporophytic maternal effect, progeny exhibit the phenotype if the mother plant is heterozygous like gametophytic maternal effects. However, in dominant sporophytic mutants, some defective seeds can be homozygous wild-type because the genotype of the surrounding sporophytic cells is critical rather than that of the gametophytes, while, for maternal gametophyte effect mutants, defective seeds always carry the mutant allele by inheritance through the haploid embryo sac.

4. Gametophytic maternal effects

Many forward genetic screens for mutants with reduced transmission rates and/or abnormal seed development have identified gametophytic maternal effect mutants (Bai et al., 2016; Chettoor et al., 2016; Gavazzi et al., 1997; Grini, Jurgens, & Hulskamp, 2002; Grossniklaus, Vielle-Calzada, Hoeppner, & Gagliano, 1998; Kohler & Grossniklaus, 2005; Pagnussat et al., 2005). Analysis of gametophytic maternal effect mutants have revealed several possible causes for the defects, including: (1) abnormal embryo sac morphology; (2) loss of embryo sac expressed factors normally stored for post-fertilization function (perdurance); (3) imprinting; or (4) dosage sensitivity in the endosperm (Chettoor et al., 2016). Interestingly, several gametophytic maternal effect mutants in maize have been identified that progress until very late in seed development, while those in Arabidopsis almost all arrest very early in seed development (e.g., (Bai et al., 2016; Chettoor et al., 2016; Pagnussat et al., 2005)). This may be a consequence of the short-lived nature of the endosperm of Arabidopsis compared to maize.

4.1 Maternal effects caused by abnormal embryo sac morphology

Many mutants have been identified that regulate embryo sac morphology prior to fertilization with hypothesized or demonstrated downstream effects on seed development (Chettoor et al., 2016; Gutierrez-Marcos, Costa, & Evans, 2006; Huang & Sheridan, 1996; Lin, 1978; Pagnussat et al., 2005; Phillips & Evans, 2011). Some mutants have extra nuclei in the central cell—changing the maternal-paternal genome ratio in the endosperm after fertilization—while in others the egg cell or central cell are abnormal (Huang & Sheridan, 1996; Ingouff, Jullien, & Berger, 2006; Lin, 1978, 1981). Maize *indeterminate gametophyte1 (ig1)* embryo sacs undergo extra rounds of free-nuclear divisions before cellularization and consequently produce extra cells and central cells with extra nuclei (Huang & Sheridan, 1996; Lin, 1978, 1981). The extra cells differentiate depending on their position along the m-c axis (Guo et al., 2004; Lin, 1978, 1981). The *rbr1* (*retinoblastoma related1*) mutant in Arabidopsis also undergoes excessive nuclear proliferation in the micropylar region of the embryo sac, resulting in supernumerary central cell nuclei (Ingouff et al., 2006). In seeds from *rbr1* mutant embryo sacs, both embryo and endosperm develop abnormally and seeds abort early (Ebel et al., 2004).

Other maternal effect mutants have been identified in Arabidopsis that fail to form endosperm and also abort early. *wyrd (wyr)*, *eostre/eda12*, *lachesis (lis)*, *agamous-like80 (agl80)*, *agl61*, *gamete cell defective1* (gcd*1)*, and *exportin1a/b* (*xpo1a/b*) all show embryo arrest at variable stages from the 1-celled zygotic stage to globular embryos (Blanvillain, Boavida, McCormick, & Ow, 2008; Gross-Hardt et al., 2007; Kirioukhova et al., 2011; Pagnussat, Yu, & Sundaresan, 2007; Portereiko et al., 2006; Steffen, Kang, Portereiko, Lloyd, & Drews, 2008; Wu et al., 2012). If synergids are miss-specified as egg cells, two eggs can be fertilized but not the central cell leading to endosperm absence. Embryos cannot develop past globular stage without the support of the endosperm tissue and so seeds abort very early in development. Twins were seen in the *eostre* mutants (Kong, Lau, & Jurgens, 2015; Pagnussat et al., 2007), but not the *wyr* or *lis* mutants (Gross-Hardt et al., 2007; Kirioukhova et al., 2011). In *agl80* and *agl61* mutants, antipodal, egg, and synergid cells are normal, but the central cell nucleus and vacuole are smaller than wild type (Portereiko et al., 2006; Steffen et al., 2008). In *agl80* and *agl61* endosperm does not form or only has a few nuclei again leading to embryo arrest.

The *xpo1a/b* mutants of Arabidopsis and *maternally reduced endosperm2* (*mrn2*) and *mrn3* mutants of maize cause embryo sac arrest, but a subset of embryo sacs can complete free-nuclear divisions, cellularize, and be fertilized but produce abnormal seeds (Blanvillain et al., 2008; Chettoor et al., 2016; Wu et al., 2012). *xpo1a/b* mutants are disrupted at various stages of development from the two-nucleus stage, to cellularization and positioning of nuclei, to final maturation. The Arabidopsis *maternal effect embryo arrest33 (mee33)/oiwa/msd1* mutant exhibits many embryo sac defects including early arrest, aberrant micropylar cell-specific gene expression, extra egg cells, and polar nuclei that fail to fuse (Martin, Fiol, Sundaresan, Zabaleta, & Pagnussat, 2013). *oiwa* antipodal patterning appears normal. Some *oiwa* embryo sacs are able to be fertilized, but arrest at the one-celled zygotic stage with one-celled endosperm development. *gcd1* mutants have all cells typically present but abnormal late stage egg cells, central cells, and antipodals with subsequent embryo and endosperm abortion at the one to two celled stages (Wu et al., 2012). In maize, two mutants with abnormal egg cells, *sans scion1* (*ssc1*) and *empty creche1* (*ecr1*), (Chettoor et al., 2016) either fail to produce an embryo or the embryos abort early, perhaps by a similar mechanism to the *gcd1* mutant.

Additional classes of embryo sac defects associated with abnormal kernel development have been identified in maize (Chettoor et al., 2016; Gutierrez-Marcos et al., 2006; Phillips & Evans, 2011). Models linking pre-fertilization defects to abnormal seed development are sufficient to explain some maternal effects. Classes include: (1) smaller embryo sacs with miniature central cells and fewer antipodal cells, *stunter1* (*stt1*), *stt2*, *stt3*, and *ecr1* (Chettoor et al., 2016; Phillips & Evans, 2011); (2) misplaced polar nuclei, *baseless1* (*bsl1*), *bsl2*, *no legacy1* (*nol1*), *heirless1* (*hrl1*), *superbase1* (*sba1*),and *mrn1* (Chettoor et al., 2016; Gutierrez-Marcos et al., 2006); (3) abnormal antipodal cell clusters (*hrl1*, *nol1*, *sba1*, *mrn1*, *stt2*, *stt3*) (Chettoor et al., 2016); and (4) extra cells or duplication of the whole embryo sac (*mrn2*, *stt3*) (Chettoor et al., 2016).

The mutants (*stt1* and *stt2*) that only reduce embryo sac size have normal BETL (based on expression of *pBET1::GUS* (Hueros et al., 1999)) and AL patterning (based on expression of *pVP1::GUS* (Cao et al., 2007)). However, both endosperms and embryos grow more slowly resulting in smaller seeds. The mutants with reduced embryo sac size plus other defects (*stt3* and *ecr1*) have some BETL patterning defects; however, the effect on other endosperm domains has not yet been examined.

Most of the mutants with abnormal polar nuclei position (except for *mrn1*) display disruptions in BETL patterning (either a reduced domain or ectopic expression of BETL markers) (Chettoor et al., 2016; Gutierrez-Marcos et al., 2006). Consistent with this data is a model in which a factor is asymmetrically distributed within the central cell, and this asymmetry is propagated within the endosperm until BETL differentiation. The localization of this factor would also share some (but not all) mechanisms in common with the localization of the polar nuclei. The BETL transports nutrients from the mother plant into the seed during development and grain fill. Consequently, these mutants have variable levels of reduced endosperm, hypothetically due to poor nutrient uptake, and occasional embryo abnormalities. In *bsl1* mutants, for example, some basal nuclei within the center of the normal BETL domain have CSE morphology rather than BETL due to mislocalization of this hypothetical factor (Gutierrez-Marcos et al., 2006). Markers for other domains of the endosperm, ESR, BIZ, CZ, SAL, and CSE have not been examined in these mutants.

4.2 Perdurance of maternal products

Some maternal effect genes appear to be expressed in the gametes with products stored for post-fertilization functions (Springer, Holding, Groover, Yordan, & Martienssen, 2000), a process referred to as perdurance. Additionally, recent results in rice and maize have demonstrated that the egg cell makes a large contribution to the zygote transcriptome (Anderson et al., 2017; Chen et al., 2017). The *PROLIFERA* (*PRL*) gene in Arabidopsis is expressed in both the central cell and the egg cell prior to fertilization and the protein products accumulate in the nucleus (Springer et al., 2000; Springer, McCombie, Sundaresan, & Martienssen, 1995). *PRL* encodes an MCM homolog that functions in yeast and mammals in a nuclear protein complex to regulate the initiation of DNA replication during the G1 phase of the cell cycle (Springer et al., 2000, 1995). In Arabidopsis, expression in the embryo sac leads to a significant maternal contribution of PRL protein to the early embryo and endosperm, which can then be used to support early nuclear divisions. Upon *prl* loss-of-function, many embryo sacs arrest at the four-nucleate stage (Springer et al., 1995). Incomplete penetrance results in some mature *prl* embryo sacs, but fertilized *prl* embryo sacs produce seeds with the embryo and the endosperm arrested (1-celled to globular stage embryos with 1–8 endosperm nuclei) (Springer et al., 2000).

Embryo Surrounding Factor1.1 (*ESF1.1*), *1.2* and *1.3* in Arabidopsis are expressed in the central cell and persist in the micropylar endosperm (MCE) to control early suspensor and proembryo patterning (Costa et al., 2014). Loss of *ESF* function results in significantly smaller, abnormal seeds with aberrant patterning of the embryo proper and a reduced number of suspensor cells with no associated effect on endosperm development (Costa et al., 2014). Further analysis of the embryo patterning defects suggests a role for *ESF1* together with the paternal effect *SHORT SUSPENSOR1* gene upstream of the *MAPKKK*, *YODA* gene, to regulate development of the basal lineage of the embryo (Costa et al., 2014). *ESF1* acts non-cell autonomously, and *ESF1* peptides are present pre-fertilization and stored in the cytoplasm of the central cell until needed post-fertilization in the endosperm to control embryo patterning.

4.3 Imprinting effects

Non-equivalence of parental alleles in seed development can be explained by parental conflict theory, leading to restriction or promotion of the growth of the endosperm by maternal and paternal alleles, respectively (Haig & Westoby, 1989). This theory is supported by the endosperm phenotypes of seeds with maternal or paternal genome excess from crosses between diploids and tetraploids (Charlton et al., 1995; Haig & Westoby, 1991; Scott et al., 1998). In maize, the BETL is particularly sensitive to maternal or paternal genome excess (Charlton et al., 1995). In Arabidopsis, loss-of-function in the pollen parent of the paternally expressed *ADMETOS*, *su(var)3–9 homolog7* (*suvh7*), *paternally expressed gene2* (*peg2*), or *peg9* genes restores viability in the presence of paternal excess (Kradolfer, Wolff, Jiang, Siretskiy, & Kohler, 2013; Wolff, Jiang, Wang, Santos-Gonzalez, & Kohler, 2015). Other models have also been proposed to explain imprinting, including generating an increase in effective allelic diversity, modulating gene dosage, or maternal regulation of growth (Bai & Settles, 2014; Dilkes & Comai, 2004; Feil & Berger, 2007; Pignatta et al., 2014). Imprinting is controlled by complex interactions involving chromatin modification, small RNAs (sRNAs), and

DNA methylation (reviewed in Feil & Berger, 2007; Martinez & Kohler, 2017; Rodrigues, Luo, Berger, & Koltunow, 2010). The silencing histone modification H3K27me3 depends on Polycomb Group Repressor Complex2 (PRC2), while DNA methylation requires the maintenance methylase *DNA METHYL TRANSFERASE1* (*MET1*) in Arabidopsis (Rodrigues & Zilberman, 2015). There is extensive demethylation of the central cell genome, dependent on the activity of *DEMETER* in Arabidopsis, leading to hypomethylation of maternal alleles of certain genes in the endosperm (Gehring, Bubb, & Henikoff, 2009; Gehring, Reik, & Henikoff, 2009; Hsieh et al., 2009; Park et al., 2016). One of these genes is *MEDEA* (*MEA*), part of the PRC2, which then suppresses the expression of the maternal alleles of many genes in the endosperm, such as *PHERES* and other Class I MADS-Box genes (Kohler et al., 2003; Kohler, Page, Gagliardini, & Grossniklaus, 2005; Zhang et al., 2018). The repressive chromatin mark H3K27me3 and binding by PRC2 are associated with silencing of maternal alleles of paternally expressed genes in maize and Arabidopsis (Moreno-Romero, Jiang, Santos-Gonzalez, & Kohler, 2016; Zhang et al., 2014).

Many sRNAs have roles in plant reproduction, including silencing of transposable elements in the gametes (Slotkin et al., 2009), and have imprinted expression in the endosperm (reviewed in Martinez & Kohler, 2017). PolIV-dependent sRNAs are maternally expressed in the endosperm and depend on maternal *NRPD1A* (Mosher et al., 2009), and the RNA-directed DNA Methylation (RdDM) pathway is important for imprinted expression of maternally expressed genes (Vu et al., 2013).

RNA-Seq has enabled the identification of hundreds of genes with parent-specific and parent-biased expression in the seeds of several plant species, in the endosperm and embryo (Autran et al., 2011; Baroux, Blanvillain, & Gallois, 2001; Baroux & Grossniklaus, 2015; Dong et al., 2017; Gehring, Missirian, & Henikoff, 2011; Grimanelli, Perotti, Ramirez, & Leblanc, 2005; Hsieh et al., 2011; Pignatta et al., 2014; Vielle-Calzada, Baskar, & Grossniklaus, 2000; Waters et al., 2011; Wolff et al., 2011; Xin et al., 2013; Zhang et al., 2011). In some cases, a gene can be imprinted in one endosperm domain but have biallelic expression in another, such as the CSE vs aleurone in maize (Zhang, Lv, Yang, Fu, & Liu, 2018). Imprinting can also be stage-specific, with uniparental expression early and biallelic expression later in development. Often, the expression is from both alleles but is unequal with maternal-biased expression being more common (Hornslien, Miller, & Grini, 2019).

A developmental function has not been verified for most of these imprinted genes. However, loss of function for the PRC2 components, *MEA*, *FERTILIZATION INDEPENDENT ENDOSPERM*, or *FERTILIZATION INDEPENDENT SEED2* cause a variety of defects in seed development in Arabidopsis (Kohler et al., 2003; Luo et al., 1999; Sørensen et al., 2001). For example, embryos arrest prematurely, endosperm development can initiate without fertilization and the endosperm has abnormal patterning with an expansion of the chalazal cyst domain of the endosperm into the PEN (Sørensen et al., 2001). Loss of the maternally expressed gene, *AtFormin Homologue5*, prevents chalazal cyst formation in *mea* mutants (FitzGerald, Hui, & Berger, 2009). Loss-of-function of the imprinted *maternally expressed gene1* (*meg1*) gene reduces the domain of the BETL in maize and consequently diminishes nutrient allocation to the seed (Costa et al., 2012). This contradicts Parent Conflict Theory and supports a role for maternal control of essential seed developmental processes unrelated to parental conflict. Moreover, the maize mutant, *floury3* (*fl3*), was long thought to be dosage-sensitive in the endosperm (Ma & Nelson, 1975), but cloning of *fl3* revealed it is actually imprinted in the endosperm, and *fl3* mutants produce smaller seeds like *meg1* (Li et al., 2017).

4.4 Maternal effects of undetermined mechanism

Many maternal effect mutants in Arabidopsis were first identified based on transmission defects and reduced fertility. Reciprocal crosses and cytological analyses revealed they had been fertilized but the endosperm and/or embryo arrested so early that no seed development was visible. The largest class of such mutants were identified by Pagnussat et al. (2005) and named *mee* for *maternal effect embryo arrest*. Of these, 56 arrested at the 1-celled zygotic stage with 1–4 endosperm nuclei, 6 arrested at the 1-celled zygotic stage but endosperm development was normal, and 8 arrested at the 2-celled or globular embryo stage with various amounts of endosperm development. These mutants confirmed that embryo development cannot proceed past the globular stage without support of the endosperm, but endosperm can develop in the absence of embryo development. The exact mechanism underlying arrest for most is unknown but the early arrest is similar to *prolifera*, which acts through perdurance of gametophyte expressed proteins (Pagnussat et al., 2005; Springer et al., 2000, 1995). The *capulet1* (*cap1*) and *cap2* maternal effect mutants may also function in this way (Grini et al., 2002). They do not display any pre-fertilization defects, and endosperms and embryos arrest 1–5 days after pollination at variable stages in development. To date it is not known what genes are disrupted in the *cap1* and *cap2* mutants.

Several maize gametophytic maternal effect mutants affect endosperm patterning or growth but the underlying mechanism is unclear (Bai et al., 2016; Chettoor et al., 2016; Evans & Kermicle, 2001; Gavazzi et al., 1997; Pan & Peterson, 1989). *Miniature-Uq* (*Mn-Uq*), *top knot1* (*tpn1*), and *no bet expression1* (*nbe1*) all have seeds with a loose pericarp due to a reduced endosperm (Chettoor et al., 2016; Pan & Peterson, 1989); *hrl2* and *maternal effect lethal1* (*mel1*) have an etched or pitted endosperm with abnormal embryos (Chettoor et al., 2016; Evans & Kermicle, 2001); and *ecr2* has a reduced endosperm with aborted embryos (Chettoor et al., 2016). All of these mutants have normal embryo sac morphology. *Mn-Uq*, *tpn1*, *nbe1*, *hrl2*, and *ecr2* show defects in the BETL marker expression patterns (Chettoor et al., 2016), while *mel1* does not (unpublished observations). The defects include complete loss of *pBET1::GUS* expression and ectopic expression of *pBET1::GUS* in extra cell layers (potentially in the BIZ or CZ domains) or in spots on the abgerminal (away from the embryo) surface of the endosperm. These mutants could affect production or function of a hypothetical patterning factor.

Two additional classes of mutants have yet to be examined for defects in embryo sac morphology: the *maternal rough endosperm* (*mre*) and *Defective aleurone pigmentation* (*Dap*) mutants. All three *mre* mutants have defects in endosperm growth and BETL development (Bai et al., 2016). The *Dap* mutants have reduced endosperm size and patches of abnormal aleurone and subaleurone tissue (Gavazzi et al., 1997). Only aleurone and subaleurone have been examined in the *Dap* mutants, and only BETL patterning has been examined in *mel1*, *Mn-Uq*, *tpn1*, *nbe1*, *hrl2*, *ecr2* and *mre* mutants (Bai et al., 2016; Chettoor et al., 2016; Evans & Kermicle, 2001; Gavazzi et al., 1997). Once these genes have been cloned it will be possible to test if they act via imprinting or perdurance or possibly by encoding, regulating, or localizing a hypothetical patterning factor(s).

5. Sporophytic maternal effects on seed development

Tissues of the maternal sporophyte have profound effects on the size and morphology of the seed. Mechanisms include: (1) disruption of maternal transfer tissues—pedicel, integuments, or other ovule tissues; or (2) the effect of non-coding RNAs including microRNA or siRNA production. The majority of the mutants studied thus far act recessively in the maternal sporophyte and have a significant impact on the early patterning of the seed coat and the embryo.

5.1 Control of seed size and composition by maternal transfer tissues

The maternal sporophyte controls seed size via transfer tissues such as the integuments, which allocate nutrients to the endosperm and deliver molecules or signals that mediate patterning of the embryo (Chen et al., 2015; Ingouff et al., 2006; Schruff et al., 2006). SWEET11, 12 and 15 are sucrose transporters in Arabidopsis that likely coordinate sucrose efflux from the inner integuments through the micropylar region of the endosperm, to the suspensor, and finally fuel growth of the embryo proper (Chen et al., 2015). When disrupted in the maternal sporophytic *sweet11;12;15* triple mutant, seeds have reduced seed weight and underdeveloped embryos (Chen et al., 2015). A set of maternal sporophyte effect mutants in barley also have a shrunken endosperm reminiscent of defects in nutrient transfer from the mother plant (Felker, Peterson, & Nelson, 1985).

Many mutants affect integument growth, which in turn has effects on seed size and shape (reviewed in Li & Li, 2015). Increased or decreased integument length can either change the size of the seed cavity allowing for more or less endosperm expansion or change nutrient transfer into the developing seed (Schruff et al., 2006). Mutants of this class include disruptions in transcription factors, the ubiquitin pathway, cytochrome P450s, G-protein signaling, and hormone signaling. Many have not been rigorously tested for sporophytic or gametophytic maternal effects. However, since they disrupt integument development, a sporophytic maternal effect is assumed.

Transcription factors of the WRKY, AP2/EREBP (ethylene responsive element binding protein), MADS-box, and C2H2 type zinc finger families, as well as the plant hormones auxin, brassinosteroid (BR), and cytokinin control seed coat and seed development (Colombo et al., 1997; Gaiser, Robinson-Beers, & Gasser, 1995; Li & Li, 2015). *Transparent Testa Glabra 2* (*TTG2*) encodes a transcription factor of the WRKY family that is expressed in the integument (Johnson, Kolevski, & Smyth, 2002). Loss of *TTG2* causes a reduction in integument length and consequent reduction in endosperm size (Garcia, Fitz Gerald, & Berger, 2005). The maternal effect of *ttg2* also modulates the phenotype of crosses between diploid and tetraploid plants (Dilkes et al., 2008). Other testa mutants of Arabidopsis, such as *aberrant testa shape* (*ats*) and *apetela2* (*ap2*), display sporophytic maternal effects on seed development (Leon-Kloosterziel, Keijzer, & Koornneef, 1994; Ohto, Fischer, Goldberg, Nakamura, & Harada, 2005). Wild-type *AP2* restricts seed size and regulates the accumulation of proteins, oils, and sugars in the seed by limiting the development of the outer integuments/seed coat (Ohto et al., 2005). Potential targets of *AP2* include signaling by the hormones, gibberellin (Jofuku, Omidyar, Gee, & Okamuro, 2005) and brassinosteroid (Li & Li, 2015), and enzymes involved in sugar metabolism, such as cell-wall-bound invertases (Ohto et al., 2005). Reduced function of the MADS-box *Floral Binding Protein7* (*FBP7*) and *FBP11* genes in petunia results in sporophytic maternal effects on seed coat development and endosperm development (Colombo et al., 1997). Specifically the endothelial cells of the seed coat, derived from the integuments, completely degenerate late in development, with negative impacts on transport into the developing endosperm (Colombo et al., 1997).

DA1 and DA1-Related1 (DAR1), two ubiquitin receptors, and DA2 and Enhancer Of Da1 (EOD1)/Big Brother (BB), two E3 ubiquitin ligases, restrict cell proliferation in the integuments (Disch et al., 2006; Li, Zheng, Corke, Smith, & Bevan, 2008; Xia et al., 2013). Both DA2 and EOD1 act synergistically with DA1 but independent of each other (Xia et al., 2013). Finally, *UBP15*, a ubiquitin specific protease, has been identified as a (likely sporophytic) maternal effect gene that functions with DA1 to regulate seed growth, but acts to promote rather than restrict cell proliferation (Du et al., 2014). Members of the *CYP78A* cytochrome P450 family act to promote integument growth and consequently affect seed size in Arabidopsis and rice, and *CYP78A6* specifically enhances the effects of the ubiquitin pathway genes above (Adamski, Anastasiou, Eriksson, O'Neill, & Lenhard, 2009; Fang, Wang, Cui, Li, & Li, 2012; Nagasawa et al., 2013). *AGG3*, a heterotrimeric G-protein complex subunit, has a positive maternal effect on seed growth (Chakravorty et al., 2011; Li et al., 2012), while two homologs in rice, *GS3* and *DEP1* negatively regulate seed size (Huang et al., 2009; Li & Li, 2015; Mao et al., 2010).

Several major developmental hormones in plants, including auxin, brassinosteroids and cytokinin, have maternal effects on seed size through effects on integument development. In *auxin response factor2* (*arf2*) mutants, seeds are larger and have an abnormal shape, due to increased cells in both the inner and outer integuments (Schruff et al., 2006). Analysis of several BR-insensitive mutants (*de-etiolated2*, *dwarf4* and *brassinosteroid insensitive1*) revealed decreased integument growth resulting in reduced elongation of seeds presumably via BR-responsive genes, such as *KLUH, AINTEGUMENTA, APETELA2, ARF2, ARABIDOPSIS HIS KINASE3 (AHK3)*, and *AHK4* (Jiang et al., 2013). The triple cytokinin receptor mutant, *ahk2;ahk3;akh4/cre1*, has larger seeds than wild type although the nature of the maternal effect is less clear (Riefler, Novak, Strnad, & Schmulling, 2006).

The *altered meristem program1* (*amp1*) mutant produces seeds with two embryos (because there are multiple egg cells) and often no endosperm, similar to some of the gametophytic maternal effectmutants mentioned above (Kong et al., 2015). However, *amp1* is required in the surrounding maternal sporophytic tissue rather than in the embryo sac itself (Kong et al., 2015). This is an example of a sporophytic maternal effect on embryo sac development which then leads to a maternal effect on seed development.

5.2 Sporophytic effects on embryo patterning

The YODA pathway regulates embryonic patterning in part through *WRKY2* and *WOX* (*WUSCHEL* related homeobox) genes (reviewed in Armenta-Medina & Gillmor, 2019). MAP KINASE KINASE4/5 (MKK4/5) and MAP KINASE6 (MPK6) are intermediate kinases in this pathway and when disrupted result in sporophytic maternal effects causing wrinkled seed coats and protruding embryos due to changes in apical-basal patterning of the embryo (Zhang et al., 2017). *mpk6* mutants display abnormal organization of the basal half of the embryo and altered distribution of auxin and the auxin transporter, PIN-FORMED1 (PIN1) (Zhang et al., 2017), suggesting that this MAPK cascade could modulate either auxin biosynthesis or transport. Previous reports show that embryos are able to synthesize their own auxin only after the globular stage when *TRYPTOPHYAN AMINOTRANSFERASE OF ARABIDOPSIS1* (*TAA1*) and *YUCCA* expression are detected in the proembryo apex of globular embryos (Robert et al., 2013). Analyses of the *mpk6* and *mkk4/5* mutants, however, show an important role for auxin prior to the globular stage to establish apical-basal polarity (Zhang et al., 2017), and thus a more direct role for sporophytic maternal control of early embryonic patterning.

HOMEODOMAIN GLABROUS11 and *12* (*HDG11* and *12*) modulate *WOX8* expression and zygote patterning (Ueda et al., 2017). Reciprocal crosses with the *hdg11/12* double mutant and wild type reveal a clear maternal effect on zygote development. Zygotes of *hdg11/12* mutants do not fully elongate and the first division is more symmetric than wild type disrupting apical-basal polarity (Ueda et al., 2017). It is unclear whether the maternal effect is sporophytic or gametophytic in nature and the authors propose two potential explanations: (1) perdurance of the *HDG11/12* transcripts or proteins from the egg cell to function in the early embryo or (2) imprinting or delayed activation of paternal copies of the gene such that only the maternal copies are transcribed in the zygote. Both possibilities would assume a gametophytic maternal effect. However, *HDG11* and *12* are both expressed in the embryo sac (egg, synergid and central cells) as well as in the integuments (Ueda et al., 2017). Other maternal effect genes converge on *WOX8* including *mkk4/5*, *mpk6*, and *ESF1*, described above (Costa et al., 2014; Zhang et al., 2017). A curious aspect of the maternal effects of these various patterning mutants is that some genes in the *YODA* pathway display gametophytic maternal effects (*ESF1*) and others sporophytic maternal effects (*mpk6* and *mkk4/5*), while *YODA* is a zygotic effect gene (Lukowitz, Roeder, Parmenter, & Somerville, 2004). All however affect embryo patterning, suggesting a complex interplay between maternal gametophytic, maternal sporophytic, and zygotic factors in the control of early embryo patterning.

5.3 Effect of DNA methylation and small RNAs on seed development

Imprinting related processes, particularly DNA methylation and small RNA pathways, have sporophytic effects as well as gametophytic effects. In Arabidopsis, loss of paternal *MET1* or *DECREASE IN DNA METHYLATION1* (*DDM1*) leads to reduced seed size, and loss of maternal *MET1* or *DDM1* leads to increased seed size (FitzGerald, Luo, Chaudhury, & Berger, 2008; Xiao et al., 2006). The maternal effect *met1* mutations appears to be sporophytic rather than gametophytic, based on frequency of seed phenotypes from *met1* mutants (FitzGerald et al., 2008). In addition to a function in gametophytic imprinting in the endosperm, small RNAs also have maternal sporophytic effects. Impairment of small RNA production or function in the maternal sporophyte of *dicer like1* (*dcl1*) mutants of Arabidopsis affects embryo patterning (Golden et al., 2002; Ray, Golden, & Ray, 1996). Mutants of some *dcl1* alleles reach maturity and have variable integument defects. Some of the morphologically normal ovules produce seeds with abnormal morphology (Golden et al., 2002; Ray et al., 1996). Embryos from these plants have a reduction in cotyledon size or number, hypocotyl reduction, and root pole abnormalities.

6. Conclusion

The gametes and other cells of the embryo sac as well as the surrounding tissues of the sporophyte all play a role in controlling the final size and pattern of the endosperm and embryo. Interactions between the three generations of the seed are necessary for regulating seed development. Some of the molecules involved have been identified but there are still many unanswered questions. There is indirect evidence for asymmetric distribution of patterning factors in the central cell but their nature and the mechanisms regulating their distribution have not yet been identified. Additionally, while some of the signaling cascades have been worked out for these maternal effects, many of the steps are still unknown. Future innovations in single cell proteomics and transcriptomics as well as live-cell imaging of the developing embryo sac and seed within the ovule will improve our understanding of maternal regulation of these processes.

Acknowledgments

We would like to thank Kathy Barton for helpful comments and years of discussions, encouragement and support. This work was supported by National Science Foundation Plant Genome Program Grant, DBI-1340050.

References

Adamski, N. M., Anastasiou, E., Eriksson, S., O'Neill, C. M., & Lenhard, M. (2009). Local maternal control of seed size by KLUH/CYP78A5-dependent growth signaling. *Proceedings of the National Academy of Sciences of the United States of America, 106*(47), 20115–20120. https://doi.org/10.1073/pnas.0907024106.

Anderson, S. N., Johnson, C. S., Chesnut, J., Jones, D. S., Khanday, I., Woodhouse, M., et al. (2017). The zygotic transition is initiated in unicellular plant zygotes with asymmetric activation of parental genomes. *Developmental Cell, 43*(3), 349–358.e344. https://doi.org/10.1016/j.devcel.2017.10.005.

Armenta-Medina, A., & Gillmor, C. S. (2019). Genetic, molecular and parent-of-origin regulation of early embryogenesis in flowering plants. *Current Topics in Developmental Biology, 131*, 497–543. https://doi.org/10.1016/bs.ctdb.2018.11.008.

Autran, D., Baroux, C., Raissig, M. T., Lenormand, T., Wittig, M., Grob, S., et al. (2011). Maternal epigenetic pathways control parental contributions to Arabidopsis early embryogenesis. *Cell, 145*(5), 707–719.

Bai, F., Daliberti, M., Bagadion, A., Xu, M., Li, Y., Baier, J., et al. (2016). Parent-of-origin-effect rough endosperm mutants in maize. *Genetics, 204*(1), 221–231. https://doi.org/10.1534/genetics.116.191775.

Bai, F., & Settles, A. M. (2014). Imprinting in plants as a mechanism to generate seed phenotypic diversity. *Frontiers in Plant Science, 5*, 780. https://doi.org/10.3389/fpls.2014.00780.

Baroux, C., Blanvillain, R., & Gallois, P. (2001). Paternally inherited transgenes are down-regulated but retain low activity during early embryogenesis in Arabidopsis. *FEBS Letters, 509*(1), 11–16.

Baroux, C., & Grossniklaus, U. (2015). The maternal-to-zygotic transition in flowering plants: Evidence, mechanisms, and plasticity. *Current Topics in Developmental Biology, 113*, 351–371. https://doi.org/10.1016/bs.ctdb.2015.06.005.

Blanvillain, R., Boavida, L. C., McCormick, S., & Ow, D. W. (2008). Exportin1 genes are essential for development and function of the gametophytes in Arabidopsis thaliana. *Genetics, 180*(3), 1493–1500. https://doi.org/10.1534/genetics.108.094896.

Cao, X., Costa, L. M., Biderre-Petit, C., Kbhaya, B., Dey, N., Perez, P., et al. (2007). Abscisic acid and stress signals induce Viviparous1 expression in seed and vegetative tissues of maize. *Plant Physiology, 143*(2), 720–731.

Ceccato, L., Masiero, S., Sinha Roy, D., Bencivenga, S., Roig-Villanova, I., Ditengou, F. A., et al. (2013). Maternal control of PIN1 is required for female gametophyte development in Arabidopsis. *PLoS One, 8*(6). e66148https://doi.org/10.1371/journal.pone.0066148.

Chakravorty, D., Trusov, Y., Zhang, W., Acharya, B. R., Sheahan, M. B., McCurdy, D. W., et al. (2011). An atypical heterotrimeric G-protein gamma-subunit is involved in guard cell K(+)-channel regulation and morphological development in *Arabidopsis thaliana. The Plant Journal, 67*(5), 840–851. https://doi.org/10.1111/j.1365-313X.2011.04638.x.

Charlton, W. L., Keen, C. L., Merriman, C., Lynch, P., Greenland, A. J., & Dickinson, H. G. (1995). Endosperm development in *Zea mays*: Implication of gametic imprinting and paternal excess in regulation of transfer layer development. *Development, 121*, 3089–3097.

Chen, L. Q., Lin, I. W., Qu, X. Q., Sosso, D., McFarlane, H. E., Londono, A., et al. (2015). A cascade of sequentially expressed sucrose transporters in the seed coat and endosperm provides nutrition for the Arabidopsis embryo. *Plant Cell, 27*(3), 607–619. https://doi.org/10.1105/tpc.114.134585.

Chen, J., Strieder, N., Krohn, N. G., Cyprys, P., Sprunck, S., Engelmann, J. C., et al. (2017). Zygotic genome activation occurs shortly after fertilization in maize. *Plant Cell, 29*(9), 2106–2125. https://doi.org/10.1105/tpc.17.00099.

Chettoor, A. M., & Evans, M. M. S. (2015). Correlation between a loss of auxin signaling and a loss of proliferation in maize antipodal cells. *Frontiers in Plant Science, 6*, 187. https://doi.org/10.3389/fpls.2015.00187.

Chettoor, A. M., Phillips, A. R., Coker, C. T., Dilkes, B., & Evans, M. M. (2016). Maternal gametophyte effects on seed development in maize. *Genetics, 204*(1), 233–248. https://doi.org/10.1534/genetics.116.191833.

Colombo, L., Franken, J., Van der Krol, A., Wittich, P. E., Dons, H. J. M., & Angenent, G. C. (1997). Down-regulation of ovule-specific MADS box genes from petunia results in maternally controlled defects in seed development. *Plant Cell, 9*, 703–715.

Costa, L. M., Marshall, E., Tesfaye, M., Silverstein, K. A., Mori, M., Umetsu, Y., et al.Gutierrez-Marcos, J. F., (2014). Central cell-derived peptides regulate early embryo patterning in flowering plants. *Science, 344*(6180), 168–172. https://doi.org/10.1126/science.1243005.

Costa, L. M., Yuan, J., Rouster, J., Paul, W., Dickinson, H., & Gutierrez-Marcos, J. F. (2012). Maternal control of nutrient allocation in plant seeds by genomic imprinting. *Current Biology, 22*(2), 160–165. https://doi.org/10.1016/j.cub.2011.11.059.

Diboll, A. G., & Larson, D. A. (1966). An electron microscopic study of the mature megagametophyte in Zea mays. *American Journal of Botany, 53*(4), 391–402.

Dilkes, B. P., & Comai, L. (2004). A differential dosage hypothesis for parental effects in seed development. *Plant Cell, 16*(12), 3174–3180. https://doi.org/10.1105/tpc.104.161230.

Dilkes, B. P., Spielman, M., Weizbauer, R., Watson, B., Burkart-Waco, D., Scott, R. J., et al. (2008). The maternally expressed WRKY transcription factor TTG2 controls lethality in interploidy crosses of Arabidopsis. *PLoS Biology, 6*(12), 2707–2720.

Disch, S., Anastasiou, E., Sharma, V. K., Laux, T., Fletcher, J. C., & Lenhard, M. (2006). The E3 ubiquitin ligase BIG BROTHER controls arabidopsis organ size in a dosage-dependent manner. *Current Biology, 16*(3), 272–279. https://doi.org/10.1016/j.cub.2005.12.026.

Dong, X., Zhang, M., Chen, J., Peng, L., Zhang, N., Wang, X., et al. (2017). Dynamic and antagonistic allele-specific epigenetic modifications controlling the expression of imprinted genes in maize endosperm. *Molecular Plant, 10*(3), 442–455. https://doi.org/10.1016/j.molp.2016.10.007.

Drews, G. N., & Yadegari, R. (2002). Development and function of the angiosperm female gametophyte. *Annual Review of Genetics, 36*, 99–124.

Du, L., Li, N., Chen, L., Xu, Y., Li, Y., Zhang, Y., et al. (2014). The ubiquitin receptor DA1 regulates seed and organ size by modulating the stability of the ubiquitin-specific protease UBP15/SOD2 in Arabidopsis. *Plant Cell, 26*(2), 665–677. https://doi.org/10.1105/tpc.114.122663.

Ebel, C., Mariconti, L., & Gruissem, W. (2004). Plant retinoblastoma homologues control nuclear proliferation in the female gametophyte. *Nature, 429*(6993), 776–780.

Evans, M. M. S., & Grossniklaus, U. (2009). The maize megagametophyte. In J. L. Bennetzen & S. Hake (Eds.), *Handbook of maize: Its biology* (pp. 79–104). New York: Springer.

Evans, M. M. S., & Kermicle, J. L. (2001). Interaction between maternal effect and zygotic effect mutations during maize seed development. *Genetics, 159*(1), 303–315.

Fang, W., Wang, Z., Cui, R., Li, J., & Li, Y. (2012). Maternal control of seed size by EOD3/CYP78A6 in *Arabidopsis thaliana*. *The Plant Journal, 70*(6), 929–939. https://doi.org/10.1111/j.1365-313X.2012.04907.x.

Feil, R., & Berger, F. (2007). Convergent evolution of genomic imprinting in plants and mammals. *Trends in Genetics, 23*(4), 192–199. https://doi.org/10.1016/j.tig.2007.02.004.

Felker, F. C., Peterson, D. M., & Nelson, O. M. (1985). Anatomy of immature grains of eight maternal effect shrunken endosperm barley mutants. *American Journal of Botany, 72*, 248–256.

FitzGerald, J., Hui, P. S., & Berger, F. (2009). Polycomb group-dependent imprinting of the actin regulator *AtFH5* regulates morphogenesis in *Arabidopsis thaliana*. *Development, 136*, 3399–3404.

FitzGerald, J., Luo, M., Chaudhury, A., & Berger, F. (2008). DNA methylation causes predominant maternal controls of plant embryo growth. *PLoS One, 3*(5) e2298.

Gaiser, J. C., Robinson-Beers, K., & Gasser, C. S. (1995). The Arabidopsis SUPERMAN gene mediates asymmetric growth of the outer integument of ovules. *Plant Cell, 7*(3), 333–345. https://doi.org/10.1105/tpc.7.3.333.

Garcia, D., Fitz Gerald, J. N., & Berger, F. (2005). Maternal control of integument cell elongation and zygotic control of endosperm growth are coordinated to determine seed size in Arabidopsis. *Plant Cell, 17*(1), 52–60. https://doi.org/10.1105/tpc.104.027136.

Gavazzi, G., Dolfini, S., Allegra, D., Castiglioni, P., Todesco, G., & Hoxha, M. (1997). *Dap (Defective aleurone pigmentation)* mutations affect maize aleurone development. *Molecular & General Genetics, 256*(3), 223–230.

Gehring, M., Bubb, K. L., & Henikoff, S. (2009). Extensive demethylation of repetitive elements during seed development underlies gene imprinting. *Science, 324*(5933), 1447–1451.

Gehring, M., Missirian, V., & Henikoff, S. (2011). Genomic analysis of parent-of-origin allelic expression in *Arabidopsis thaliana* seeds. *PLoS One*, *6*(8). e23687. https://doi.org/10.1371/journal.pone.0023687.

Gehring, M., Reik, W., & Henikoff, S. (2009). DNA demethylation by DNA repair. *Trends in Genetics*, *25*(2), 82–90.

Golden, T. A., Schauer, S. E., Lang, J. D., Pien, S., Mushegian, A. R., Grossniklaus, U., et al. (2002). Short INTEGUMENTS1/SUSPENSOR1/CARPEL FACTORY, a Dicer homolog, is a maternal effect gene required for embryo development in Arabidopsis. *Plant Physiology*, *130*(2), 808–822.

Grimanelli, D., Perotti, E., Ramirez, J., & Leblanc, O. (2005). Timing of the maternal-to-zygotic transition during early seed development in maize. *Plant Cell*, *17*(4), 1061–1072.

Grini, P. E., Jurgens, G., & Hulskamp, M. (2002). Embryo and endosperm development is disrupted in the female gametophytic *capulet* mutants of *Arabidopsis*. *Genetics*, *162*(4), 1911–1925.

Gross-Hardt, R., Kagi, C., Baumann, N., Moore, J. M., Baskar, R., Gagliano, W. B., et al. (2007). LACHESIS restricts gametic cell fate in the female gametophyte of Arabidopsis. *PLoS Biology*, *5*(3) e47.

Grossniklaus, U., & Schneitz, K. (1998). The molecular and genetic basis of ovule and megagametophyte development. *Seminars in Cell & Developmental Biology*, *9*(2), 227–238.

Grossniklaus, U., Vielle-Calzada, J. P., Hoeppner, M. A., & Gagliano, W. B. (1998). Maternal control of embryogenesis by MEDEA, a polycomb group gene in Arabidopsis. *Science*, *280*(5362), 446–450.

Guo, F., Huang, B. Q., Han, Y., & Zee, S. Y. (2004). Fertilization in maize indeterminate gametophyte1 mutant. *Protoplasma*, *223*(2–4), 111–120.

Gutierrez-Marcos, J. F., Costa, L. M., & Evans, M. M. S. (2006). Maternal gametophytic baseless1 is required for development of the central cell and early endosperm patterning in maize (Zea mays). *Genetics*, *174*, 317–329.

Hackenberg, D., & Twell, D. (2019). The evolution and patterning of male gametophyte development. *Current Topics in Developmental Biology*, *131*, 257–298. https://doi.org/10.1016/bs.ctdb.2018.10.008.

Haig, D., & Westoby, M. (1989). Parent specific gene expression and the triploid endosperm. *The American Naturalist*, *134*, 147–155.

Haig, D., & Westoby, M. (1991). Genomic imprinting in endosperm: Its effect on seed development in crosses between species, and between different ploidies of the same species, and its implications for the evolution of apomixis. *Philosophical Transactions of the Royal Society of London. Series B, Biological Sciences*, *333*(1266), 1–13.

Hornslien, K. S., Miller, J. R., & Grini, P. E. (2019). Regulation of parent-of-origin allelic expression in the endosperm. *Plant Physiology*, *180*(3), 1498–1519. https://doi.org/10.1104/pp.19.00320.

Hsieh, T. F., Ibarra, C. A., Silva, P., Zemach, A., Eshed-Williams, L., Fischer, R. L., et al. (2009). Genome-wide demethylation of Arabidopsis endosperm. *Science*, *324*(5933), 1451–1454.

Hsieh, T. F., Shin, J., Uzawa, R., Silva, P., Cohen, S., Bauer, M. J., et al. (2011). Regulation of imprinted gene expression in Arabidopsis endosperm. *Proceedings of the National Academy of Sciences of the United States of America*, *108*(5), 1755–1762. https://doi.org/10.1073/pnas.1019273108.

Huang, X., Qian, Q., Liu, Z., Sun, H., He, S., Luo, D., et al. (2009). Natural variation at the DEP1 locus enhances grain yield in rice. *Nature Genetics*, *41*(4), 494–497. https://doi.org/10.1038/ng.352.

Huang, B. Q., & Sheridan, W. F. (1994). Female gametophyte development in maize: Microtubular organization and embryo sac polarity. *Plant Cell*, *6*(6), 845–861.

Huang, B. Q., & Sheridan, W. F. (1996). Embryo sac development in the maize indeterminate gametophyte1 mutant: Abnormal nuclear behavior and defective microtubule organization. *Plant Cell*, *8*(8), 1391–1407.

Hueros, G., Gomez, E., Cheikh, N., Edwards, J., Weldon, M., Salamini, F., et al. (1999). Identification of a promoter sequence from the BETL1 gene cluster able to confer transfer-cell-specific expression in transgenic maize. *Plant Physiology*, *121*(4), 1143–1152.

Hulskamp, M., Schneitz, K., & Pruitt, R. E. (1995). Genetic evidence for a long-range activity that directs pollen tube guidance in Arabidopsis. *Plant Cell*, 7(1), 57–64.

Ingouff, M., Jullien, P. E., & Berger, F. (2006). The female gametophyte and the endosperm control cell proliferation and differentiation of the seed coat in Arabidopsis. *Plant Cell*, *18*(12), 3491–3501. https://doi.org/10.1105/tpc.106.047266.

Jiang, W. B., Huang, H. Y., Hu, Y. W., Zhu, S. W., Wang, Z. Y., & Lin, W. H. (2013). Brassinosteroid regulates seed size and shape in Arabidopsis. *Plant Physiology*, *162*(4), 1965–1977. https://doi.org/10.1104/pp.113.217703.

Jofuku, K. D., Omidyar, P. K., Gee, Z., & Okamuro, J. K. (2005). Control of seed mass and seed yield by the floral homeotic gene APETALA2. *Proceedings of the National Academy of Sciences of the United States of America*, *102*(8), 3117–3122. https://doi.org/10.1073/pnas.0409893102.

Johnson, C. S., Kolevski, B., & Smyth, D. R. (2002). TRANSPARENT TESTA GLABRA2, a trichome and seed coat development gene of Arabidopsis, encodes a WRKY transcription factor. *Plant Cell*, *14*(6), 1359–1375. https://doi.org/10.1105/tpc.001404.

Kirioukhova, O., Johnston, A. J., Kleen, D., Kagi, C., Baskar, R., Moore, J. M., et al. (2011). Female gametophytic cell specification and seed development require the function of the putative Arabidopsis INCENP ortholog WYRD. *Development*, *138*(16), 3409–3420. https://doi.org/10.1242/dev.060384.

Kohler, C., & Grossniklaus, U. (2005). Seed development and genomic imprinting in plants. *Progress in Molecular and Subcellular Biology*, *38*, 237–262.

Kohler, C., Hennig, L., Bouveret, R., Gheyselinck, J., Grossniklaus, U., & Gruissem, W. (2003). Arabidopsis MSI1 is a component of the MEA/FIE Polycomb group complex and required for seed development. *The EMBO Journal*, *22*(18), 4804–4814.

Kohler, C., Hennig, L., Spillane, C., Pien, S., Gruissem, W., & Grossniklaus, U. (2003). The Polycomb-group protein MEDEA regulates seed development by controlling expression of the MADS-box gene PHERES1. *Genes & Development*, *17*(12), 1540–1553.

Kohler, C., Page, D. R., Gagliardini, V., & Grossniklaus, U. (2005). The *Arabidopsis thaliana* MEDEA Polycomb group protein controls expression of PHERES1 by parental imprinting. *Nature Genetics*, *37*(1), 28–30.

Kong, J., Lau, S., & Jurgens, G. (2015). Twin plants from supernumerary egg cells in Arabidopsis. *Current Biology*, *25*(2), 225–230. https://doi.org/10.1016/j.cub.2014.11.021.

Kradolfer, D., Wolff, P., Jiang, H., Siretskiy, A., & Kohler, C. (2013). An imprinted gene underlies postzygotic reproductive isolation in *Arabidopsis thaliana*. *Developmental Cell*, *26*(5), 525–535. https://doi.org/10.1016/j.devcel.2013.08.006.

Leon-Kloosterziel, K. M., Keijzer, C. J., & Koornneef, M. (1994). A seed shape mutant of Arabidopsis that is affected in integument development. *Plant Cell*, *6*(3), 385–392. https://doi.org/10.1105/tpc.6.3.385.

Leroux, B. M., Goodyke, A. J., Schumacher, K. I., Abbott, C. P., Clore, A. M., Yadegari, R., et al. (2014). Maize early endosperm growth and development: From fertilization through cell type differentiation. *American Journal of Botany*, *101*(8), 1259–1274. https://doi.org/10.3732/ajb.1400083.

Li, J. (2017). Endosperm differentiation. In B. Thomas, B. G. Murray, & D. J. Murphy (Eds.), *Vol. 1. Encyclopedia of applied plant sciences* (pp. 497–503). Amsterdam: Elsevier.

Li, N., & Li, Y. (2015). Maternal control of seed size in plants. *Journal of Experimental Botany, 66*(4), 1087–1097. https://doi.org/10.1093/jxb/eru549.

Li, S., Liu, Y., Zheng, L., Chen, L., Li, N., Corke, F., et al. (2012). The plant-specific G protein gamma subunit AGG3 influences organ size and shape in *Arabidopsis thaliana*. *The New Phytologist, 194*(3), 690–703. https://doi.org/10.1111/j.1469-8137.2012.04083.x.

Li, G., Wang, D., Yang, R., Logan, K., Chen, H., Zhang, S., et al. (2014). Temporal patterns of gene expression in developing maize endosperm identified through transcriptome sequencing. *Proceedings of the National Academy of Sciences of the United States of America, 111*(21), 7582–7587. https://doi.org/10.1073/pnas.1406383111.

Li, Q., Wang, J., Ye, J., Zheng, X., Xiang, X., Li, C., et al. (2017). The maize imprinted gene Floury3 encodes a PLATZ protein required for tRNA and 5S rRNA transcription through interaction with RNA polymerase III. *Plant Cell, 29*(10), 2661–2675. https://doi.org/10.1105/tpc.17.00576.

Li, Y., Zheng, L., Corke, F., Smith, C., & Bevan, M. W. (2008). Control of final seed and organ size by the DA1 gene family in *Arabidopsis thaliana*. *Genes & Development, 22*(10), 1331–1336. https://doi.org/10.1101/gad.463608.

Lin, B.-Y. (1978). Structural modifications of the female gametophyte associated with the indeterminate gametophyte (ig) mutant in maize. *Canadian Journal of Genetics and Cytology, 20*, 249–257.

Lin, B.-Y. (1981). Megagametogenetic alterations associated with the indeterminate gametophyte (ig) mutation in maize. *Revista Brasileira de Biologia, 41*, 557–563.

Lituiev, D. S., Krohn, N. G., Müller, B., Jackson, D., Hellriegel, B., Dresselhaus, T., et al. (2013). Theoretical and experimental evidence indicates that there is no detectable auxin gradient in the angiosperm female gametophyte. *Development, 140*(22), 4544–4553. https://doi.org/10.1242/dev.098301.

Lukowitz, W., Roeder, A., Parmenter, D., & Somerville, C. (2004). A MAPKK kinase gene regulates extra-embryonic cell fate in Arabidopsis. *Cell, 116*(1), 109–119. https://doi.org/10.1016/s0092-8674(03)01067-5.

Luo, M., Bilodeau, P., Koltunow, A., Dennis, E. S., Peacock, W. J., & Chaudhury, A. M. (1999). Genes controlling fertilization-independent seed development in *Arabidopsis thaliana*. *Proceedings of the National Academy of Sciences of the United States of America, 96*(1), 296–301.

Ma, Y., & Nelson, O. E. (1975). Amino acid composition and storage proteins in two high lysine mutants in maize. *Cereal Chemistry, 52*, 412–419.

Mao, H., Sun, S., Yao, J., Wang, C., Yu, S., Xu, C., et al. (2010). Linking differential domain functions of the GS3 protein to natural variation of grain size in rice. *Proceedings of the National Academy of Sciences of the United States of America, 107*(45), 19579–19584. https://doi.org/10.1073/pnas.1014419107.

Martin, M. V., Fiol, D. F., Sundaresan, V., Zabaleta, E. J., & Pagnussat, G. C. (2013). Oiwa, a female gametophytic mutant impaired in a mitochondrial manganese-superoxide dismutase, reveals crucial roles for reactive oxygen species during embryo sac development and fertilization in Arabidopsis. *Plant Cell, 25*(5), 1573–1591. https://doi.org/10.1105/tpc.113.109306.

Martinez, G., & Kohler, C. (2017). Role of small RNAs in epigenetic reprogramming during plant sexual reproduction. *Current Opinion in Plant Biology, 36*, 22–28. https://doi.org/10.1016/j.pbi.2016.12.006.

Moreno-Romero, J., Jiang, H., Santos-Gonzalez, J., & Kohler, C. (2016). Parental epigenetic asymmetry of PRC2-mediated histone modifications in the Arabidopsis endosperm. *The EMBO Journal, 35*(12), 1298–1311. https://doi.org/10.15252/embj.201593534.

Mosher, R. A., Melnyk, C. W., Kelly, K. A., Dunn, R. M., Studholme, D. J., & Baulcombe, D. C. (2009). Uniparental expression ofPolIV-dependent siRNAs in developing endosperm of Arabidopsis. *Nature*, *460*(7252), 283–286.

Nagasawa, N., Hibara, K., Heppard, E. P., Vander Velden, K. A., Luck, S., Beatty, M., et al. (2013). GIANT EMBRYO encodes CYP78A13, required for proper size balance between embryo and endosperm in rice. *The Plant Journal*, *75*(4), 592–605. https://doi.org/10.1111/tpj.12223.

Nardmann, J., & Werr, W. (2009). Patterning of the maize embryo and the perspective of evolutionary developmental biology. In J. L. Bennetzen & S. Hake (Eds.), *Handbook of maize: Its biology* (pp. 105–119). New York: Springer.

Ohto, M. A., Fischer, R. L., Goldberg, R. B., Nakamura, K., & Harada, J. J. (2005). Control of seed mass by APETALA2. *Proceedings of the National Academy of Sciences of the United States of America*, *102*(8), 3123–3128. https://doi.org/10.1073/pnas.0409858102.

Olsen, O. A. (2004). Nuclear endosperm development in cereals and *Arabidopsis thaliana*. *Plant Cell*, *16*(Suppl), S214–S227.

Olsen, O. A., Linnestad, C., & Nichols, S. E. (1999). Developmental biology of the cereal endosperm. *Trends in Plant Science*, *4*(7), 253–257.

Pagnussat, G. C., Yu, H. J., Ngo, Q. A., Rajani, S., Mayalagu, S., Johnson, C. S., et al. (2005). Genetic and molecular identification of genes required for female gametophyte development and function in Arabidopsis. *Development*, *132*(3), 603–614.

Pagnussat, G. C., Yu, H. J., & Sundaresan, V. (2007). Cell-fate switch of synergid to egg cell in Arabidopsis eostre mutant embryo sacs arises from misexpression of the BEL1-like homeodomain gene BLH1. *Plant Cell*, *19*(11), 3578–3592.

Pan, Y. B., & Peterson, P. A. (1989). Tagging of a maize gene involved in kernel development by an activated Uq transposable element. *Molecular & General Genetics*, *219*(1–2), 324–327.

Park, K., Kim, M. Y., Vickers, M., Park, J. S., Hyun, Y., Okamoto, T., et al. (2016). DNA demethylation is initiated in the central cells of Arabidopsis and rice. *Proceedings of the National Academy of Sciences of the United States of America*, *113*(52), 15138–15143. https://doi.org/10.1073/pnas.1619047114.

Phillips, A. R., & Evans, M. M. (2011). Analysis of stunter1, a maize mutant with reduced gametophyte size and maternal effects on seed development. *Genetics*, *187*(4), 1085–1097.

Pignatta, D., Erdmann, R. M., Scheer, E., Picard, C. L., Bell, G. W., & Gehring, M. (2014). Natural epigenetic polymorphisms lead to intraspecific variation in Arabidopsis gene imprinting. *eLife3.* , e03198https://doi.org/10.7554/eLife.03198.

Portereiko, M. F., Lloyd, A., Steffen, J. G., Punwani, J. A., Otsuga, D., & Drews, G. N. (2006). AGL80 is required for central cell and endosperm development in Arabidopsis. *Plant Cell*, *18*(8), 1862–1872.

Randolph, L. F. (1936). Developmental morphology of the caryopsis in maize. *Journal of Argicultural Research*, *53*(12), 881–916.

Ray, S., Golden, T., & Ray, A. (1996). Maternal effects of the short integument mutation on embryo development in Arabidopsis. *Developmental Biology*, *180*(1), 365–369.

Riefler, M., Novak, O., Strnad, M., & Schmulling, T. (2006). Arabidopsis cytokinin receptor mutants reveal functions in shoot growth, leaf senescence, seed size, germination, root development, and cytokinin metabolism. *Plant Cell*, *18*(1), 40–54. https://doi.org/10.1105/tpc.105.037796.

Robert, H. S., Grones, P., Stepanova, A. N., Robles, L. M., Lokerse, A. S., Alonso, J. M., et al. (2013). Local auxin sources orient the apical-basal axis in Arabidopsis embryos. *Current Biology*, *23*(24), 2506–2512. https://doi.org/10.1016/j.cub.2013.09.039.

Rodrigues, J. C., Luo, M., Berger, F., & Koltunow, A. M. (2010). Polycomb group gene function in sexual and asexual seed development in angiosperms. *Sexual Plant Reproduction*, *23*(2), 123–133.

Rodrigues, J. A., & Zilberman, D. (2015). Evolution and function of genomic imprinting in plants. *Genes & Development*, *29*(24), 2517–2531. https://doi.org/10.1101/gad.269902.115.

Russell, S. D. (1979). Fine structure of megagametophyte development in *Zea mays*. *Canadian Journal of Botany*, *57*(10), 1093–1110.

Schruff, M. C., Spielman, M., Tiwari, S., Adams, S., Fenby, N., & Scott, R. J. (2006). The AUXIN RESPONSE FACTOR 2 gene of Arabidopsis links auxin signalling, cell division, and the size of seeds and other organs. *Development*, *133*(2), 251–261. https://doi.org/10.1242/dev.02194.

Scott, R. J., Spielman, M., Bailey, J., & Dickinson, H. G. (1998). Parent-of-origin effects on seed development in *Arabidopsis thaliana*. *Development*, *125*, 3329–3341.

Serbes, I. E., Palovaara, J., & Gross-Hardt, R. (2019). Chapter fifteen—Development and function of the flowering plant female gametophyte. *Current Topics in Developmental Biology*, *131*, 401–434.

Slotkin, R. K., Vaughn, M., Borges, F., Tanurdzic, M., Becker, J. D., Feijo, J. A., et al. (2009). Epigenetic reprogramming and small RNA silencing of transposable elements in pollen. *Cell*, *136*(3), 461–472. https://doi.org/10.1016/j.cell.2008.12.038.

Song, X., Yuan, L., & Sundaresan, V. (2014). Antipodal cells persist through fertilization in the female gametophyte of Arabidopsis. *Plant Reproduction*, *27*(4), 197–203. https://doi.org/10.1007/s00497-014-0251-1.

Sørensen, M. B., Chaudhury, A. M., Robert, H., Bancharel, E., & Berger, F. (2001). Polycomb group genes control pattern formation in plant seed. *Current Biology*, *11*, 277–281.

Springer, P. S., Holding, D. R., Groover, A., Yordan, C., & Martienssen, R. A. (2000). The essential Mcm7 protein PROLIFERA is localized to the nucleus of dividing cells during the G(1) phase and is required maternally for early Arabidopsis development. *Development*, *127*(9), 1815–1822.

Springer, P. S., McCombie, W. R., Sundaresan, V., & Martienssen, R. A. (1995). Gene trap tagging of PROLIFERA, an essential MCM2-3-5-like gene in Arabidopsis. *Science*, *268*(5212), 877–880.

Steffen, J. G., Kang, I. H., Portereiko, M. F., Lloyd, A., & Drews, G. N. (2008). AGL61 interacts with AGL80 and is required for central cell development in Arabidopsis. *Plant Physiology*, *148*(1), 259–268.

Sundaresan, V., & Alandete-Saez, M. (2010). Pattern formation in miniature: The female gametophyte of flowering plants. *Development*, *137*(2), 179–189.

Tekleyohans, D. G., Nakel, T., & Gross-Hardt, R. (2017). Patterning the female gametophyte of flowering plants. *Plant Physiology*, *173*(1), 122–129. https://doi.org/10.1104/pp.16.01472.

Ueda, M., Aichinger, E., Gong, W., Groot, E., Verstraeten, I., Vu, L. D., et al. (2017). Transcriptional integration of paternal and maternal factors in the Arabidopsis zygote. *Genes & Development*, *31*(6), 617–627. https://doi.org/10.1101/gad.292409.116.

Vielle-Calzada, J. P., Baskar, R., & Grossniklaus, U. (2000). Delayed activation of the paternal genome during seed development. *Nature*, *404*(6773), 91–94.

Vu, T. M., Nakamura, M., Calarco, J. P., Susaki, D., Lim, P. Q., Kinoshita, T., et al. (2013). RNA-directed DNA methylation regulates parental genomic imprinting at several loci in Arabidopsis. *Development*, *140*(14), 2953–2960. https://doi.org/10.1242/dev.092981.

Walbot, V., & Evans, M. M. S. (2003). Unique features of the plant life cycle and their consequences. *Nature Reviews Genetics*, *4*(5), 369–379.

Waters, A. J., Makarevitch, I., Eichten, S. R., Swanson-Wagner, R. A., Yeh, C. T., Xu, W., et al. (2011). Parent-of-origin effects on gene expression and DNA methylation in the maize endosperm. *Plant Cell*, *23*(12), 4221–4233. https://doi.org/10.1105/tpc.111.092668.

Wolff, P., Jiang, H., Wang, G., Santos-Gonzalez, J., & Kohler, C. (2015). Paternally expressed imprinted genes establish postzygotic hybridization barriers in *Arabidopsis thaliana*. *eLife4*, e10074. https://doi.org/10.7554/eLife.10074.

Wolff, P., Weinhofer, I., Seguin, J., Roszak, P., Beisel, C., Donoghue, M. T., et al. (2011). High-resolution analysis of parent-of-origin allelic expression in the Arabidopsis endosperm. *PLoS Genetics*, 7(6). e1002126. https://doi.org/10.1371/journal.pgen.1002126.

Wu, J. J., Peng, X. B., Li, W. W., He, R., Xin, H. P., & Sun, M. X. (2012). Mitochondrial GCD1 dysfunction reveals reciprocal cell-to-cell signaling during the maturation of Arabidopsis female gametes. *Developmental Cell*, *23*(5), 1043–1058. https://doi.org/10.1016/j.devcel.2012.09.011.

Xia, T., Li, N., Dumenil, J., Li, J., Kamenski, A., Bevan, M. W., et al. (2013). The ubiquitin receptor DA1 interacts with the E3 ubiquitin ligase DA2 to regulate seed and organ size in Arabidopsis. *Plant Cell*, *25*(9), 3347–3359. https://doi.org/10.1105/tpc.113.115063.

Xiao, W., Brown, R. C., Lemmon, B. E., Harada, J. J., Goldberg, R. B., & Fischer, R. L. (2006). Regulation of seed size by hypomethylation of maternal and paternal genomes. *Plant Physiology*, *142*(3), 1160–1168.

Xin, M., Yang, R., Li, G., Chen, H., Laurie, J., Ma, C., et al. (2013). Dynamic expression of imprinted genes associates with maternally controlled nutrient allocation during maize endosperm development. *Plant Cell*, *25*(9), 3212–3227. https://doi.org/10.1105/tpc.113.115592.

Zhang, M., Lv, R., Yang, W., Fu, T., & Liu, B. (2018). Imprinted gene expression in maize starchy endosperm and aleurone tissues of reciprocal F1 hybrids at a defined developmental stage. *Genes & Genomics*, *40*(1), 99–107. https://doi.org/10.1007/s13258-017-0613-9.

Zhang, S., Wang, D., Zhang, H., Skaggs, M. I., Lloyd, A., Ran, D., et al. (2018). FERTILIZATION-INDEPENDENT SEED-Polycomb repressive complex 2 plays a dual role in regulating type I MADS-box genes in early endosperm development. *Plant Physiology*, *177*(1), 285–299. https://doi.org/10.1104/pp.17.00534.

Zhang, M., Wu, H., Su, J., Wang, H., Zhu, Q., Liu, Y., et al. (2017). Maternal control of embryogenesis by MPK6 and its upstream MKK4/MKK5 in Arabidopsis. *The Plant Journal*, *92*(6), 1005–1019. https://doi.org/10.1111/tpj.13737.

Zhang, M., Xie, S., Dong, X., Zhao, X., Zeng, B., Chen, J., et al. (2014). Genome-wide high resolution parental-specific DNA and histone methylation maps uncover patterns of imprinting regulation in maize. *Genome Research*, *24*(1), 167–176. https://doi.org/10.1101/gr.155879.113.

Zhang, M., Zhao, H., Xie, S., Chen, J., Xu, Y., Wang, K., et al. (2011). Extensive, clustered parental imprinting of protein-coding and noncoding RNAs in developing maize endosperm. *Proceedings of the National Academy of Sciences of the United States of America*, *108*(50), 20042–20047. https://doi.org/10.1073/pnas.1112186108.

CHAPTER TEN

Maternal factors regulating symmetry breaking and dorsal–ventral axis formation in the sea urchin embryo

Maria Dolores Molina, Thierry Lepage*
Institut de Biologie Valrose, Université Côte d'Azur, Nice, France
*Corresponding author: e-mail address: thierry.lepage@unice.fr

Contents

1. Introduction 284
2. Maternal determination of the D/V axis in sea urchin embryos: The heritage from experimental embryology 285
3. Nodal signaling initiates D/V axis specification of the sea urchin embryo 288
4. Studies on the Nodal promoter identify maternal transcription factors and signaling molecules required for the initiation of Nodal expression 291
5. Yan, a maternal ETS domain transcription factor represses the *nodal* promoter 294
6. The maternal function of the transcriptional repressor Yan/Tel is essential for spatial regulation of *nodal* 295
7. Stability of sea urchin Yan/Tel is regulated by MAPK and maternal GSK3 and β-TRCP 297
8. Activity of the maternal type I BMP receptors Alk1/2 and Alk3/6 is required for the early spatial restriction of *nodal* 300
9. Identification of Panda as a factor required to restrict *nodal* expression during dorsal–ventral axis formation in the sea urchin embryo 302
10. *panda* mRNA is distributed asymmetrically in the oocyte and unfertilized egg 303
11. Increasing or decreasing the levels of Panda locally orients the D/V axis 304
12. Spatially restricted Panda signaling specifies the dorsal–ventral axis 305
13. Panda and Yan/Tel act upstream of the Lefty-dependent reaction–diffusion mechanism to initiate the spatial restriction of *nodal* 306
14. Panda's mechanism of action 306
    14.1 Panda does not directly promote phosphorylation of Smad1/5/8 306
    14.2 Yan/Tel acts downstream of Panda to restrict *nodal* expression 309
15. Conclusion: Maternal determinants of D/V axis formation and developmental plasticity of the early blastomeres 310

Acknowledgments 312
References 312

Current Topics in Developmental Biology, Volume 140
ISSN 0070-2153
https://doi.org/10.1016/bs.ctdb.2019.10.007

Abstract

Specification of the main axes of polarity of the embryo is an essential process during embryonic development. In many species, this process is achieved by the localization of maternal factors into discrete regions of the egg. However, in other animals, like in amniotes and in echinoderms, the considerable plasticity of the early blastomeres seems to preclude the existence of maternal determinants and the mechanisms by which the radial symmetry of the egg is broken remain largely enigmatic. In this chapter, we review recent progress on the identification of maternal components involved in symmetry breaking and dorsal–ventral (D/V) axis formation of the sea urchin embryo. We will first review some key experiments on D/V axis formation from classical embryologists that provided evidence for a weak maternal D/V prepattern. We will then detail more recent molecular analyses that established the critical role played by Nodal signaling in allocating cell fates along the secondary axis and led to the discovery that maternal transcription factors such as the Sry-related HMG box B1 (SoxB1), the Octamer binding factor1/2 (Oct1/2), the T-cell factor/Lymphoid enhancer-binding *factor* (TCF/LEF) and the Erythroblastosis virus E26 Oncogene Homolog (ETS) domain transcriptional repressor Translocation-Ets-Leukemia virus protein (Yan/Tel) as well as maternal signaling molecules like Univin are essential for the initiation of *nodal* expression. Finally, we will describe recent advances that uncovered a role in symmetry breaking and dorsal–ventral axis orientation for the transforming growth factor beta (TGF-beta)-like factor Panda, which appears to be both necessary and sufficient for D/V axis orientation. Therefore, even in the highly regulative sea urchin embryo, the activity of localized maternal factors provides the embryo with a blueprint of the D/V axis.

1. Introduction

In bilaterians, specification of the dorsal–ventral axis is a crucial event during embryogenesis to establish the correct body plan. In many species, this process relies on gene products translated from maternal mRNAs deposited in the egg. For example, in *Drosophila*, specification of the dorsal–ventral axis of the embryo is initiated by the product of the *gurken* gene, which is active in the oocyte nucleus during oogenesis and encodes a member of the EGF superfamily that acts as a secreted dorsalizing signal (Roth, Stein, & Nusslein-Volhard, 1989; Rushlow, Han, Manley, & Levine, 1989; Schupbach, 1987; Steward, 1989). Similarly, in *Xenopus* and zebrafish, although the D/V axis is not preformed in the unfertilized egg, dorsal determinants are localized to the vegetal pole of the egg (Jesuthasan & Stahle, 1997; Mizuno, Yamaha, Wakahara, Kuroiwa, & Takeda, 1996; Ober & Schulte-Merker, 1999; Tao et al., 2005). Fertilization breaks the radial symmetry of the egg and triggers the asymmetric transport of these determinants from the vegetal pole to the future dorsal side where they activate the canonical Wnt pathway (Langdon & Mullins, 2011; Weaver & Kimelman, 2004). While maternal information is clearly important for specification of the dorsal–ventral axis in a number of species, in contrast, there is very little evidence for the presence of maternal determinants of axis formation in the oocytes of mammals consistent with the idea that the embryonic axes are specified entirely by cell interactions (Arnold & Robertson, 2009). Accordingly, it has been argued that the regulative abilities of the first blastomeres of the mouse embryo rule out the possibility that maternal determinants influence axis specification (Papaioannou, Mkandawire, & Biggers, 1989) (reviewed in (Frankenberg & Zernicka-Goetz, 2004)). Similarly, in the sea urchin embryo, the tremendous plasticity of the early blastomeres seemed to preclude the possibility that maternal determinants regulate specification of the D/V axis in the blastomeres that inherit them (Henry, 1998; Hörstadius, 1973). Recent studies on axis specification in the sea urchin have identified the first maternal determinant of the D/V axis. This chapter will explain how this factor was discovered and describe its properties. The concept that maternal determination of the secondary axis does not necessarily conflict with the plasticity of cell fates and the possibility that similar maternal determinants exist in amniotes will be discussed.

2. Maternal determination of the D/V axis in sea urchin embryos: The heritage from experimental embryology

Whether the dorsal–ventral axis of the sea urchin embryo is pre-established by maternal factors localized in the unfertilized egg or if it is specified de novo after fertilization has long been a question that preoccupied experimental embryologists. Driesch dissociated the early blastomeres by treating two and four cell-stage embryos with calcium free sea water and gentle shaking. He reported that the isolated fragments developed into miniature but complete pluteus larvae (Driesch, 1892) a result later confirmed by Boveri and Hörstadius (Boveri, 1908; Hörstadius & Wolsky, 1936) (Fig. 1A). This experiment, which became famous since it provided the foundation for the analysis of regulative development, argued against a predetermination of the dorsal–ventral axis. That four pluteus larvae could develop from the four blastomeres isolated at the four-cell stage was taken as evidence against the presence of localized maternal determinants. The argument at the basis of this conclusion was that if each blastomere has the same

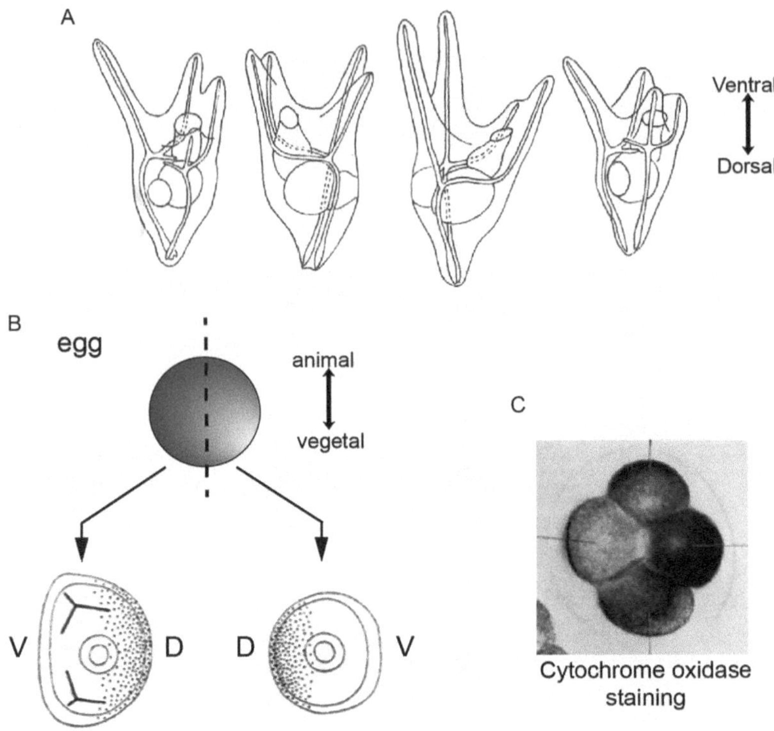

Fig. 1 Sea urchin regulative development. (A) Separation of the first four blastomeres of a sea urchin embryo can give rise to four well-formed pluteus larvae according to Hörstadius and Wolsky (1936). (B) Bisection experiment from Hörstadius and Wolsky (1936). (C) Cytochrome oxidase staining at the eight-cell stage reveals an asymmetry in the distribution of mitochondria in the early embryo (Czihack from Rous'x archive fûr Entwicklungsmechanik 154, 272–292 (1963)). *Panels A and B: Image from Hörstadius, S. (1973). Experimental embryology of echinoderms. Oxford: Clarendon Press.*

potential to differentiate D/V structures then, cell fates are unlikely to be differentially allocated along the D/V axis. Using *Paracentrotus lividus* zygotes, Hörstadius and Wolsky searched for a correlation between the plane of first cleavage and the secondary axis but found that this plane was oriented randomly with respect to the D/V axis. In many urchin species, however, a correlation between the plane of the first cleavage and the orientation of the D/V axis exists but because this relationship varies between the species considered, it reinforced the idea that specification of the D/V axis is not strongly associated with the plane of first cleavage (Henry, 1998; Kominami, 1988).

Despite the amazing plasticity of the early blastomeres and the lack of strong correlation of the D/V axis with the plane of first cleavage, a whole body of evidence led Hörstadius to nevertheless conclude that "there is no doubt as to the existence of a ventro-dorsal axis in the unfertilized sea urchin egg." First, Hörstadius noticed that like in the case of the animal-vegetal axis, for which there is in *P. lividus*, a natural marker in the form of a subequatorial pigment band that allows visualization of the A/V axis, the eggs of two sea cucumbers and a starfish are bilaterally symmetrical, a feature that could be interpreted as evidence for a predetermined dorsal–ventral axis (Hörstadius, 1973). The conclusion that a significant level of organization preexists in the unfertilized egg comes from the results of experiments in which Hörstadius bisected unfertilized eggs and subsequently fertilized the meridional halves. The two halves were reared in pairs and their development was analyzed. Hörstadius observed that development of the pairs of half larvae could be interpreted as development of presumptive left and right or dorsal and ventral egg halves, suggesting that bisection had revealed asymmetries of D/V determination present in the egg. Intriguingly, vital dye staining of the bisection plane revealed that in presumptive dorsal halves the orientation of the D/V axis was reversed suggesting that if there is some maternal pre-organization it can be easily perturbed (Hörstadius, 1973). Moreover, unlike the animal vegetal axis which is fixed rigidly, the D/V axis is quite labile and various physical or chemical treatments including stretching of eggs by placing them in a micropipette, centrifugation, ligature as well as treatments with a wide variety of compounds such as detergents, lithium chloride and methyl chloroxanthines can alter the D/V organization of the egg resulting in inversions, duplications or suppression of the D/V axis (Reviewed Hörstadius, 1973). There is a large body of evidence correlating formation of the dorsal–ventral axis with the activity of redox gradients and with the asymmetric distribution of mitochondria in the unfertilized sea urchin egg. Classical experiments performed by Child, Pease and Czihak in the thirties and sixties showed that it is possible to bias the dorsal–ventral axis by treating embryos with steep gradients of respiratory inhibitors and that the activity of the mitochondrial enzyme cytochrome oxidase can predict the dorsal–ventral axis as early as the eight-cell stage, with the presumptive ventral side being more oxidizing than the dorsal side (Child, 1941; Cinquin, 2006; Czihak, 1963; Pease, 1941) (Fig. 1C). Until recently, this asymmetry of mitochondria activity was the first known manifestation of dorsal–ventral polarity. Several recent studies by Coffman and colleagues addressed the question of causality between this early asymmetry and the orientation of the dorsal–ventral axis (Coffman, Coluccio, Planchart, & Robertson, 2009; Coffman & Denegre, 2007; Coffman, McCarthy, Dickey-Sims, & Robertson, 2004).

Although these studies provided evidence that the dorsal–ventral axis can be entrained by centrifugation or by microinjection of purified mitochondria or by overexpressing a form of catalase targeted to the mitochondria, the correlations obtained remained modest and in no case were these perturbations shown to efficiently orient the dorsal–ventral axis (Coffman et al., 2004, 2009; Coffman & Denegre, 2007). Furthermore, perturbations that were expected to influence D/V axis formation, such as overexpression of a mitochondrially-targeted form of superoxide dismutase, which generates the strong oxidizing component H_2O_2 and that would be predicted to efficiently orient the axis, did not show any effect on the orientation of the dorsal–ventral axis (Coffman et al., 2009). Similarly, Chang et al. used this axis induction assay to analyze the function of Hypoxia induced factor alpha (HIF-1alpha), the master regulator of Hypoxia, in dorsal–ventral specification in *Strongylocentrotus purpuratus* (Chang et al., 2017). In that study, overexpression of *HIF1-alpha* mRNA into one blastomere at the two-cell stage modestly increased by 10% the proportion of dorsal clones suggesting that this factor may possibly have a role in specification of the dorsal–ventral axis in that species. However, blocking the function of HIF1 with morpholino did not strongly impact D/V axis specification (Chang et al., 2017). Therefore, the redox gradient model of dorsal–ventral axis formation, though very interesting, clearly needs further experimental validation to determine the biological significance of the early asymmetry of mitochondria and of redox gradients and to clarify their relation to the prospective dorsal–ventral axis.

In conclusion, there is a rich heritage of experimental embryology regarding the maternal factors regulating D/V axis formation. There appears to be a certain level of predetermination of the D/V axis in the unfertilized egg but this maternal prepattern is labile and the axis can easily be re-specified.

3. Nodal signaling initiates D/V axis specification of the sea urchin embryo

In the sea urchin embryo, specification of the dorsal–ventral (D/V) axis critically relies on zygotic expression of the gene encoding the TGF-β family member Nodal in the presumptive ventral ectoderm (Fig. 2A and B) (Duboc et al., 2004). *nodal* is the earliest zygotic gene displaying a restricted expression pattern along the dorsal–ventral axis during sea urchin development and Nodal function is absolutely required for establishment

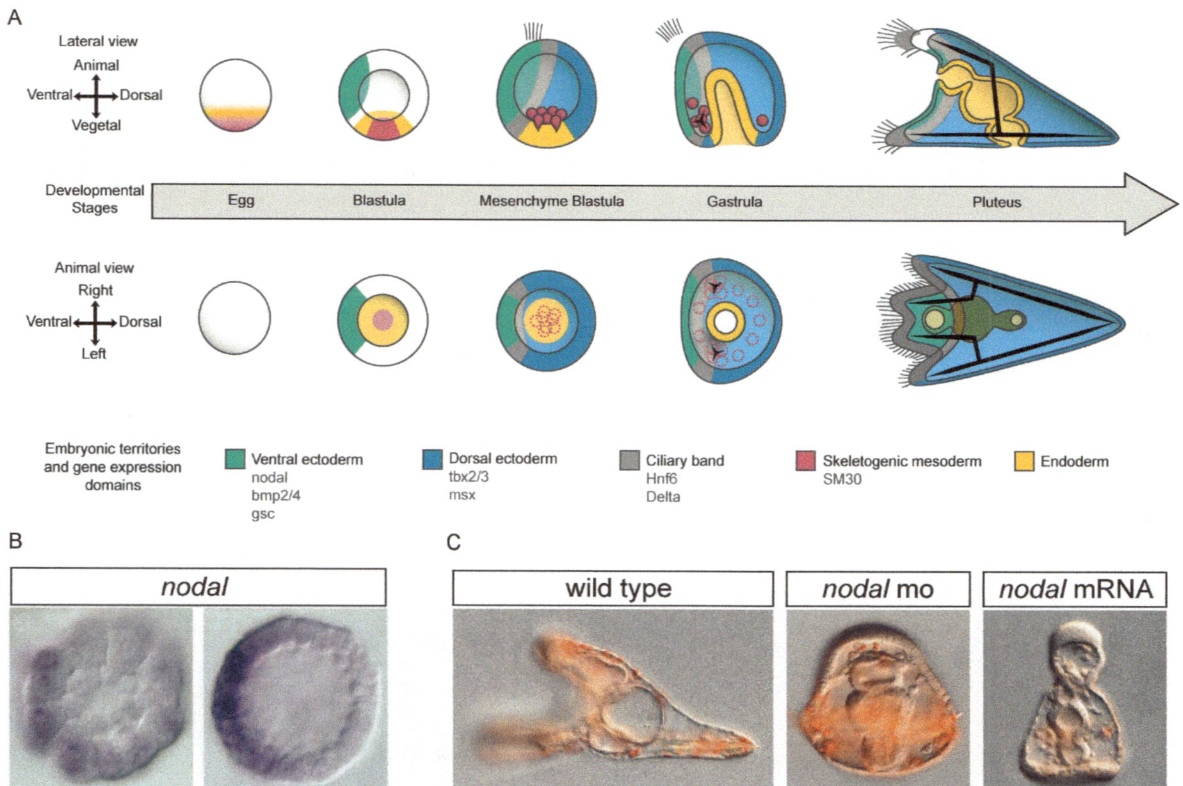

Fig. 2 See legend on next page.

of dorsal–ventral polarity (Fig. 2C) (Duboc et al., 2004). Nodal signaling promotes specification of the ventral ectoderm and triggers expression of BMP2/4, which acts as a long-range morphogen that specifies and patterns the dorsal region (Lapraz, Besnardeau, & Lepage, 2009a). Sea urchin embryos lacking Nodal function develop with a strongly radialized phenotype and fail to express both ventral and dorsal marker genes (Duboc et al., 2004; Saudemont et al., 2010). However, injection of *nodal* mRNA into a single cell can rescue a complete dorsal–ventral axis in these *nodal* morpholino injected embryos (Duboc et al., 2004). Therefore, expression of *nodal* is a key event that launches the gene regulatory network that specifies the D/V axis and understanding how *nodal* expression is regulated is essential to understand how secondary axis specification is established (Molina, de Croze, Haillot, & Lepage, 2013).

nodal expression is initiated around the 32/64 cell-stage in a domain that is initially rather broad, encompassing most cells of the presumptive ectoderm, but starting at the early blastula stage, *nodal* expression is progressively restricted to a smaller domain that corresponds to the presumptive ventral ectoderm (Fig. 2B) (Duboc et al., 2004). The progressive restriction of *nodal* expression in the sea urchin embryo is thought to rely on the ability of Nodal

Fig. 2 Early development of the sea urchin embryo and phenotypes resulting from the perturbation of the Nodal pathway. (A) Scheme describing early development of the sea urchin embryo together with the fate maps and gene expression territories. D/V polarity first becomes apparent at the beginning of gastrulation by the flattening of the presumptive ventral side, the bending of the gut, and the asymmetrical positioning of the skeletogenic mesenchyme cells that build the spicules. D/V polarity is further accentuated when four arms grow on the ventral side while the mouth opens on the ventral ectoderm and the larva lengthens along the dorsal side to become the pluteus larva. (B) In situ hybridization with *nodal* probe showing broad *nodal* expression at 32-cell stage (left panel) and its progressive restriction to the ventral ectoderm at blastula stage (right panel). (C) Perturbations of Nodal signaling strongly affect dorsal–ventral polarity. Inhibition of Nodal function with antisense morpholino oligonucleotides (mo) eliminates all manifestation of dorsal–ventral polarity, giving rise to radialized embryos covered with a ciliary band-like ectoderm, harboring a straight archenteron, lacking a mouth, and displaying a striking excess of mesodermal cells called pigment cells. In contrast, misexpression of *nodal* mRNA ventralizes the embryo and induces formation of a large proboscis at the animal pole (Duboc, Rottinger, Besnardeau, & Lepage, 2004). Lateral views. *Panel A: Images from Lapraz, F., Besnardeau, L., & Lepage, T. (2009). Dorsal-ventral patterning in echinoderms: Insights into the evolution of the BMP-chordin signaling network.* PLoS Biology, *7, 1–25; Panel B: Images from Duboc, V., Rottinger, E., Besnardeau, L., & Lepage, T. (2004). Nodal and BMP2/4 signaling organizes the oral-aboral axis of the sea urchin embryo.* Developmental Cell, *6, 397–410.*

to stimulate its own expression as well as that of the *lefty* gene, which encodes a diffusible long-range inhibitor of Nodal signaling (Meno et al., 1996, 1998, 1999; Thisse & Thisse, 1999). In the absence of Lefty, *nodal* expression is not restricted ventrally and expands toward the dorsal ectoderm (Duboc, Lapraz, Besnardeau, & Lepage, 2008). Together, the auto-activation of Nodal and the long-range inhibition of Lefty constitute the basis for a reaction-diffusion mechanism, i.e., a system with self-organizing properties, able to amplify weak initial anisotropies and to generate sharp boundaries within a field of cells (Chen & Schier, 2002; Duboc et al., 2008; Juan & Hamada, 2001; Meinhardt & Gierer, 2000; Solnica-Krezel, 2003; Turing, 1952). Therefore, understanding the molecular mechanisms responsible for the initiation of *nodal* expression is crucial for understanding how the D/V axis is established.

4. Studies on the Nodal promoter identify maternal transcription factors and signaling molecules required for the initiation of Nodal expression

Several components of the Nodal signaling pathway, such as the Nodal receptor Alk4/5/7 and the coreceptor Cripto are provided maternally and could potentially function as maternal determinants of the D/V axis (Range et al., 2007; Wei, Angerer, & Angerer, 2006). However, in situ hybridization analysis revealed that mRNA encoding these essential components of the pathway are distributed uniformly in the egg and are unlikely to fulfill the role of maternal determinants of the axis unless they are translated in a spatially restricted manner. Alternatively, maternal determinants of the D/V axis could encode transcription factors that regulate *nodal* expression.

Cis-regulatory analysis of the *nodal* promoter led to the identification of a proximal cis-regulatory module called the R-module (Range et al., 2007). The R-module contains consensus binding sites for Sox, Oct and TCF transcription factors (Fig. 3A), suggesting that these factors could play a role in initiating *nodal* expression (Range et al., 2007; Wei et al., 2006). This idea was supported by the observation that all three factors are expressed maternally and broadly distributed in the early embryo (Huang et al., 2000; Kenny, Oleksyn, Newman, Angerer, & Angerer, 2003; Range & Lepage, 2011). Interfering with SoxB1 function using antisense morpholino oligonucleotides severely affected dorsal–ventral patterning, causing embryos to develop with a strongly radialized phenotype (Fig. 3B) (Kenny et al., 2003).

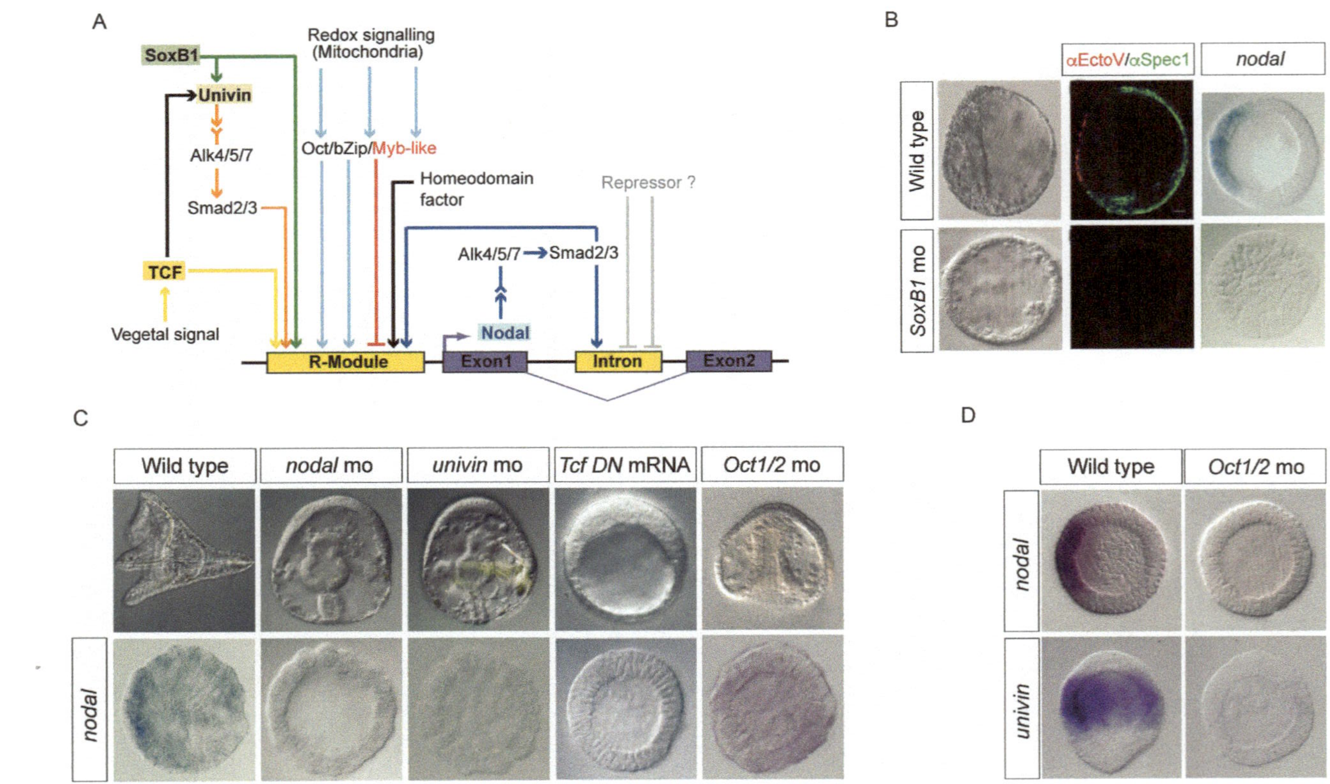

Fig. 3 See legend on opposite page.

Examination of endogenous *nodal* expression in the SoxB1 morpholino injected embryos at early blastula stages revealed that *nodal* expression was abolished in these embryos (Fig. 3B) (Range et al., 2007). Also, the activity of the R-module driven reporter gene decreased to 18% of its normal value in embryos injected with the SoxB1 morpholino at hatched blastula stage (Range et al., 2007). Similarly, co-injection of RNA encoding a dominant negative version of TCF reduced the activity of the reporter gene to 11% of its original value. Moreover, inhibition of Oct1/2 mRNA translation produced embryos with severe axial defects similar to those observed following inhibition of Nodal function (Range & Lepage, 2011). Perturbing Oct1/2 function specifically disrupted specification of the ventral and dorsal ectodermal regions by preventing *nodal* expression early in development (Fig. 3C). So maternal transcription factors clearly play important roles in D/V axis formation through the regulation of *nodal* expression.

In addition to maternal transcription factors, maternal ligands also play crucial roles in D/V axis formation. One of these factors is the TGF-β superfamily member Vg1/Univin. Vg1/Univin is the sea urchin orthologue of Vg1/GDF1, which likely functions as a heterodimerization partner for Nodal (Tanaka, Sakuma, Nakamura, Hamada, & Saijoh, 2007). *Vg1/Univin* is first expressed uniformly and maternally, then zygotic *Vg1/Univin* transcripts accumulate in a belt of ectodermal cells surrounding the equatorial region of the embryo (Lapraz et al., 2006; Zito et al., 2003).

Fig. 3 TCF, SoxB1, Oct1 and Univin are required for the initiation of Nodal expression. (A) Integration of signaling and maternal transcription factor inputs by cis-regulatory elements of the *nodal* gene. (B) Down-regulation of SoxB1 results in radialized embryos that loose *nodal* expression as well as the staining for EctoV (red) and Spec1 (green), which label differentiated ventral and dorsal ectoderm, respectively. (C) Overexpression of TCF-DN mRNA or inhibition of Univin, SoxB1 or Oct1 by morpholino injection phenocopies the loss-of function of *nodal* and results in radialized embryos that fail to express *nodal* (D) Zygotic expression of *nodal* and *vg1/univin* requires maternal Oct1/2. *Panel C: Images from Duboc, V., Rottinger, E., Besnardeau, L., & Lepage, T. (2004). Nodal and BMP2/4 signaling organizes the oral-aboral axis of the sea urchin embryo.* Developmental Cell, 6, *397–410; Range, R., Lapraz, F., Quirin, M., Marro, S., Besnardeau, L., & Lepage, T. (2007). Cis-regulatory analysis of nodal and maternal control of dorsal-ventral axis formation by Univin, a TGF-{beta} related to Vg1.* Development, 134, *3649–3664; Range, R., & Lepage, T. (2011). Maternal Oct1/2 is required for nodal and Vg1/Univin expression during dorsal-ventral axis specification in the sea urchin embryo.* Developmental Biology, 357, *440–449. Panel D: Images from Range, R., & Lepage, T. (2011). Maternal Oct1/2 is required for nodal and Vg1/Univin expression during dorsal-ventral axis specification in the sea urchin embryo.* Developmental Biology, 357, *440–449.*

Thus, both maternal and zygotic expression of *vg1/univin* overlaps with early *nodal* expression. Strikingly, blocking translation of the *vg1/univin* mRNA abolishes *nodal* expression and mimics the Nodal loss-of-function phenotype (Fig. 3C) (Range et al., 2007). Therefore, in the sea urchin, Vg1/Univin plays a role just as important as that played by Nodal during D/V axis formation. Both cWnt and Vg1 signaling work together to activate and/or maintain the levels of *nodal* expression necessary to specify and pattern the D/V axis of sea urchin embryos, a situation which is remarkably similar to what is observed in vertebrate embryos (please add refs). Interestingly, the zygotic expression of sea urchin *vg1/univin* was also found to be completely dependent on Oct1/2 (Fig. 3D) (Range & Lepage, 2011), implying that the maternal Oct1/2 protein plays an early and essential role in D/V axis specification by initiating the expression of *nodal* and *vg1/univin*, two genes that act at the top of the D/V ectoderm gene regulatory network.

In conclusion, analysis of the *nodal* promoter allowed to identify several maternal transcription factors such as TCF, SoxB1 and Oct1/2 that have essential functions in initiating and maintaining high levels of *nodal* expression, possibly through direct binding to the R-module.

5. Yan, a maternal ETS domain transcription factor represses the *nodal* promoter

Although initial dissection of the *nodal* promoter led to the identification of several transcription factors important for the temporal expression of *nodal*, disappointingly, it failed to identify transcription factors regulating the spatial expression of this gene. A key spatial regulator of *nodal* expression was nevertheless later identified on the basis of its repressive action on the 5′ proximal module of the *nodal* promoter (please add refs). In addition to bZIP, Sox, Oct, homeobox and Smad binding sites, the 5′ proximal module of the *nodal* promoter contains several conserved 5′-GGAA/T-3’ ETS binding motifs (Range et al., 2007) the function of which had not been analyzed in the initial study of this promoter. Interestingly, mutational analysis of these ETS motifs combined with reporter analysis revealed that the ETS motifs are bound by a transcriptional repressor (Molina et al., 2018). This suggested that members of the ETS family with transcriptional repressor activity negatively control *nodal* expression by binding to the proximal module of its promoter. The relevant repressor was subsequently identified as the transcriptional repressor Yan/Tel (Molina et al., 2018).

The overall structure of sea urchin Yan/Tel protein is conserved when compared to that of vertebrate Tel or to that of Yan from *Drosophila*, with an

N-terminal SAM domain, a central co-repressor binding domain and a C-terminal ETS DNA binding domain (Fig. 4A) (Molina et al., 2018). The N-terminal SAM domain of Yan and Tel oligomerizes in a head to tail manner and this polymerization has been shown to be essential for repression (Jousset et al., 1997; Kim et al., 2001; Qiao et al., 2004). Interestingly, most of the hydrophobic residues that make up the interfaces between two monomers of Yan from *Drosophila* or Tel from vertebrates are conserved in the sea urchin Yan/Tel protein, suggesting that sea urchin Yan/Tel may also form higher order polymers as do its vertebrate and fly orthologs (Molina et al., 2018). Further experiments are required to test this hypothesis.

6. The maternal function of the transcriptional repressor Yan/Tel is essential for spatial regulation of *nodal*

yan/tel is a maternal transcript expressed abundantly and ubiquitously during cleavage (Fig. 4B). To downregulate the function of Yan/Tel, Molina et al. used antisense morpholino oligonucleotides and found that embryos injected with morpholinos directed against the translation start site of the *yan/tel* transcript fail to initiate dorsal–ventral polarity and resemble embryos partially ventralized by treatment with recombinant Nodal or by overexpression of low doses of *nodal* mRNA (Fig. 4C). In that work the PMCs of *yan/tel* morphants remained arranged in a ring around the archenteron (arrowheads) and the embryos conserved a rounded shape (Molina et al., 2018). This apparent lack of dorsal–ventral polarity persisted at the late gastrula stage: when in control embryos, the archenteron bent toward the presumptive ventral ectoderm, the *yan/tel* morphants remained radialized as evidenced by the straight position of the archenteron at the center of the blastocoel and by the radial arrangement of the PMCs (Fig. 4C arrowheads). Consistent with their radialized phenotype, *yan/tel* morphants displayed a dramatic ectopic expression of *nodal* (Molina et al., 2018) (Fig. 4C) as well as of *chordin, bmp2/4* and *lefty,* three direct downstream targets of Nodal signaling (Saudemont et al., 2010). In that work, injection of a morpholino oligonucleotide targeting a splice junction of *yan/tel,* in contrast, did not affect establishment of the dorsal–ventral axis and did not prevent formation of the bilateral PMC clusters implying that the maternal function of Yan/Tel but not its zygotic function is required for the spatial regulation of *nodal* expression and the establishment of the dorsal–ventral axis.

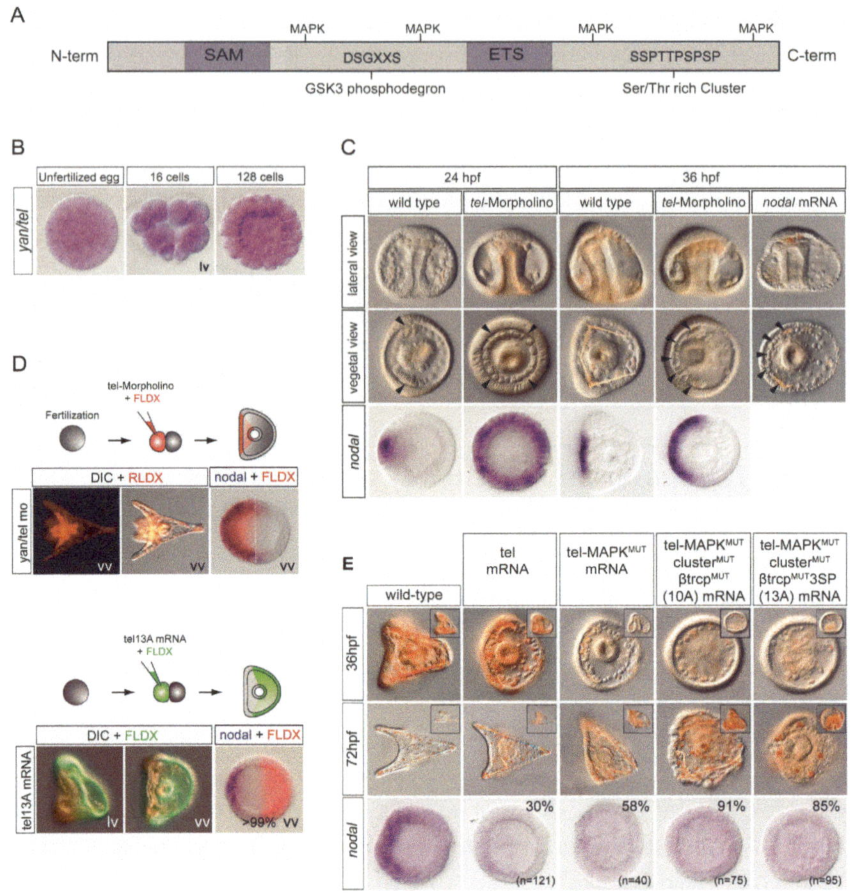

Fig. 4 The transcriptional repressor Yan/Tel controls the spatial expression of *nodal*. (A) Structure of sea urchin Yan/Tel. The Yan/Tel protein contains a SAM domain, an ETS binding site domain and several MAPK consensus phosphorylation sites as well as a GSK3 phosphodegron and a cluster rich in serine/threonine residues. (B) Maternal ubiquitous expression of *yan/tel*. lv, lateral view. (C) Inhibition of maternal Yan/Tel function disrupts dorsal–ventral axis formation and causes massive ectopic or expanded *nodal* expression at blastula and gastrula stages, respectively. Note the radial arrangement of the primary mesenchymal cells (PMCs, arrowheads) at gastrula stage (24 hpf) and the rounded shape of the embryo at prism stage (36 hpf). hpf, hours post-fertilization. (D) Yan/Tel efficiently orients the D/V axis. In embryos injected randomly with the *yan/tel* morpholino into one blastomere at the two or four-cell stage, *nodal* expression (blue) overlaps with the progeny of the injected blastomere (red), which will later develop the ventral region of the pluteus larva. On the other hand, *nodal* expression (blue) at swimming blastula stage is systematically found on the side opposite to the injection clone (red) after random injection of a phosphorylation mutant form of Yan/Tel. vv, vegetal view; lv, lateral view, with animal to the top, and ventral to the left. (E) Overexpression of a wild-type form of Yan/Tel causes little effects on development of

Because *yan* morphants ectopically express *nodal*, local inhibition of *yan/tel* mRNA translation was predicted to orient the dorsal–ventral axis (Haillot, Molina, Lapraz, & Lepage, 2015). Since in *Paracentrotus* the D/V axis is not related to the plane of first cleavage, the *yan/tel* morpholino was injected randomly into one blastomere of embryos at the two or four cell-stage and the position of the clone of injected cells was scored at pluteus stage (Fig. 4D) (Duboc et al., 2004; Haillot et al., 2015). Strikingly, at pluteus stage, the authors found that the progeny of the injected cells were always found on the ventral side of the larva. Remarkably, in most of the injected embryos, early *nodal* expression was found in a territory either congruent or overlapping with the clone of cells derived from the injected blastomere. Notably, the boundaries of the clone were always contained within the *nodal* expressing territory and at least one boundary precisely aligned with the border of the *nodal* expression domain (Fig. 4D). Therefore, blocking translation of Yan/Tel mRNA caused the cell-autonomous expression of *nodal*, irrespective of the position of the clone, strongly suggesting a model of *nodal* activation by release of Yan/Tel-mediated transcriptional repression (Molina et al., 2018).

7. Stability of sea urchin Yan/Tel is regulated by MAPK and maternal GSK3 and β-TRCP

Despite Yan's role as a transcriptional repressor of *nodal* expression, overexpression of wild-type *yan/tel* mRNA had only moderate effects on development of sea urchin embryos (Fig. 4E) (Molina et al., 2018). Even when injected with high doses of this mRNA, the embryos develop into pluteus larvae with a normal D/V axis (Molina et al., 2018). This result, that may seem surprising at first glance, is consistent with previous studies of

the embryos, while overexpression of phosphorylation mutant forms of Yan/Tel progressively radializes the embryos and increases the percentage of embryos that fail to express *nodal* at swimming blastula stage. The phosphorylation mutant *yan/tel10A* produces a phenotype similar to the *nodal* loss-of-function phenotype: note the radial arrangement of the PMCs, the abundance of pigment cells, the straight archenteron and the thickened ectoderm. Overexpression of the phosphorylation mutant form *yan/tel13A* fully radializes the embryo and disrupts gastrulation. All images are vegetal views at pluteus stage. Lateral views of the same embryo are shown in the upper corner. *All images from Molina, M. D., Quirin, M., Haillot, E., De Croze, N., Range, R., Rouel, M., et al. (2018). MAPK and GSK3/ss-TRCP-mediated degradation of the maternal Ets domain transcriptional repressor Yan/Tel controls the spatial expression of nodal in the sea urchin embryo.* PLoS Genetics, *14, e1007621.*

Drosophila Yan and vertebrate Tel which demonstrated that regulation is not at the transcriptional level but instead is post-transcriptionally controlled by MAP kinase phosphorylation. MAPK dependent phosphorylation of Yan and Tel is the central mechanism that regulates the activity and stability of these transcriptional repressors (Lai & Rubin, 1992; Maki et al., 2004; O'Neill, Rebay, Tjian, & Rubin, 1994; Poirel et al., 1997; Rebay & Rubin, 1995) (reviewed in (Tootle & Rebay, 2005)). Specific phosphorylation events triggered by either ERK, JNK or p38 downregulate the transcriptional repressor function of *Drosophila* Yan or of vertebrate Tel, leading to their nuclear export, promoting their degradation, and/or reducing DNA binding interactions. Consequently, it has been shown that mutations that convert the serine and threonine residues normally phosphorylated by MAPK into non-phosphorylatable residues transform these factors into constitutively active repressors (Lai & Rubin, 1992; Maki et al., 2004; O'Neill et al., 1994; Poirel et al., 1997; Rebay & Rubin, 1995). Conversely, mutations that convert them into phospho-mimetic residues promote degradation of these factors. Interestingly, the sea urchin Yan/Tel also contains three canonical PXS/TP phosphorylation sites for MAP kinases, plus one additional LSTP site (Fig. 4A, see also Molina et al., 2018) that has been shown to be efficiently phosphorylated by ERK (Cruzalegui, Cano, & Treisman, 1999). Intriguingly, although the positions of these sites do not appear to be conserved with those of vertebrates and *Drosophila*, replacing the four putative MAPK consensus phosphorylation sites with alanine increased the ability of sea urchin Yan/Tel to repress *nodal* at blastula stage and caused a partial radialization of the embryos at gastrula stages (Molina et al., 2018), suggesting that sea urchin Yan/Tel is also regulated by MAPK phosphorylation. However, these embryos progressively recovered their dorsal–ventral polarity as shown by their pluteus-like morphology at 72 h (Fig. 4E), indicating that additional mechanisms regulate the activity of sea urchin Yan/Tel factor.

Molina et al. found that in addition to MAPK sites, the sequence of sea urchin Yan/Tel contains at least two regions that may be phosphorylated by GSK3 (Fig. 4A), which targets several proteins for degradation including β-catenin and Snail (Ikeda et al., 1998; Sutherland, 2011; Yost et al., 1996; Zhou et al., 2004). The first motif DSGHSS conforms to the degradation box recognized by the E3 Ubiquitin ligase β-TRCP that recognizes the phosphorylated residues within the motif $DpSGX_{(1-4)}pS$ and promotes ubiquitinylation by the Skp1-Cul1-Fbox complex (Fuchs, Spiegelman, & Kumar, 2004). The second putative GSK3 phosphorylation motif is a short region rich in serines, threonines and prolines that is located in the C-terminal region (Molina et al., 2018). Intriguingly, in the sea urchin, inhibition of GSK3 causes a dose dependent stabilization of Yan/Tel. Furthermore, in vivo, overexpression of GSK3 strongly destabilizes wild-type Yan/Tel but does not affect the stability of Yan13A, a phosphorylation mutant of Yan/Tel (Molina et al., 2018). Finally, mutation of the β-TRCP and S/T rich cluster motifs markedly stabilizes Yan/Tel and consistent with these in vivo effects both the β-TRCP motif and the S/T rich cluster are phosphorylated in vitro by purified GSK3 (Molina et al., 2018). Therefore, sea urchin Yan/Tel can be added to the list of GSK3 substrates. Intriguingly, neither Yan from *Drosophila* nor Tel from vertebrates contain a β-TRCP motif in their sequence. Indeed, in flies and vertebrates, phosphorylation by MAP kinases, not GSK3, is the main mode of regulation of Yan or Tel, which both have been shown to be degraded primarily following ubiquitination by the ubiquitin conjugating enzyme FBL6 (Roukens, Alloul-Ramdhani, Moghadasi, Op den Brouw, & Baker, 2008), not by β-TRCP. The regulation of sea urchin Yan/Tel by a GSK3/β-TRCP pathway therefore appears to be a novel feature of this ETS family member. It is interesting to note that in echinoderm embryos *nodal* is expressed in the ectoderm and that *nodal* expression requires inhibition of animal pole fates by Wnt signals emanating from the vegetal pole. Therefore, this novel mode of regulation of Yan/Tel in the sea urchin may be linked to the need to coordinate patterning along the dorsal–ventral axis and *nodal* expression with patterning along the animal-vegetal axis.

Finally, Molina et al. showed that random injection of the phosphorylation mutant *yan/tel13A* mRNA into one blastomere at the two-cell stage is sufficient to orient the dorsal–ventral axis (Fig. 4D). Yan/Tel 13A is a constitutive active form of Yan/Tel that cannot be regulated by phosphorylation and that acts as a constitutive repressor (ref). Indeed, in all the *yan/tel13A* injected embryos, *nodal* expression was found in a discrete region on the opposite side of the cells expressing Yan/Tel13A and the progeny of the injected blastomere later occupied a territory that precisely coincided with the dorsal ectoderm (Fig. 4D). Taken together these findings revealed that Yan/Tel acts as a major negative regulator of *nodal* expression. The stability of Yan/Tel is itself regulated by GSK3, which is active in the ectoderm and which targets Yan/Tel for degradation. Since GSK3 regulates β-catenin stability and animal-vegetal patterning, it is tempting to speculate that Yan/Tel acts at the crossroads of both animal-vegetal and dorsal–ventral patterning to coordinate patterning along the two axes.

8. Activity of the maternal type I BMP receptors Alk1/2 and Alk3/6 is required for the early spatial restriction of *nodal*

It has recently been shown that specification of the ventral territory is not independent of BMP signaling, as previously thought (Lapraz, Besnardeau, & Lepage, 2009b; Saudemont et al., 2010). In addition to specifying the dorsal region at the onset of gastrulation, signaling from the two BMP receptors Alk3/6 and Alk1/2 is critically required during or before blastula stages to restrict *nodal* expression to the ventral side (Haillot et al., 2015). Both *alk3/6* and *alk1/2* are expressed maternally and ubiquitously during cleavage and blastula stages (Haillot et al., 2015; Lapraz et al., 2009b). Interestingly, blocking *alk1/2* mRNA translation disrupts dorsal–ventral axis formation causing a weak ventralization, a phenotype stronger than that resulting from inhibition of Alk3/6, which mainly disrupts the specification of the dorsal territory (Fig. 5A) (Haillot et al., 2015; Lapraz et al., 2009b). Strikingly, blocking the function of both *alk1/2+alk3/6* produces completely radialized embryos that develop with a prominent proboscis in the animal pole region (white arrowhead in Fig. 5A) and with an ectopic ciliary band surrounding the vegetal pole region (black arrowhead in Fig. 5A) (Haillot et al., 2015). These features are typical of the strongly ventralized phenotype observed in *nodal* overexpressing or nickel treated embryos and correlate with a massive ectopic expression of *nodal* (Fig. 5A) and of its downstream target genes *chordin* and *foxA* in the presumptive ectoderm at blastula stage (Haillot et al., 2015).

The finding that double inactivation of Alk1/2 and Alk3/6 produced an extreme radialization due to the unrestricted expression of *nodal* raised novel questions (Haillot et al., 2015). Since this phenotype was much stronger than the *bmp2/4* morphant phenotype (which similarly to *alk3/6* morphants mainly disrupts the specification of the dorsal territory) and since the effects of abolishing BMP signaling on *nodal* expression could be observed well before the onset of *bmp2/4* expression, the inescapable conclusion was that another TGF-β ligand acting through these two TGF-β receptors was cooperating with BMP2/4 during dorsal–ventral axis formation (Haillot et al., 2015).

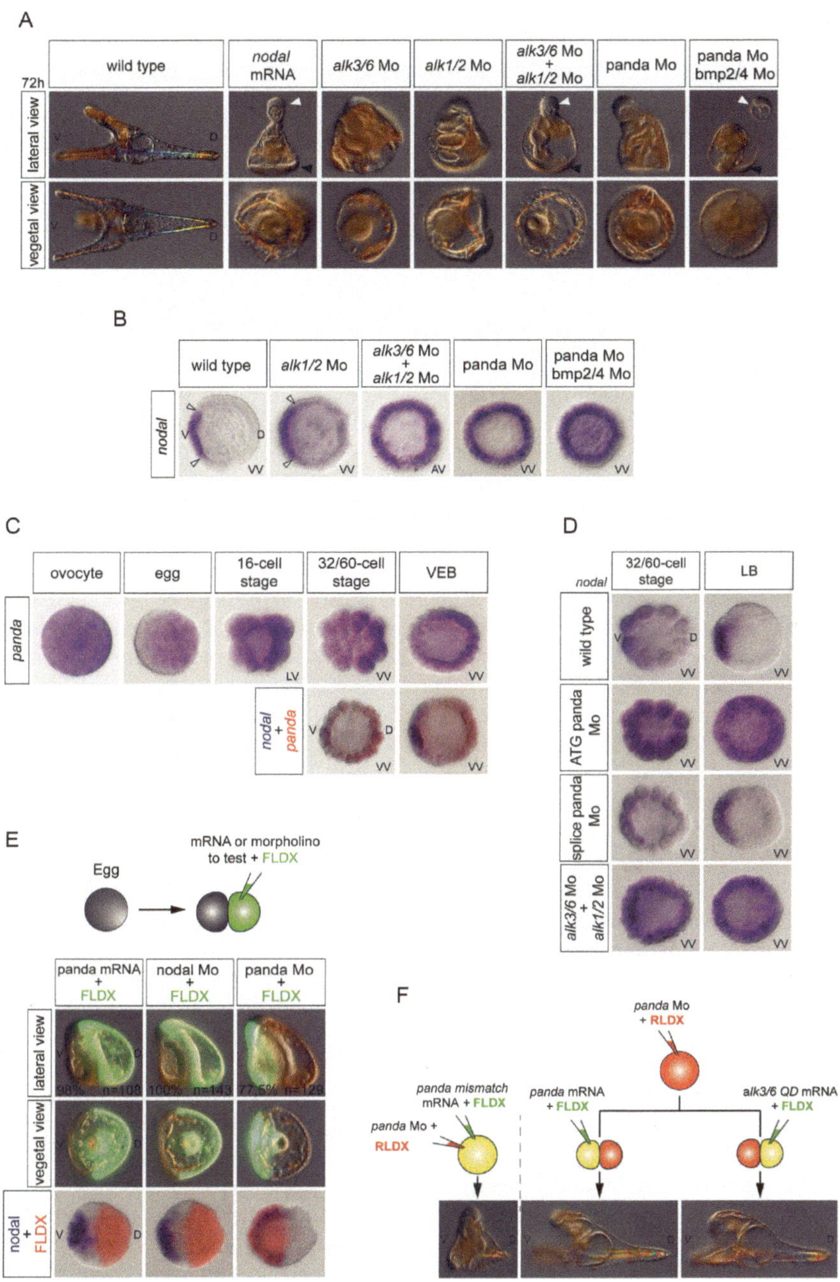

Fig. 5 BMP receptors and Panda restrict *nodal* expression during D/V patterning. (A) Down-regulation of *panda* completely radializes the embryos during the first 48 h, but a partial recovery of D/V polarity occurs afterwards, as evidenced by the *(Continued)*

9. Identification of Panda as a factor required to restrict *nodal* expression during dorsal–ventral axis formation in the sea urchin embryo

Haillot et al. screened the TGF-β ligands encoded in the sea urchin genome that were expressed during early development. Using a double morpholino injection assay, the authors identified a TGF-β protein related to Maverick from *Drosophila* and GDF15 from vertebrates, as a promising candidate. Simultaneous inactivation of both this TGF-β and *bmp2/4* caused

Fig. 5—Cont'd formation of a short dorsal apex. In contrast, simultaneous down-regulation of Alk1/2 and Alk3/6 or of *panda* and *bmp2/4* causes a strong ventralization similar to that resulting from overexpression of *nodal*. Note the presence of a ciliary band in the vegetal pole region (black arrowheads) and the prominent proboscis (white arrowheads) in the animal pole region in the double morphants and in *nodal*-overexpressing embryos. V, ventral. D, dorsal. h, hours. (B) In situ hybridization reveals that the strong ventralization observed in *panda* or double *panda*+ *bmp2/4* or *alk1/2* +*alk3/6* morphants is presaged by the massive ectopic expression of *nodal* at blastula stages. V, ventral; D, dorsal; VV, vegetal view; AV, animal view. (C) Whole mount in situ hybridizations with *panda* alone or with *panda* (red) and *nodal* (blue) probes. A gradient of maternal *panda* mRNA is detected in immature oocytes and to a lesser extent in the unfertilized mature eggs, whereas during the cleavage and blastula stages, *panda* mRNA is detected in a shallow D/V gradient. VEB, very early blastula (about 120 cells); V, ventral; D, dorsal; LV, lateral view; VV, vegetal view. (D) In situ hybridizations to detect *nodal* transcript at early stages. In the absence of maternal but not of zygotic Panda, a massive ectopic expression of *nodal* is observed starting at the 60-cell stage. Massive and early ectopic expression of *nodal* is also observed in the double *alk1/2*+*alk3/6* morphants. VEB, very early blastula (about 120 cells). LB, late blastula; V, ventral; D, dorsal; VV, vegetal view. (E) Effects of local overexpression or down-regulation of *panda* and *nodal* on the orientation of the D/V axis. Injection of panda into one blastomere at the two-cell stage imposes a dorsal identity to the progeny of the injected cell in nearly 100% of the injected embryos. Local down-regulation of *nodal* also imposes a dorsal identity. Conversely, down-regulation of *panda* strongly biases the orientation of the D/V axis and promotes ventral fates. V, ventral; D, dorsal. Consistent with this strong effect on the orientation of the D/V axis, in embryos injected with *panda* mRNAs or with the *nodal* morpholino, at the blastula stage, *nodal* was expressed in a sector of the embryo located on the opposite side of the clone of injected cells. V, ventral. D, dorsal. (F) Panda activity must be provided locally to efficiently rescue *panda* morphants. While providing Panda activity into the egg does not rescue D/V polarity, providing *panda* mRNA or low doses (50 μg/mL) of mRNA encoding the activated form of Alk3/6 (Alk3/6QD) into one blastomere at the two-cell stage fully rescues D/V polarity of *panda* morphants. Lateral views; V, ventral; D, dorsal. *Panel A–F: Images from Haillot, E., Molina, M. D., Lapraz, F., & Lepage, T. (2015). The maternal maverick/GDF15-like TGF-beta ligand panda directs dorsal-ventral Axis formation by restricting nodal expression in the sea urchin embryo.* PLoS Biology, 13, *e1002247.*

a massive ectopic expression of *nodal* starting at early stages leading to strong ventralization, similar to that caused by the double inactivation of Alk1/2 and Alk3/6 (Fig. 5A and B) (Haillot et al., 2015). Strikingly, the ventralization induced by the double inactivation of this TGF-β and *bmp2/4* was so strong that it frequently led to bisection of the embryos due to formation of a circular stomodeum that separated the animal pole-derived proboscis from the vegetal part of the larva that contains the gut (Fig. 5A) (Haillot et al., 2015). On the basis of its repressive effect on *nodal* expression, this TGF-β protein was named Panda (*Paracentrotus Anti-Nodal Dorsal Activity*).

Interestingly, phylogenetic analysis showed that Panda, Maverick and GDF15 do not belong to any known subclass of canonical BMP ligands (Haillot et al., 2015). Consistent with this conclusion, Panda, Maverick and GDF15 share with Inhibins beta chains, TGF-β and Myostatins a pattern of nine cysteines in the ligand domain, a pattern that is not shared by any prototypical BMP ligand (Haillot et al., 2015). Therefore, Panda, Maverick and GDF15-like sequences define a distinct subclass of TGF-β ligands within a larger branch of the TGF-β superfamily that comprises Inhibins beta chains, Lefty factors, Myostatins and TGF-β *sensu stricto* (Haillot et al., 2015).

10. *panda* mRNA is distributed asymmetrically in the oocyte and unfertilized egg

Previous studies of sea urchin *maverick/panda* had failed to detect expression of this gene by in situ hybridization (Lapraz et al., 2006). However, using an oligonucleotide microarray, very weak expression was detected in 2h zygotes and in 72h pluteus larvae (Wei et al., 2006). Remarkably, a recent re-analysis by Haillot et al. detected a graded distribution of *panda* transcripts in the subcortical region of immature oocytes, eggs and during early stages, with one side of the embryo showing a slightly stronger staining than the other, reinforcing the idea that this factor plays an early role in D/V axis formation (Fig. 5C). Panda is therefore one of the very few transcripts known to be spatially localized in sea urchin eggs. Furthermore, double labeling with *nodal* revealed that the side with the highest concentration of mRNA was the dorsal side, opposite to the side of *nodal* expression and consistent with the idea that Panda is a factor that cooperates with BMP2/4 to restrict *nodal* expression (Fig. 5C) (Haillot et al., 2015).

Haillot et al. further showed that Panda function is required early to restrict *nodal* expression. In the absence of Panda, ventral fates are expanded at the expense of dorsal fates. Molecular analysis revealed that knocking down *panda* causes a strong ventralization accompanied by massive and early ectopic expression of *nodal* (Fig. 5A and B) and *chordin,* which appear expressed throughout most of the ectoderm at mesenchyme blastula stage, and a concomitant loss of the dorsal marker gene *hox7* (Haillot et al., 2015). The observed ventralization was, however, more severe during blastula and gastrula stages, and the embryos progressively recovered dorsal–ventral polarity to some extent after 48 h (Fig. 5A) (see Haillot et al., 2015). Noticeably, Panda, as well as the activities of Alk1/2 and Alk3/6 are required very early to restrict *nodal* expression to the ventral side (Haillot et al., 2015). Consistent with this notion, *nodal* expression is not restricted and remains radialized in embryos injected with the morpholino targeting the translation start site of *panda* mRNA, presumed to block both maternal and zygotic *panda* transcripts, or with a combination of the *alk1/2* and *alk3/6* morpholinos (Fig. 5D) (Haillot et al., 2015). Interestingly, in contrast, *nodal* expression is unaffected in embryos injected with a morpholino targeting the splice junction of *panda* and blocking zygotic *panda* function does not noticeably perturb development of the embryos (Fig. 5D) (Haillot et al., 2015), suggesting that the maternal function of Panda, but not zygotic Panda, is required to restrict early *nodal* expression to the ventral side.

11. Increasing or decreasing the levels of Panda locally orients the D/V axis

Panda, like Lefty and Yan, is critically required for the correct spatial restriction of *nodal* to the ventral side during early stages. Surprisingly, however, overexpression of *panda* in the egg does not perturb establishment of the D/V axis suggesting that unlike Lefty, Panda alone is not capable of suppressing Nodal signaling (Haillot et al., 2015). This raised the possibility that rather than inhibiting Nodal signaling, the function of Panda may instead be to bias early Nodal signaling, perhaps by simply attenuating Nodal signaling on the dorsal side (Haillot et al., 2015). Accordingly, local overexpression of *panda* mRNA into one blastomere at the two-cell stage imposes a dorsal identity to the progeny of the injected cells (Fig. 5E). Interestingly, local overexpression of a constitutively active version of Alk3/6 (Alk3/6QD), local overexpression of the constitutive repressor Yan/Tel13A or local inhibition of Nodal signaling by a morpholino oligonucleotide mimics the effects of local overexpression of *panda*, efficiently orienting the D/V axis in injected embryos (Fig. 4D and 5E)

(Haillot et al., 2015). Consistent with this strong effect on the orientation of the D/V axis, in embryos injected with *panda* mRNA, with *alk3/6QD* mRNA, with *yan/tel 13A* mRNA or with the *nodal* morpholino, at blastula stage *nodal* is expressed in a sector of the embryo located on the opposite side of the clone of injected cells (Fig. 4D and 5E) (Haillot et al., 2015; Molina et al., 2018). On the other hand, removing the function of *panda* from part of the early embryo is also sufficient to orient the D/V axis (Fig. 5E) (Haillot et al., 2015). Indeed, injecting the *panda* morpholino randomly into one blastomere at the two-cell stage significantly biases the orientation of the D/V axis, with most embryos showing a clone of fluorescently labeled cells in the ventral region (Haillot et al., 2015). These results demonstrated that upregulating or downregulating the levels of Panda, Alk3/6 and Yan in a localized manner in early embryos strongly impacts the orientation of the D/V axis, partially mimicking manipulation of Nodal signaling levels (Haillot et al., 2015; Molina et al., 2018).

12. Spatially restricted Panda signaling specifies the dorsal–ventral axis

A very important control when working with morpholino oligonucleotides is to demonstrate that a wild-type mRNA immune to the injected morpholino can rescue normal development. Surprisingly, in the case of Panda, introducing a synthetic *panda* mRNA fails to rescue the severe defects of dorsal–ventral polarity caused by the *panda* morpholino (Fig. 5F) (Haillot et al., 2015). Since Panda is required to restrict *nodal* expression and since the endogenous *panda* mRNA is enriched on the presumptive dorsal side, Haillot et al. reasoned that Panda activity may need to be provided locally in order to mimic the distribution of endogenous *panda* mRNA and to rescue dorsal–ventral polarity of *panda* morphants. This conclusion was supported by the following observations. Although injection of *panda* mRNA into eggs did not rescue the D/V axis, injection of *panda* mRNA into one blastomere completely rescued dorsal–ventral polarity of embryos previously injected with the *panda* morpholino, such that the embryos developed into perfectly normal pluteus larvae with the dorsal side corresponding to the *panda* expressing clone (Fig. 5F) (Haillot et al., 2015). This experiment demonstrated that the activity of exogenous Panda must be spatially restricted to rescue the lack of maternal Panda function, consistent with the idea that the activity of endogenous Panda is spatially restricted in the early embryo. Similarly, injection of subdorsalizing doses of *alk3/6QD* mRNA into one blastomere completely rescued dorsal–ventral polarity of embryos previously injected with *panda* morpholino, consistent with previous results showing that local misexpression of an activated form of Alk3/6 is sufficient to antagonize *nodal* expression and to orient the D/V axis (Fig. 5F) (Haillot et al., 2015). Down-regulation of Nodal signaling by Panda or by type I BMP receptor signaling is therefore essential to spatially restrict *nodal* expression.

13. Panda and Yan/Tel act upstream of the Lefty-dependent reaction–diffusion mechanism to initiate the spatial restriction of *nodal*

As already mentioned, in the sea urchin, like in vertebrates, the spatial restriction of *nodal* expression relies on the early establishment of a reaction-diffusion mechanism between Nodal and Lefty starting at the early blastula stage (Duboc et al., 2004, 2008; Yaguchi, Yaguchi, & Burke, 2006). Interestingly, *lefty* morphant embryos at 60-cell stage display localized *nodal* expression indicating that spatial restriction of *nodal* is initiated normally in these embryos (Fig. 6A) (Molina et al., 2018). Nevertheless, in the absence of Lefty, spatial restriction of *nodal* expression is not maintained at later stages and most embryos eventually show a massive ectopic expression of *nodal* (Duboc et al., 2008). In contrast, in the absence of either maternal *yan/tel* or *panda* transcripts, no restricted expression of *nodal* is visible at any stage and instead the injected embryos show a massive ectopic expression of *nodal* from early stages (Fig. 6A) (Molina et al., 2018). These results suggest that the function of Panda and Yan/Tel is required before that of Lefty to initiate restriction of *nodal* expression, while the function of Lefty appears to be required only later for maintenance of the spatially restricted expression of *nodal*.

14. Panda's mechanism of action

14.1 Panda does not directly promote phosphorylation of Smad1/5/8

The finding that knocking down Panda causes a phenotype similar to that caused by knocking down the two BMP type I receptors Alk1/2 and Alk3/6, specifically early ectopic expression of *nodal*, and the observation that local expression of *alk3/6QD* efficiently rescues D/V polarity in *panda* morphants indicates that Panda most likely uses Alk1/2 and Alk3/6 to signal

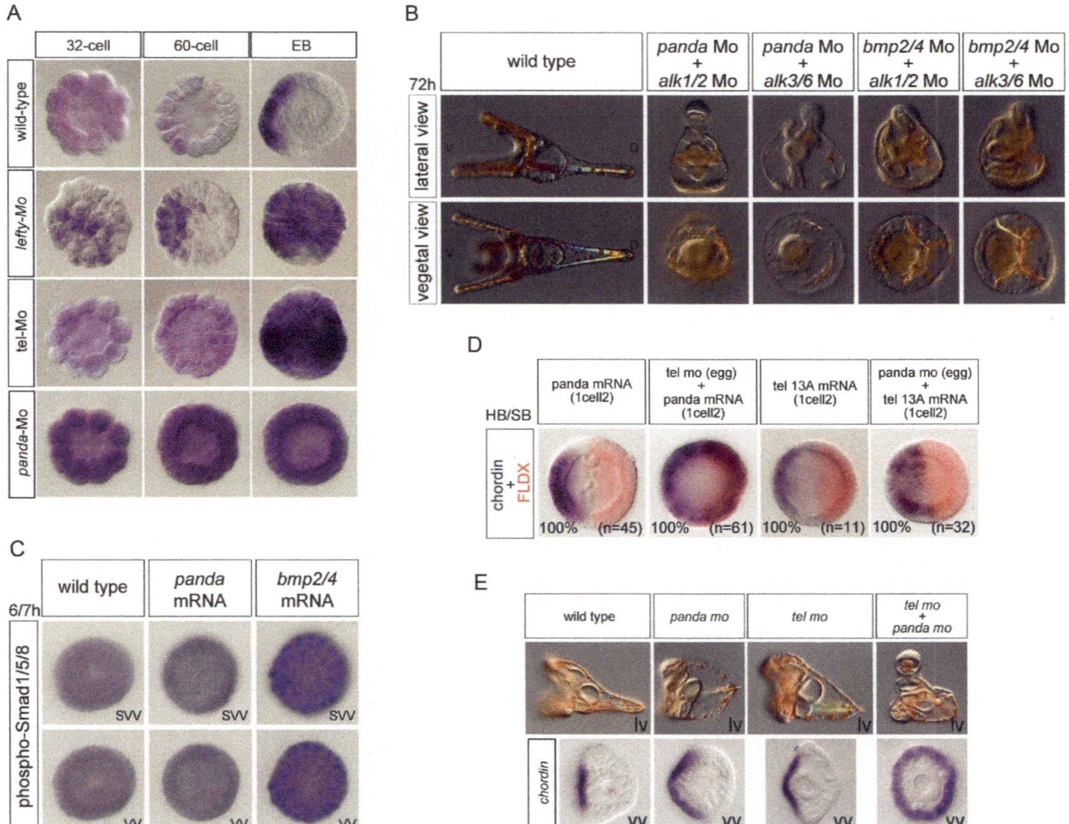

Fig. 6 See legend on next page.

(Haillot et al., 2015). To further address the question of the specificity of the ligands regarding the receptors, Haillot et al. used an assay based on double knockdown and showed that double inactivation of *panda* and *alk1/2*, of *panda* and *alk3/6*, of *bmp2/4* and *alk1/2* or of *bmp2/4* and *alk3/6* causes strong ventralization, similar to that caused by the double inactivation of *panda* and *bmp2/4 or of alk1/2 and alk3/6* (Haillot et al., 2015), consistent with the idea that the activities of Alk1/2 and Alk3/6 are required to transduce both BMP2/4 and Panda signals (Fig. 6B).

Haillot et al. further tested the hypothesis that Panda requires the BMP type I receptors to orient the D/V axis. In an axis induction assay, the *alk3/6*

Fig. 6 Mechanism of action of Panda. (A) Time-course of *nodal* expression in control, *lefty* morpholino, *yan/tel* morpholino and *panda* morpholino injected embryos. Note that ectopic expression of *nodal* is detected in *yan/tel* and *panda* morphants earlier than in the *lefty* morphants. (EB), early blastula. (B) Co-injection of *panda* and *alk1/2* morpholinos or of *panda* and *alk3/6* morpholinos causes a strong ventralization, as does the double inactivation of *bmp2/4+alk1/2* or of *bmp2/4* and *alk3/6*. V, ventral; D, Dorsal. (C) Phospho-Smad1/5/8 immunostaining at very early or early blastula stages in control embryos and in embryos overexpressing *panda* or *bmp2/4* mRNA. The highly sensitive alkaline phosphatase-based detection of pSmad1/5/8 does not allow detection of Smad1/5/8 signaling at early stages in control embryos. In contrast, following overexpression of *bmp2/4* strong nuclear phosphoSmad1/5/8 immunostaining is easily detected at early blastula. This pSmad1/5/8 immunoreactivity is not detected following injection of *panda* mRNA. h, hours; SVV, Surface vegetal view; VV, vegetal view. (D) *panda* mRNA overexpression into one cell at the two-cell stage (FLDX clone developed in red) does not restrict early *nodal* expression (blue) to the ventral side in the absence of Yan/Tel. Instead, Yan/Tel mRNA overexpression in one cell at the two-cell stage (FLDX clone developed in red) confines *nodal* expression (blue) to the ventral side even in the absence of Panda. HB, hatching blastula; SB, swimming blastula; Vegetal views, ventral to the left. (E) Embryos injected with suboptimal doses of Panda or Yan/Tel morpholinos develop into pluteus larvae, while double Panda+Yan/Tel morphants appear partially ventralized and show radial chordin expression at late gastrula stage (LG). lv, lateral view, ventral to the left. vv, vegetal view. *Panel A: Images from Haillot, E., Molina, M. D., Lapraz, F., & Lepage, T. (2015). The maternal maverick/GDF15-like TGF-beta ligand panda directs dorsal-ventral axis formation by restricting nodal expression in the sea urchin embryo.* PLoS Biology, 13, *e1002247; Molina, M. D., Quirin, M., Haillot, E., De Croze, N., Range, R., Rouel, M., et al. (2018). MAPK and GSK3/ss-TRCP-mediated degradation of the maternal Ets domain transcriptional repressor Yan/Tel controls the spatial expression of nodal in the sea urchin embryo.* PLoS Genetics, 14, *e1007621; Panel B and C: Images from Haillot, E., Molina, M. D., Lapraz, F., & Lepage, T. (2015). The maternal maverick/GDF15-like TGF-beta ligand panda directs dorsal-ventral Axis formation by restricting nodal expression in the sea urchin embryo.* PLoS Biology, 13, *e1002247; Panels D and E: Images from Molina, M. D., Quirin, M., Haillot, E., De Croze, N., Range, R., Rouel, M., et al. (2018). MAPK and GSK3/ss-TRCP-mediated degradation of the maternal Ets domain transcriptional repressor Yan/Tel controls the spatial expression of nodal in the sea urchin embryo.* PLoS Genetics, 14, *e1007621.*

morpholino was first injected into the egg, then, at the two-cell stage, *panda* mRNA was injected into one blastomere. When injected alone, *panda* mRNA efficiently oriented the D/V axis, but previous injection of the *alk3/6* morpholino into the egg abolished the ability of *panda* mRNA to orient the D/V axis suggesting that the Alk3/6 receptor is required for the activity of Panda (Haillot et al., 2015). Taken together, these results strongly suggested that Panda requires the activity of the BMP type I receptors to orient the D/V axis; however, these findings do not distinguish which signal transduction pathway is activated by this ligand. Indeed, although Panda requires the BMP type I receptor Alk3/6, it does not seem to activate canonical BMP signaling as endogenous phospho-Smad1/5/8 remains below the level of detection during cleavage stages and overexpression of *panda* does not detectably increases the level of phosphorylated Smad1/5/8 at early stages (Fig. 6C) (Haillot et al., 2015).

14.2 Yan/Tel acts downstream of Panda to restrict *nodal* expression

The similarity of the phenotypes caused by inactivation of Panda and Yan/Tel strongly suggested the possibility that Yan/Tel may act downstream of Panda in a common pathway. According to this model, Panda signaling may regulate *nodal* expression by inhibiting phosphorylation of Yan/Tel and stabilizing this factor. In support of this idea, overexpression of Panda was shown to affect the phosphorylation of Yan/Tel by Western blot and caused Yan/Tel to migrate predominantly as a fast migrating (presumably stabilized and non-phosphorylated) isoform, consistent with the idea that Panda acts at least in part by regulating the phosphorylation state and therefore the stability and/or the activity of the transcription factor Yan/Tel (Molina et al., 2018). In addition, the fact that both Yan/Tel and Panda are required early to restrict *nodal* expression supported this idea (Haillot et al., 2015; Molina et al., 2018). Furthermore, not only does inactivation of either Yan/Tel or Panda cause ectopic expression of *nodal* but they both appear to do so in a cell-autonomous manner (Fig. 4D and 5E) (Molina et al., 2018).

To further test if Panda and Yan Tel act in the same pathway and determine if Panda acts through stabilization of Yan/Tel, the ability of Panda to orient the D/V axis in the absence of Yan/Tel and reciprocally, the ability of Yan/Tel to orient the axis in the absence of Panda were tested using co-injection assays (Molina et al., 2018). In Yan morphants the progeny of the clones that were also injected with Panda mRNA were found predominantly on the dorsal side at prism stage, indicating that Panda can orient the D/V axis of Yan morphants (Molina et al., 2018). However, Yan morphants injected with Panda mRNA into one blastomere failed to restrict early *nodal* expression at early stages, strongly suggesting that Yan is epistatic to Panda, or in other words, that Yan acts downstream of Panda (Molina et al., 2018). Consistent with this idea, local overexpression of activated *yan/tel* efficiently restricted early *nodal* expression and oriented the D/V axis in *panda* morphants (Fig. 6D) (Molina et al., 2018). That Yan and Panda act in the same pathway during D/V axis formation, is further suggested by the results of experiments testing for synergistic effects. Assays using suboptimal doses of either the Yan or the Panda morpholino resulted in a spatially restricted expression of the nodal target gene *chordin* at gastrula stage (Molina et al., 2018). In contrast, co-injection of the Yan and Panda morpholinos at these suboptimal doses caused a dramatic expansion of *chordin* expression and resulted in a strong ventralization (Fig. 6E) (Molina et al., 2018). These results strongly suggest that Yan/Tel and Panda likely work in the same pathway and synergize to restrict *nodal* expression.

Cumulatively, these findings suggest a model wherein Panda acts upstream of Yan/Tel to restrict early *nodal* expression, possibly by antagonizing phosphorylation and degradation of Yan/Tel on the dorsal side (Fig. 7). By integrating information along the animal-vegetal and dorsal–ventral axis, Yan/Tel may therefore act as a factor that coordinates patterning along these two orthogonal patterning systems.

In conclusion, recent studies suggest that in addition to the zygotic function of Lefty, spatial restriction of *nodal* expression critically requires the activity of the maternal TGF-β ligand Panda (Molina et al., 2018). As discussed above, maternal Panda mRNA is differentially expressed in the egg and Panda function is required very early and locally to spatially restrict *nodal* expression and is sufficient to orient the axis when locally overexpressed (Molina et al., 2018). Taken together these properties strongly suggest that Panda may act as a *bona fide* maternal determinant of dorsal–ventral axis formation in the sea urchin embryo. However, how Panda works is still unclear and elucidating the mechanism by which Panda restricts *nodal* expression will require additional experiments.

15. Conclusion: Maternal determinants of D/V axis formation and developmental plasticity of the early blastomeres

The sea urchin embryo is well known for its remarkable developmental plasticity, the best example of this flexibility being the ability of each

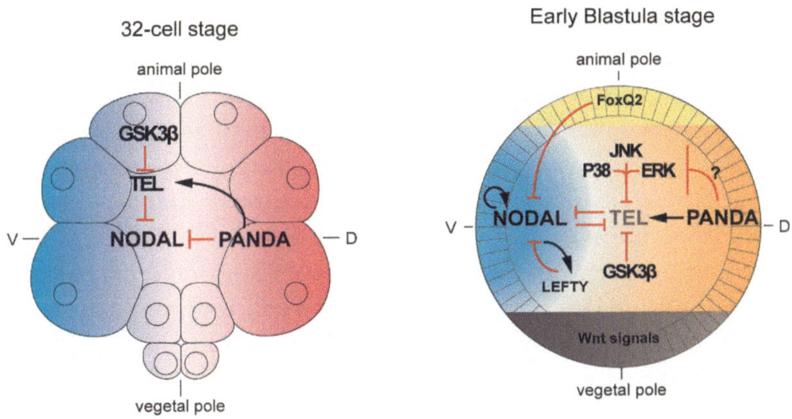

Fig. 7 Model of the regulation of *nodal* expression by GSK3, Yan/Tel and Panda. Starting at the 32-cell stage GSK3 activity in the animal hemisphere starts to target Yan/Tel for degradation, thereby releasing *nodal* from repression by Yan/Tel and allowing *nodal* expression to be initiated. On the dorsal side of the embryo, Panda creates an asymmetry of *nodal* expression by antagonizing *nodal* expression by an incompletely elucidated mechanism that may rely on the function of Yan/Tel. At early blastula stage, *nodal* expression in the animal pole is repressed by the presence of FoxQ2. In the rest of the ectoderm, MAP kinases, GSK3 and possibly Nodal signaling contribute to *nodal* expression by promoting phosphorylation and degradation of Yan/Tel. On the contrary, Panda signaling on the dorsal side may prevent degradation of Yan/Tel contributing to repression of *nodal* expression on the dorsal side. *From Molina, M. D., Quirin, M., Haillot, E., De Croze, N., Range, R., Rouel, M., et al. (2018). MAPK and GSK3/ss-TRCP-mediated degradation of the maternal Ets domain transcriptional repressor Yan/Tel controls the spatial expression of nodal in the sea urchin embryo.* PLoS Genetics, 14, *e1007621*.

blastomere of the four-cell stage to regulate and to develop into smaller but normally patterned pluteus larvae (Hörstadius, 1973). This property deeply influenced ideas about how the dorsal–ventral axis may be specified in this embryo and led to the commonly accepted view that dorsal–ventral patterning of the sea urchin embryo relies on cell interactions in the zygote and not on asymmetrically distributed maternal determinants (Henry, 1998). On the other hand, classical experiments of Hörstadius using bisected unfertilized eggs showed that artificially activated meridional halves frequently differentiate with a complementary symmetry. These observations led Hörstadius to think that "the D/V axis was already specified in the unfertilized egg." (Hörstadius, 1973). As discussed, the observation that the spatially restricted activity of the maternal factor Panda directs dorsal–ventral axis formation strongly supports this conclusion (Haillot et al., 2015). However, the finding that the spatially restricted activity of maternal *panda* mRNA directs the orientation of the dorsal–ventral axis may seem at odds with the results of the Driesch experiment. How can we reconcile the fact that the first blastomeres show an equivalent potential to re-establish a secondary axis with the graded activity of a maternal factor controlling formation the D/V axis in the early embryo? Following dissociation, each blastomere is expected to inherit a portion of the gradient of Panda activity. One possibility is therefore that the portion of the gradient of activity of Panda inherited by each blastomere following dissociation is sufficient to re-establish the secondary axis. Indeed, with a reduced gradient of Panda activity, the reaction-diffusion mechanism between Nodal and Lefty may in some cases be sufficient to amplify an initial asymmetry in the expression of *nodal* or *lefty* thereby restricting *nodal* expression to enable re-establishment of the secondary axis.

In conclusion, experiments carried out by early experimental embryologists like Child, Pease, Hörstadius and Czihak had provided some evidence for a possible maternal determination of the D/V axis, however, the molecular nature of the prepattern responsible for initiating D/V axis formation had remained elusive. Recent molecular analysis of this process identified a new pathway comprising the secreted factor Panda and the transcription factor Yan/Tel acting in the early steps of D/V axis specification. These new findings illustrate how in the highly regulative sea urchin embryo, the secondary axis is already "penciled in" by the graded maternal information deposited into the egg in the form of a gradient of *panda* mRNA. Since *nodal* plays a key role in specification of the proximal distal axis in mammals and in specification of the secondary and left–right axes in a number of species, this raises the question as to whether maternal factors of the Panda/Maverick/GDF15 family also provide a blueprint of axial development in these embryos.

Acknowledgments

This work was supported by the Agence Nationale de la Recherche (ANR) (ANR-14-CE11-0006-01 to T.L.), the Fondation ARC pour la Recherche sur le Cancer (Projet 20171206616 to T.L.) and the Fondation pour la Recherche Médicale (FRM), (Projet DEQ20180339195 to T.L.). We acknowledge previous members of the lab, particularly to Emmanuel Haillot, for their contribution to the identification of the maternal factors for dorsal–ventral specification in the sea urchin.

References

Arnold, S. J., & Robertson, E. J. (2009). Making a commitment: Cell lineage allocation and axis patterning in the early mouse embryo. *Nature Reviews. Molecular Cell Biology*, *10*, 91–103.

Boveri, T. H. (1908). Zellenstudien VI. Die Entwicklung dispermer Seeigeleier. Ein Beitrag zur Befruchtungslehre und zur Theorie des Kerns. *Jena Z. Naturwiss.*, *43*, 1.

Chang, W. L., Chang, Y. C., Lin, K. T., Li, H. R., Pai, C. Y., Chen, J. H., et al. (2017). Asymmetric distribution of hypoxia-inducible factor α regulates dorsoventral axis establishment in the early sea urchin embryo. *Development*, *144*, 2940–2950.
Chen, Y., & Schier, A. F. (2002). Lefty proteins are long-range inhibitors of squint-mediated nodal signaling. *Current Biology: CB*, *12*, 2124–2128.
Child, C. M. (1941). Formation and reduction of indophenol blue in development of an echinoderm. *Proceedings of the National Academy of Sciences of the United States of America*, *27*, 523–528.
Cinquin, O. (2006). Fast-tracking morphogen diffusion. *Journal of Theoretical Biology*, *238*, 532–540.
Coffman, J. A., Coluccio, A., Planchart, A., & Robertson, A. J. (2009). Oral-aboral axis specification in the sea urchin embryo III. Role of mitochondrial redox signaling via H_2O_2. *Developmental Biology*, *330*, 123–130.
Coffman, J. A., & Denegre, J. M. (2007). Mitochondria, redox signaling and axis specification in metazoan embryos. *Developmental Biology*, *308*, 266–280.
Coffman, J. A., McCarthy, J. J., Dickey-Sims, C., & Robertson, A. J. (2004). Oral-aboral axis specification in the sea urchin embryo II. Mitochondrial distribution and redox state contribute to establishing polarity in *Strongylocentrotus purpuratus*. *Developmental Biology*, *273*, 160–171.
Cruzalegui, F. H., Cano, E., & Treisman, R. (1999). ERK activation induces phosphorylation of Elk-1 at multiple S/T-P motifs to high stoichiometry. *Oncogene*, *18*, 7948–7957.
Czihak, G. (1963). Entwicklungsphysiologische Untersuchungen an Echininiden (Verteilung und bedeutung der Cytochomoxydase). *Wilhelm Roux' Archiv für Entwicklungsmechanik der Organismen*, *154*, 272–292.
Driesch, H. (1892). *The potency of the first two cells in echinoderm development.* New York: Hafner.
Duboc, V., Lapraz, F., Besnardeau, L., & Lepage, T. (2008). Lefty acts as an essential modulator of nodal activity during sea urchin oral-aboral axis formation. *Developmental Biology*, *320*, 49–59.
Duboc, V., Rottinger, E., Besnardeau, L., & Lepage, T. (2004). Nodal and BMP2/4 signaling organizes the oral-aboral axis of the sea urchin embryo. *Developmental Cell*, *6*, 397–410.
Frankenberg, S., & Zernicka-Goetz, M. (2004). Breaking radial symmetry. In C. Stern (Ed.), *Gastrulation*. New York: Cold Spring Harbor Laboratory Press.
Fuchs, S. Y., Spiegelman, V. S., & Kumar, K. G. (2004). The many faces of beta-TrCP E3 ubiquitin ligases: Reflections in the magic mirror of cancer. *Oncogene*, *23*, 2028–2036.
Haillot, E., Molina, M. D., Lapraz, F., & Lepage, T. (2015). The maternal maverick/GDF15-like TGF-beta ligand panda directs dorsal-ventral Axis formation by restricting nodal expression in the sea urchin embryo. *PLoS Biology*, *13* e1002247.
Henry, J. (1998). The development of dorsoventral and bilateral axial properties in sea urchin embryos. *Seminars in Cell and Developmental Biology*, *9*, 43–52.
Hörstadius, S. (1973). *Experimental embryology of echinoderms.* Oxford: Clarendon Press.
Hörstadius, S., & Wolsky, A. (1936). Studien über die Determination der Bilateralsymmetrie des jungen Seeigelkeimes. *Wilhelm Roux's Archives of Developmental Biology*, *135*, 69–113.
Huang, L., Li, X., El-Hodiri, H. M., Dayal, S., Wikramanayake, A. H., & Klein, W. H. (2000). Involvement of Tcf/Lef in establishing cell types along the animal-vegetal axis of sea urchins. *Development Genes and Evolution*, *210*, 73–81.
Ikeda, S., Kishida, S., Yamamoto, H., Murai, H., Koyama, S., & Kikuchi, A. (1998). Axin, a negative regulator of the Wnt signaling pathway, forms a complex with GSK-3beta and beta-catenin and promotes GSK-3beta-dependent phosphorylation of beta-catenin. *The EMBO Journal*, *17*, 1371–1384.

Jesuthasan, S., & Stahle, U. (1997). Dynamic microtubules and specification of the zebrafish embryonic axis. *Current Biology: CB*, *7*, 31–42.

Jousset, C., Carron, C., Boureux, A., Quang, C. T., Oury, C., Dusanter-Fourt, I., et al. (1997). A domain of TEL conserved in a subset of ETS proteins defines a specific oligomerization interface essential to the mitogenic properties of the TEL-PDGFR beta oncoprotein. *The EMBO Journal*, *16*, 69–82.

Juan, H., & Hamada, H. (2001). Roles of nodal-lefty regulatory loops in embryonic patterning of vertebrates. *Genes to Cells*, *6*, 923–930.

Kenny, A. P., Oleksyn, D. W., Newman, L. A., Angerer, R. C., & Angerer, L. M. (2003). Tight regulation of SpSoxB factors is required for patterning and morphogenesis in sea urchin embryos. *Developmental Biology*, *261*, 412–425.

Kim, C. A., Phillips, M. L., Kim, W., Gingery, M., Tran, H. H., Robinson, M. A., et al. (2001). Polymerization of the SAM domain of TEL in leukemogenesis and transcriptional repression. *The EMBO Journal*, *20*, 4173–4182.

Kominami, T. (1988). Determination of dorso-ventral axis in early embryos of the sea urchin, *Hemicentrotus pulcherrimus*. *Developmental Biology*, *127*, 187–196.

Lai, Z. C., & Rubin, G. M. (1992). Negative control of photoreceptor development in Drosophila by the product of the yan gene, an ETS domain protein. *Cell*, *70*, 609–620.

Langdon, Y. G., & Mullins, M. C. (2011). Maternal and zygotic control of zebrafish dorsoventral axial patterning. *Annual Review of Genetics*, *45*, 357–377.

Lapraz, F., Besnardeau, L., & Lepage, T. (2009a). Dorsal-ventral patterning in echinoderms: Insights into the evolution of the BMP-chordin signaling network. *PLoS Biology*, *7*, 1–25.

Lapraz, F., Besnardeau, L., & Lepage, T. (2009b). Patterning of the dorsal-ventral axis in echinoderms: Insights into the evolution of the BMP-chordin signaling network. *PLoS Biology*, *7*, e1000248.

Lapraz, F., Rottinger, E., Duboc, V., Range, R., Duloquin, L., Walton, K., et al. (2006). RTK and TGF-beta signaling pathways genes in the sea urchin genome. *Developmental Biology*, *300*, 132–152.

Maki, K., Arai, H., Waga, K., Sasaki, K., Nakamura, F., Imai, Y., et al. (2004). Leukemia-related transcription factor TEL is negatively regulated through extracellular signal-regulated kinase-induced phosphorylation. *Molecular and Cellular Biology*, *24*, 3227–3237.

Meinhardt, H., & Gierer, A. (2000). Pattern formation by local self-activation and lateral inhibition. *BioEssays*, *22*, 753–760.

Meno, C., Gritsman, K., Ohishi, S., Ohfuji, Y., Heckscher, E., Mochida, K., et al. (1999). Mouse Lefty2 and zebrafish antivin are feedback inhibitors of nodal signaling during vertebrate gastrulation. *Molecular Cell*, *4*, 287–298.

Meno, C., Saijoh, Y., Fujii, H., Ikeda, M., Yokoyama, T., Yokoyama, M., et al. (1996). Left-right asymmetric expression of the TGF beta-family member lefty in mouse embryos. *Nature*, *381*, 151–155.

Meno, C., Shimono, A., Saijoh, Y., Yashiro, K., Mochida, K., Ohishi, S., et al. (1998). Lefty-1 is required for left-right determination as a regulator of lefty-2 and nodal. *Cell*, *94*, 287–297.

Mizuno, T., Yamaha, E., Wakahara, M., Kuroiwa, A., & Takeda, H. (1996). Mesoderm induction in zebrafish. *Nature*, *383*, 131–132.

Molina, M. D., de Croze, N., Haillot, E., & Lepage, T. (2013). Nodal: Master and commander of the dorsal-ventral and left-right axes in the sea urchin embryo. *Current Opinion in Genetics & Development*, *23*, 445–453.

Molina, M. D., Quirin, M., Haillot, E., De Croze, N., Range, R., Rouel, M., et al. (2018). MAPK and GSK3/ss-TRCP-mediated degradation of the maternal Ets domain transcriptional repressor Yan/Tel controls the spatial expression of nodal in the sea urchin embryo. *PLoS Genetics*, *14* e1007621.

Ober, E. A., & Schulte-Merker, S. (1999). Signals from the yolk cell induce mesoderm, neuroectoderm, the trunk organizer, and the notochord in zebrafish. *Developmental Biology, 215*, 167–181.

O'Neill, E. M., Rebay, I., Tjian, R., & Rubin, G. M. (1994). The activities of two Ets-related transcription factors required for Drosophila eye development are modulated by the Ras/MAPK pathway. *Cell, 78*, 137–147.

Papaioannou, V. E., Mkandawire, J., & Biggers, J. D. (1989). Development and phenotypic variability of genetically identical half mouse embryos. *Development, 106*, 817–827.

Pease, D. C. (1941). Echinoderm bilateral determination in chemical concentration gradients I. the effects of cyanide, ferricyanide, iodoacetate, picrate, dinitrophenol, urethane, iodine, malonate, etc. *The Journal of Experimental Zoology, 86*, 381–405.

Poirel, H., Oury, C., Carron, C., Duprez, E., Laabi, Y., Tsapis, A., et al. (1997). The TEL gene products: Nuclear phosphoproteins with DNA binding properties. *Oncogene, 14*, 349–357.

Qiao, F., Song, H., Kim, C. A., Sawaya, M. R., Hunter, J. B., Gingery, M., et al. (2004). Derepression by depolymerization; structural insights into the regulation of Yan by Mae. *Cell, 118*, 163–173.

Range, R., Lapraz, F., Quirin, M., Marro, S., Besnardeau, L., & Lepage, T. (2007). Cis-regulatory analysis of nodal and maternal control of dorsal-ventral axis formation by Univin, a TGF-{beta} related to Vg1. *Development, 134*, 3649–3664.

Range, R., & Lepage, T. (2011). Maternal Oct1/2 is required for nodal and Vg1/ Univin expression during dorsal-ventral axis specification in the sea urchin embryo. *Developmental Biology, 357*, 440–449.

Rebay, I., & Rubin, G. M. (1995). Yan functions as a general inhibitor of differentiation and is negatively regulated by activation of the Ras1/MAPK pathway. *Cell, 81*, 857–866.

Roth, S., Stein, D., & Nusslein-Volhard, C. (1989). A gradient of nuclear localization of the dorsal protein determines dorsoventral pattern in the Drosophila embryo. *Cell, 59*, 1189–1202.

Roukens, M. G., Alloul-Ramdhani, M., Moghadasi, S., Op den Brouw, M., & Baker, D. A. (2008). Downregulation of vertebrate Tel (ETV6) and Drosophila Yan is facilitated by an evolutionarily conserved mechanism of F-box-mediated ubiquitination. *Molecular and Cellular Biology, 28*, 4394–4406.

Rushlow, C. A., Han, K., Manley, J. L., & Levine, M. (1989). The graded distribution of the dorsal morphogen is initiated by selective nuclear transport in Drosophila. *Cell, 59*, 1165–1177.

Saudemont, A., Haillot, E., Mekpoh, F., Bessodes, N., Quirin, M., Lapraz, F., et al. (2010). Ancestral regulatory circuits governing ectoderm patterning downstream of Nodal and BMP2/4 revealed by gene regulatory network analysis in an echinoderm. *PLoS Genetics, 6*, e1001259.

Schupbach, T. (1987). Germ line and soma cooperate during oogenesis to establish the dorsal-ventral pattern of egg shell and embryo in *Drosophila melanogaster*. *Cell, 49*, 699–707.

Solnica-Krezel, L. (2003). Vertebrate development: Taming the nodal waves. *Current Biology: CB, 13*, R7–R9.

Steward, R. (1989). Relocalization of the dorsal protein from the cytoplasm to the nucleus correlates with its function. *Cell, 59*, 1179–1188.

Sutherland, C. (2011). What are the bona fide GSK3 substrates? *International Journal of Alzheimer's Disease, 2011*, 505607.

Tanaka, C., Sakuma, R., Nakamura, T., Hamada, H., & Saijoh, Y. (2007). Long-range action of nodal requires interaction with GDF1. *Genes & Development, 21*, 3272–3282.

Tao, Q., Yokota, C., Puck, H., Kofron, M., Birsoy, B., Yan, D., et al. (2005). Maternal wnt11 activates the canonical wnt signaling pathway required for axis formation in Xenopus embryos. *Cell, 120*, 857–871.

Thisse, C., & Thisse, B. (1999). Antivin, a novel and divergent member of the TGFbeta superfamily, negatively regulates mesoderm induction. *Development, 126*, 229–240.

Tootle, T. L., & Rebay, I. (2005). Post-translational modifications influence transcription factor activity: A view from the ETS superfamily. *BioEssays, 27*, 285–298.

Turing, A. (1952). The chemical basis of morphogenesis. *Philosophical Transactions of the Royal Society of London, 237*, 37–72.

Weaver, C., & Kimelman, D. (2004). Move it or lose it: Axis specification in Xenopus. *Development, 131*, 3491–3499.

Wei, Z., Angerer, R. C., & Angerer, L. M. (2006). A database of mRNA expression patterns for the sea urchin embryo. *Developmental Biology, 300*, 476–484.

Yaguchi, S., Yaguchi, J., & Burke, R. D. (2006). Specification of ectoderm restricts the size of the animal plate and patterns neurogenesis in sea urchin embryos. *Development, 133*, 2337–2346.

Yost, C., Torres, M., Miller, J. R., Huang, E., Kimelman, D., & Moon, R. T. (1996). The axis-inducing activity, stability, and subcellular distribution of beta-catenin is regulated in Xenopus embryos by glycogen synthase kinase 3. *Genes & Development, 10*, 1443–1454.

Zhou, B. P., Deng, J., Xia, W., Xu, J., Li, Y. M., Gunduz, M., et al. (2004). Dual regulation of Snail by GSK-3beta-mediated phosphorylation in control of epithelial-mesenchymal transition. *Nature Cell Biology, 6*, 931–940.

Zito, F., Costa, C., Sciarrino, S., Poma, V., Russo, R., Angerer, L. M., et al. (2003). Expression of univin, a TGF-beta growth factor, requires ectoderm-ECM interaction and promotes skeletal growth in the sea urchin embryo. *Developmental Biology, 264*, 217–227.

CHAPTER ELEVEN

Maternal factors regulating preimplantation development in mice

Di Wu*, Jurrien Dean*
Laboratory of Cellular and Developmental Biology, NIDDK, National Institutes of Health, Bethesda, MD, United States
*Corresponding authors: e-mail address: di.wu2@nih.gov; jurrien.dean@nih.gov

Contents

1. Introduction 318
2. Initiation of embryonic development 318
3. Transition to independence: Zygotic genome activation 322
   3.1 Maternal transcription factors prime ZGA 323
   3.2 Preparing chromatin for transcription by maternal protein modifiers 324
   3.3 Other maternal factors governing ZGA 325
4. Other preimplantation events regulated by maternal factors 327
   4.1 Cell cleavage 327
   4.2 Cell fate specification 328
5. Maternal factors require timely clearance 330
6. Concluding remarks 333
Acknowledgments 333
References 333

Abstract

Mammalian embryogenesis depends on maternal factors accumulated in eggs prior to fertilization and on placental transfers later in gestation. In this review, we focus on initial events when the organism has insufficient newly synthesized embryonic factors to sustain development. These maternal factors regulate preimplantation embryogenesis both uniquely in pronuclear formation, genome reprogramming and cell fate determination and more universally in regulating cell division, transcription and RNA metabolism. Depletion, disruption or inappropriate persistence of maternal factors can result in developmental defects in early embryos. To better understand the origins of these maternal effects, we include oocyte maturation processes that are responsible for their production. We focus on recent publications and reference comprehensive reviews that include earlier scientific literature of early mouse development.

Current Topics in Developmental Biology, Volume 140
ISSN 0070-2153
https://doi.org/10.1016/bs.ctdb.2019.10.006

1. Introduction

Preimplantation development begins with fertilization in which two haploid gametes fuse to form one-cell, diploid zygotes that divide and eventually form mature blastocysts. Although implantation is unique to mammals, the window of preimplantation development shares many similarities with non-mammalian animals and takes ~4–5 days in mice. The one-cell zygote remodels parental chromatin, activates embryonic transcription of its genome, defines early polarity and establishes different cell lineages. After implantation on the uterine wall, specified lineages—epiblast, primitive endoderm and trophectoderm—further differentiate and define the three germ layers of the developing embryo (Fig. 1).

Although many maternal mutants arrest at the 2-cell stage of development, others have defects later in preimplantation embryogenesis. Maternal factor deficiency leads to embryonic arrest either directly or indirectly. Directly, a maternal factor can regulate specific activities at a particular developmental stage, the lack of which immediately blocks development. Indirectly, a maternal factor can cause non-lethal defects that accumulate to create cellular stress that impairs development at a later stage. We will dissect the direct and indirect roles of maternal factors during mouse preimplantation by following the sequence of events in early development.

2. Initiation of embryonic development

The life of mammals begins at fertilization when sperm and egg meet in the oviductal ampulla of the female reproductive tract. Gametic fusion

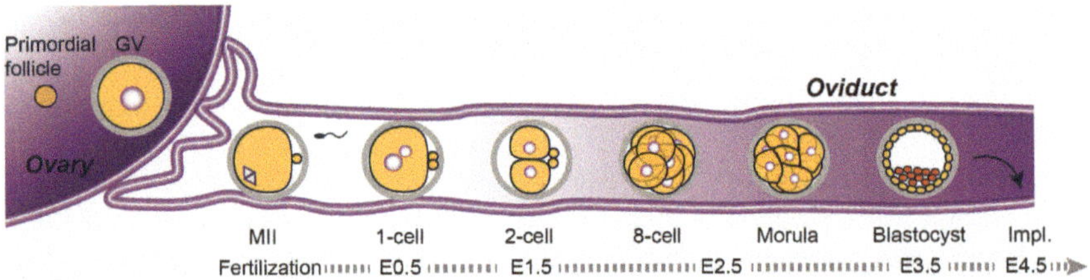

Fig. 1 An outline of mouse oogenesis and preimplantation development. Oocytes grow and undergo meiotic maturation within the ovary prior to ovulation into the oviduct where they are fertilized by capacitated sperm to form a zygote (embryonic day 1, E1). The 1-cell zygote divides into two totipotent blastomeres and continues to cleave (~E1.5). At the 8-cell stage (~E2.5), the embryo compacts to form the morula. After 2–3 additional rounds of cell division (~E3.5), a blastocoel forms, transforming the embryo into a blastocyst that implants on the wall of the uterus ~E4.5. Impl., Implantation.

begins with membrane binding and can be visualized by sperm internalization that happens within minutes to hours of gamete mixing. However, interdigitating nuclear syngamy of the two haploid gametes does not occur prior to a sequence of key events that take place following sperm entry. In the absence of zygotic gene transcription, maternal factors actively participate in sperm protamine-to-histone transition, pronuclear formation, epigenetic remodeling, genomic imprinting protection and establishment of higher-order chromatin structures, the consequence of which collectively prepare the zygote for nuclear union.

Sperm chromatin undergoes dramatic remodeling after penetrating into the egg. Although protamine packaging of genomic DNA is critical for spermiogenesis, after fertilization, protamines must be replaced with somatic histones during the decondensation and recondensation phases of the sperm-to-paternal pronucleus transition (Balhorn, 2007). Decondensation happens right after sperm enter the egg cortical zone and requires reduction of intra- and inter-protamine disulfide bonds (Caglar et al., 2005). In mice, maternal NPM2 (nucleoplasmin 2) can induce sperm chromatin decondensation by regulating levels of histone acetylation (Burns et al., 2003; Inoue, Ogushi, Saitou, Suzuki, & Aoki, 2011), which is consistent with the documented role of *Xenopus* oocyte-supplied NPM in removing sperm protamine and facilitating nucleosome assembly (Philpott & Leno, 1992). Purified mouse NPM2 protein interacts with mouse P1/P2 protamines and displaces them from DNA (Ellard, Serpa, Petrotchenko, Borchers, & Ausio, 2016; Philpott & Leno, 1992). Genetic ablation of mouse *Npm2* results in reduced female fertility due to disorganized chromatin, including the absence of nucleoli and heterochromatin (Burns et al., 2003; Ogushi et al., 2017). In the subsequent recondensation phase, oocyte-supplied histone 3.3 (H3.3), but not H3.1 or H3.2, can access and reorganize sperm DNA (van der Heijden et al., 2005). This H3.3 preference may differentiate paternal from maternal pronuclei and account for early transcription activation of the paternal genome (Kong et al., 2018) (Fig. 2A).

After sperm chromatin decondensation and recondensation, both sperm and egg haploid genomes form pronuclei and move toward each other whereupon their membranes interdigitate and disperse to facilitate nuclear union. During this process, active DNA demethylation (removal of 5mC) in paternal chromatin is mediated by maternal TET1/2/3 through a series of oxidative transformation to 5hmC (5-hydroxymethylcytosine), 5fC (5-formylcytosine) and 5caC (5-carboxylcytosine) (Wu & Zhang, 2014). All intermediate products can be converted to cytosine by base excision DNA repair or by 5caC decarboxylation (Wu & Zhang, 2010). In contrast,

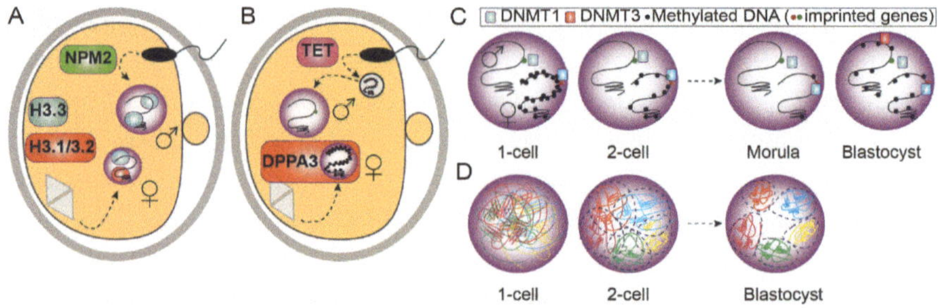

Fig. 2 Maternal factors function in chromatin remodeling post-fertilization. (A) Paternal pronucleus formation by maternal NPM2 that mediates protamine removal and H3.3 that mediates recondensation. Maternal chromatin is occupied by H3.3 and H3.1/3.2. (B) Paternal DNA demethylation is mediated by maternal TET proteins while maternal DNA methylation persists due to the masking by DPPA3. (C) DNA methylation differs in paternal and maternal alleles during preimplantation development. Paternal DNA is actively and rapidly demethylated at the 1-cell stage whereas maternal DNA methylation decreases passively through cell division. Both paternal and maternal DNA then undergo remethylation by DNMT3A/3B beginning in blastocysts. Imprinted genes remain unaffected by DNMT1 during preimplantation. (D) Establishment of topologically associating domains (TADs) during preimplantation. While there are no clearly defined TADs at the 1-cell stage, they are gradually established during preimplantation development.

maternal DNA is largely spared from active demethylation by DPPA3 (developmental pluripotency associated 3) masking, which is recruited by dimethylation of Lys9 on maternal H3 (H3K9me2) (Nakamura et al., 2012). Maternal DNA demethylation occurs passively through replication-dependent dilution during subsequent embryonic cell cleavage (Mayer, Niveleau, Walter, Fundele, & Haaf, 2000). The three maternal TETs are partially redundant (Dawlaty et al., 2011; Gu et al., 2011; Kang et al., 2015; Li et al., 2011, 2016; Shen et al., 2014). Depletion of maternal TET3 impairs mouse paternal DNA demethylation in the 5mC to 5hmC conversion, which results in repression of paternal derived embryonic stem cell genes *Oct4* and *Nanog* and increases embryonic lethality (Gu et al., 2011). However, TET3 depletion does not affect zygotic gene activation or preimplantation development (Shen et al., 2014). Maternal TET1 or TET2 depletion is associated with normal or subfertility (Dawlaty et al., 2011; Li et al., 2011). Double knockouts of maternal *Tet1*; *Tet3* results in delayed or aborted preimplantation development due to increased transcriptome variability (Kang et al., 2015); and triple maternal knockout of *Tet1–3* does not further worsen the phenotype (Li et al., 2016). The minor defects associated with *Tet* gene ablation may reflect the greater role exerted by passive DNA demethylation that occurs concurrently by dilution due to cell cleavage (Messerschmidt, Knowles, & Solter, 2014). Moreover, defects in zygotic transcription may not necessarily lead to an immediate phenotype because of maternal compensation (Fig. 2B).

Despite global DNA demethylation, gene imprinting established in the parental germline remains unchanged by the maternal DNA methyltransferase DNMT1 (Hirasawa et al., 2008). DNMT1 targets hemi-methylated DNA to maintain the methylation pattern during replication (Jeltsch & Jurkowska, 2014). Mouse maternal DNMT1 depletion results in partial reduction of methylation at parental differentially methylated regions (DMRs), including paternal *H19* and *Rasgrf1* DMRs, and maternal *Peg3* and *Snrpn* DMRs (Howell et al., 2001). In addition to DNMT1, the oocyte-specific isoform DNMT1o also contributes to maintaining allele-specific DNA methylation (Howell et al., 2001). Unlike DNMT1, DNMT3A/3B performs *de novo* DNA methylation (Okano, Bell, Haber, & Li, 1999). Oocyte-specific knockout of *Dnmt3* causes hypomethylation of maternal DNA and recurrent post-implantation embryonic loss (Kaneda et al., 2010) (Fig. 2C). Similar to TET deficiency, even though dysregulated DNA methylation directly perturbs zygotic gene transcription, the physiological defect is manifest after implantation, which suggests proper maternal support can bypass certain zygotic deficiencies in this developmental window.

Recent studies have identified higher-order chromatin structure in mouse embryos and documented establishment of topologically associating domains (TADs) during preimplantation (Ke et al., 2017). In mouse oocytes or 1-cell zygotes, clearly defined TADs are not present; in sperm, however, somatic cell-like TADs have been identified, but are quickly lost at fertilization and the higher-order chromatin structures become obscure as in unfertilized eggs (Ke et al., 2017). TADs are reestablished at the 2-cell stage and maternal CTCF (CCCTC-binding factor) appears involved in this process: CTCF occupancy is enriched at the TAD boundaries together with chromatin structural proteins, active histone modifications including H3K4me3, H3K27Ac, H3K9Ac, H3K4me1 and SMC1 (Ke et al., 2017). The interactions between CTCF and other transcription factors persist during preimplantation development to maintain this chromatin configuration (Jung et al., 2019) (Fig. 2D). However, TAD formation relies on DNA replication and not zygotic gene transcription as previously documented in *Drosophila* (Hug, Grimaldi, Kruse, & Vaquerizas, 2017; Ke et al., 2017). Thus, as yet unknown maternal proteins/RNAs must be required to establish TADs during preimplantation development. In summary, most of the basic cellular activities and specific events in the oocyte-to-embryo transition are performed by oocyte loaded executers.

3. Transition to independence: Zygotic genome activation

Zygotic genome activation (ZGA) provides a blueprint for the independent future of embryos. As a highly conserved process from sea urchin to humans, ZGA refers to a huge number of genes transitioning from quiescence to transcription in a short time window after fertilization. In mice, the first evidence of ZGA comes from transcription of the paternal genome and emergence of transcripts unavailable in the egg. Increased mRNA levels have been described both for the entire transcriptome and for single gene loci at the 2-cell stage of development (Schultz, 1993). Later, the ZGA splits into two phases: minor and major. Minor ZGA happens specifically in the male pronucleus, both earlier and to a lesser extent than major ZGA (Aoki, Worrad, & Schultz, 1997) (Fig. 3A). The increasingly consensus is that ZGA contains many coordinated waves of gene transcription that form a cascade which differs from the old "burst" view of zygote transcription (Jukam, Shariati, & Skotheim, 2017). During ZGA, maternal factors regulate transcription, both directly and indirectly.

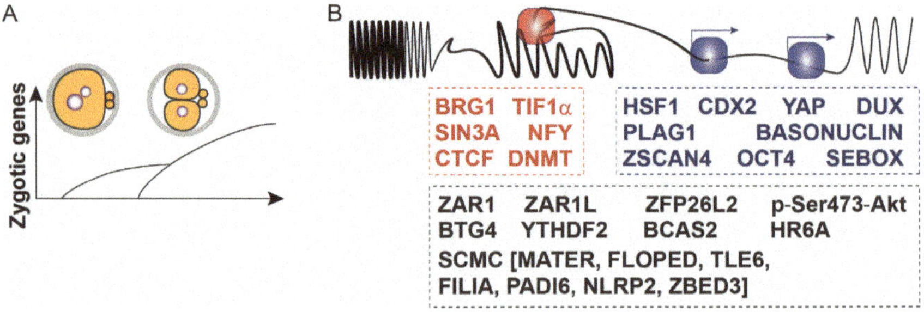

Fig. 3 Maternal factors affect zygotic gene activation (ZGA). (A) Minor and major waves of ZGA in the 1- and 2-cell embryos. (B) Maternal factors can directly regulate ZGA by affecting transcription (red) or chromatin structure (blue), or indirectly regulate ZGA by affecting other cellular activities including RNA metabolism, DNA repair and cytoskeleton (black).

3.1 Maternal transcription factors prime ZGA

Many maternal transcription factors are loaded into the zygote as RNA which requires active translation to produce functional proteins including E2F transcription factor 1, RNA polymerase 1–3, MYST histone acetyltransferase 1 and SWI/SNF-related proteins (Potireddy, Vassena, Patel, & Latham, 2006). Most of the identified factors are not ZGA-specific, but are ubiquitously required for transcription and their zygotic ablation causes defects later in development (Maslon et al., 2019; Tsai et al., 2008).

Loss-of-function studies of maternal transcription factors have been achieved by gene ablation in late oocytes or in 1-cell embryos, which most immediately affects ZGA. Oocyte-specific knockout of *Hsf1* (heat shock factor 1), which binds and controls expression of the minor ZGA gene *Hspa1b*, results in extensive defects and developmental arrest at the 2-cell stage (Bierkamp et al., 2010; Wilkerson, Murphy, & Sarge, 2008). Maternal-zygotic knockout of *Cdx2*, a well-known trophectoderm marker, leads to an earlier embryonic lethality compared to zygotic knockout, suggesting a unique role of maternal *Cdx2* (Blij, Frum, Akyol, Fearon, & Ralston, 2012; Jedrusik, Cox, Wicher, Glover, & Zernicka-Goetz, 2015). Genetic depletion of maternal YAP (yes-associated protein) in oocytes prolongs the 2-cell stage, delays progression to 4-cells and causes concomitant transcriptional disruption of zygotic genes *Rpl13* and *Rrm2* (Yu, Ji, Dang, et al., 2016; Yu, Ji, Sha, et al., 2016). DUX (double-homeodomain proteins) is known to activate ZGA genes, and when depleted zygotically results in preimplantation arrest in *ex vivo* cultured embryos (De Iaco et al., 2017). Zygotic or maternal-zygotic knockout of *Dux* can have subtle defects in ZGA (Chen & Zhang, 2019) which raises the possibility that occupancy of DUX at actively transcribed genes could be compensated by other factors. PLAG1 (Pleomorphic adenoma gene 1) has been found to directly promote the transcription of a few zygotic genes having PLAG1 binding motifs (Madissoon et al., 2019). Heterozygous embryos lacking maternal PLAG1 delay the 2-cell stage development (Madissoon et al., 2019). Maternal transcription factors thus activate zygotic gene transcription both cooperatively and independently.

In addition to genetic ablation, siRNA or morpholinos have been used to degrade targeted transcripts or block translation for transient loss-of-function in early embryonic stages (Falco et al., 2007; Ma, Zeng, Schultz, & Tseng, 2006; Park et al., 2015; Wu & Scholer, 2014). An advantage of these approaches is that they target both maternal and zygotic RNA which can lead to a more severe phenotype than genetic perturbation of the maternal gene. Lack of BASONUCLIN in transgenic RNAi mouse models blocks embryonic development at the 2-cell stage by perturbing both RNA polymerase I and II mediated transcription (Ma et al., 2006). ZSCAN4 (zinc finger and SCAN domain-containing protein 4), a potential transcription factor possessing zinc finger domains, is expressed ubiquitously at the 2–4-cell stage. siRNA mediated depletion of ZSCAN4 in 1-cell zygotes causes arrest at the 2-cell stage (Falco et al., 2007). Although an oocyte-specific knockout of maternal *Oct4* has little effect, injection of *Oct4* morpholinos can disrupt embryogenesis at the morula stage, raising the possibility that zygotic OCT4 can quickly compensate for the loss of maternal OCT4 (Wu & Scholer, 2014). Eliminating maternal SEBOX (skin-embryo-brain-oocyte homeobox) by pronuclear injection of double-stranded RNA results in embryonic arrest at the 2–8-cell stage and the absence of several known zygotic transcripts including *Mt1a*, *Rpl23*, *Ube2a* and *Wee1* (Park et al., 2015). In summary, oocyte-provided transcription factors can activate early embryonic genes by directly participating in the zygotic gene transcription network (Fig. 3B).

3.2 Preparing chromatin for transcription by maternal protein modifiers

In addition to directly controlling transcription, maternal factors can prepare chromatin for transcription. As discussed below, maternal factors can modify chromatin activity, mediate transcription factor binding to the genome, dissociate histones from the genome and demethylate genomic DNA (Bultman et al., 2006; Jimenez et al., 2015; Lu et al., 2016; Torres-Padilla & Zernicka-Goetz, 2006).

One of the best studied maternal chromatin modifiers is BRG1, a catalytic subunit of a chromatin remodeling complex (Kim, Bresnick, & Bultman, 2009). Zygotic null homozygotes of *Brg1* are embryonic lethal at the blastocyst stage while oocyte-depletion by Cre-loxP or 1-cell depletion by RNAi results in 2-cell arrest with reduced dimethyl-H3K4 (Bultman et al., 2006). BRG1 chromosomal localization during minor ZGA seems to be controlled by another transcription intermediary factor TIF1α, and ablation of TIF1α through RNAi or antibody blocking leads to mis-localization of BRG1 and RNA polymerase (Torres-Padilla & Zernicka-Goetz, 2006). Due to limited biological material, BRG1-binding sites in preimplantation embryos have not been identified. Knowing target genes of maternal BRG1 at the 2-cell stage would add greatly to our understanding of ZGA. Another factor, SIN3A, is a member of the HDAC1/2 family that functions during the oocyte to embryo transition (Yang & Seto, 2008). Disruption of oocyte SIN3A causes 2-cell arrest with dysregulated ZGA (Jimenez et al., 2015). Maternal NFY binds to open chromatin structures and can contribute to DNase I-hypersensitive sites formation (Lu et al., 2016). NFY depletion at the 2-cell stage disrupts zygotic gene expression (Lu et al., 2016).

In addition to local remodeling of chromatin structure for transcription, establishing higher-order structure can also affect ZGA (Wan et al., 2008). CTCF can facilitate or repress interactions of different genomic domains (Ong & Corces, 2014). Depletion of oocyte CTCF by transgenic RNAi causes defects in meiosis and mitosis and adversely affects ZGA which leads to delayed and failed development (Wan et al., 2008). Though TADs formation is independent of ZGA, genomic CTCF footprints remain constant from oocytes to preimplantation embryos which suggests that these sites require additional factors to regulate developmentally appropriate transcription (Jung et al., 2019). It may be that CTCF can recruit different interacting factors to achieve different regulatory roles, which would separate activation from repression, and TAD establishment from ZGA. Along with the development of low-input sequencing technology and single-cell omics, we anticipate further molecular clarity of CTCF binding dynamics and its relationship with other chromatin modifications in ZGA regulation (Fig. 3B).

3.3 Other maternal factors governing ZGA

In addition to specific transcription factors or chromatin modifiers, maternal factors can affect ZGA through multiple indirect pathways, such as controlled RNA splicing, translation and transcript stability. *Zar1*, one of the first maternal effect genes identified by mouse genetics, has been studied for more than 10 years (Wu et al., 2003). *Zar1*-depleted oocytes appear normal, but embryos derived from null females arrest at the 1-cell stage with marked reduction of ZGA (Wu et al., 2003). ZAR1-like (ZAR1L) has a similar protein sequence to ZAR1, and also induces 2-cell arrest by ectopic expression of its C terminus, with concomitant defects of histone modification, RNA synthesis and RNA polymerase II (Hu et al., 2010). The co-localization of ZAR1L and RNA granules indicate a role in RNA regulation (Hu et al., 2010). Meanwhile, the similarity between the carboxyl-termini of ZAR1 and ZAR1L suggests that ZAR1 may also regulate RNA stability to indirectly affect ZGA (Hu et al., 2010). ZFP36L2, known to bind the 3′ UTR of mRNA and participate in the deadenylation complex to trigger mRNA decay, leads to 2-cell arrest when the maternal ZFP36L2 protein is truncated at its amino-terminus (Ramos et al., 2004). Oocyte-specific depletion of ZFP36L2 stabilizes a group of histone modification genes normally degraded during oocyte maturation, which results in insufficient transcription termination and poor oocyte quality (Dumdie et al., 2018). It is possible that ZFP36L2 could adopt a similar strategy in targeting chromatin modifiers during ZGA. pSer473-Akt (phosphorylated Ser473-Akt), present in 1–2 cell nuclei, leads to impaired mRNA levels of some zygotic genes by its enzymatic inhibition (Chen et al., 2016). BTG4, a scaffold protein of the deadenylation-translation complex, can regulate RNA stability and translation across the oocyte-to-embryo transition (Wu & Dean, 2016). *Btg4* knockout females produce morphologically normal eggs, but are infertile because embryos arrest after fertilization at 2-cell stage with perturbed transcriptomes (Liu et al., 2016; Yu, Ji, Dang, et al., 2016; Yu, Ji, Sha, et al., 2016). YTHDF2, an RNA N6-methyladenosine (m6A)-binding and stabilization enzyme, is essential to ZGA since its maternal disruption impairs degradation of its target genes and is also associated with reduced female fecundity (Ivanova et al., 2017).

In addition to post-transcription modification, the DNA repair system also indirectly affects ZGA. Maternal BCAS2 (breast carcinoma amplified sequence 2) deficiency compromises the DNA damage response and leads to 2–4-cell stage arrest accompanied by accumulation of damaged DNA, micronuclei and disrupted zygotic gene expression (Xu et al., 2015). Oocytes, depleted of HR6A, a ubiquitin-conjugating DNA repair enzyme homologous to yeast RAD6, have normal oogenesis; but the derived embryos fail to develop beyond the 2-cell stage (Roest et al., 2004). The ZGA defects described above could be caused indirectly by either cellular abnormalities or mis-regulated transcription factors. Some of the defects, especially those involved in maternal RNA metabolism, may generate deleterious deposits that impair oocyte quality prior to preimplantation development (discussed below).

The mammalian subcortical maternal complex (SCMC) plays essential roles in early embryonic development. The SCMC contains a group of proteins localizing in the cortical region of eggs and embryos and its components include MATER (official name, NLRP5), FLOPED (OOEP), TLE6, FILIA (KHDC3), PADI6, NLRP2 and ZBED3 (Gao et al., 2018; Li, Baibakov, & Dean, 2008; Lim & Knowles, 2015; Mahadevan et al., 2017; Zheng & Dean, 2009). The function of the SCMC begins as early as oocyte maturation by positioning the meiotic spindle and localizing cortical vesicles for exocytosis (Vogt et al., 2019). After fertilization, the SCMC continues to regulate cytoplasmic F-actin meshwork, RNA translation, epigenetic reprogramming and organelle redistribution (Bebbere, Masala, Albertini, & Ledda, 2016; Lu, Gao, Qin, & Li, 2017; Vogt et al., 2019). Depletion of SCMC components results in cleavage-stage embryonic arrest and maternal infertility (Li, Zheng, & Dean, 2010) (Fig. 3B). The concomitant ZGA failure in SCMC-deficient embryos is more likely to be a secondary effect of cellular stress accumulated through extensive dysregulation.

4. Other preimplantation events regulated by maternal factors

During preimplantation development, both maternal and zygotic factors regulate multiple cellular activities, including cell cleavage, embryonic polarity and cell fate specification, which are both highly conserved across the animal kingdom and have unique order-specific characteristics.

4.1 Cell cleavage

Cell cleavage during preimplantation development gradually molds embryonic polarity that guides further cell fate determination. The first cleavage of the mouse embryo generates two totipotent blastomeres (Tarkowski, 1959). Their totipotency has been documented by the two embryos that develop from the physically separated blastomeres, and by each compensating for the loss of the other (Tarkowski, 1959). Before fertilization, the egg develops asymmetry with its meiosis II spindle close to the cortical area as well as concomitant asymmetry of maternal PAR proteins and an actin meshwork. The first division cleaves the zygote longitudinally by passing through the site of the previous meiotic division and along the axis of animal-vegetal pole (Plusa, Grabarek, Piotrowska, Glover, & Zernicka-Goetz, 2002). Unique animal-pole maternal components are suggested to transmit meiotic pole localization information to the first cleavage plane, a model that has been confirmed in an animal-pole transplantation assay (Plusa et al., 2002). However, the molecular basis of these observations remains to be determined. The asymmetric localization of PAR proteins in the egg disappears during cleavage-stage development and only re-emerges after the 8-cell stage (Vinot et al., 2005). Thus, it is unlikely that PAR proteins have direct association with the regulation of the first mitotic division. Moreover, the newly transcribed zygotic genes during the minor ZGA may play an important role in building the mitotic machinery for the first cleavage.

Perturbation of many maternal factors essential for ZGA could potentially affect cell cleavage and result in 2-cell arrest (as discussed in the previous section). Additionally, several maternal proteins regulate cell cleavage independent of ZGA and are more associated with the mitotic apparatus. Maternal CENPC (centromere protein C) contributes to the first several mitotic divisions in mice and its depletion results in morula stage lethality (Kalitsis, Fowler, Earle, Hill, & Choo, 1998). Activation of maternal PLK1 (Polo-like kinase 1) by phosphorylation precedes pronuclear fusion in 1-cell zygotes, and, when inhibited, perturbs microtubule assembly and prevents completion of the first mitotic division (Baran, Solc, Kovarikova, Rehak, & Sutovsky, 2013). Oocyte-specific depletion of ERK1/2 can result in embryonic arrest at the 1–2-cell stage (Sha et al., 2017) due to insufficient translation activation in oocytes. Maternal SETDB1 deficient embryos arrest during preimplantation due to impaired chromosome segregation (Eymery, Liu, Ozonov, Stadler, & Peters, 2016).

4.2 Cell fate specification

Although each of the two blastomeres formed by the first mitotic cleavage is totipotent, thereafter they gradually gain different fates directing them toward distinct cell lineages. The first blastomere to divide will contribute to the embryonic portion of the blastocyst whereas the later dividing blastomere will contribute to the abembryo (Piotrowska, Wianny, Pedersen, & Zernicka-Goetz, 2001). The loss of totipotency is supported by the bimodal gene expression pattern at the late 2-cell and early 4-cell stages (Biase, Cao, & Zhong, 2014). The bimodal distribution of gene expression is reproducible among embryos and overlaps with differential expression of ICM (inner cell mass) and TE (trophectoderm) genes at the blastocyst stage (Biase et al., 2014). A recent report suggests that the earliest polarity may form within 6 h after the first mitotic division (Casser et al., 2017). In these experiments, early 2-cell embryos were separated and cultured to blastocysts.

Consistently, the investigators observed different abilities of the pairs to generate the epiblast (Casser et al., 2017). The loss of totipotency is likely due to coordinated regulation of both maternal and zygotic factors. However, maternal proteins are most likely involved in this process since parthenogenetic embryos can develop up to embryonic day 10 (E10) which is well-past the time of implantation (Niwa et al., 2004). Moreover, the differences of the transcriptomes in the two initial blastomeres are likely to be further modulated by localized translational regulation that can be traced back to the polarity of the egg (Plusa et al., 2002).

The next major differentiation happens at the morula-to-blastocyst transition during which an inner *vs.* outer polarization underlies the future ICM and TE lineages, respectively, and contributes to the dichotomy of the embryo and placenta. The earliest marker for inner cells is *Sox2*, while the earliest marker for outer cells is *Cdx2*, and the separation of SOX2- and CDX2-expressing cells serves as the foundation of inner and outer lineages (Avilion et al., 2003; Strumpf et al., 2005). Both *Sox2* and *Cdx2* have maternal and zygotic contributions, and the two genes are regulated dynamically in parallel (Avilion et al., 2003; Wicklow et al., 2014). Maternal-zygotic HIPPO signaling pathway is central for establishing SOX2-CDX2 polarity: HIPPO intracellular mediators YAP1/TAZ/TEAD4 directly activate *Cdx2* expression in outer cells where they moderately inhibit *Sox2* expression (Wicklow et al., 2014). The outer HIPPO activity is further modulated by Rho-ROCK pathway to limit expansion of outer fate (Shi et al., 2017). Notch signaling can also function synergistically with HIPPO in activating *Cdx2* in the outer cells (Rayon et al., 2014). On the other hand, HIPPO in the inner cells is inhibited by LATS1/2 though YAP phosphorylation which is further modulated by NF2/MERLIN (Cockburn, Biechele, Garner, & Rossant, 2013). Phosphorylated YAP loses its nuclear localization and normal regulatory roles. Maternal *Sox2* knockouts seem to have little effect on preimplantation development, but its overexpression can cause 2-cell arrest (Campolo et al., 2013; Pan & Schultz, 2011) (Fig. 4). In summary, the allocation of different signaling pathways breaks the balance of SOX2 and CDX2 to initiate inner-outer embryo polarity (Frum, Murphy, & Ralston, 2018).

The inner-outer polarity is further stabilized by gene networks. SOX2 can form a regulatory core with OCT4 and NANOG, all of which are stemness markers that dominate the ICM to activate self-renewal genes (Niwa, 2007). Meanwhile, CDX2 efficiently represses *Oct4* and *Nanog* expression in the TE (Strumpf et al., 2005). During further lineage differentiation, NANOG and GATA6 prime the epiblast and primitive endoderm formation of the ICM which are modulated by FGF-ERK signaling (Frankenberg et al., 2011). Maternal *Fgf4* mutant embryos exhibit no obvious defects and maternal/zygotic *Fgf4* mutants have defects similar to the zygotic *Fgf4* mutant alone, suggesting that maternal FGF4 is not essential for later lineage specification at implantation (Kang, Piliszek, Artus, & Hadjantonakis, 2013). Taken together, these observations suggest that maternal signaling pathways interact with each other to initiate and advance cell fate specification but may not be important for further lineage differentiation.

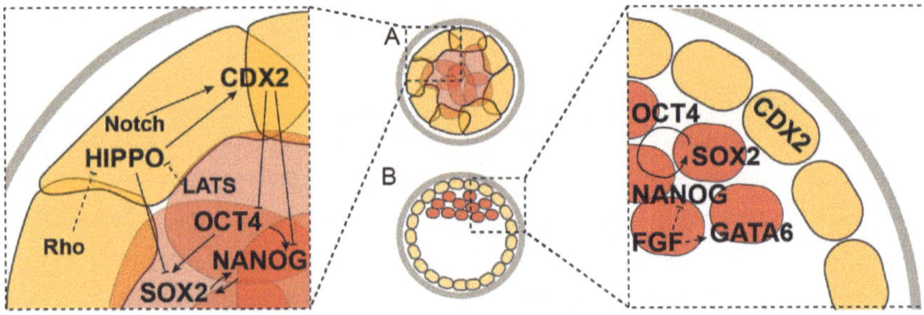

Fig. 4 Gene network in cell fate specification and differentiation in morula-to-blastocyst transition. (A) SOX2 and CDX2 determine inner-outer polarity at the morula stage. HIPPO pathway modulates the SOX2 and CDX2 balance through interacting with Notch and Rho pathways. The inner-outer polarity is maintained by a SOX2-NANOG-OCT4 network in the inner cell mass (ICM), which is inhibited by CDX2 in the trophectoderm (TE). (B) The established inner-outer polarity persists and the ICM differentiates by priming NANOG and GATA6 to form the epiblast and primitive endoderm which are regulated by zygotic FGF signaling.

5. Maternal factors require timely clearance

Not only the presence but also the timely degradation of maternal factors is essential for early development. Over-expression of cRNA encoding several maternal genes leads to embryonic defects (Gazdag et al., 2009; Jimenez et al., 2015; Pan & Schultz, 2011; Peng et al., 2012; Zeng et al., 2018). Both maternal RNA and proteins are down regulated in early embryos by degradation at the transcript level, but there are also mechanisms to promote maternal protein clearance independent of the stability of cognate RNA.

Cellular RNA level decreases during oocyte maturation (from GV to MII) and from the 1- to 2-cell stage due to reduced transcription and selective RNA degradation (Yu, Ji, Dang, et al., 2016; Yu, Ji, Sha, et al., 2016). Maternal RNA degradation refers to all RNA degradation happening before and after fertilization (Dumdie et al., 2018; Ma, Flemr, Strnad, Svoboda, & Schultz, 2013; Ma, Fukuda, & Schultz, 2015; Piko & Clegg, 1982; Yu, Ji, Dang, et al., 2016; Yu, Ji, Sha, et al., 2016). To better specify the function of maternal RNA degradation, we separate the term by the time point of fertilization into maternal transcriptome shaping (before gamete fusion) and maternal RNA clearance (after gamete fusion). The goal of maternal transcriptome shaping is to produce a fertilizable egg with the proper complement of RNA, and the consequence of maternal RNA clearance is to transfer the control of development from maternal to embryonic programs.

Impaired maternal transcriptome shaping results in poor oocyte quality. Several known oocyte-specific knockouts of genes that regulate RNA metabolism exhibit coordinated dysregulation in RNA degradation, translation and deadenylation (Yu, Ji, Dang, et al., 2016; Yu, Ji, Sha, et al., 2016; Zeng et al., 2018). Special cytoplasmic polyadenylation elements (CPEs) in the 3′ UTR of RNA molecules can switch the RNA translational activity by interactions with either inactivating or activating CPE-binding proteins; thus the number and distance between several CPEs, CPE-like elements and poly(A) signal sequences in one mRNA can define a combinatorial code to control translation and stability (Groisman et al., 2006; Pique, Lopez, Foissac, Guigo, & Mendez, 2008; Sha et al., 2017; Yang et al., 2017). The CCR4-NOT complex, bridging deadenylation and translation of mRNA, is also critical in maternal transcriptome shaping (Collart & Panasenko, 2017). Mutations disrupting members of this complex result in meiotic defects and reduced fertility (Sha et al., 2018; Yu, Ji, Dang, et al., 2016; Yu, Ji, Sha, et al., 2016). A unique characteristic of oocytes is the discordance of RNA deadenylation and degradation—several known deadenylated RNA are not degraded but are protected in RNA-containing granules (Paynton, Rempel, & Bachvarova, 1988) (Fig. 5A). RNA-containing granules in oocytes control access of translation/degradation

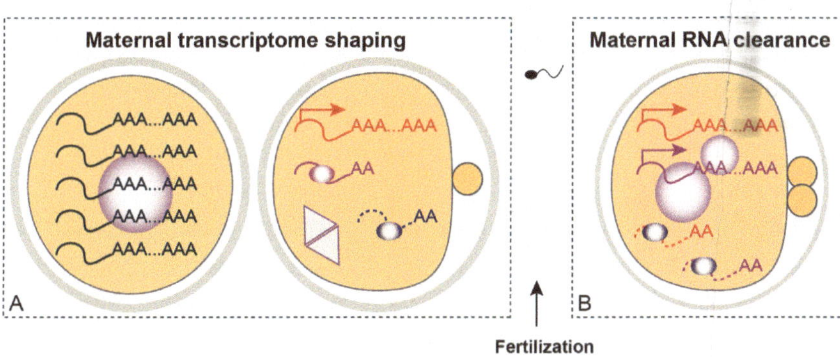

Fig. 5 Maternal transcriptome shaping and maternal RNA clearance. (A) Maternal transcriptome shaping happens during oocyte maturation from GV to MII stages. The protein coding transcripts can be either polyadenylated for translation (red), or deadenylated. The deadenylated RNAs can be degraded (blue) or protected in RNA-containing granules (magenta). (B) Maternal RNA clearance happens post-fertilization to ensure zygotic control of subsequent developmental programs. The egg-deposited RNAs can be either polyadenylated for translation, or deadenylated for degradation.

machinery to certain RNA molecules, i.e., DDX1-association with *Aog2*, *Zar1*, *Tle6*, *Floped* and *Tif1α* RNAs protects them from degradation (Hildebrandt et al., 2019). In the absence of DDX1, the embryo arrests at the 2–4-cell stage of development (Hildebrandt et al., 2019). However, the molecular mechanisms that determine whether RNAs will be selectively degraded or protected during oocyte maturation remains unclear. Thus, protein coding RNAs can be either polyadenylated for translation or deadenylated during oocyte maturation. The deadenylated RNAs can be degraded further or protected in cytoplasmic granules. Once delivered into the zygotes, the RNAs can be released from granule protection for polyadenylation, translation or degradation (Fig. 5B).

Maternal RNA clearance in mice is even more poorly understood. Unlike zebrafish in which micro-RNA 430 dramatically mediates RNA decay during a brief mid-blastula window (Giraldez et al., 2006), 1-cell mouse embryos take a much longer time to divide and micro-RNAs are not required for transcript degradation (Suh et al., 2010). Maternal-zygotic depletion of DGCR8 (required for miRNA processing) had no effect on preimplantation development and maternal mRNAs targeted by the miRNAs do not degrade more significantly compared to non-miRNA targeted mRNAs (Suh et al., 2010; Yang et al., 2016). On the other hand, DICER (required for the processing of both miRNA and siRNA) is essential for oocyte maturation (Murchison et al., 2007). Thus, it is possible that *endo*-siRNA (endogenous small interfering RNA) plays more important roles in mouse maternal RNA metabolism. At this moment, many of the known mutants identified with maternal RNA clearance defects reflect poor oocyte quality due to impaired maternal transcriptome shaping (Wu et al., 2003; Yu, Ji, Dang, et al., 2016; Yu, Ji, Sha, et al., 2016). Improved strategies for depleting maternal components after fertilization will be required to more conclusively investigate the role of maternal factors in early embryogenesis.

Other than clearing proteins by getting rid of their cognate mRNA, maternal proteins also can be enzymatically degraded. Maternal Cullin-ring finger ligase-4 (CRL4) complexes can target proteins for polyubiquitination and proteasomal degradation. This facilitates oocyte development and zygotic genome reprogramming and when disrupted results in female infertility (Yu et al., 2013; Zhang et al., 2018). Oocytes deficient in SENP7, a SUMOylation enzyme, experience meiotic arrest with cyclin downregulation and spindle defects (Huang et al., 2017). In summary, maternal factors clearance, both active through selective targeting and passive through insufficient transcription, is an essential process for preimplantation development.

6. Concluding remarks

Maternal factors selectively participate in preimplantation development. Their metabolism, function and clearance are dynamically regulated to guarantee successful embryogenesis. For mechanistic investigations, both genetics and transient assays have advantages and disadvantages. Genetic perturbation produces more consistent and reliable phenotypes necessary for drawing firm conclusions. However, genetic manipulations can adversely affect oocyte quality and inadvertently perturb post-fertilization embryonic development. Although transient assays performed at the 1-cell stage provide a clean background to study loss-of-function, there remain concerns of specificity, toxicity and artifacts introduced by in vitro culture systems. In addition, transient assays often fail to separate maternal and zygotic functions of a gene product. For future investigations, approaches of rapid protein clearance, including Trim-Away and auxin-inducible degron systems, should provide an ideal way to study the function of maternal factors in early embryogenesis (Clift et al., 2017; Holland, Fachinetti, Han, & Cleveland, 2012). Meanwhile, the application of low-input, high throughput sequencing technology will inevitably benefit our understanding of maternal regulation of embryonic development.

Acknowledgments

We apologize to the authors whose work could not be cited because of space constraints. This research was supported by the Intramural Research Program of the NIH, National Institute of Diabetes and Digestive and Kidney Disease.

References

Aoki, F., Worrad, D. M., & Schultz, R. M. (1997). Regulation of transcriptional activity during the first and second cell cycles in the preimplantation mouse embryo. *Developmental Biology*, *181*, 296–307.

Avilion, A. A., Nicolis, S. K., Pevny, L. H., Perez, L., Vivian, N., & Lovell-Badge, R. (2003). Multipotent cell lineages in early mouse development depend on SOX2 function. *Genes & Development*, *17*, 126–140.

Balhorn, R. (2007). The protamine family of sperm nuclear proteins. *Genome Biology*, *8*, 227.

Baran, V., Solc, P., Kovarikova, V., Rehak, P., & Sutovsky, P. (2013). Polo-like kinase 1 is essential for the first mitotic division in the mouse embryo. *Molecular Reproduction and Development*, *80*, 522–534.

Bebbere, D., Masala, L., Albertini, D. F., & Ledda, S. (2016). The subcortical maternal complex: Multiple functions for one biological structure? *Journal of Assisted Reproduction and Genetics*, *33*, 1431–1438.

Biase, F. H., Cao, X., & Zhong, S. (2014). Cell fate inclination within 2-cell and 4-cell mouse embryos revealed by single-cell RNA sequencing. *Genome Research*, *24*, 1787–1796.

Bierkamp, C., Luxey, M., Metchat, A., Audouard, C., Dumollard, R., & Christians, E. (2010). Lack of maternal heat shock factor 1 results in multiple cellular and developmental defects, including mitochondrial damage and altered redox homeostasis, and leads to reduced survival of mammalian oocytes and embryos. *Developmental Biology, 339*, 338–353.

Blij, S., Frum, T., Akyol, A., Fearon, E., & Ralston, A. (2012). Maternal cdx2 is dispensable for mouse development. *Development, 139*, 3969–3972.

Bultman, S. J., Gebuhr, T. C., Pan, H., Svoboda, P., Schultz, R. M., & Magnuson, T. (2006). Maternal BRG1 regulates zygotic genome activation in the mouse. *Genes & Development, 20*, 1744–1754.

Burns, K. H., Viveiros, M. M., Ren, Y., Wang, P., DeMayo, F. J., Frail, D. E., et al. (2003). Roles of npm2 in chromatin and nucleolar organization in oocytes and embryos. *Science, 300*, 633–636.

Caglar, G. S., Hammadeh, M., Asimakopoulos, B., Nikolettos, N., Diedrich, K., & Al-Hassani, S. (2005). In vivo and in vitro decondensation of human sperm and assisted reproduction technologies. *In Vivo, 19*, 623–630.

Campolo, F., Gori, M., Favaro, R., Nicolis, S., Pellegrini, M., Botti, F., et al. (2013). Essential role of sox2 for the establishment and maintenance of the germ cell line. *Stem Cells, 31*, 1408–1421.

Casser, E., Israel, S., Witten, A., Schulte, K., Schlatt, S., Nordhoff, V., et al. (2017). Totipotency segregates between the sister blastomeres of two-cell stage mouse embryos. *Scientific Reports*, 7, 8299.

Chen, J., Lian, X., Du, J., Xu, S., Wei, J., Pang, L., et al. (2016). Inhibition of phosphorylated ser473-akt from translocating into the nucleus contributes to 2-cell arrest and defective zygotic genome activation in mouse preimplantation embryogenesis. *Development, Growth & Differentiation, 58*, 280–292.

Chen, Z., & Zhang, Y. (2019). Loss of dux causes minor defects in zygotic genome activation and is compatible with mouse development. *Nature Genetics, 51*, 947–951.

Clift, D., McEwan, W. A., Labzin, L. I., Konieczny, V., Mogessie, B., James, L. C., et al. (2017). A method for the acute and rapid degradation of endogenous proteins. *Cell, 171*, 1692–1706.e1618.

Cockburn, K., Biechele, S., Garner, J., & Rossant, J. (2013). The hippo pathway member Nf2 is required for inner cell mass specification. *Current Biology, 23*, 1195–1201.

Collart, M. A., & Panasenko, O. O. (2017). The ccr4-not complex: Architecture and structural insights. *Sub-Cellular Biochemistry, 83*, 349–379.

Dawlaty, M. M., Ganz, K., Powell, B. E., Hu, Y. C., Markoulaki, S., Cheng, A. W., et al. (2011). Tet1 is dispensable for maintaining pluripotency and its loss is compatible with embryonic and postnatal development. *Cell Stem Cell, 9*, 166–175.

De Iaco, A., Planet, E., Coluccio, A., Verp, S., Duc, J., & Trono, D. (2017). Dux-family transcription factors regulate zygotic genome activation in placental mammals. *Nature Genetics, 49*, 941–945.

Dumdie, J. N., Cho, K., Ramaiah, M., Skarbrevik, D., Mora-Castilla, S., Stumpo, D. J., et al. (2018). Chromatin modification and global transcriptional silencing in the oocyte mediated by the mRNA decay activator ZFP36L2. *Developmental Cell, 44*, 392–402. e397.

Ellard, K., Serpa, J. J., Petrotchenko, E. V., Borchers, C. H., & Ausio, J. (2016). Expression and purification of the full murine NPM2 and study of its interaction with protamines and histones. *Biochemistry and Biophysics Reports, 6*, 165–171.

Eymery, A., Liu, Z., Ozonov, E. A., Stadler, M. B., & Peters, A. H. (2016). The methyltransferase setdb1 is essential for meiosis and mitosis in mouse oocytes and early embryos. *Development, 143*, 2767–2779.

Falco, G., Lee, S. L., Stanghellini, I., Bassey, U. C., Hamatani, T., & Ko, M. S. (2007). Zscan4: A novel gene expressed exclusively in late 2-cell embryos and embryonic stem cells. *Developmental Biology*, *307*, 539–550.

Frankenberg, S., Gerbe, F., Bessonnard, S., Belville, C., Pouchin, P., Bardot, O., et al. (2011). Primitive endoderm differentiates via a three-step mechanism involving Nanog and RTK signaling. *Developmental Cell*, *21*, 1005–1013.

Frum, T., Murphy, T. M., & Ralston, A. (2018). Hippo signaling resolves embryonic cell fate conflicts during establishment of pluripotency in vivo. *eLife*, *7*, e42298.

Gao, Z., Zhang, X., Yu, X., Qin, D., Xiao, Y., Yu, Y., et al. (2018). Zbed3 participates in the subcortical maternal complex and regulates the distribution of organelles. *Journal of Molecular Cell Biology*, *10*, 74–88.

Gazdag, E., Santenard, A., Ziegler-Birling, C., Altobelli, G., Poch, O., Tora, L., et al. (2009). TBP2 is essential for germ cell development by regulating transcription and chromatin condensation in the oocyte. *Genes & Development*, *23*, 2210–2223.

Giraldez, A. J., Mishima, Y., Rihel, J., Grocock, R. J., Van Dongen, S., Inoue, K., et al. (2006). Zebrafish MiR-430 promotes deadenylation and clearance of maternal mRNAs. *Science*, *312*, 75–79.

Groisman, I., Ivshina, M., Marin, V., Kennedy, N. J., Davis, R. J., & Richter, J. D. (2006). Control of cellular senescence by CPEB. *Genes & Development*, *20*, 2701–2712.

Gu, T. P., Guo, F., Yang, H., Wu, H. P., Xu, G. F., Liu, W., et al. (2011). The role of tet3 DNA dioxygenase in epigenetic reprogramming by oocytes. *Nature*, *477*, 606–610.

Hildebrandt, M. R., Wang, Y., Li, L., Yasmin, L., Glubrecht, D. D., & Godbout, R. (2019). Cytoplasmic aggregation of DDX1 in developing embryos: Early embryonic lethality associated with DDX1 knockout. *Developmental Biology*. https://doi.org/10.1016/j.ydbio.2019.07.014 [Epub ahead of print].

Hirasawa, R., Chiba, H., Kaneda, M., Tajima, S., Li, E., Jaenisch, R., et al. (2008). Maternal and zygotic Dnmt1 are necessary and sufficient for the maintenance of DNA methylation imprints during preimplantation development. *Genes & Development*, *22*, 1607–1616.

Holland, A. J., Fachinetti, D., Han, J. S., & Cleveland, D. W. (2012). Inducible, reversible system for the rapid and complete degradation of proteins in mammalian cells. *Proceedings of the National Academy of Sciences of the United States of America*, *109*, E3350–E3357.

Howell, C. Y., Bestor, T. H., Ding, F., Latham, K. E., Mertineit, C., Trasler, J. M., et al. (2001). Genomic imprinting disrupted by a maternal effect mutation in the Dnmt1 gene. *Cell*, *104*, 829–838.

Hu, J., Wang, F., Zhu, X., Yuan, Y., Ding, M., & Gao, S. (2010). Mouse ZAR1-like (XM_359149) colocalizes with mRNA processing components and its dominant-negative mutant caused two-cell-stage embryonic arrest. *Developmental Dynamics*, *239*, 407–424.

Huang, C. J., Wu, D., Jiao, X. F., Khan, F. A., Xiong, C. L., Liu, X. M., et al. (2017). Maternal SENP7 programs meiosis architecture and embryo survival in mouse. *Biochimica et Biophysica Acta. Molecular Cell Research*, *1864*, 1195–1206.

Hug, C. B., Grimaldi, A. G., Kruse, K., & Vaquerizas, J. M. (2017). Chromatin architecture emerges during zygotic genome activation independent of transcription. *Cell*, *169*, 216–228.e219.

Inoue, A., Ogushi, S., Saitou, M., Suzuki, M. G., & Aoki, F. (2011). Involvement of mouse nucleoplasmin 2 in the decondensation of sperm chromatin after fertilization. *Biology of Reproduction*, *85*, 70–77.

Ivanova, I., Much, C., Di Giacomo, M., Azzi, C., Morgan, M., Moreira, P. N., et al. (2017). The RNA m(6)a reader YTHDF2 is essential for the post-transcriptional regulation of the maternal transcriptome and oocyte competence. *Molecular Cell*, *67*, 1059–1067.e1054.

Jedrusik, A., Cox, A., Wicher, K. B., Glover, D. M., & Zernicka-Goetz, M. (2015). Maternal-zygotic knockout reveals a critical role of Cdx2 in the morula to blastocyst transition. *Developmental Biology, 398*, 147–152.

Jeltsch, A., & Jurkowska, R. Z. (2014). New concepts in DNA methylation. *Trends in Biochemical Sciences, 39*, 310–318.

Jimenez, R., Melo, E. O., Davydenko, O., Ma, J., Mainigi, M., Franke, V., et al. (2015). Maternal SIN3A regulates reprogramming of gene expression during mouse preimplantation development. *Biology of Reproduction, 93*, 89.

Jukam, D., Shariati, S. A. M., & Skotheim, J. M. (2017). Zygotic genome activation in vertebrates. *Developmental Cell, 42*, 316–332.

Jung, Y. H., Kremsky, I., Gold, H. B., Rowley, M. J., Punyawai, K., Buonanotte, A., et al. (2019). Maintenance of CTCF- and transcription factor-mediated interactions from the gametes to the early mouse embryo. *Molecular Cell, 75*, 154–171.e155.

Kalitsis, P., Fowler, K. J., Earle, E., Hill, J., & Choo, K. H. (1998). Targeted disruption of mouse centromere protein c gene leads to mitotic disarray and early embryo death. *Proceedings of the National Academy of Sciences of the United States of America, 95*, 1136–1141.

Kaneda, M., Hirasawa, R., Chiba, H., Okano, M., Li, E., & Sasaki, H. (2010). Genetic evidence for Dnmt3a-dependent imprinting during oocyte growth obtained by conditional knockout with Zp3-cre and complete exclusion of Dnmt3b by chimera formation. *Genes to Cells, 15*, 169–179.

Kang, J., Lienhard, M., Pastor, W. A., Chawla, A., Novotny, M., Tsagaratou, A., et al. (2015). Simultaneous deletion of the methylcytosine oxidases Tet1 and Tet3 increases transcriptome variability in early embryogenesis. *Proceedings of the National Academy of Sciences of the United States of America, 112*, E4236–E4245.

Kang, M., Piliszek, A., Artus, J., & Hadjantonakis, A. K. (2013). FGF4 is required for lineage restriction and salt-and-pepper distribution of primitive endoderm factors but not their initial expression in the mouse. *Development, 140*, 267–279.

Ke, Y., Xu, Y., Chen, X., Feng, S., Liu, Z., Sun, Y., et al. (2017). 3D chromatin structures of mature gametes and structural reprogramming during mammalian embryogenesis. *Cell, 170*, 367–381.e320.

Kim, S. I., Bresnick, E. H., & Bultman, S. J. (2009). BRG1 directly regulates nucleosome structure and chromatin looping of the alpha globin locus to activate transcription. *Nucleic Acids Research, 37*, 6019–6027.

Kong, Q., Banaszynski, L. A., Geng, F., Zhang, X., Zhang, J., Zhang, H., et al. (2018). Histone variant H3.3-mediated chromatin remodeling is essential for paternal genome activation in mouse preimplantation embryos. *The Journal of Biological Chemistry, 293*, 3829–3838.

Li, L., Baibakov, B., & Dean, J. (2008). A subcortical maternal complex essential for preimplantation mouse embryogenesis. *Developmental Cell, 15*, 416–425.

Li, Z., Cai, X., Cai, C. L., Wang, J., Zhang, W., Petersen, B. E., et al. (2011). Deletion of Tet2 in mice leads to dysregulated hematopoietic stem cells and subsequent development of myeloid malignancies. *Blood, 118*, 4509–4518.

Li, X., Yue, X., Pastor, W. A., Lin, L., Georges, R., Chavez, L., et al. (2016). Tet proteins influence the balance between neuroectodermal and mesodermal fate choice by inhibiting wnt signaling. *Proceedings of the National Academy of Sciences of the United States of America, 113*, E8267–E8276.

Li, L., Zheng, P., & Dean, J. (2010). Maternal control of early mouse development. *Development, 137*, 859–870.

Lim, A. K., & Knowles, B. B. (2015). Controlling endogenous retroviruses and their chimeric transcripts during natural reprogramming in the oocyte. *The Journal of Infectious Diseases, 212*(Suppl. 1), S47–S51.

Liu, Y., Lu, X., Shi, J., Yu, X., Zhang, X., Zhu, K., et al. (2016). BTG4 is a key regulator for maternal mrna clearance during mouse early embryogenesis. *Journal of Molecular Cell Biology*, *8*, 366–368.

Lu, X., Gao, Z., Qin, D., & Li, L. (2017). A maternal functional module in the mammalian oocyte-to-embryo transition. *Trends in Molecular Medicine*, *23*, 1014–1023.

Lu, F., Liu, Y., Inoue, A., Suzuki, T., Zhao, K., & Zhang, Y. (2016). Establishing chromatin regulatory landscape during mouse preimplantation development. *Cell*, *165*, 1375–1388.

Ma, J., Flemr, M., Strnad, H., Svoboda, P., & Schultz, R. M. (2013). Maternally recruited DCP1A and DCP2 contribute to messenger RNA degradation during oocyte maturation and genome activation in mouse. *Biology of Reproduction*, *88*, 11.

Ma, J., Fukuda, Y., & Schultz, R. M. (2015). Mobilization of dormant Cnot7 mRNA promotes deadenylation of maternal transcripts during mouse oocyte maturation. *Biology of Reproduction*, *93*, 48.

Ma, J., Zeng, F., Schultz, R. M., & Tseng, H. (2006). Basonuclin: A novel mammalian maternal-effect gene. *Development*, *133*, 2053–2062.

Madissoon, E., Damdimopoulos, A., Katayama, S., Krjutskov, K., Einarsdottir, E., Mamia, K., et al. (2019). Pleomorphic adenoma gene 1 is needed for timely zygotic genome activation and early embryo development. *Scientific Reports*, *9*, 8411.

Mahadevan, S., Sathappan, V., Utama, B., Lorenzo, I., Kaskar, K., & Van den Veyver, I. B. (2017). Maternally expressed NLRP2 links the subcortical maternal complex (SCMC) to fertility, embryogenesis and epigenetic reprogramming. *Scientific Reports*, *7*, 44667.

Maslon, M. M., Braunschweig, U., Aitken, S., Mann, A. R., Kilanowski, F., Hunter, C. J., et al. (2019). A slow transcription rate causes embryonic lethality and perturbs kinetic coupling of neuronal genes. *The EMBO Journal*, *38*, e101244.

Mayer, W., Niveleau, A., Walter, J., Fundele, R., & Haaf, T. (2000). Demethylation of the zygotic paternal genome. *Nature*, *403*, 501–502.

Messerschmidt, D. M., Knowles, B. B., & Solter, D. (2014). DNA methylation dynamics during epigenetic reprogramming in the germline and preimplantation embryos. *Genes & Development*, *28*, 812–828.

Murchison, E. P., Stein, P., Xuan, Z., Pan, H., Zhang, M. Q., Schultz, R. M., et al. (2007). Critical roles for dicer in the female germline. *Genes & Development*, *21*, 682–693.

Nakamura, T., Liu, Y. J., Nakashima, H., Umehara, H., Inoue, K., Matoba, S., et al. (2012). PGC7 binds histone H3K9me2 to protect against conversion of 5mC to 5hmC in early embryos. *Nature*, *486*, 415–419.

Niwa, H. (2007). How is pluripotency determined and maintained? *Development*, *134*, 635–646.

Niwa, K., Takano, R., Obata, Y., Hiura, H., Komiyama, J., Ogawa, H., et al. (2004). Nuclei of oocytes derived from mouse parthenogenetic embryos are competent to support development to term. *Biology of Reproduction*, *71*, 1560–1567.

Ogushi, S., Yamagata, K., Obuse, C., Furuta, K., Wakayama, T., Matzuk, M. M., et al. (2017). Reconstitution of the oocyte nucleolus in mice through a single nucleolar protein, NPM2. *Journal of Cell Science*, *130*, 2416–2429.

Okano, M., Bell, D. W., Haber, D. A., & Li, E. (1999). DNA methyltransferases Dnmt3a and Dnmt3b are essential for de novo methylation and mammalian development. *Cell*, *99*, 247–257.

Ong, C. T., & Corces, V. G. (2014). CTCF: An architectural protein bridging genome topology and function. *Nature Reviews. Genetics*, *15*, 234–246.

Pan, H., & Schultz, R. M. (2011). SOX2 modulates reprogramming of gene expression in two-cell mouse embryos. *Biology of Reproduction*, *85*, 409–416.

Park, M. W., Kim, K. H., Kim, E. Y., Lee, S. Y., Ko, J. J., & Lee, K. A. (2015). Associations among sebox and other MEGs and its effects on early embryogenesis. *PLoS One*, *10*, e0115050.

Paynton, B. V., Rempel, R., & Bachvarova, R. (1988). Changes in state of adenylation and time course of degradation of maternal mRNAs during oocyte maturation and early embryonic development in the mouse. *Developmental Biology, 129*, 304–314.

Peng, H., Chang, B., Lu, C., Su, J., Wu, Y., Lv, P., et al. (2012). NLRP2, a maternal effect gene required for early embryonic development in the mouse. *PLoS One, 7*, e30344.

Philpott, A., & Leno, G. H. (1992). Nucleoplasmin remodels sperm chromatin in xenopus egg extracts. *Cell, 69*, 759–767.

Piko, L., & Clegg, K. B. (1982). Quantitative changes in total RNA, total poly(a), and ribosomes in early mouse embryos. *Developmental Biology, 89*, 362–378.

Piotrowska, K., Wianny, F., Pedersen, R. A., & Zernicka-Goetz, M. (2001). Blastomeres arising from the first cleavage division have distinguishable fates in normal mouse development. *Development, 128*, 3739–3748.

Pique, M., Lopez, J. M., Foissac, S., Guigo, R., & Mendez, R. (2008). A combinatorial code for CPE-mediated translational control. *Cell, 132*, 434–448.

Plusa, B., Grabarek, J. B., Piotrowska, K., Glover, D. M., & Zernicka-Goetz, M. (2002). Site of the previous meiotic division defines cleavage orientation in the mouse embryo. *Nature Cell Biology, 4*, 811–815.

Potireddy, S., Vassena, R., Patel, B. G., & Latham, K. E. (2006). Analysis of polysomal mRNA populations of mouse oocytes and zygotes: Dynamic changes in maternal mRNA utilization and function. *Developmental Biology, 298*, 155–166.

Ramos, S. B., Stumpo, D. J., Kennington, E. A., Phillips, R. S., Bock, C. B., Ribeiro-Neto, F., et al. (2004). The CCCH tandem zinc-finger protein Zfp36l2 is crucial for female fertility and early embryonic development. *Development, 131*, 4883–4893.

Rayon, T., Menchero, S., Nieto, A., Xenopoulos, P., Crespo, M., Cockburn, K., et al. (2014). Notch and hippo converge on Cdx2 to specify the trophectoderm lineage in the mouse blastocyst. *Developmental Cell, 30*, 410–422.

Roest, H. P., Baarends, W. M., de Wit, J., van Klaveren, J. W., Wassenaar, E., Hoogerbrugge, J. W., et al. (2004). The ubiquitin-conjugating DNA repair enzyme HR6A is a maternal factor essential for early embryonic development in mice. *Molecular and Cellular Biology, 24*, 5485–5495.

Schultz, R. M. (1993). Regulation of zygotic gene activation in the mouse. *BioEssays, 15*, 531–538.

Sha, Q. Q., Dai, X. X., Dang, Y., Tang, F., Liu, J., Zhang, Y. L., et al. (2017). A MAPK cascade couples maternal mRNA translation and degradation to meiotic cell cycle progression in mouse oocytes. *Development, 144*, 452–463.

Sha, Q. Q., Yu, J. L., Guo, J. X., Dai, X. X., Jiang, J. C., Zhang, Y. L., et al. (2018). CNOT6L couples the selective degradation of maternal transcripts to meiotic cell cycle progression in Mouse oocyte. *The EMBO Journal, 37*, e99333.

Shen, L., Inoue, A., He, J., Liu, Y., Lu, F., & Zhang, Y. (2014). Tet3 and DNA replication mediate demethylation of both the maternal and paternal genomes in mouse zygotes. *Cell Stem Cell, 15*, 459–471.

Shi, X., Yin, Z., Ling, B., Wang, L., Liu, C., Ruan, X., et al. (2017). Rho differentially regulates the hippo pathway by modulating the interaction between Amot and Nf2 in the blastocyst. *Development, 144*, 3957–3967.

Strumpf, D., Mao, C. A., Yamanaka, Y., Ralston, A., Chawengsaksophak, K., Beck, F., et al. (2005). Cdx2 is required for correct cell fate specification and differentiation of trophectoderm in the mouse blastocyst. *Development, 132*, 2093–2102.

Suh, N., Baehner, L., Moltzahn, F., Melton, C., Shenoy, A., Chen, J., et al. (2010). Microrna function is globally suppressed in mouse oocytes and early embryos. *Current Biology, 20*, 271–277.

Tarkowski, A. K. (1959). Experiments on the development of isolated blastomers of mouse eggs. *Nature, 184*, 1286–1287.

Torres-Padilla, M. E., & Zernicka-Goetz, M. (2006). Role of TIF1alpha as a modulator of embryonic transcription in the mouse zygote. *The Journal of Cell Biology, 174*, 329–338.
Tsai, S. Y., Opavsky, R., Sharma, N., Wu, L., Naidu, S., Nolan, E., et al. (2008). Mouse development with a single E2F activator. *Nature, 454*, 1137–1141.
van der Heijden, G. W., Dieker, J. W., Derijck, A. A., Muller, S., Berden, J. H., Braat, D. D., et al. (2005). Asymmetry in histone H3 variants and lysine methylation between paternal and maternal chromatin of the early mouse zygote. *Mechanisms of Development, 122*, 1008–1022.
Vinot, S., Le, T., Ohno, S., Pawson, T., Maro, B., & Louvet-Vallee, S. (2005). Asymmetric distribution of par proteins in the mouse embryo begins at the 8-cell stage during compaction. *Developmental Biology, 282*, 307–319.
Vogt, E. J., Tokuhiro, K., Guo, M., Dale, R., Yang, G., Shin, S. W., et al. (2019). Anchoring cortical granules in the cortex ensures trafficking to the plasma membrane for post-fertilization exocytosis. *Nature Communications, 10*, 2271.
Wan, L. B., Pan, H., Hannenhalli, S., Cheng, Y., Ma, J., Fedoriw, A., et al. (2008). Maternal depletion of CTCF reveals multiple functions during oocyte and preimplantation embryo development. *Development, 135*, 2729–2738.
Wicklow, E., Blij, S., Frum, T., Hirate, Y., Lang, R. A., Sasaki, H., et al. (2014). Hippo pathway members restrict SOX2 to the inner cell mass where it promotes ICM fates in the mouse blastocyst. *PLoS Genetics, 10*, e1004618.
Wilkerson, D. C., Murphy, L. A., & Sarge, K. D. (2008). Interaction of HSF1 and HSF2 with the Hspa1b promoter in mouse epididymal spermatozoa. *Biology of Reproduction, 79*, 283–288.
Wu, D., & Dean, J. (2016). BTG4, a maternal mRNA cleaner. *Journal of Molecular Cell Biology, 8*, 369–370.
Wu, G., & Scholer, H. R. (2014). Role of Oct4 in the early embryo development. *Cell Regeneration (London), 3*, 7.
Wu, X., Viveiros, M. M., Eppig, J. J., Bai, Y., Fitzpatrick, S. L., & Matzuk, M. M. (2003). Zygote arrest 1 (Zar1) is a novel maternal-effect gene critical for the oocyte-to-embryo transition. *Nature Genetics, 33*, 187–191.
Wu, S. C., & Zhang, Y. (2010). Active DNA demethylation: Many roads lead to Rome. *Nature Reviews. Molecular Cell Biology, 11*, 607–620.
Wu, H., & Zhang, Y. (2014). Reversing DNA methylation: Mechanisms, genomics, and biological functions. *Cell, 156*, 45–68.
Xu, Q., Wang, F., Xiang, Y., Zhang, X., Zhao, Z. A., Gao, Z., et al. (2015). Maternal BCAS2 protects genomic integrity in mouse early embryonic development. *Development, 142*, 3943–3953.
Yang, Q., Lin, J., Liu, M., Li, R., Tian, B., Zhang, X., et al. (2016). Highly sensitive sequencing reveals dynamic modifications and activities of small RNAs in mouse oocytes and early embryos. *Science Advances, 2*, e1501482.
Yang, X. J., & Seto, E. (2008). The Rpd3/Hda1 family of lysine deacetylases: From bacteria and yeast to mice and men. *Nature Reviews Molecular Cell Biology, 9*, 206–218.
Yang, Y., Yang, C. R., Han, S. J., Daldello, E. M., Cho, A., Martins, J. P. S., et al. (2017). Maternal mRNAs with distinct 3′ UTRs define the temporal pattern of Ccnb1 synthesis during mouse oocyte meiotic maturation. *Genes & Development, 31*, 1302–1307.
Yu, C., Ji, S. Y., Dang, Y. J., Sha, Q. Q., Yuan, Y. F., Zhou, J. J., et al. (2016). Oocyte-expressed yes-associated protein is a key activator of the early zygotic genome in mouse. *Cell Research, 26*, 275–287.
Yu, C., Ji, S. Y., Sha, Q. Q., Dang, Y., Zhou, J. J., Zhang, Y. L., et al. (2016). BTG4 is a meiotic cell cycle-coupled maternal-zygotic-transition licensing factor in oocytes. *Nature Structural & Molecular Biology, 23*, 387–394.

Yu, C., Zhang, Y. L., Pan, W. W., Li, X. M., Wang, Z. W., Ge, Z. J., et al. (2013). CRL4 complex regulates mammalian oocyte survival and reprogramming by activation of TET proteins. *Science*, *342*, 1518–1521.

Zeng, J., Jiang, M., Wu, X., Diao, F., Qiu, D., Hou, X., et al. (2018). SIRT4 is essential for metabolic control and meiotic structure during mouse oocyte maturation. *Aging Cell*, *17*, e12789.

Zhang, Y. L., Zhao, L. W., Zhang, J., Le, R., Ji, S. Y., Chen, C., et al. (2018). DCAF13 promotes pluripotency by negatively regulating SUV39H1 stability during early embryonic development. *The EMBO Journal*, *37*, e98981.

Zheng, P., & Dean, J. (2009). Role of Filia, a maternal effect gene, in maintaining euploidy during cleavage-stage mouse embryogenesis. *Proceedings of the National Academy of Sciences of the United States of America*, *106*, 7473–7478.

CHAPTER TWELVE

The maternal coordinate system: Molecular-genetics of embryonic axis formation and patterning in the zebrafish

Ricardo Fuentes[a,*], Benjamin Tajer[b], Manami Kobayashi[b], Jose L. Pelliccia[b], Yvette Langdon[c], Elliott W. Abrams[d], Mary C. Mullins[b,*]
[a]Departamento de Biología Celular, Facultad de Ciencias Biológicas, Universidad de Concepción, Concepción, Chile
[b]Department of Cell and Developmental Biology, University of Pennsylvania, Perelman School of Medicine, Philadelphia, PA, United States
[c]Millsaps College, Jackson, MS, United States
[d]Department of Biology, Purchase College, State University of New York, Harrison, NY, United States
*Corresponding authors: e-mail address: ricfuentes@udec.cl; mullins@pennmedicine.upenn.edu

Contents

1. Polarity and cytoplasmic reorganization prime embryonic axis formation 342
 1.1 Cytoplasmic domain formation in oocytes: Models of cytoskeleton- and phase separation-based polarity establishment 343
 1.2 Cytoplasmic domain formation in eggs: Models for calcium-induced and cell cycle-dependent cytoplasmic flow to direct axis specification 346
 1.3 Single-cell genotype-phenotype maps: Revealing positional patterning signatures during the oocyte-to-embryo transition 349
2. Maternal control of DV axial specification 351
 2.1 Maternal control of dorsal cell fates 351
 2.2 Signaling and dorsal cell fates 353
 2.3 Maternal factors and ventral cell fates 355
3. Signaling pathways and the patterning of the embryonic axes 357
 3.1 The activation of anterior-posterior patterning 357
 3.2 Maternal and zygotic Wnt signaling in dorsal-ventral and anterior-posterior patterning 359
 3.3 The maternal and zygotic expression of Wnt pathway components 362
 3.4 Wnt and Fgf signaling during organizer specification and AP patterning 364
 3.5 Integration of maternal and zygotic Fgf signaling components in patterning 364
 3.6 The activation of dorsal-ventral patterning 365
4. Nodal signaling in germ layer formation and left-right patterning 368

Current Topics in Developmental Biology, Volume 140
ISSN 0070-2153
https://doi.org/10.1016/bs.ctdb.2020.05.002

4.1 Nodal-Vg1 heterodimer specifies the mesoderm and endoderm germ layers 368
4.2 Maternal Vg1/Gdf3 control of left-right axis formation 371
5. Concluding remarks 373
Acknowledgments 373
References 374

Abstract

Axis specification of the zebrafish embryo begins during oogenesis and relies on proper formation of well-defined cytoplasmic domains within the oocyte. Upon fertilization, maternally-regulated cytoplasmic flow and repositioning of dorsal determinants establish the coordinate system that will build the structure and developmental body plan of the embryo. Failure of specific genes that regulate the embryonic coordinate system leads to catastrophic loss of body structures. Here, we review the genetic principles of axis formation and discuss how maternal factors orchestrate axis patterning during zebrafish early embryogenesis. We focus on the molecular identity and functional contribution of genes controlling critical aspects of oogenesis, egg activation, blastula, and gastrula stages. We examine how polarized cytoplasmic domains form in the oocyte, which set off downstream events such as animal-vegetal polarity and germ line development. After gametes interact and form the zygote, cytoplasmic segregation drives the animal-directed reorganization of maternal determinants through calcium- and cell cycle-dependent signals. We also summarize how maternal genes control dorsoventral, anterior-posterior, mesendodermal, and left-right cell fate specification and how signaling pathways pattern these axes and tissues during early development to instruct the three-dimensional body plan. Advances in reverse genetics and phenotyping approaches in the zebrafish model are revealing positional patterning signatures at the single-cell level, thus enhancing our understanding of genotype-phenotype interactions in axis formation. Our emphasis is on the genetic interrogation of novel and specific maternal regulatory mechanisms of axis specification in the zebrafish.

1. Polarity and cytoplasmic reorganization prime embryonic axis formation

In most animals, formation of distinct, yet topologically-connected cytoplasmic domains is crucial to generate an egg with a polarized coordinate system that then directs axial patterning during embryonic development. In the zebrafish, these spatially defined cytoplasmic domains form during oogenesis when distinct gene products localize to different domains within the oocyte to establish polarity. Upon fertilization, the cytoplasm reorganizes itself; a process which continues through the first cleavage divisions and ultimately determines the positions of the embryonic axes.

The organization of cytoplasmic domains in the egg is conferred by maternally supplied molecular messages, which direct initial axis formation during the oocyte-to-embryo transition through tightly regulated processes. First, transcripts and proteins segregate into polarized cytoplasmic domains during oogenesis. Next, fertilization activates calcium- and cell cycle-dependent pathways, orchestrating the animal-directed reorganization of cytoplasm. Finally, the cytoskeleton ensures that the gene products in the oocyte, zygote and embryo are distributed correctly, establishing the basis for the embryonic coordinate system, which will culminate in the events of germline development and axis specification. However, our understanding of how these processes are regulated and the functions of the individual genetic factors remains limited, as only a few maternal genes orchestrating these processes have been identified.

Several pharmacological and genetic approaches have been developed to identify maternal functions contributing to embryo development. For example, these approaches have revealed that both calcium (Ca^{2+}) ions and the cytoskeleton play important roles in the massive cytoplasmic transport initiated following egg activation that is essential to determining the embryo architecture and axes. Since these dynamics are directed by maternally expressed genes, maternal-effect mutants are valuable genetic resources to decipher the molecular mechanisms and symmetry-breaking events that underlie axis formation and embryonic patterning. Current and emerging DNA-, RNA- and protein expression-modifying technologies and phenotyping projects are providing unprecedented tools to enhance our understanding of the genetic underpinnings of the vertebrate maternally-driven coordinate system.

1.1 Cytoplasmic domain formation in oocytes: Models of cytoskeleton- and phase separation-based polarity establishment

In zebrafish, the animal-vegetal axis of the egg is established during early oogenesis and is important for the later formation of the dorsal organizer and, thus, the establishment of the embryonic axes (Fig. 1). It is well-known that maternal dorsal determinants first localize to the vegetal pole of the egg and, after fertilization, are transported toward the future dorsal side. Specifically, the mRNAs *huluwa, syntabulin* and *grip2a*, which are required for dorsal organizer formation, localize to the vegetal pole during early oogenesis, and this localization is dependent on a structure called the Balbiani body (Bb) (Ge et al., 2014; Hino et al., 2018; Nojima et al., 2010; Varga, Maegawa, Bellipanni, & Weinberg, 2007; Yan et al., 2018).

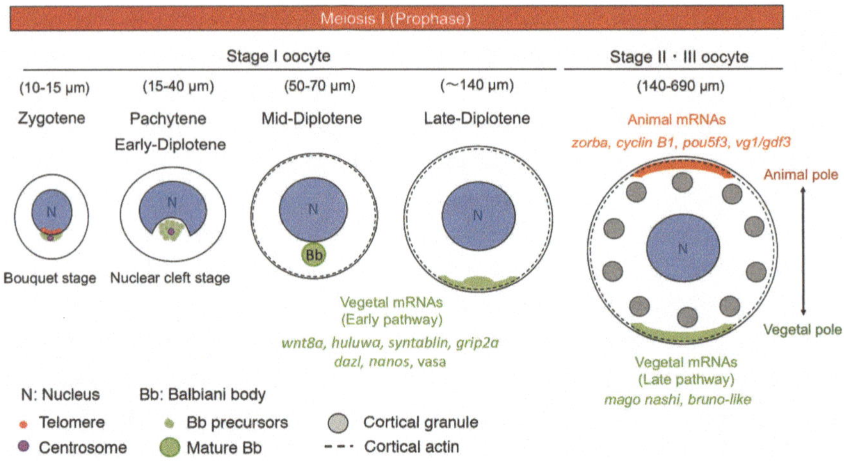

Fig. 1 Schematic of animal-vegetal cytoplasmic domain formation and spatiotemporal mRNA localization during zebrafish oogenesis. Animal (orange) and vegetal (green) transcripts are shown. The early stage I oocyte polarizes and the nucleus becomes asymmetric as Balbiani body (Bb) components first begin to aggregate in the zygotene bouquet stage of meiosis. Bb components continue to aggregate in the nuclear cleft during the pachytene and early diplotene stage of meiosis. The nuclear cleft then recedes and the mature Bb forms. Later in the diplotene stage, the Bb disassembles at the oocyte cortex, with vegetal mRNAs becoming anchored to the cortex (early pathway), thus forming a vegetal cytoplasmic domain. In stage II and III oocytes, additional mRNAs localize to the vegetal pole (late pathway). The localization of transcripts to the opposite pole establishes an animal cytoplasmic domain.

The Bb, also called the mitochondrial cloud in *Xenopus*, is a large, cytoplasmic, electron-dense structure conserved from mammals to insects, which contains mitochondria, ER and Golgi (Jamieson-Lucy & Mullins, 2019; Kloc, Bilinski, & Etkin, 2004). In addition to these organelles, the Bb of zebrafish and *Xenopus* also contains mRNA-protein complexes (mRNPs) or granules, which are essential for both dorsal-ventral (DV) axis and germline formation (Chang et al., 2004; Heasman, Quarmby, & Wylie, 1984; Kloc et al., 2001; Kosaka, Kawakami, Sakamoto, & Inoue, 2007; Wilk, Bilinski, Dougherty, & Kloc, 2005).

Initially, components of the Bb precursor start to gather at the meiotic zygotene bouquet stage, adjacent to the clustered telomeres, and then aggregate in a nuclear cleft at the pachytene and early diplotene stages of stage I oocytes (Elkouby, Jamieson-Lucy, & Mullins, 2016). By the mid-diplotene stage, the Bb matures at the periphery of the nucleus. By late stage I of oogenesis, the Bb has relocated and disassembles at the oocyte cortex, asymmetrically localizing mRNPs that specify the vegetal pole of the oocyte (Fig. 1). This Bb-dependent localization is called the "early pathway" in zebrafish or the METRO pathway in *Xenopus* (Kloc & Etkin, 1995; Kosaka et al., 2007). After the Bb disassembles, in stage II and III oocytes, "late pathway" mRNAs, such as *Vg1* in *Xenopus*, and *bruno-like* and *mago nashi* in zebrafish, localize to the vegetal pole (Fig. 1), while the transcripts *cyclin B1*, *pou5f3*, and *Vg1* in zebrafish localize to the animal pole (Bally-Cuif, Schatz, & Ho, 1998; Howley & Ho, 2000; Kloc & Etkin, 1995; Kosaka et al., 2007; Marlow & Mullins, 2008).

A forward genetic screen in zebrafish identified *bucky ball* (*buc*) and *macf1* (previously called *magellan*) as key regulators of the Bb-dependent localization pathway (Dosch et al., 2004; Gupta et al., 2010; Marlow & Mullins, 2008). *buc* is the first and only gene essential for Bb formation. In the *buc* mutant, the Bb does not form and Bb mRNAs do not localize to the vegetal cortex of the oocyte, while animally-anchored mRNAs become radially localized and these oocytes form ectopic micropyles, a single one of which normally localizes to the animal pole (Bontems et al., 2009; Marlow & Mullins, 2008). On the other hand, in the *macf1* mutant the Bb forms and accumulates its components, but fails to disassemble and the mRNAs remain trapped in an enlarged Bb. Like the *buc* mutant, the *macf1* mutant also fails to localize vegetal pole mRNAs in oocytes and embryonic development fails (Escobar-Aguirre, Zhang, Jamieson-Lucy, & Mullins, 2017; Gupta et al., 2010). Macf1 (microtubule actin cross-linking factor) is required to dock Bb components to the vegetal pole localized cortical actin of the oocyte (Escobar-Aguirre et al., 2017). Macf1 is a member of the spectraplakin family, which interacts with cytoskeletal components, including intermediate filaments (IF), microtubules (MTs) and actin in various cellular contexts, via its multiple cytoskeletal interaction domains (Karakesisoglou, Yang, & Fuchs, 2000; Lin, Chen, Leung, Parry, & Liem, 2005). The IF cytokeratin is enriched in the zebrafish and *Xenopus* Bb, and abrogating actin filaments mimics the *macf1* mutant phenotype (Escobar-Aguirre et al., 2017). Analysis of a *macf1* CRISPR mutant that specifically deletes the actin-binding domain demonstrates that this domain is essential for Bb disassembly, while deletion of the IF binding domain shows that it is dispensable (Escobar-Aguirre et al., 2017).

Though it contains ER and mitochondria, the Bb can also be considered as a large membrane-free mRNP granule with a hydrogel-like structure. A hydrogel is a three-dimensional polymer network that undergoes phase transition. Intrinsically disordered proteins (IDPs) can form liquid droplets and hydrogel phases, and even stable amyloid or prion aggregations under certain conditions, via their highly flexible (disordered) regions (Banani, Lee, Hyman, & Rosen, 2017). Intriguingly, Buc and the *Xenopus* homolog Xvelo are predicted to be IDPs and the N-terminal region of Xvelo contains a conserved prion-like domain (Boke et al., 2016; Bontems et al., 2009; Toretsky & Wright, 2014). This domain is required to form amyloid-like structures in vitro and is necessary and sufficient for a GFP-tagged Xvelo to localize to the Bb in the oocyte (Boke et al., 2016).

The mechanisms by which specific mRNAs localize into discrete cytoplasmic domains during oogenesis are not fully understood. However, research in *Xenopus* and zebrafish has revealed *cis*-acting localizing elements in the 3′UTR of mRNAs, trans-acting RNA binding proteins, and cytoskeleton-dependent localization mechanisms (Betley, Frith, Graber, Choo, & Deshler, 2002; Chabanon, Mickleburgh, & Hesketh, 2004; Kosaka et al., 2007; Takahashi, Ishii, & Yamashita, 2018; Yoon & Mowry, 2004; Zhou & King, 1996). In the late pathway, MTs are required for localization of RNAs to the vegetal pole, whereas in the early pathway, the localization of RNAs into the Bb and the movement of the Bb to the vegetal pole of the oocyte are independent of MTs and microfilaments. Since Bb localization of RNAs is cytoskeleton independent and is associated with organelles in the Bb, "a diffusion and entrapment" mechanism has been suggested (Chang et al., 2004). In this model, the Bb mRNAs diffuse in the cytoplasm and are trapped in the organelles enriched in the Bb, which is likely conferred by a Bucky ball amyloid-like network.

1.2 Cytoplasmic domain formation in eggs: Models for calcium-induced and cell cycle-dependent cytoplasmic flow to direct axis specification

From invertebrates to vertebrates, establishing cytoplasmic domains accompanies the nuclear and cytoplasmic events of egg activation to prepare the zygote for embryogenesis. In telolecithal organisms, the physical separation of the ooplasm (cytoplasm derived from the oocyte) from the vitelloplasm (yolk globules) reorganizes the cytoplasm, while changes in the cytoskeletal network regulate polarized cytoplasmic flows, which together redistribute maternal mRNAs, proteins, and organelles. This highly dynamic process of cytoplasmic segregation has important morphogenetic and embryonic body plan functions (reviewed in Fuentes, Mullins, & Fernandez, 2018). Remarkably, cell-wide, cytoskeleton-based cytoplasmic flows have been proposed as morphokinetic parameters to be evaluated in fertilized human oocytes during infertility treatment (Ajduk et al., 2011; Ajduk & Zernicka-Goetz, 2016; Milewski, Szpila, & Ajduk, 2018), highlighting the importance of this process in reproductive success.

Unlike many other metazoan eggs, visualization of cytoplasmic segregation in zebrafish eggs is facilitated by a massive displacement of cytoplasm toward the animal pole, resulting in the growth of a prominent cytoplasmic domain, the blastodisc, that will give rise to the embryo (Fig. 2) (Fuentes et al., 2018). Upon egg activation following fertilization, the release of a Ca^{2+} wave throughout the egg initiates both a spatial reorganization and the flow of cytoplasm, forming a subcortical, ooplasm-rich animal region in two main phases: slow and fast flows, 20–30 and 30–40 min post fertilization (mpf), respectively (Fuentes & Fernandez, 2010; Sharma & Kinsey, 2008). MTs and actin filaments play a significant role in the animal-directed transport of maternal determinants after egg activation (Fig. 2). Central and cortical cytoplasmic components move animally along arrays of Ca^{2+}-dependent polymerizing MTs, translocating them to the blastodisc (Fuentes et al., 2018). Meanwhile, a central, but not cortical, actin cytoskeleton forms transportation pathways or streamers, and Ca^{2+}-induced central cytoplasmic segregation in a cell cycle-dependent manner (Fuentes & Fernandez, 2010; Shamipour et al., 2019; Sharma & Kinsey, 2008). Importantly, defects in cytoplasmic flow disrupt axis patterning and embryo formation (Dosch et al., 2004; Li-Villarreal et al., 2015; Mei, Lee, Marlow, Miller, & Mullins, 2009; R. Fuentes and M.C. Mullins, unpublished).

In recent years, the zebrafish oocyte, egg, and zygote have emerged as excellent experimental systems to study cytoplasmic segregation, allowing rigorous analysis of the mechanisms behind cytoplasmic reorganization, a maternally-dependent process (Fernandez, Valladares, Fuentes, & Ubilla, 2006; Fuentes & Fernandez, 2010; Fuentes et al., 2018; Shamipour et al., 2019). However, the mechanisms by which these maternal regulators control cytoplasmic domain formation and maintenance remain largely unknown.

Egg activation events, which lead to the reorganization of the cytoplasm of the egg, are carried out by the spatiotemporal regulation of Ca^{2+} dynamics. The maternal-effect *brom bones* (*brb*) mutant has provided insight into the genetic regulation of egg activation in zebrafish. The *brb* mutant displays altered cytoplasmic segregation and a subsequent axis formation defect (Mei et al., 2009). The mutated gene encodes the RNA binding protein Ptbp1a (also called HnRNP I), which regulates intracellular Ca^{2+} release

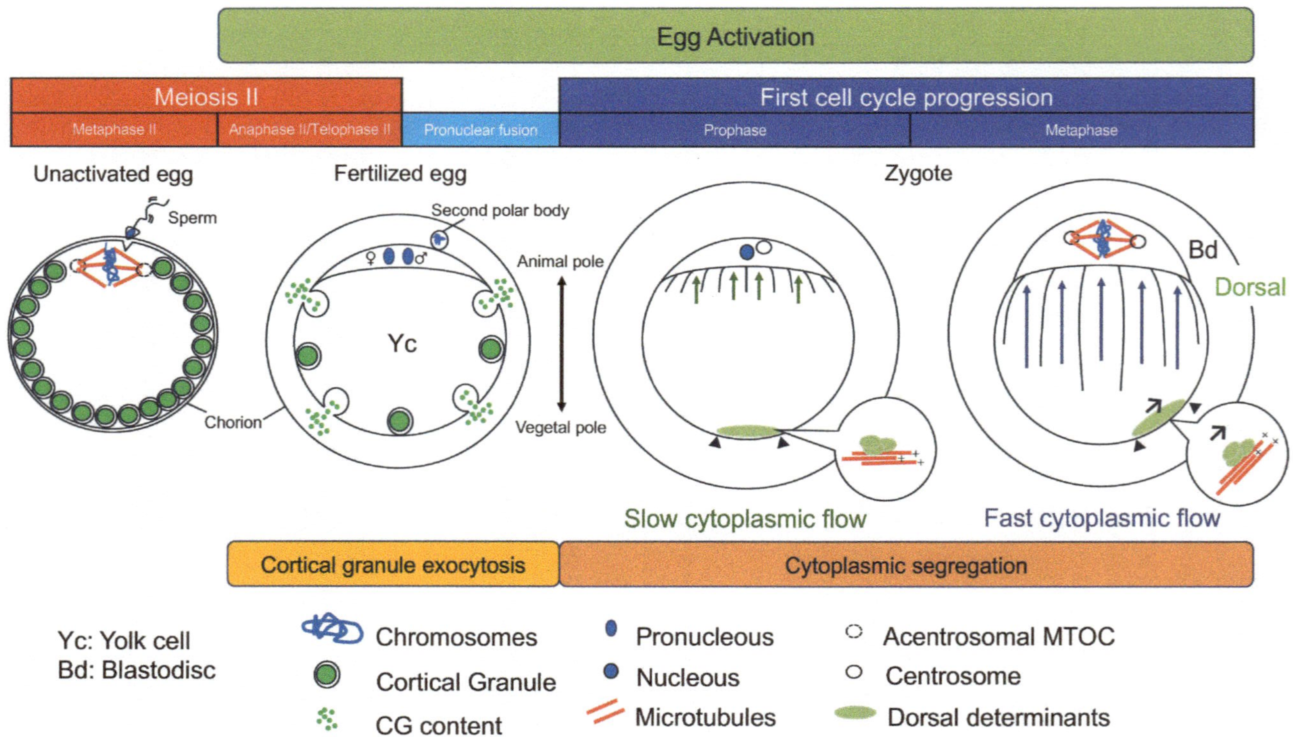

Fig. 2 Schematic of egg activation, cytoplasmic segregation, and dorsal determinant translocation during zebrafish early embryogenesis. Following egg laying and fertilization, egg activation occurs and cortical granule exocytosis is initiated. This process proceeds as meiosis II completes, the pronuclei fuse, and the chorion elevates. A brief animal-directed cytoplasmic segregation starts during the first prophase progression (slow cytoplasmic flow) and dorsal determinants anchor to a forming parallel array of vegetal microtubules. Then, blastodisc growth is accompanied by a long and massive animal-directed streaming of cytoplasm (fast cytoplasmic flow) from the yolk cell (yc), which correlates with the first metaphase. During this process, an animally-directed transport of dorsal determinants along microtubules or a so-called cortical rotation-like mechanism (black arrow heads) is activated. Green and blue arrows indicate the direction of low- and high-speed cytoplasm displacement, respectively. Black arrow indicates translocation of vegetally localized dorsal determinants.

during egg activation (Mei et al., 2009). The Ca^{2+} ion plays a fundamental role in many cellular processes, including those associated with early embryogenesis (Webb & Miller, 2006; Whitaker, 2006). The versatility of Ca^{2+} signaling is remarkable as different egg activation events and stimuli produce different Ca^{2+} responses, with some stimuli prompting short-lived Ca^{2+} waves and others generating more sustained signals. Some Ca^{2+} waves are cortically localized, while others invade the central region of the cell (Sharma & Kinsey, 2008). Each Ca^{2+} response has a distinct kinetic profile and known mutants disrupt different aspects of this Ca^{2+} signaling, with divergent effects on axis specification (Li-Villarreal et al., 2015; Mei et al., 2009). Currently, chemical indicators and genetic approaches are the main means of studying Ca^{2+} dynamics during oogenesis and embryonic development, allowing the modeling of intracellular organization and regulation of this ion (Chen, Xia, Bruchas, & Solnica-Krezel, 2017; Webb, Chan, & Miller, 2013; Webb, Fluck, & Miller, 2011).

Using cellular and physical modeling methodologies, Shamipour et al. (2019) have demonstrated in a recent elegant study that a highly dynamic bulk actin-driven flow in the egg pulls the cytoplasm animally, which combined with actin filaments pushing on yolk granules vegetally acts in cytoplasmic segregation to the animal pole. Polarized ooplasmic comet-like actin structures on the yolk globules actively push them toward the vegetal pole. The authors show that these pulling and pushing forces are generated by actin/myosin-based machinery in the early embryo. The authors also identified a cell cycle-dependent mechanism of actin polymerization and contraction, which regulates how the cytoplasm separates from the yolk cell (Shamipour et al., 2019). This is consistent with the idea that the events which initiate the first mitosis also trigger the progression of cytoplasmic segregation (Fig. 2), as previously proposed (Fuentes & Fernandez, 2010). The mechanisms driving these cytoplasmic flows during egg activation and their subsequent impact on body plan program execution will be important areas of future research.

1.3 Single-cell genotype-phenotype maps: Revealing positional patterning signatures during the oocyte-to-embryo transition

As discussed in Section 1.1, maternally-loaded dorsal determinants accumulate in the Bb of stage I oocytes and localize to the vegetal cortex of stage II and stage III oocytes. These axis inducing transcripts remain in the vegetal region of the egg and zygote, as was elegantly demonstrated by removing the vegetal-most region of the zygote, which produces embryos with reduced dorsal tissues (Mizuno, Yamaha, Kuroiwa, & Takeda, 1999). Thus, maternal transcripts are asymmetrically localized over the course of oocyte development and establish polar tissue coordinates in the single-cell female gamete.

High-resolution fluorescent in situ hybridization and imaging analysis has shown that at the vegetal pole cortex of the zygote, the dorsal axis determining gene *grip2a* mRNA is differentially positioned when compared with the germline transcript *dazl* (Welch & Pelegri, 2014). Interestingly, *grip2a* and *wnt8a* (also a presumptive dorsal-acting factor) exhibit a more cortical location than *dazl,* which localizes more internally at the vegetal pole. Following egg activation, the vegetally localized *dazl* transcript is translocated animally in an actin-dependent manner (Welch & Pelegri, 2014). In contrast, the more cortically localized *grip2a* and *wnt8a* are transported along MTs through a *Xenopus* cortical rotation-like mechanism to the presumptive dorsal side of the embryo after fertilization (Ge et al., 2014; Lu, Thisse, & Thisse, 2011; Tran et al., 2012). Thus, the precise localization depth or sub-domain at the vegetal pole may lead to dorsal-determining transcripts being translocated dorsally by MTs versus germ line-destined transcripts being translocated animally by actin filaments. Intriguingly, *grip2a, wnt8a,* and *dazl* transcripts all localize during oogenesis to the Bb and then to the vegetal pole following Bb disassembly. Thus, it is possible that these different subdomains arise due to distinct localization signatures or mechanisms in their recruitment to the Bb, their cortical anchoring following Bb dissociation, or processes at later stages in oogenesis or in the egg. These subdomains may be key to their distinct redistribution following egg activation and distinct functions in the embryo.

In cleavage stage embryos, MTs also act to dorsally enrich the maternal transcripts of *squint* (also called *nodal-related 1*), which can serve as a scaffold that binds and translocates maternal factors to the presumptive dorsal side (Gore & Sampath, 2002; Lim et al., 2012). However, a recently generated deletion of the *squint* gene indicates that maternal *squint* is not essential for dorsal axis formation (Goudarzi, Berg, Pieper, & Schier, 2019).

To investigate the mechanisms which allocate gene expression products to specific subcellular cytoplasmic domains, a better characterization of protein and mRNA dynamics during and after the oocyte-to-embryo transition is needed. To this end, Sun, Yan, Shen, and Meng (2018) have demonstrated that the incorporation of cytoplasmic components into the blastodisc of the zebrafish zygote is controlled by limiting the global translation of proteins during oocyte maturation and egg activation (Sun et al., 2018).

Hence, translocation of dorsal determinants would rely first, on the polar subcellular RNA localization in the oocyte, and second, on the enhanced local translation of these transcripts in the egg's vegetal cytoplasmic domain after fertilization (Nojima et al., 2010). Analysis of vegetal RNA distribution and translocation in mutants with defective protein translation, such as the *ybx1* mutant (Kumari et al., 2013; Sun et al., 2018), supports this hypothesis.

Recently, single-cell RNA-sequencing (RNA-seq) and 3-D spatial mapping have confirmed known regulators of the DV axis and implicated previously unknown genes in this process (Junker et al., 2014; Satija, Farrell, Gennert, Schier, & Regev, 2015; Wagner et al., 2018). In addition, informative genome-wide transcriptomics and proteomics data for oocytes of different stages, eggs and one-cell stage embryos is available and fundamental to the characterization of the genetic program that directs the oocyte-to-embryo transition in zebrafish (Cheung, Patinote, Guiguen, & Bobe, 2018; Sun et al., 2018; Yilmaz et al., 2017; Zayed, Qi, & Peng, 2019; Zhu, Pardeshi, Chen, & Ge, 2018). Genetic analysis will be warranted to reveal maternally-regulated mechanisms that specifically control mRNA localization, translation, and novel regulators of these processes.

2. Maternal control of DV axial specification

2.1 Maternal control of dorsal cell fates

The zebrafish egg contains mRNAs, dorsal determinants, and cytoskeletal components that are critical, first to establish the dorsal side of the embryo, and then to pattern the dorsoventral axis. It is well established that specification of dorsal cell fates requires a Wnt/β-Catenin signaling pathway and begins with translocation of a dorsal determinant(s) along a microtubule network to the dorsal blastomeres. This process culminates with the nuclear localization of β-catenin and activation of dorsal genes (Langdon & Mullins, 2011).

2.1.1 The vegetal microtubule (MT) network that translocates dorsal determinants

Dorsal axis determining factors are transported to the dorsal side of the embryo via a vegetal pole MT network and the cargo linker protein Syntabulin (Fig. 3A). In vivo imaging of formation of the vegetal parallel MT array demonstrates its transient nature, forming only between ~15 and ~30 mpf, and importantly, its directionality toward the future dorsal side of the embryo (see Fig. 2) (Tran et al., 2012). The requirement for

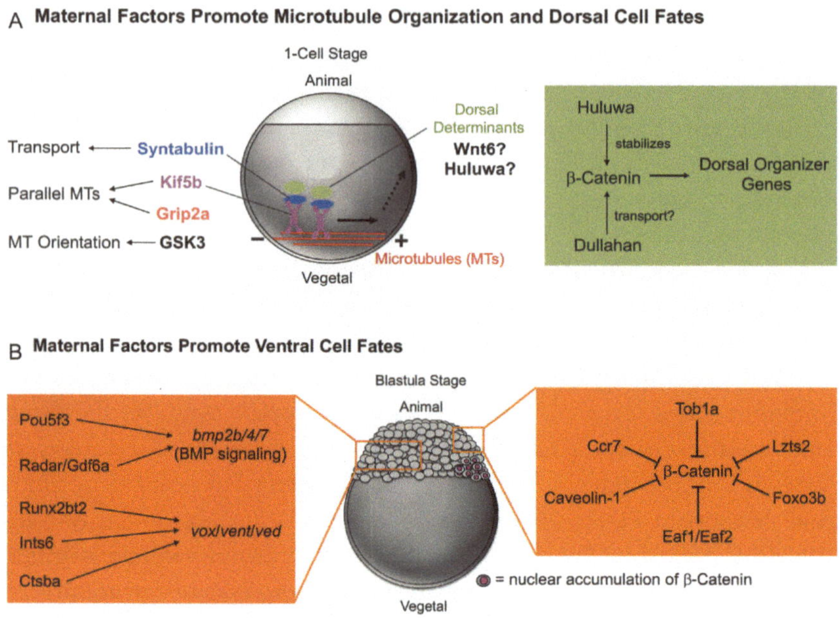

Fig. 3 Maternal factors required for dorsal and ventral cell fate specification. (A) Maternal Kif5b, Grip2a, and GSK3 are required for proper vegetal pole microtubule organization, while Syntabulin functions as a linker molecule that mediates dorsal determinant (Huluwa and possibly Wnt6) transport to the prospective dorsal side of the embryo (black arrow). Lateral MTs then translocate the dorsal determinants animally toward the blastomeres (dotted arrow). Subsequently, Huluwa leads to the stabilization of β-catenin on the dorsal side of the embryo. Dullahan functions to promote dorsal organizer target gene expression and dorsal cell fates (right). (B) Maternal Pou5f3 and Radar/Gdf6a promote BMP signaling ventrally, while Runx2bt2, Inst6, and Ctsba regulate Vox/Vent/Ved expression in ventrolateral blastomeres. Maternal factors Tob1a, Lzts2, Foxo3b, Eaf1, Eaf2, Caveolin-1, and CCr7 function to inhibit β-catenin nuclear accumulation in non-dorsal blastomeres.

Syntabulin was revealed by analysis of the ventralized maternal-effect *tokkaebi* (*syntabulin*) mutant, which was shown to disrupt dorsal determinant transport (Mei et al., 2009; Nojima et al., 2010, 2004). While it was suggested that Syntabulin binds the motor protein Kinesin 1, as it co-immunoprecipitates with Kif5b, the heavy chain of Kinesin 1, the role for maternal *kif5b* during dorsal cell fate specification and axis formation was still unclear (Nojima et al., 2010). Interestingly, generation of maternal *kif5Ba* mutants demonstrates its role in regulating MT dorsal determinant translocation. Maternal *kif5Ba* mutant embryos are ventralized and fail to translocate Syntabulin protein or *wnt8a* transcript but surprisingly display normal *grip2a* transcript translocation (Campbell, Heim, Smith, & Marlow, 2015). Moreover, maternal *kif5Ba* mutant embryos have disrupted vegetal parallel MT arrays, whereas their lateral cortex MTs are normal. This suggests that the primary function of Kif5Ba may be to regulate parallel MT array formation during dorsal specification (Fig. 3A) (Campbell et al., 2015).

Another maternal factor, Grip2a, is also required to organize and bundle MTs that promote the symmetry-breaking events required for dorsal specification (Ge et al., 2014). In *hecate* (*grip2a*) mutants, vegetal MT arrays are disorganized and do not form bundles. Interestingly, in *hecate* mutants, at 20 mpf Syntabulin protein is localized to the vegetal pole, as in wild-type embryos, but by 30 mpf, Syntabulin expression is completely lost (Ge et al., 2014). Like *kif5Ba* mutants, *grip2a* mutant embryos also exhibit normal lateral cortical MTs, which were shown to function normally in translocating fluorescent beads animally. Thus loss of Grip2a specifically affects formation of the parallel vegetal MT array, which transports dorsal axis determining factors to the lateral MT arrays for further translocation animally toward the future dorsal blastomeres (Ge et al., 2014). Since Kif5Ba and Grip2a are critical for both parallel MT array formation and dorsal translocation (Fig. 3A), it will be interesting to determine whether they function in the same or in independent pathways. Finally, evidence also suggests that GSK-3 functions between 0 and 10 mpf to promote the correct orientation of vegetal MTs independent of its later role in canonical Wnt signaling (Fig. 3A) (Shao et al., 2012).

2.2 Signaling and dorsal cell fates

As the most upstream component of the Wnt/β-Catenin signaling pathway, Wnt ligands have been implicated as the dorsal determinant in Xenopus and zebrafish. Many Wnt ligands are maternally expressed in zebrafish, including *wnt8a*, *wnt11*, *wnt6a*, *wnt5a*, *wnt9b*, *wnt10a*, *wnt10b*, *wnt2*, *wnt4a*, and *wnt4b* (Hino et al., 2018; Lu et al., 2011). Of these, only *wnt8a* and *wnt6a* can activate canonical Wnt signaling in the zebrafish embryo (Hino et al., 2018; Lu et al., 2011). Intriguingly, maternal *wnt8a* transcript localizes to the vegetal pole of the egg in a Bb-dependent manner during oogenesis. Furthermore, *wnt8a* transcript translocates asymmetrically after egg activation via microtubules, all consistent with expected properties of the dorsal determinant. Inhibition of Wnt signaling results in ventralization, while Wnt8a over expression causes dorsalization (Lu et al., 2011).

Recently, however, it was found that maternal *wnt8a* mutants exhibit no defects in dorsal axis formation, and maternal-zygotic mutants exhibit anteriorization defects that are stronger than those observed in zygotic *wnt8a* mutants alone (Hino et al., 2018). This suggests that maternal *wnt8a* is unnecessary for dorsal determination, but functions later, with zygotic *wnt8a* in posteriorizing tissue (Hino et al., 2018). Finally, *wnt6a* has now been identified as a new dorsal determinant candidate that is vegetally localized in the one-cell zebrafish embryo with overexpression causing mild dorsalization, similar to that shown for *wnt8a* (Hino et al., 2018). Further studies are necessary to determine whether maternal *wnt6a* is the zebrafish dorsal determinant.

Recently, a spontaneous zebrafish maternal-effect mutant *huluwa* was identified that reveals an exciting new factor acting upstream of maternal β-Catenin in dorsal axis determination (Yan et al., 2018). Maternal *huluwa* mutant embryos fail to form the dorsal organizer or dorsoanterior structures, exhibiting a strongly ventralized phenotype, whereas misexpression of *huluwa* does the opposite, causing the induction of a secondary dorsal axis (Yan et al., 2018). The *huluwa* transcript localizes to the vegetal pole of the egg through the Bb-dependent localization pathway in oogenesis (Yan et al., 2018). Like *wnt8a* and *grip2a* transcripts, the *huluwa* transcript is also translocated asymmetrically by the four-cell stage in a cortical rotation-like manner. Huluwa is a novel transmembrane protein that localizes to the plasma membrane of dorsal blastomeres. Through direct binding to Axin, a key component of the β-Catenin destruction complex, and promotion of Axin degradation, Huluwa stabilizes β-Catenin, which can then relocalize to the nucleus to induce dorsal gene expression and dorsal axis formation (Fig. 3A) (Yan et al., 2018). Interestingly, Huluwa can induce a dorsal axis independent of Wnt ligand and receptors. These and other Wnt antagonist overexpression studies suggest the exciting possibility that a Wnt ligand may not act in dorsal axis formation and Huluwa could be the dorsal determinant in zebrafish (Shinya, Eschbach, Clark, Lehrach, & Furutani-Seiki, 2000). Additional studies are required to investigate that hypothesis further.

Wnt/β-catenin signaling establishes dorsal cell fates through the nuclear localization of β-Catenin in prospective dorsal blastomeres, which activates expression of a number of dorsal-acting genes (Kelly, Chin, Leatherman, Kozlowski, & Weinberg, 2000; Langdon & Mullins, 2011; Schneider, Steinbeisser, Warga, & Hausen, 1996). Specifically, the nuclear translocation of β-Catenin activates genes required for dorsal organizer formation including *bozozok, goosecoid,* and *chordin* (Dixon Fox & Bruce, 2009;

Melby, Beach, Mullins, & Kimelman, 2000; Schulte-Merker, Lee, McMahon, & Hammerschmidt, 1997; Stachel, Grunwald, & Myers, 1993; Yamanaka et al., 1998). Zebrafish express two maternal *β-catenin* genes, *β-catenin-1* and *β-catenin-2*. The maternal-effect *ichabod* mutant is deficient in β-Catenin-2, which results in ventralization (Bellipanni et al., 2006). It has been shown that β-Catenin-1 and β-Catenin-2 function similarly to promote dorsal axis formation, but that β-Catenin-1 localization and increased Axin expression may account for the inability of β-catenin-1 to rescue the *ichabod* mutant (Valenti et al., 2015).

β-Catenin is regulated by multiple additional maternal factors. Caveolin-1 and Transducer of Erbb2 1 (Tob1) both can bind and inhibit β-Catenin. Caveolin-1 binding inhibits β-Catenin nuclear translocation, while Tob1 blocks β-Catenin binding to LEF1, thus inhibiting dorsal cell fate induction (Mo et al., 2010; Xiong et al., 2006). Maternal β-catenin is also regulated by G-protein coupled receptor (GPCR) signaling. GPCR (C-C motif chemokine receptor 7) Ccr7 signaling functions by reducing the levels of β-Catenin, disrupting its ability to translocate into the nucleus, and preventing it from activating dorsal genes (Fig. 3B) (Wu, Shin, Sepich, & Solnica-Krezel, 2012). While the majority of maternal factors identified to promote dorsal cell fates do so via maternal β-Catenin, knockdown of maternal R-Spondin 3 had no effect on *bozozok* expression, a direct readout of maternal β-Catenin (Rong et al., 2014). Instead R-Spondin 3 regulates dorsal cell fate specification by inhibiting ventralizing zygotic Wnt signaling (Rong et al., 2014).

Maternal factors acting in dorsal axis formation still remain to be discovered. For example, the maternal-effect mutant *dullahan* was recently identified in a chemical mutagenesis screen in zebrafish (Fig. 3A) (Abrams et al., 2020). Embryos from *dullahan* homozygous mutant mothers are significantly ventralized, with the expression of dorsal organizer genes *goosecoid* and *chordin* either absent or reduced (Abrams et al., 2020). Thus, the future cloning of the corresponding *dullahan* gene will certainly add to our molecular understanding of dorsal organizer induction.

2.3 Maternal factors and ventral cell fates

As with dorsal cell fate specification, maternal factors actively promote ventral cell fates. The first maternal factors identified to promote ventral cell fates were *radar/gdf6a* and *pou5f3* (also called *pou2* and *oct4*), which act upstream of BMP signaling, and *runx2b type2* (*runx2bt2*), which is a transcriptional activator of the ventral genes *vox*, *vent*, and *ved* (Goutel, Kishimoto, Schulte-Merker, & Rosa, 2000; Langdon & Mullins, 2011; Sidi, Goutel, Peyrieras, & Rosa, 2003). Together, Vox, Vent, and Ved transcription factors promote ventral cell fates by inhibiting the ventral and lateral expression of dorsal-acting genes, and through the maintenance of ventral gene expression including *bmp* expression (Fig. 3B) (Imai et al., 2001; Ramel & Lekven, 2004; Shimizu et al., 2002). The expression of *vox, vent,* and *ved* is in turn maintained by zygotic Wnt8a signaling.

Another important maternal factor regulating the specification of ventral cell fates is Integrator Complex subunit 6 (Ints6). Maternal-effect *ints6* mutant embryos exhibit a unique strongly dorsalized phenotype, where secondary dorsal axes form in the strongest dorsalized mutant embryos (Kapp, Abrams, Marlow, & Mullins, 2013). The dorsalization defect can be rescued by suppressing dorsal organizer gene function or forcing *bmp* gene expression (Kapp et al., 2013). Expression of *bmp* genes is not maintained, and *vox* and *ved* expression is strongly reduced in *ints6* deficient embryos. Ints6 functions to promote ventral cell fates by restricting the dorsal organizer from expanding ventrally at late blastula stages, possibly by modifying the response to Nodal signaling in non-axial cells or by regulating Wnt8, *vox, vent,* and *ved* (Fig. 3B) (Kapp et al., 2013). Ints6 is a subunit of the Integrator Complex, which functions in cleaving nascent mRNA in various processes, including snRNA 3′ UTR processing and transcripts associated with paused RNA polymerase (Baillat et al., 2005; Elrod et al., 2019; Ezzeddine et al., 2011; Lai, Gardini, Zhang, & Shiekhattar, 2015; Schmidt et al., 2018; Tatomer et al., 2019). Whether Ints6 functions as part of the integrator complex or independently in regulating ventral cell fate specification is still unknown. Future studies will be required to determine its molecular mode of action in embryonic patterning.

Additionally, Leucine zipper tumor suppressor 2 (Lzts2), Forkhead box O transcription factor 3 (Foxo3b), and ELL associated factor 1 and 2 (Eaf1 and Eaf2) have been suggested as maternal factors that directly bind to maternal β-Catenin to promote ventral cell fates, as knockdown of each resulted in increased dorsal gene expression (Fig. 3B) (Li, Li, Long, & Cui, 2011; Liu et al., 2018; Xie, Liu, Hu, & Xiao, 2011). Finally, maternal *cathepsin Ba* has been identified to function upstream of BMP signaling to promote ventral cell fates as the dorsalization defect observed in the *cathepsin Ba* (*split top)* maternal-effect mutant is rescued by BMP signaling (Fig. 3B) (Langdon et al., 2016). Interestingly, *cathepsin Ba* mutants also display defects in morphogenesis due to alterations of the yolk cell actin and microtubule network.

Intriguingly, Cathepsin Ba is a lysosomal endopeptidase but how it regulates DV patterning and morphogenesis mechanistically, and the molecular pathway involved has yet to be determined (Langdon et al., 2016).

3. Signaling pathways and the patterning of the embryonic axes

During the blastula and gastrula stages of zebrafish development, the Wnt and Fgf signaling pathways pattern the embryonic anterior-posterior (AP) axis, while the BMP signaling pathway patterns the embryonic DV axis. Though most of this activity is zygotic, multiple maternally expressed genes play key roles in these processes and define the initial regions of ligand and antagonist expression (Fig. 4A) (Langdon & Mullins, 2011; Tuazon & Mullins, 2015; Zinski, Tajer, & Mullins, 2018). Crucially, maternally expressed factors define the dorsal organizer, which plays a key antagonistic role in both AP and DV patterning. Moreover, the embryo uses the inherited animal-vegetal polarity of the egg as a basis for axial patterning, as the marginal region of the embryo adjacent to the yolk cell forms an important posteriorizing center (Fig. 4A) (Langdon & Mullins, 2011; Thisse & Thisse, 2015; Tuazon & Mullins, 2015; Zinski et al., 2018). Finally, many receptors, transcription factors, other cytoplasmic intermediaries, and even some ligands are maternally provided (White et al., 2017). Indeed, by the midblastula transition (MBT), both the AP and the DV axes are already defined; subsequent zygotic signaling serves to both reinforce these identities, and to establish a graded coordinate system along these axes, by which cells can properly establish their location, and assume their appropriate fates (Fig. 4B) (Langdon & Mullins, 2011; Schier & Talbot, 2005; Tuazon & Mullins, 2015).

3.1 The activation of anterior-posterior patterning

During gastrulation Wnt and Fgf act as posteriorizing signals to pattern the AP axis (Fig. 5A) (Green, Whitener, Mohanty, & Lekven, 2015; Tuazon & Mullins, 2015). Loss of both Wnt and Fgf function leads to the loss of posterior structures, such as the tail, somites, notochord, and hindbrain, and the expansion of anterior structures such as the forebrain (Fig. 4B) (Bellipanni et al., 2006; Lekven, Thorpe, Waxman, & Moon, 2001; Shimizu, Bae, Muraoka, & Hibi, 2005; Shinya et al., 2000), while gain of function leads to the expansion of posterior structures at the expense of anterior ones

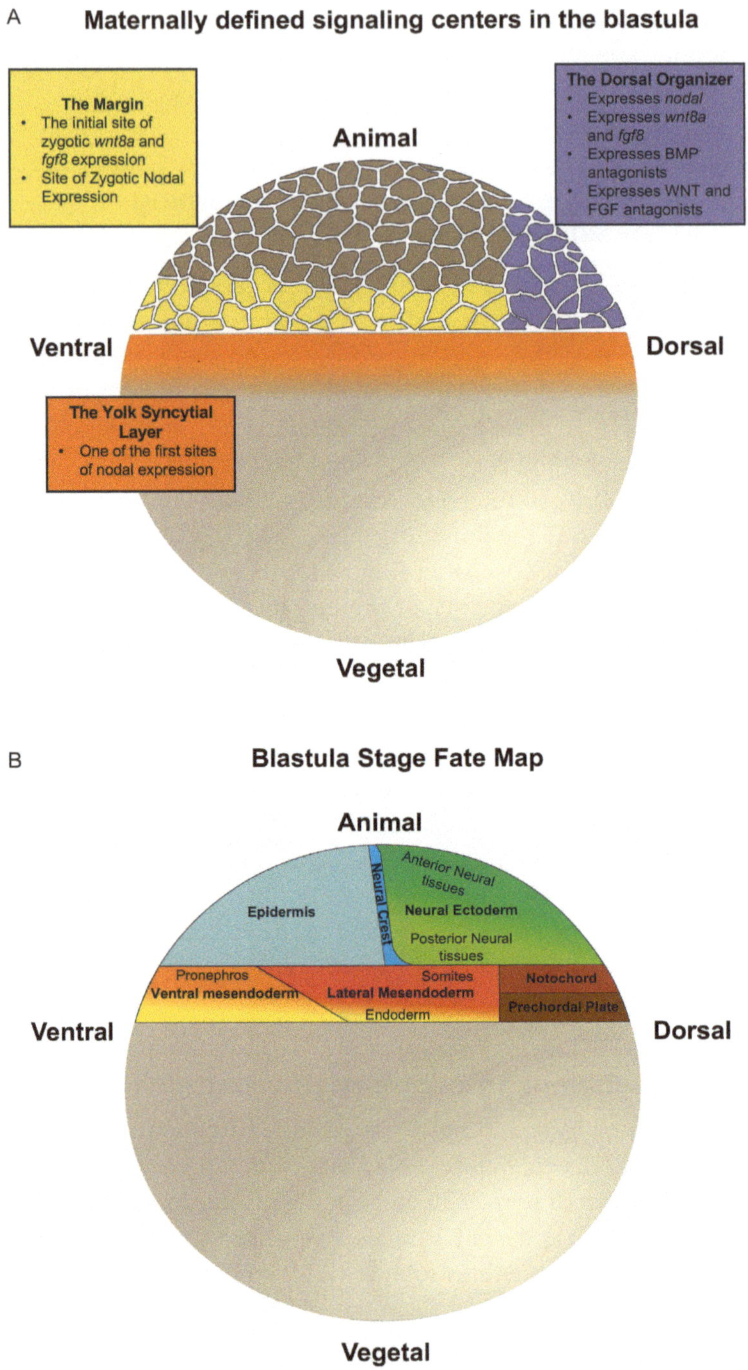

Fig. 4 See figure legend on opposite page.

(Dorsky, Itoh, Moon, & Chitnis, 2003; Kelly, Greenstein, Erezyilmaz, & Moon, 1995; Kim et al., 2000; Stachel et al., 1993). During this process, the two pathways act synergistically, with Wnt signaling activating FGF via Nodal and both signaling pathways being required for proper AP patterning (Fig. 5A) (Dyer et al., 2014; Green et al., 2015; Maegawa, Varga, & Weinberg, 2006; Tuazon & Mullins, 2015; Zinski et al., 2018). The zygotic expression of both ligands initiates in the organizer and margin of the embryo (Fig. 4) (Langdon & Mullins, 2011; Tuazon & Mullins, 2015). These same domains also express the FGF antagonists *sprouty2*, *sprouty4*, and *sef1*, as well as the Wnt antagonist *dkk1b* (Fodor et al., 2013; Zinski et al., 2018). The interaction between the ligands and their antagonists establishes a high Wnt/Fgf signaling region along the margin, with levels decreasing toward the animal pole, translating the initial animal-vegetal polarity of the oocyte into the AP axis of the embryo (Green et al., 2015; Tuazon & Mullins, 2015).

3.2 Maternal and zygotic Wnt signaling in dorsal-ventral and anterior-posterior patterning

While the relationship between Wnt and FGF signaling during AP patterning is complex, Wnt/β-Catenin can be said to be upstream, as maternal β-Catenin indirectly activates zygotic Fgf expression in both the organizer and the margin via Nodal signaling (Dougan, Warga, Kane, Schier, & Talbot, 2003; Dyer et al., 2014; Maegawa et al., 2006; Tanaka, Hosokawa, Weinberg, & Maegawa, 2017; Tuazon & Mullins, 2015; Zinski et al., 2018). Wnt pathway activity has two seemingly contradictory roles during early development: Wnt/β-Catenin signaling can be both dorsalizing and ventralizing in different contexts (Kapp et al., 2013; Langdon & Mullins, 2011; Tuazon & Mullins, 2015; Zinski et al., 2018).

Fig. 4 Maternally-defined signaling centers and fate map at the midblastula stage (A) By MBT, maternal factors have already defined three important signaling regions for axial patterning. The Dorsal Organizer (blue), which is defined by maternal β-Catenin, expresses *wnt8a*, *fgf8*, *ndr1*, and *ndr2*. The dorsal organizer also forms an antagonistic center for both AP and DV patterning. The dorsal organizer opposes posteriorizing activity by expressing WNT and FGF antagonists, and opposes ventralizing activity by expressing BMP antagonists. The Margin (yellow), formed of the cells closest to the yolk cell, is an important posteriorizing center, expressing *wnt8a* and *fgf8*. The Margin, along with the adjacent Yolk Syncytial layer (orange), is also a site of *ndr1* and *ndr2* expression. (B) Blastula stage fate map showing the cell types specified by the various signaling centers.

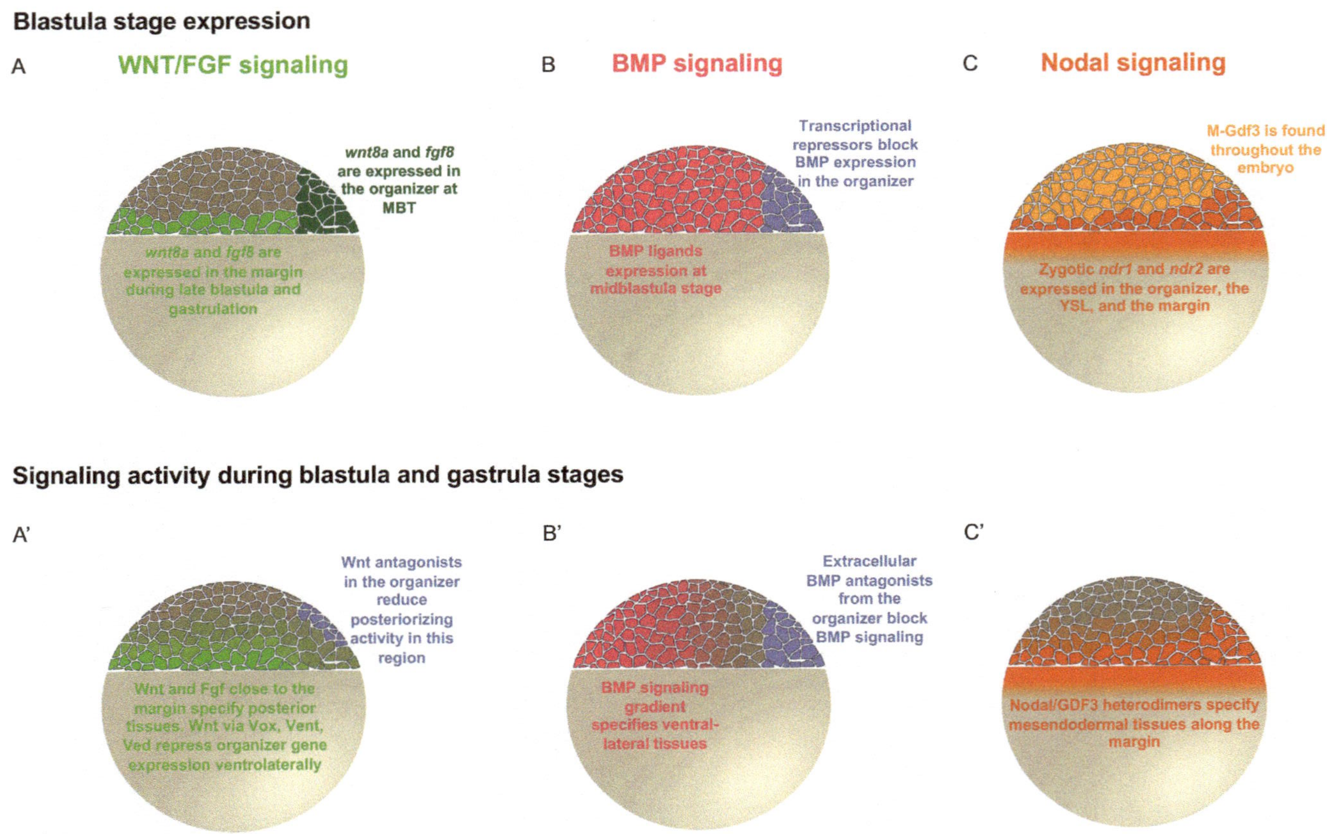

Fig. 5 See figure legend on opposite page.

As mentioned in Section 2.2, maternal, dorsally-localized nuclear β-Catenin specifies the dorsal organizer (Bellipanni et al., 2006; Kelly et al., 2000; Langdon & Mullins, 2011; Schneider et al., 1996). Reflecting this role, several Wnt pathway gain-of-function experiments show an expansion of the dorsal organizer (Kelly et al., 1995; Schneider et al., 1996; Shao et al., 2012; Speer et al., 2019; Stachel et al., 1993; Yan et al., 2018), whereas some Wnt pathway loss-of-function experiments show a reduction in organizer tissue (Bellipanni et al., 2006; Lu et al., 2011). The organizer expresses several dorsalizing BMP antagonists such as *chordin*, *noggin*, *bozozok*, and *goosecoid* (Fig. 5B) (Langdon & Mullins, 2011; Zinski et al., 2018). However, outside of the organizer, in ventrolateral regions, zygotic Wnt signaling drives the expression of the pro-ventral genes *vox*, *vent*, and *ved*, which in turn repress dorsal organizer markers such as *fgf8a*, and synergize with BMP signaling to confine the organizer (Fig. 5A′) (Langdon & Mullins, 2011; Tuazon & Mullins, 2015; Zinski et al., 2018). Thus, while maternal Wnt pathway activation driven by the asymmetric nuclear localization of β-Catenin is dorsalizing, the later, zygotic signaling of secreted Wnt ligands is ventralizing and posteriorizing (Fig. 5A and A′).

In addition to activating BMP antagonists, maternal β-Catenin activates several extracellular Wnt antagonists such as *dkk1*, *srfp1*, and *frzb* that limit the impact of subsequent zygotic signaling in this region (Fig. 5B)

Fig. 5 Expression signaling activities in blastula and gastrula stages. (A) The zygotic expression of *wnt8a* and *fgf8* initiates in the dorsal organizer at MBT but shifts to the margin in the later blastula. (A′) Wnt and Fgf signaling forms a gradient with the highest levels of signaling at the margin and decreasing levels toward the animal pole. The organizer also expresses Wnt and Fgf antagonists which somewhat limit signaling in this region. Higher levels of Wnt and Fgf signaling specify posterior tissues, such as the hindbrain, and the absence of these signals specifies more anterior tissues like the forebrain. (B) BMP ligands are initially expressed throughout the embryo, except for the organizer, where maternal β-Catenin drives the expression of transcriptional repressors of *bmp* transcription. (B′) Extracellular BMP antagonists expressed from the organizer block BMP signaling dorsally in the embryo. BMP signals in a gradient with high BMP signaling levels specifying ventral tissues, like epidermis and ventral mesendoderm, while lower levels specify lateral tissues, like neural crest, and the absence of BMP specifying dorsal tissues, like neural ectoderm. (C) Maternal *gdf3* RNA is distributed throughout the embryo, while the YSL, organizer, and margin also express zygotic *ndr1* and *ndr2*. (C′) Nodal signaling is strongest at the margin and diminishes toward the animal pole. High and moderate levels of Nodal signaling specify mesendodermal tissues, while the absence of Nodal signaling allows ectodermal tissue specification.

(Hashimoto et al., 2000; Lu et al., 2011; Nojima et al., 2004; Peng & Westerfield, 2006; Pezeron et al., 2006; Seiliez, Thisse, & Thisse, 2006; Shinya et al., 2000; Tendeng & Houart, 2006). *Dkk1* in particular is initially expressed in the organizer, expanding to the margin by 30% epiboly, but retreating again to the dorsal shield by mid-gastrulation (Hashimoto et al., 2000; Shinya et al., 2000). This raises the interesting possibility that dorsalizing, maternal Wnt signaling actually antagonizes ventralizing and posteriorizing, zygotic Wnt signaling. Outside of the organizer, in the margin, the maternally-deposited transcription factor *Kzp* initiates zygotic *wnt8a* transcription (Yao et al., 2010), but this pathway remains relatively uncharacterized.

3.3 The maternal and zygotic expression of Wnt pathway components

During canonical Wnt signaling, Wnt ligands bind Frizzled receptors and LRP5/6 coreceptors at the surface of the cell, which recruits the intracellular protein Disheveled (Nusse & Clevers, 2017; Yamaguchi, 2001). The combinatorial activation of Frizzled, LRP5/6, and Disheveled in turn inhibits a β-Catenin degrading complex containing Axin, GSK3, and APC (Nusse & Clevers, 2017; Yamaguchi, 2001). This inhibition leads to the accumulation of β-Catenin, which can then localize to the nucleus where it activates target genes by partnering with the TCF1/LEF1 transcription factors, and inactivates other target genes by pairing with HIPK2, which phosphorylates and displaces TCF3 transcription factors (Nusse & Clevers, 2017; Yamaguchi, 2001).

As discussed in Section 2.1, several Wnt ligands are maternally expressed, but these ligands are not involved in the pre-MBT process of dorsal organizer specification. Though maternal *wnt8a* is not required for organizer specification, zygotic *wnt8a* is necessary at a later stage to specify posterior tissues. Maternal-zygotic *wnt8a* mutants show more severe anteriorization phenotypes than zygotic mutants (Hino et al., 2018). Maternal only mutants, however, show no phenotype at all.

Several Frizzled receptors and Lrp5/6 coreceptors are maternally expressed, including *frizzled3b*, *frizzled5*, *frizzled7a*, *frizzled7b*, *frizzled8b*, *lrp5*, and *lrp6* (White et al., 2017). While the contributions of these different receptors and coreceptors to early zebrafish development remains unexplored, a recent study has thoroughly characterized the differential contributions of five zebrafish *disheveled* (*dvl*) genes (Xing et al., 2018).

This study found that only *dvl2* and *dvl3a* are expressed maternally. Through an innovative use of CRISPR, the authors generated mosaic adults, capable of producing maternal-zygotic *dvl2, dvl3a* double mutants. These animals were strongly anteriorized and cyclopic, supporting the requirement of *dvl2* and *dvl3a* in AP patterning. They also found that dorsal organizer marker expression was unaffected in these mutants. These results support the idea that Wnt signaling components upstream of β-Catenin, e.g., Wnt, Frizzled, and Disheveled may not act during dorsal organizer specification and instead Huluwa is key to stabilizing β-Catenin to act in this process (see Section 2.2).

The other intracellular components Axin, APC, GSK3, β-Catenin, and TCF1/LEF1 are also maternally supplied (Bellipanni et al., 2006; Cadigan & Waterman, 2012; Dorsky et al., 1999; Kelly et al., 2000; Nojima et al., 2004; Shimizu et al., 2000; Valenti et al., 2015; Veien, Grierson, Saund, & Dorsky, 2005; White et al., 2017). Of these components, maternal β-*catenin2* mutants lack a dorsal organizer, reflecting the role of maternal β-Catenin in organizer specification (Bellipanni et al., 2006; Kelly et al., 2000). Zygotic *axin1* (*masterblind*) mutants have overactive Wnt signaling, reflecting Axin's role in β-Catenin degradation. This leads to the loss of anterior tissues such as the eyes and forebrain (Heisenberg et al., 1996, 2001; Kapsimali, Caneparo, Houart, & Wilson, 2004; van de Water et al., 2001). Loss of function experiments with morpholinos suggest that this Axin is not required for organizer specification but is required for AP patterning (Schneider, Slusarski, & Houston, 2012). As these morpholino experiments eliminate both maternal and zygotic expression, it is not clear if maternal expression is specifically required. One study also suggests that maternal axin overexpression is partially responsible for the β-*catenin ichabod* mutant phenotype (Valenti et al., 2015).

While there is no reported analysis of APC, LEF1 or GSK3 maternal mutants in organizer formation, various loss of function experiments indicate their involvement in organizer specification and AP patterning. As these components negatively regulate β-Catenin, their inhibition results in an overactivation of Wnt signaling. Morpholino studies with APC and LEF1 cause posteriorization of the zebrafish brain (Paridaen et al., 2009). Lithium inhibition of GSK3 expands the organizer prior to the MBT (Stachel et al., 1993), but posteriorizes the embryo post-MBT (van de Water et al., 2001), reflecting the different roles of Wnt signaling at these two stages of development. Ultimately more experiments are needed to determine the maternal importance of these components.

3.4 Wnt and Fgf signaling during organizer specification and AP patterning

Though maternal Wnt signaling initiates zygotic expression of FGF, in the absence of FGF signaling, Wnt is incapable of specifying the dorsal organizer and patterning the AP axis (Tuazon & Mullins, 2015; Zinski et al., 2018). As such, Fgf signaling is essential for both the dorsalizing activity of maternal β-Catenin, and the posteriorizing activity of zygotic Wnt8a. Zygotic *fgf8a* expression initiates alongside Wnt in the organizer, and extends with Wnt to the margin in the late blastula stage (Langdon & Mullins, 2011; Tuazon & Mullins, 2015). Fgf also activates the expression of its own antagonists *sprouty2*, *sprouty4*, and *sef* in the organizer and the margin (Zinski et al., 2018), though *sef* is also initially maternally expressed ubiquitously (Tsang, Friesel, Kudoh, & Dawid, 2002; White et al., 2017). Fgf signaling is required for the specification of several posterior neuronal fates such as the midbrain/hind brain boundary, the hind brain, and the spinal cord, and posterior mesodermal fates such as trunk and tail mesoderm, and may act as a morphogen (Draper, Stock, & Kimmel, 2003; Green et al., 2015; Griffin & Kimelman, 2003; Scholpp & Brand, 2004; Tuazon & Mullins, 2015; Yu et al., 2009).

During dorsal organizer specification, Fgf signaling plays a crucial role in defining dorsal identity (Fig. 5A and A′). Fgf signaling is required for the expression of the BMP antagonist *chordin* (Kuo, Lam, Hewitt, & Scotting, 2013; Maegawa et al., 2006; Varga et al., 2007). Fgf signaling can also antagonize BMP signaling independently of Chordin by both directly inhibiting the transcription of BMP (Furthauer, Van Celst, Thisse, & Thisse, 2004; Zinski et al., 2018), and by inducing the expression of *goosecoid*, which inhibits expression of the ventral gene *vent* (Fig. 5C and C′) (Joore et al., 1996; Kawahara, Wilm, Solnica-Krezel, & Dawid, 2000; Kuo et al., 2013; Maegawa et al., 2006). Curiously, the maternal factor Pou5f3, which exhibits a strongly dorsalized maternal-effect phenotype, restricts organizer activity by inhibiting *fgf8a* transcription, rather than by disrupting Wnt signaling (Belting et al., 2011; Reim & Brand, 2006).

3.5 Integration of maternal and zygotic Fgf signaling components in patterning

Fgf ligands signal by assembling a complex of two FGFR receptor tyrosine kinases. These receptors trans-phosphorylate each other and can activate a wide variety of downstream pathways including the canonical

RAS-MAPK pathway, as well as the PI3K-AKT, PLCγ, and STAT pathways (Ornitz & Itoh, 2015). All five of the canonical zebrafish FGF receptors, *fgfr1a*, *fgfr1b*, *fgfr2*, *fgfr3*, and *fgfr4*, are expressed maternally except for possibly *fgfr2* (Challa & Chatti, 2013; Rohner et al., 2009; White et al., 2017). Disruption of these receptors expectedly anteriorizes the embryo, while constitutive active versions posteriorize the embryo (Ota, Tonou-Fujimori, & Yamasu, 2009). One version of constitutive active *fgfr2* lacking a regulatory Valine-Threonine dipeptide additionally dorsalizes the embryo. Whether these processes require maternal activation remains unexplored, but, as Fgf signaling is required during organizer specification (Maegawa et al., 2006), it seems likely that at least some receptors will be required maternally. Of the 30 zebrafish Fgf ligands, 13 are expressed maternally before the MBT: *fgf1b*, *fgf2*, *fgf4*, *fgf6a*, *fgf8a*, *fgf10b*, *fgf11a*, *fgf13a*, *fgf13b*, *fgf14*, *fgf17b*, *fgf18a*, and *fgf20b* (Cao et al., 2004; Lee et al., 2010; Reifers et al., 1998; Songhet, Adzic, Reibe, & Rohr, 2007; White et al., 2017; Yamauchi, Miyakawa, Miyake, & Itoh, 2009; Yan, Zheng, & Gong, 2015). So far the roles of individual Fgfs in early development, and their relative maternal and zygotic importance remains relatively unexplored: zygotic *fgf8a* mutants show disruption to posterior brain structures (Brand et al., 1996), and the overexpression of *fgf17b* both dorsalizes and posteriorizes embryos in different contexts (Cao et al., 2004).

3.6 The activation of dorsal-ventral patterning

While Wnt and Fgf signaling pattern tissues along the AP axis, BMP signaling patterns structures along the DV axis (De Robertis & Sasai, 1996; Tuazon & Mullins, 2015; Zinski et al., 2018). Prior to the MBT, maternal factors play an important role in both specifying and confining the dorsal organizer. The dorsal organizer expresses both extracellular BMP antagonists such as *chordin*, *noggin*, and *follistatin*, and transcriptional repressors, such as *bozozok* and *goosecoid* (Fig. 5B and B′) (Zinski et al., 2018). Outside of the dorsal organizer, maternal factors *radar/gdf6a* and *pou5f3* activate the expression of BMPs (Goutel et al., 2000; Langdon & Mullins, 2011; Reim & Brand, 2006; Sidi et al., 2003; Wilm & Solnica-Krezel, 2003), discussed more in Section 2.1. The results of these interactions create a quantifiable morphogen gradient of BMP signaling along the DV axis, with high levels of BMP specifying ventral tissues and lower levels specifying more lateral tissues (Fig. 5B′) (Pomreinke et al., 2017; Ramel & Hill, 2013; Tuazon & Mullins, 2015; Zinski et al., 2017).

BMP ligands are members of the TGF-β superfamily and signal by assembling tetrameric complexes at the cell surface (Hata & Chen, 2016). These complexes contain two type II receptors, which phosphorylate and activate two type I receptors. The activated type I receptors in turn, phosphorylate receptor (R-)Smad1/5/9 transcription factors (Hata & Chen, 2016). Upon phosphorylation, two R-Smads trimerize with the co-Smad, Smad4, and accumulate in the nucleus where they activate the expression of BMP target genes (Hata & Chen, 2016; Hill, 2016).

In contrast to the BMP ligands, which are largely expressed zygotically, most BMP receptors are expressed maternally. The type I BMP receptors *acvr1l*, *bmpr1aa*, *bmpr1ab*, *bmpr1ba*, and *bmpr1bb* are all expressed both maternally and zygotically (Bauer, Lele, Rauch, Geisler, & Hammerschmidt, 2001; Little & Mullins, 2009; Mintzer et al., 2001; Mullins et al., 1996; Nikaido, Tada, & Ueno, 1999; White et al., 2017; Woods et al., 2005). Notably, zygotic *acvr1l* mutants are mildly dorsalized, while maternal-zygotic mutants lack all BMP signaling (Bauer et al., 2001; Mintzer et al., 2001; Mullins et al., 1996). Maternal loss alone, however, yields all wild-type phenotype embryos (Mintzer et al., 2001). Thus maternal and zygotic *acvr1l* are largely redundant with each other, similar to the Nodal co-receptor Oep (Gritsman et al., 1999). Morpholino knockdown and some mutant experiments suggest that all four zebrafish Bmpr1 receptors are required, somewhat redundantly for DV patterning (Little & Mullins, 2009; Smith et al., 2011). Acvr1l and Bmpr1 class receptors function non-redundantly during DV patterning, with both receptors involved in the BMP2/7 heterodimer signaling complex (Little & Mullins, 2009).

BMP type II receptors are much less characterized than the type I receptors. All four zebrafish Acvr2 receptors, *acvr2aa*, *acvr2ab*, *acvr2ba*, and *acvr2bb*, are expressed maternally and zygotically (Albertson, Payne-Ferreira, Postlethwait, & Yelick, 2005; Thisse and Thisse, 2004; White et al., 2017), yet the role of these receptors in DV patterning remains uncharacterized. It is also unclear whether the Bmpr2 class type II receptors, *bmpr2a* and *bmpr2b*, are expressed during early zebrafish development. In situ hybridization and RT-PCR data from one publication suggest both maternal and zygotic expression, and morpholino experiments suggest that these receptors are required for both DV patterning (Monteiro et al., 2008) and left-right patterning. More recent RNA-seq data, however, fails to detect these receptors prior to somitogenesis (White et al., 2017).

All three zebrafish BMP *R-smads smad1, smad5*, and *smad9* are expressed zygotically during gastrulation, but only *smad5* is expressed maternally (Wei, Wang, Zhu, & Sun, 2014; White et al., 2017). The *smad5* gene is also expressed at much higher levels than the other BMP R-*smads* throughout early embryogenesis (Wei et al., 2014; White et al., 2017). Moreover, *smad1* and *smad9* knockdown have relatively minor phenotypes that are only apparent later in development, and the combinatorial knockdown of all three Smads does not enhance the *smad5* knockdown phenotype (Wei et al., 2014). One *smad5* allele "*somitabun*" or *dtc24* acts as an antimorphic allele and has a dominant maternal phenotype, with all offspring from heterozygous females being strongly dorsalized, while embryos from wild-type females crossed to heterozygous males have a mild or wild-type phenotype (Kramer et al., 2002; Mullins et al., 1996). Homozygous mutant females of a null allele of *smad5* also display a 100% penetrant strongly dorsalized, maternal phenotype in their progeny (Kramer et al., 2002). These results show that the maternal contribution of *smad5* is essential and may be more important than the zygotic contribution during DV patterning. The zebrafish co-smad *smad4* is also expressed maternally (White et al., 2017).

The BMP signaling gradient is also modulated by several maternal agonists, including intracellular factors such as Pou5f3 and Ints6, the extracellular Chordin-cleaving proteases Tolloid and Bmp1a (Langdon & Mullins, 2011; Tuazon & Mullins, 2015; Zinski et al., 2018). *Pou5f3* is maternally expressed ubiquitously in the embryo, and enhances BMP expression possibly by inhibiting FGF signaling from the organizer, this has the effect of confining the organizer and inhibiting the expression of many BMP antagonists (Belting et al., 2011; Lippok, Song, & Driever, 2014; Onichtchouk et al., 2010; Reim & Brand, 2006). The extracellular BMP agonist *bmp1a* is maternally and ubiquitously expressed (Muraoka et al., 2006; White et al., 2017), while *tolloid* is expressed zygotically in the ventral region of the embryo (Connors, Trout, Ekker, & Mullins, 1999). Of these two BMP agonists, *tolloid* mutants are mildly dorsalized, while *bmp1a* mutants have a later, non-DV patterning phenotype (Connors et al., 1999; van Eeden et al., 1996). The simultaneous knockdown of both genes results in complete dorsalization (Muraoka et al., 2006). While it is clear that these two genes are partially redundant, it is not clear to what extent the maternal expression of *bmp1a* is relevant.

4. Nodal signaling in germ layer formation and left-right patterning

4.1 Nodal-Vg1 heterodimer specifies the mesoderm and endoderm germ layers

During the earliest stages of embryonic development, maternal factors are instrumental in forming the three germ layers, the ectoderm, mesoderm, and endoderm (Fig. 4). The Nodal signaling pathway plays a critical role in germ layer formation by specifying mesoderm and endoderm (mesendoderm) in vertebrates (Figs. 4 and 5C, C′) (Conlon et al., 1994; Feldman et al., 1998; Iannaccone, Zhou, Khokha, Boucher, & Kuehn, 1992; Jones, Kuehn, Hogan, Smith, & Wright, 1995). Recent studies show that Nodal does not act as a homodimer and instead functions as an obligate heterodimer with Gdf3 (also called Vg1 and Gdf1). Signaling is activated when the TGF-β ligand Nodal-Gdf3 binds and assembles a heteromeric complex of receptors consisting of the Type I receptor Acvr1b (Gu et al., 1998) and the Type II receptor Acvr2a or Acvr2b (Song et al., 1999). Nodal-Gdf3 binding also requires the presence of the EGF-CFC co-receptor, (called Oep and Tdgf1 in zebrafish, CRIPTO and CRYPTIC in mouse, and FRL-1, Xcr2, and Xcr3 in Xenopus) (Chu & Shen, 2010; Ding et al., 1998; Dorey & Hill, 2006; Gritsman et al., 1999; Yeo & Whitman, 2001). Upon assembly of the Nodal receptor complex the Type II receptor phosphorylates and activates the Type I receptor. The Type I receptor in turn phosphorylates Smad2 and/or Smad3, which then associate with Smad4 and accumulate in the nucleus (Hata & Chen, 2016). The Smad complex then interacts with transcription co-factors to activate the expression of genes involved in the formation of mesoderm and endoderm (Hata & Chen, 2016; Hoodless et al., 2001; Kofron et al., 2004; Slagle, Aoki, & Burdine, 2011). In the zebrafish, Nodal signaling relies on multiple maternally supplied components of this pathway, including Gdf3 (Vg1), the co-receptor Oep, and Smad2, among other components, as discussed below.

The function of the TGF-β ligand Vg1 (also called Gdf1 and Gdf3) in mesendoderm induction was a conundrum in the field for many years until recently. It was first identified in *Xenopus* and named Vg1 due to its vegetal pole localization in the egg and oocyte (Melton, 1987; Weeks & Melton, 1987). Misexpression of Vg1 had no developmental consequences and failed to be secreted, except when chimeric forms were generated using the prodomains of other TGF-β ligands, when it could then robustly induce ectopic mesendoderm formation (Andersson, Bertolino, & Ibanez, 2007; Birsoy, Kofron, Schaible, Wylie, & Heasman, 2006; Chen et al., 2006; Dohrmann, Kessler, & Melton, 1996; Kessler & Melton, 1995; Thomsen & Melton, 1993; Wall, Craig, Labosky, & Kessler, 2000). These chimeric Vg1 variants signaled through the same receptors as Nodal and the EGF-CFC co-receptor to activate Nodal target genes (Andersson et al., 2007; Chen et al., 2006; Cheng, Olale, Bennett, Brivanlou, & Schier, 2003).

Recent work in the zebrafish by three independent labs has further elucidated the role of Vg1 in Nodal signaling and germ layer specification (Bisgrove, Su, & Yost, 2017; Montague & Schier, 2017; Pelliccia, Jindal, & Burdine, 2017). The *Vg1* ortholog in zebrafish is called *gdf3*. The *gdf3* mRNA is also localized asymmetrically during oogenesis, but instead of localizing to the vegetal pole as in *Xenopus*, it localizes to the animal pole (Bally-Cuif et al., 1998; Marlow & Mullins, 2008). Maternal *gdf3* (M-*gdf3*) is ubiquitously present through early stages of development until late in gastrulation (Dohrmann et al., 1996; Helde & Grunwald, 1993; Pelliccia et al., 2017; Peterson, Wang, & Yost, 2013). M-*gdf3* expression overlaps from the late blastula stage with the more restricted *nodal* expression domain around the margin of the embryo (Figs. 4 and 5C, C′). In contrast to M-*gdf3* expression, zygotic *gdf3* (Z-*gdf3*) has a more defined expression pattern that is detected at the bud stage (end of gastrulation) in the left lateral plate mesoderm (Pelliccia et al., 2017; Peterson et al., 2013). With the advent of efficient gene targeting methods in the zebrafish, three groups made mutations in the zebrafish *gdf3* gene. All groups found that loss of Z-*gdf3* causes no morphological defects in the embryo, whereas loss of M-*gdf3* causes a severe loss of mesendoderm formation (Bisgrove et al., 2017; Montague & Schier, 2017; Pelliccia et al., 2017).

The loss of M-*gdf3* causes mesendoderm defects very similar to phenotypes seen in zygotic double mutants for the zebrafish *nodal* ligand orthologs *sqt/ndr1* and *cyc/ndr2* (Feldman et al., 1998) and maternal-zygotic (MZ) mutant embryos for *oep* (Gritsman et al., 1999). The MZ- *gdf3* loss-of-function phenotype is no more severe than the maternal phenotype, suggesting that zygotic *gdf3* is dispensable for mesendoderm formation. This is a notable difference between Nodal and Gdf3. While zygotic Nodal is critical for mesoderm and endoderm specification (Bennett et al., 2007; Feldman et al., 1998), maternal Gdf3, and not zygotic Gdf3, is the critical Gdf3 component that functions to form the germ layers.

Like in *Xenopus*, Gdf3 homodimers in zebrafish are not secreted, nor is Gdf3 processed from its prodomain; hence, it remains intracellularly and is thus not functional (Bisgrove et al., 2017; Dohrmann et al., 1996; Montague & Schier, 2017). In contrast, when Gdf3 is co-expressed with Nodal, then it is secreted as a heterodimer with Nodal (Montague & Schier, 2017). Nodal can be secreted in the absence of Gdf3, but it is not functional at physiological levels unless it heterodimerizes with Gdf3. Because *gdf3* is expressed ubiquitously in the early embryo, ectopic *nodal* expression induces ectopic mesendoderm to form. In a M-*gdf3* mutant, however, physiological levels of ectopic *nodal* expression are no longer active (Bisgrove et al., 2017; Montague & Schier, 2017), because only Nodal-Gdf3 heterodimers are active in signaling, due to a still unknown mechanism. Thus, in the zebrafish embryo maternal Gdf3 is critical for Nodal signaling and proper germ layer formation. Maternal Gdf3 is present prior to *nodal* expression in the zebrafish embryo (Dohrmann et al., 1996) but it remains inactive. Nodal is made in the zygote, heterodimerizes with maternal Gdf3, and is released to the extracellular space where it specifies mesoderm and endoderm. This mechanism is conserved in the mouse as well, although *Gdf3* and *Nodal* are both expressed zygotically. Like in zebrafish, *Gdf3* mutant embryos in the mouse lack mesendoderm similar to *Nodal* mutant embryos (Conlon et al., 1994; Iannaccone et al., 1992; Lowe, Yamada, & Kuehn, 2001; Norris, Brennan, Bikoff, & Robertson, 2002), consistent with Nodal-Gdf1 heterodimers acting in the mouse as well.

Other components of the Nodal signaling pathway are also supplied maternally in mesendoderm specification in the zebrafish. In addition to the *acvr1b* and *acvr2* receptors (White et al., 2017), *smad2* is also expressed maternally and zygotically, both of which are required for mesendoderm development (Dubrulle et al., 2015). Like *gdf3* zygotic mutants, *smad2* zygotic mutants are not deficient in mesendoderm; however, unlike *gdf3* zygotic mutants, which are viable, *smad2* zygotic mutants are larval lethal (Dubrulle et al., 2015). Thus, MZ-*smad2* mutants were generated by making germline clones of *smad2* deficient cells in otherwise wild-type embryos, which are then raised to adults (Ciruna et al., 2002; Dubrulle et al., 2015). Such *smad2 −/−* germline mutant adult females when crossed to heterozygous *smad2* males, produce 50% MZ-*smad2* mutant embryos. MZ-*smad2* mutant embryos display an identical phenotype to double *nodal* ligand mutants (*sqt/ndr1*; *cyc/ndr2)* (Feldman et al., 1998), MZ-*oep* mutant embryos (Gritsman et al., 1999), and M-*gdf3* mutant embryos, displaying a severe deficiency in mesendoderm. Furthermore, misexpression of Nodal in MZ-*smad2* mutant embryos fails to induce mesendoderm, indicating that Smad2 is essential for all Nodal signaling activity.

The formation of mesoderm and endoderm is also regulated by the maternal transcription factor Eomesodermin (Eomes) (Arnold, Hofmann, Bikoff, & Robertson, 2008; Bjornson et al., 2005; Bruce et al., 2003; Russ et al., 2000; Ryan, Garrett, Mitchell, & Gurdon, 1996). Eomes is both maternally and zygotically expressed in the zebrafish embryo, and overexpression of *eomes* in the embryo induces ectopic expression of mesoderm and endoderm associated genes and can lead to the formation of a secondary axis (Bjornson et al., 2005; Bruce et al., 2003). Defects in mesoderm formation due to loss of FoxH1 are exacerbated by the loss of Eomes function (Slagle et al., 2011). Likewise, knockdown of FoxH1 in *MZ-eomes* mutant embryos leads to a greater loss of mesendoderm marker expression (Nelson et al., 2014). Notably, the loss of maternal *eomes* leads to delays in the initiation of epiboly (Du, Draper, Mione, Moens, & Bruce, 2012) and to defects in the temporal expression of genes involved in mesendoderm specification (Xu et al., 2014).

Thus, like DV axis induction and patterning, mesendodermal specification also relies heavily on maternal regulation of not only signaling but transcriptional components that are functionally intertwined with zygotic pathway components to elaborate the body plan.

4.2 Maternal Vg1/Gdf3 control of left-right axis formation

Vertebrates have an asymmetric left-right (L-R) arrangement of their organs within the body. The L-R axial placement of organs within the vertebrate body is regulated by Nodal signaling (Collignon, Varlet, & Robertson, 1996; Levin, Johnson, Stern, Kuehn, & Tabin, 1995; Long, Ahmad, & Rebagliati, 2003; Lowe et al., 1996; Sampath, Cheng, Frisch, & Wright, 1997). During L-R patterning, *nodal* must be expressed in the left lateral plate mesoderm (LPM) and restricted from the right LPM. If Nodal is lost from the left LPM or misexpressed in the right LPM, the embryo will develop with irregular organ laterality. The process that ensures *nodal* is correctly activated in the LPM is regulated by the left-right organizer (LRO), known as Kupffer's vesicle in zebrafish, the node in mouse, and the gastrocoel roof plate in *Xenopus* (Grimes & Burdine, 2017). Cilia in the LRO generate fluid flow, which is responsible for the right-biased expression of the Nodal inhibitor Dand5 (Dasgupta & Amack, 2016;

Hashimoto et al., 2004; Hojo et al., 2007; Marques et al., 2004; Nakamura et al., 2012; Schweickert et al., 2010). Dand5 prevents Nodal that emanates from around the LRO from activating the Nodal signaling pathway in the right LPM, restricting the initiation of *nodal* expression to the left LPM. (Hashimoto et al., 2004; Hojo et al., 2007; Marques et al., 2004; Montague, Gagnon, & Schier, 2018; Nakamura et al., 2012; Sampaio et al., 2014; Schweickert et al., 2010). Nodal activity is then dampened by the Nodal inhibitor known as Lefty1 (Lft1) (Shiratori & Hamada, 2014), which prevents Nodal from activating its own expression in the right LPM. Nodal expressed in the left LPM ultimately communicates positional information to the organ primordia (Grimes & Burdine, 2017).

Recent work in zebrafish has revealed that Vg1 is also required for L-R patterning in the zebrafish embryo (Pelliccia et al., 2017; Peterson et al., 2013), which is a conserved function in the mouse and *Xenopus* (Hanafusa, Masuyama, Kusakabe, Shibuya, & Nishida, 2000; Hyatt, Lohr, & Yost, 1996; Rankin, Bunton, Lawler, & Lee, 2000; Wall et al., 2000). Depletion of Gdf3 using morpholino-mediated knockdown adversely affects development of KV and leads to irregular expression of Dand5 (Pelliccia et al., 2017). Knockdown of Gdf3 also reduces the expression of the Nodal ortholog *spaw* in the left LPM (Pelliccia et al., 2017; Peterson et al., 2013). However, homozygous *Z-gdf3* mutant embryos do not exhibit defects in L-R patterning (Bisgrove et al., 2017; Montague & Schier, 2017; Pelliccia et al., 2017). Although homozygous M-*gdf3* and MZ-*gdf3* mutant embryos exhibit severe defects in mesendoderm formation, these gross morphological defects can be rescued to an extent that allows the investigation of maternal Gdf3 in L-R patterning. Interestingly, rescue experiments using increasing amounts of Gdf3 in homozygous M-*gdf3* and MZ-*gdf3* mutant embryos revealed that maternal Gdf3 is critical for L-R patterning in zebrafish (Pelliccia et al., 2017). Both KV development and *dand5* expression was rescued, in addition to correct *spaw* expression in the left LPM. These experiments also showed that zygotic Gdf3 can competently function, but it cannot overcome the loss of maternal Gdf3 in L-R patterning.

Maternal Gdf3 functions at multiple levels during L-R patterning. Nodal first specifies the dorsal forerunner cells (DFCs), which later form KV (Oteiza, Koppen, Concha, & Heisenberg, 2008). Nodal is also needed for the zebrafish embryo to form a KV with proper structure and morphology (Compagnon et al., 2014), which is required for correct L-R organ placement (Okabe, Xu, & Burdine, 2008; Sampaio et al., 2014;

Wang et al., 2011; Wang, Manning, & Amack, 2012). Maternal Gdf3 subsequently heterodimerizes with the Nodal ortholog Spaw during somitogenesis to activate Nodal signaling in L-R patterning (Montague et al., 2018; Pelliccia et al., 2017).

It is remarkable that maternal Gdf3 plays such a crucial role in a developmental process like L-R patterning, which occurs many hours after the embryo begins zygotic transcription. The zebrafish embryo also transcribes *gdf3* a few hours prior to the expression of *nodal* in the left LPM in zebrafish. Yet, maternal Gdf3 can compensate for the loss of this zygotic Gdf3 expression and allows L-R patterning to proceed without adverse effects. Gdf3 produced later by the embryo likely acts as a buttress, a redundant system for L-R patterning in case maternal Gdf3 was diminished. Even so, maternal Gdf3 is clearly doing the "heavy lifting" during zebrafish development.

5. Concluding remarks

Unlocking the molecular code to early vertebrate patterning is fundamental to our understanding of how the embryo ultimately takes shape. The zebrafish model system has been integral in advancing this effort, serving both as a powerful genetic tool and an optically-accessible organism during the key early stages of development. We have indeed come a long way in our understanding, from the initial ground-breaking zygotic mutant screens reported in the mid-1990s to maternal-effect screens conducted throughout the 2000s and beyond. More recently, our genetic approaches in the zebrafish have become more sophisticated and with the recent explosion of genomic and single-cell sequencing data, combined with the advancement of genome editing techniques, reverse genetic approaches are rapidly filling in the gaps. These advancements aided by innovations in biochemical and imaging techniques have certainly fueled the exciting era we are currently witnessing with respect to expanding our knowledge of the molecular genetic mechanisms of early vertebrate patterning.

Acknowledgments

We thank Francesca Tuazon for valuable comments on the manuscript. We are grateful for NIH grant funding, R35-GM131908 and R21- HD094096 to M.C.M., and a University of Pennsylvania Provost's Postdoctoral fellowship to J.L.P.

References

Abrams, E. W., Fuentes, R., Marlow, F. L., Kobayashi, M., Zhang, H., Lu, S., et al. (2020). Molecular genetics of maternally-controlled cell divisions. *PLoS Genetics*, *16*(4), e1008652. https://doi.org/10.1371/journal.pgen.1008652.

Ajduk, A., Ilozue, T., Windsor, S., Yu, Y., Seres, K. B., Bomphrey, R. J., et al. (2011). Rhythmic actomyosin-driven contractions induced by sperm entry predict mammalian embryo viability. *Nature Communications*, *2*, 417. https://doi.org/10.1038/ncomms1424.

Ajduk, A., & Zernicka-Goetz, M. (2016). Polarity and cell division orientation in the cleavage embryo: From worm to human. *Molecular Human Reproduction*, *22*(10), 691–703. https://doi.org/10.1093/molehr/gav068.

Albertson, R. C., Payne-Ferreira, T. L., Postlethwait, J., & Yelick, P. C. (2005). Zebrafish acvr2a and acvr2b exhibit distinct roles in craniofacial development. *Developmental Dynamics*, *233*(4), 1405–1418. https://doi.org/10.1002/dvdy.20480.

Andersson, O., Bertolino, P., & Ibanez, C. F. (2007). Distinct and cooperative roles of mammalian Vg1 homologs GDF1 and GDF3 during early embryonic development. *Developmental Biology*, *311*(2), 500–511. https://doi.org/10.1016/j.ydbio.2007.08.060.

Arnold, S. J., Hofmann, U. K., Bikoff, E. K., & Robertson, E. J. (2008). Pivotal roles for eomesodermin during axis formation, epithelium-to-mesenchyme transition and endoderm specification in the mouse. *Development*, *135*(3), 501–511. https://doi.org/10.1242/dev.014357.

Baillat, D., Hakimi, M. A., Naar, A. M., Shilatifard, A., Cooch, N., & Shiekhattar, R. (2005). Integrator, a multiprotein mediator of small nuclear RNA processing, associates with the C-terminal repeat of RNA polymerase II. *Cell*, *123*(2), 265–276. https://doi.org/10.1016/j.cell.2005.08.019.

Bally-Cuif, L., Schatz, W. J., & Ho, R. K. (1998). Characterization of the zebrafish orb/CPEB-related RNA binding protein and localization of maternal components in the zebrafish oocyte. *Mechanisms of Development*, 77(1), 31–47. https://doi.org/10.1016/s0925-4773(98)00109-9.

Banani, S. F., Lee, H. O., Hyman, A. A., & Rosen, M. K. (2017). Biomolecular condensates: Organizers of cellular biochemistry. *Nature Reviews Molecular Cell Biology*, *18*(5), 285–298. https://doi.org/10.1038/nrm.2017.7.

Bauer, H., Lele, Z., Rauch, G. J., Geisler, R., & Hammerschmidt, M. (2001). The type I serine/threonine kinase receptor Alk8/lost-a-fin is required for Bmp2b/7 signal transduction during dorsoventral patterning of the zebrafish embryo. *Development*, *128*(6), 849–858.

Bellipanni, G., Varga, M., Maegawa, S., Imai, Y., Kelly, C., Myers, A. P., et al. (2006). Essential and opposing roles of zebrafish beta-catenins in the formation of dorsal axial structures and neurectoderm. *Development*, *133*(7), 1299–1309. https://doi.org/10.1242/dev.02295.

Belting, H. G., Wendik, B., Lunde, K., Leichsenring, M., Mossner, R., Driever, W., et al. (2011). Pou5f1 contributes to dorsoventral patterning by positive regulation of vox and modulation of fgf8a expression. *Developmental Biology*, *356*(2), 323–336. https://doi.org/10.1016/j.ydbio.2011.05.660.

Bennett, J. T., Stickney, H. L., Choi, W. Y., Ciruna, B., Talbot, W. S., & Schier, A. F. (2007). Maternal nodal and zebrafish embryogenesis. *Nature*, *450*(7167), E1–E2. discussion E2-4. https://doi.org/10.1038/nature06314.

Betley, J. N., Frith, M. C., Graber, J. H., Choo, S., & Deshler, J. O. (2002). A ubiquitous and conserved signal for RNA localization in chordates. *Current Biology*, *12*(20), 1756–1761. https://doi.org/10.1016/s0960-9822(02)01220-4.

Birsoy, B., Kofron, M., Schaible, K., Wylie, C., & Heasman, J. (2006). Vg1 is an essential signaling molecule in Xenopus development. *Development, 133*(1), 15–20. https://doi.org/10.1242/dev.02144.

Bisgrove, B. W., Su, Y. C., & Yost, H. J. (2017). Maternal Gdf3 is an obligatory cofactor in nodal signaling for embryonic axis formation in zebrafish. *eLife, 6*, e28534. https://doi.org/10.7554/eLife.28534.

Bjornson, C. R., Griffin, K. J., Farr, G. H., 3rd, Terashima, A., Himeda, C., Kikuchi, Y., et al. (2005). Eomesodermin is a localized maternal determinant required for endoderm induction in zebrafish. *Developmental Cell, 9*(4), 523–533. https://doi.org/10.1016/j.devcel.2005.08.010.

Boke, E., Ruer, M., Wuhr, M., Coughlin, M., Lemaitre, R., Gygi, S. P., et al. (2016). Amyloid-like self-assembly of a cellular compartment. *Cell, 166*(3), 637–650. https://doi.org/10.1016/j.cell.2016.06.051.

Bontems, F., Stein, A., Marlow, F., Lyautey, J., Gupta, T., Mullins, M. C., et al. (2009). Bucky ball organizes germ plasm assembly in zebrafish. *Current Biology, 19*(5), 414–422. https://doi.org/10.1016/j.cub.2009.01.038.

Brand, M., Heisenberg, C. P., Jiang, Y. J., Beuchle, D., Lun, K., Furutani-Seiki, M., et al. (1996). Mutations in zebrafish genes affecting the formation of the boundary between midbrain and hindbrain. *Development, 123*, 179–190.

Bruce, A. E., Howley, C., Zhou, Y., Vickers, S. L., Silver, L. M., King, M. L., et al. (2003). The maternally expressed zebrafish T-box gene eomesodermin regulates organizer formation. *Development, 130*(22), 5503–5517. https://doi.org/10.1242/dev.00763.

Cadigan, K. M., & Waterman, M. L. (2012). TCF/LEFs and Wnt signaling in the nucleus. *Cold Spring Harbor Perspectives in Biology, 4*(11), a007906. https://doi.org/10.1101/cshperspect.a007906.

Campbell, P. D., Heim, A. E., Smith, M. Z., & Marlow, F. L. (2015). Kinesin-1 interacts with Bucky ball to form germ cells and is required to pattern the zebrafish body axis. *Development, 142*(17), 2996–3008. https://doi.org/10.1242/dev.124586.

Cao, Y., Zhao, J., Sun, Z., Zhao, Z., Postlethwait, J., & Meng, A. (2004). fgf17b, a novel member of Fgf family, helps patterning zebrafish embryos. *Developmental Biology, 271*(1), 130–143. https://doi.org/10.1016/j.ydbio.2004.03.032.

Chabanon, H., Mickleburgh, I., & Hesketh, J. (2004). Zipcodes and postage stamps: mRNA localisation signals and their trans-acting binding proteins. *Briefings in Functional Genomics & Proteomics, 3*(3), 240–256. https://doi.org/10.1093/bfgp/3.3.240.

Challa, A. K., & Chatti, K. (2013). Conservation and early expression of zebrafish tyrosine kinases support the utility of zebrafish as a model for tyrosine kinase biology. *Zebrafish, 10*(3), 264–274. https://doi.org/10.1089/zeb.2012.0781.

Chang, P., Torres, J., Lewis, R. A., Mowry, K. L., Houliston, E., & King, M. L. (2004). Localization of RNAs to the mitochondrial cloud in Xenopus oocytes through entrapment and association with endoplasmic reticulum. *Molecular Biology of the Cell, 15*(10), 4669–4681. https://doi.org/10.1091/mbc.e04-03-0265.

Chen, C., Ware, S. M., Sato, A., Houston-Hawkins, D. E., Habas, R., Matzuk, M. M., et al. (2006). The Vg1-related protein Gdf3 acts in a nodal signaling pathway in the pre-gastrulation mouse embryo. *Development, 133*(2), 319–329. https://doi.org/10.1242/dev.02210.

Chen, J., Xia, L., Bruchas, M. R., & Solnica-Krezel, L. (2017). Imaging early embryonic calcium activity with GCaMP6s transgenic zebrafish. *Developmental Biology, 430*(2), 385–396. https://doi.org/10.1016/j.ydbio.2017.03.010.

Cheng, S. K., Olale, F., Bennett, J. T., Brivanlou, A. H., & Schier, A. F. (2003). EGF-CFC proteins are essential coreceptors for the TGF-beta signals Vg1 and GDF1. *Genes & Development, 17*(1), 31–36. https://doi.org/10.1101/gad.1041203.

Cheung, C. T., Patinote, A., Guiguen, Y., & Bobe, J. (2018). foxr1 is a novel maternal-effect gene in fish that is required for early embryonic success. *PeerJ*, *6*, e5534. https://doi.org/10.7717/peerj.5534.

Chu, J., & Shen, M. M. (2010). Functional redundancy of EGF-CFC genes in epiblast and extraembryonic patterning during early mouse embryogenesis. *Developmental Biology*, *342*(1), 63–73. https://doi.org/10.1016/j.ydbio.2010.03.009.

Ciruna, B., Weidinger, G., Knaut, H., Thisse, B., Thisse, C., Raz, E., et al. (2002). Production of maternal-zygotic mutant zebrafish by germ-line replacement. *Proceedings of the National Academy of Sciences of the United States of America*, *99*(23), 14919–14924. https://doi.org/10.1073/pnas.222459999.

Collignon, J., Varlet, I., & Robertson, E. J. (1996). Relationship between asymmetric nodal expression and the direction of embryonic turning. *Nature*, *381*(6578), 155–158. https://doi.org/10.1038/381155a0.

Compagnon, J., Barone, V., Rajshekar, S., Kottmeier, R., Pranjic-Ferscha, K., Behrndt, M., et al. (2014). The notochord breaks bilateral symmetry by controlling cell shapes in the zebrafish laterality organ. *Developmental Cell*, *31*(6), 774–783. https://doi.org/10.1016/j.devcel.2014.11.003.

Conlon, F. L., Lyons, K. M., Takaesu, N., Barth, K. S., Kispert, A., Herrmann, B., et al. (1994). A primary requirement for nodal in the formation and maintenance of the primitive streak in the mouse. *Development*, *120*(7), 1919–1928.

Connors, S. A., Trout, J., Ekker, M., & Mullins, M. C. (1999). The role of tolloid/mini fin in dorsoventral pattern formation of the zebrafish embryo. *Development*, *126*(14), 3119–3130.

Dasgupta, A., & Amack, J. D. (2016). Cilia in vertebrate left-right patterning. *Philosophical Transactions of the Royal Society of London. Series B, Biological Sciences*, *371*(1710), 20150410. https://doi.org/10.1098/rstb.2015.0410.

De Robertis, E. M., & Sasai, Y. (1996). A common plan for dorsoventral patterning in bilateria. *Nature*, *380*(6569), 37–40. https://doi.org/10.1038/380037a0.

Ding, J., Yang, L., Yan, Y. T., Chen, A., Desai, N., Wynshaw-Boris, A., et al. (1998). Cripto is required for correct orientation of the anterior-posterior axis in the mouse embryo. *Nature*, *395*(6703), 702–707. https://doi.org/10.1038/27215.

Dixon Fox, M., & Bruce, A. E. (2009). Short- and long-range functions of Goosecoid in zebrafish axis formation are independent of chordin, noggin 1 and follistatin-like 1b. *Development*, *136*(10), 1675–1685. https://doi.org/10.1242/dev.031161.

Dohrmann, C. E., Kessler, D. S., & Melton, D. A. (1996). Induction of axial mesoderm by zDVR-1, the zebrafish orthologue of Xenopus Vg1. *Developmental Biology*, *175*(1), 108–117. https://doi.org/10.1006/dbio.1996.0099.

Dorey, K., & Hill, C. S. (2006). A novel Cripto-related protein reveals an essential role for EGF-CFCs in nodal signalling in Xenopus embryos. *Developmental Biology*, *292*(2), 303–316. https://doi.org/10.1016/j.ydbio.2006.01.006.

Dorsky, R. I., Itoh, M., Moon, R. T., & Chitnis, A. (2003). Two tcf3 genes cooperate to pattern the zebrafish brain. *Development*, *130*(9), 1937–1947. https://doi.org/10.1242/dev.00402.

Dorsky, R. I., Snyder, A., Cretekos, C. J., Grunwald, D. J., Geisler, R., Haffter, P., et al. (1999). Maternal and embryonic expression of zebrafish lef1. *Mechanisms of Development*, *86*(1–2), 147–150. https://doi.org/10.1016/s0925-4773(99)00101-x.

Dosch, R., Wagner, D. S., Mintzer, K. A., Runke, G., Wiemelt, A. P., & Mullins, M. C. (2004). Maternal control of vertebrate development before the midblastula transition: Mutants from the zebrafish I. *Developmental Cell*, *6*(6), 771–780. https://doi.org/10.1016/j.devcel.2004.05.002.

Dougan, S. T., Warga, R. M., Kane, D. A., Schier, A. F., & Talbot, W. S. (2003). The role of the zebrafish nodal-related genes squint and cyclops in patterning of mesendoderm. *Development*, *130*(9), 1837–1851. https://doi.org/10.1242/dev.00400.

Draper, B. W., Stock, D. W., & Kimmel, C. B. (2003). Zebrafish fgf24 functions with fgf8 to promote posterior mesodermal development. *Development*, *130*(19), 4639–4654. https://doi.org/10.1242/dev.00671.

Du, S., Draper, B. W., Mione, M., Moens, C. B., & Bruce, A. (2012). Differential regulation of epiboly initiation and progression by zebrafish Eomesodermin A. *Developmental Biology*, *362*(1), 11–23. https://doi.org/10.1016/j.ydbio.2011.10.036.

Dubrulle, J., Jordan, B. M., Akhmetova, L., Farrell, J. A., Kim, S. H., Solnica-Krezel, L., et al. (2015). Response to Nodal morphogen gradient is determined by the kinetics of target gene induction. *eLife*, *4*, e05042. https://doi.org/10.7554/eLife.05042.

Dyer, C., Blanc, E., Hanisch, A., Roehl, H., Otto, G. W., Yu, T., et al. (2014). A bi-modal function of Wnt signalling directs an FGF activity gradient to spatially regulate neuronal differentiation in the midbrain. *Development*, *141*(1), 63–72. https://doi.org/10.1242/dev.099507.

Elkouby, Y. M., Jamieson-Lucy, A., & Mullins, M. C. (2016). Oocyte polarization is coupled to the chromosomal bouquet, a conserved polarized nuclear configuration in meiosis. *PLoS Biology*, *14*(1), e1002335. https://doi.org/10.1371/journal.pbio.1002335.

Elrod, N. D., Henriques, T., Huang, K. L., Tatomer, D. C., Wilusz, J. E., Wagner, E. J., et al. (2019). The integrator complex attenuates promoter-proximal transcription at protein-coding genes. *Molecular Cell*, *76*(5), 738–752.e737. https://doi.org/10.1016/j.molcel.2019.10.034.

Escobar-Aguirre, M., Zhang, H., Jamieson-Lucy, A., & Mullins, M. C. (2017). Microtubule-actin crosslinking factor 1 (Macf1) domain function in Balbiani body dissociation and nuclear positioning. *PLoS Genetics*, *13*(9), e1006983. https://doi.org/10.1371/journal.pgen.1006983.

Ezzeddine, N., Chen, J., Waltenspiel, B., Burch, B., Albrecht, T., Zhuo, M., et al. (2011). A subset of drosophila integrator proteins is essential for efficient U7 snRNA and spliceosomal snRNA 3′-end formation. *Molecular and Cellular Biology*, *31*(2), 328–341. https://doi.org/10.1128/MCB.00943-10.

Feldman, B., Gates, M. A., Egan, E. S., Dougan, S. T., Rennebeck, G., Sirotkin, H. I., et al. (1998). Zebrafish organizer development and germ-layer formation require nodal-related signals. *Nature*, *395*(6698), 181–185. https://doi.org/10.1038/26013.

Fernandez, J., Valladares, M., Fuentes, R., & Ubilla, A. (2006). Reorganization of cytoplasm in the zebrafish oocyte and egg during early steps of ooplasmic segregation. *Developmental Dynamics*, *235*(3), 656–671. https://doi.org/10.1002/dvdy.20682.

Fodor, E., Zsigmond, A., Horvath, B., Molnar, J., Nagy, I., Toth, G., et al. (2013). Full transcriptome analysis of early dorsoventral patterning in zebrafish. *PLoS One*, *8*(7), e70053. https://doi.org/10.1371/journal.pone.0070053.

Fuentes, R., & Fernandez, J. (2010). Ooplasmic segregation in the zebrafish zygote and early embryo: Pattern of ooplasmic movements and transport pathways. *Developmental Dynamics*, *239*(8), 2172–2189. https://doi.org/10.1002/dvdy.22349.

Fuentes, R., Mullins, M. C., & Fernandez, J. (2018). Formation and dynamics of cytoplasmic domains and their genetic regulation during the zebrafish oocyte-to-embryo transition. *Mechanisms of Development*, *154*, 259–269. https://doi.org/10.1016/j.mod.2018.08.001.

Furthauer, M., Van Celst, J., Thisse, C., & Thisse, B. (2004). Fgf signalling controls the dorsoventral patterning of the zebrafish embryo. *Development*, *131*(12), 2853–2864. https://doi.org/10.1242/dev.01156.

Ge, X., Grotjahn, D., Welch, E., Lyman-Gingerich, J., Holguin, C., Dimitrova, E., et al. (2014). Hecate/Grip2a acts to reorganize the cytoskeleton in the symmetry-breaking event of embryonic axis induction. *PLoS Genetics*, *10*(6), e1004422. https://doi.org/10.1371/journal.pgen.1004422.

Gore, A. V., & Sampath, K. (2002). Localization of transcripts of the zebrafish morphogen squint is dependent on egg activation and the microtubule cytoskeleton. *Mechanisms of Development*, *112*(1–2), 153–156. https://doi.org/10.1016/s0925-4773(01)00622-0.

Goudarzi, M., Berg, K., Pieper, L. M., & Schier, A. F. (2019). Individual long non-coding RNAs have no overt functions in zebrafish embryogenesis, viability and fertility. *eLife, 8*, e40815. https://doi.org/10.7554/eLife.40815.

Goutel, C., Kishimoto, Y., Schulte-Merker, S., & Rosa, F. (2000). The ventralizing activity of radar, a maternally expressed bone morphogenetic protein, reveals complex bone morphogenetic protein interactions controlling dorso-ventral patterning in zebrafish. *Mechanisms of Development, 99*(1–2), 15–27. https://doi.org/10.1016/s0925-4773(00)00470-6.

Green, D., Whitener, A. E., Mohanty, S., & Lekven, A. C. (2015). Vertebrate nervous system posteriorization: Grading the function of Wnt signaling. *Developmental Dynamics, 244*(3), 507–512. https://doi.org/10.1002/dvdy.24230.

Griffin, K. J., & Kimelman, D. (2003). Interplay between FGF, one-eyed pinhead, and T-box transcription factors during zebrafish posterior development. *Developmental Biology, 264*(2), 456–466. https://doi.org/10.1016/j.ydbio.2003.09.008.

Grimes, D. T., & Burdine, R. D. (2017). Left-right patterning: Breaking symmetry to asymmetric morphogenesis. *Trends in Genetics, 33*(9), 616–628. https://doi.org/10.1016/j.tig.2017.06.004.

Gritsman, K., Zhang, J., Cheng, S., Heckscher, E., Talbot, W. S., & Schier, A. F. (1999). The EGF-CFC protein one-eyed pinhead is essential for nodal signaling. *Cell, 97*(1), 121–132. https://doi.org/10.1016/s0092-8674(00)80720-5.

Gu, Z., Nomura, M., Simpson, B. B., Lei, H., Feijen, A., van den Eijnden-van Raaij, J., et al. (1998). The type I activin receptor ActRIB is required for egg cylinder organization and gastrulation in the mouse. *Genes & Development, 12*(6), 844–857. https://doi.org/10.1101/gad.12.6.844.

Gupta, T., Marlow, F. L., Ferriola, D., Mackiewicz, K., Dapprich, J., Monos, D., et al. (2010). Microtubule actin crosslinking factor 1 regulates the Balbiani body and animal-vegetal polarity of the zebrafish oocyte. *PLoS Genetics, 6*(8), e1001073. https://doi.org/10.1371/journal.pgen.1001073.

Hanafusa, H., Masuyama, N., Kusakabe, M., Shibuya, H., & Nishida, E. (2000). The TGF-beta family member derriere is involved in regulation of the establishment of left-right asymmetry. *EMBO Reports, 1*(1), 32–39. https://doi.org/10.1093/embo-reports/kvd008.

Hashimoto, H., Itoh, M., Yamanaka, Y., Yamashita, S., Shimizu, T., Solnica-Krezel, L., et al. (2000). Zebrafish Dkk1 functions in forebrain specification and axial mesendoderm formation. *Developmental Biology, 217*(1), 138–152. https://doi.org/10.1006/dbio.1999.9537.

Hashimoto, H., Rebagliati, M., Ahmad, N., Muraoka, O., Kurokawa, T., Hibi, M., et al. (2004). The Cerberus/Dan-family protein Charon is a negative regulator of nodal signaling during left-right patterning in zebrafish. *Development, 131*(8), 1741–1753. https://doi.org/10.1242/dev.01070.

Hata, A., & Chen, Y. G. (2016). TGF-beta signaling from receptors to Smads. *Cold Spring Harbor Perspectives in Biology, 8*(9), a022061. https://doi.org/10.1101/cshperspect.a022061.

Heasman, J., Quarmby, J., & Wylie, C. C. (1984). The mitochondrial cloud of Xenopus oocytes: The source of germinal granule material. *Developmental Biology, 105*(2), 458–469. https://doi.org/10.1016/0012-1606(84)90303-8.

Heisenberg, C. P., Brand, M., Jiang, Y. J., Warga, R. M., Beuchle, D., van Eeden, F. J., et al. (1996). Genes involved in forebrain development in the zebrafish, Danio rerio. *Development, 123*, 191–203.

Heisenberg, C. P., Houart, C., Take-Uchi, M., Rauch, G. J., Young, N., Coutinho, P., et al. (2001). A mutation in the Gsk3-binding domain of zebrafish Masterblind/Axin1 leads to a fate transformation of telencephalon and eyes to diencephalon. *Genes & Development, 15*(11), 1427–1434. https://doi.org/10.1101/gad.194301.

Helde, K. A., & Grunwald, D. J. (1993). The DVR-1 (Vg1) transcript of zebrafish is maternally supplied and distributed throughout the embryo. *Developmental Biology, 159*(2), 418–426. https://doi.org/10.1006/dbio.1993.1252.

Hill, C. S. (2016). Transcriptional control by the SMADs. *Cold Spring Harbor Perspectives in Biology, 8*(10), a022079. https://doi.org/10.1101/cshperspect.a022079.

Hino, H., Nakanishi, A., Seki, R., Aoki, T., Yamaha, E., Kawahara, A., et al. (2018). Roles of maternal wnt8a transcripts in axis formation in zebrafish. *Developmental Biology, 434*(1), 96–107. https://doi.org/10.1016/j.ydbio.2017.11.016.

Hojo, M., Takashima, S., Kobayashi, D., Sumeragi, A., Shimada, A., Tsukahara, T., et al. (2007). Right-elevated expression of charon is regulated by fluid flow in medaka Kupffer's vesicle. *Development, Growth & Differentiation, 49*(5), 395–405. https://doi.org/10.1111/j.1440-169X.2007.00937.x.

Hoodless, P. A., Pye, M., Chazaud, C., Labbe, E., Attisano, L., Rossant, J., et al. (2001). FoxH1 (fast) functions to specify the anterior primitive streak in the mouse. *Genes & Development, 15*(10), 1257–1271. https://doi.org/10.1101/gad.881501.

Howley, C., & Ho, R. K. (2000). mRNA localization patterns in zebrafish oocytes. *Mechanisms of Development, 92*(2), 305–309.

Hyatt, B. A., Lohr, J. L., & Yost, H. J. (1996). Initiation of vertebrate left-right axis formation by maternal Vg1. *Nature, 384*(6604), 62–65. https://doi.org/10.1038/384062a0.

Iannaccone, P. M., Zhou, X., Khokha, M., Boucher, D., & Kuehn, M. R. (1992). Insertional mutation of a gene involved in growth regulation of the early mouse embryo. *Developmental Dynamics, 194*(3), 198–208. https://doi.org/10.1002/aja.1001940305.

Imai, Y., Gates, M. A., Melby, A. E., Kimelman, D., Schier, A. F., & Talbot, W. S. (2001). The homeobox genes vox and vent are redundant repressors of dorsal fates in zebrafish. *Development, 128*(12), 2407–2420.

Jamieson-Lucy, A., & Mullins, M. C. (2019). The vertebrate Balbiani body, germ plasm, and oocyte polarity. *Current Topics in Developmental Biology, 135*, 1–34. https://doi.org/10.1016/bs.ctdb.2019.04.003.

Jones, C. M., Kuehn, M. R., Hogan, B. L., Smith, J. C., & Wright, C. V. (1995). Nodal-related signals induce axial mesoderm and dorsalize mesoderm during gastrulation. *Development, 121*(11), 3651–3662.

Joore, J., Fasciana, C., Speksnijder, J. E., Kruijer, W., Destree, O. H., van den Eijnden-van Raaij, A. J., et al. (1996). Regulation of the zebrafish goosecoid promoter by mesoderm inducing factors and Xwnt1. *Mechanisms of Development, 55*(1), 3–18. https://doi.org/10.1016/0925-4773(95)00481-5.

Junker, J. P., Noel, E. S., Guryev, V., Peterson, K. A., Shah, G., Huisken, J., et al. (2014). Genome-wide RNA tomography in the zebrafish embryo. *Cell, 159*(3), 662–675. https://doi.org/10.1016/j.cell.2014.09.038.

Kapp, L. D., Abrams, E. W., Marlow, F. L., & Mullins, M. C. (2013). The integrator complex subunit 6 (Ints6) confines the dorsal organizer in vertebrate embryogenesis. *PLoS Genetics, 9*(10), e1003822. https://doi.org/10.1371/journal.pgen.1003822.

Kapsimali, M., Caneparo, L., Houart, C., & Wilson, S. W. (2004). Inhibition of Wnt/Axin/beta-catenin pathway activity promotes ventral CNS midline tissue to adopt hypothalamic rather than floorplate identity. *Development, 131*(23), 5923–5933. https://doi.org/10.1242/dev.01453.

Karakesisoglou, I., Yang, Y., & Fuchs, E. (2000). An epidermal plakin that integrates actin and microtubule networks at cellular junctions. *The Journal of Cell Biology, 149*(1), 195–208. https://doi.org/10.1083/jcb.149.1.195.

Kawahara, A., Wilm, T., Solnica-Krezel, L., & Dawid, I. B. (2000). Functional interaction of vega2 and goosecoid homeobox genes in zebrafish. *Genesis, 28*(2), 58–67. https://doi.org/10.1002/1526-968x(200010)28:2<58::aid-gene30>3.0.co;2-n.

Kelly, C., Chin, A. J., Leatherman, J. L., Kozlowski, D. J., & Weinberg, E. S. (2000). Maternally controlled (beta)-catenin-mediated signaling is required for organizer formation in the zebrafish. *Development*, *127*(18), 3899–3911.

Kelly, G. M., Greenstein, P., Erezyilmaz, D. F., & Moon, R. T. (1995). Zebrafish wnt8 and wnt8b share a common activity but are involved in distinct developmental pathways. *Development*, *121*(6), 1787–1799.

Kessler, D. S., & Melton, D. A. (1995). Induction of dorsal mesoderm by soluble, mature Vg1 protein. *Development*, *121*(7), 2155–2164.

Kim, C. H., Oda, T., Itoh, M., Jiang, D., Artinger, K. B., Chandrasekharappa, S. C., et al. (2000). Repressor activity of headless/Tcf3 is essential for vertebrate head formation. *Nature*, *407*(6806), 913–916. https://doi.org/10.1038/35038097.

Kloc, M., Bilinski, S., Chan, A. P., Allen, L. H., Zearfoss, N. R., & Etkin, L. D. (2001). RNA localization and germ cell determination in Xenopus. *International Review of Cytology*, *203*, 63–91. https://doi.org/10.1016/s0074-7696(01)03004-2.

Kloc, M., Bilinski, S., & Etkin, L. D. (2004). The Balbiani body and germ cell determinants: 150 years later. *Current Topics in Developmental Biology*, *59*, 1–36. https://doi.org/10.1016/S0070-2153(04)59001-4.

Kloc, M., & Etkin, L. D. (1995). Two distinct pathways for the localization of RNAs at the vegetal cortex in Xenopus oocytes. *Development*, *121*(2), 287–297.

Kofron, M., Puck, H., Standley, H., Wylie, C., Old, R., Whitman, M., et al. (2004). New roles for FoxH1 in patterning the early embryo. *Development*, *131*(20), 5065–5078. https://doi.org/10.1242/dev.01396.

Kosaka, K., Kawakami, K., Sakamoto, H., & Inoue, K. (2007). Spatiotemporal localization of germ plasm RNAs during zebrafish oogenesis. *Mechanisms of Development*, *124*(4), 279–289. https://doi.org/10.1016/j.mod.2007.01.003.

Kramer, C., Mayr, T., Nowak, M., Schumacher, J., Runke, G., Bauer, H., et al. (2002). Maternally supplied Smad5 is required for ventral specification in zebrafish embryos prior to zygotic bmp signaling. *Developmental Biology*, *250*(2), 263–279.

Kumari, P., Gilligan, P. C., Lim, S., Tran, L. D., Winkler, S., Philp, R., et al. (2013). An essential role for maternal control of nodal signaling. *eLife*, *2*, e00683. https://doi.org/10.7554/eLife.00683.

Kuo, C. L., Lam, C. M., Hewitt, J. E., & Scotting, P. J. (2013). Formation of the embryonic organizer is restricted by the competitive influences of Fgf signaling and the SoxB1 transcription factors. *PLoS One*, *8*(2), e57698. https://doi.org/10.1371/journal.pone.0057698.

Lai, F., Gardini, A., Zhang, A., & Shiekhattar, R. (2015). Integrator mediates the biogenesis of enhancer RNAs. *Nature*, *525*(7569), 399–403. https://doi.org/10.1038/nature14906.

Langdon, Y. G., Fuentes, R., Zhang, H., Abrams, E. W., Marlow, F. L., & Mullins, M. C. (2016). Split top: A maternal cathepsin B that regulates dorsoventral patterning and morphogenesis. *Development*, *143*(6), 1016–1028. https://doi.org/10.1242/dev.128900.

Langdon, Y. G., & Mullins, M. C. (2011). Maternal and zygotic control of zebrafish dorsoventral axial patterning. *Annual Review of Genetics*, *45*, 357–377. https://doi.org/10.1146/annurev-genet-110410-132517.

Lee, H. O., Choe, H., Seo, K., Lee, H., Lee, J., & Kim, J. (2010). Fgfbp1 is essential for the cellular survival during zebrafish embryogenesis. *Molecules and Cells*, *29*(5), 501–507. https://doi.org/10.1007/s10059-010-0062-7.

Lekven, A. C., Thorpe, C. J., Waxman, J. S., & Moon, R. T. (2001). Zebrafish wnt8 encodes two wnt8 proteins on a bicistronic transcript and is required for mesoderm and neurectoderm patterning. *Developmental Cell*, *1*(1), 103–114. https://doi.org/10.1016/s1534-5807(01)00007-7.

Levin, M., Johnson, R. L., Stern, C. D., Kuehn, M., & Tabin, C. (1995). A molecular pathway determining left-right asymmetry in chick embryogenesis. *Cell*, *82*(5), 803–814. https://doi.org/10.1016/0092-8674(95)90477-8.

Li, Y., Li, Q., Long, Y., & Cui, Z. (2011). Lzts2 regulates embryonic cell movements and dorsoventral patterning through interaction with and export of nuclear beta-catenin in zebrafish. *The Journal of Biological Chemistry*, *286*(52), 45116–45130. https://doi.org/10.1074/jbc.M111.267328.

Lim, S., Kumari, P., Gilligan, P., Quach, H. N., Mathavan, S., & Sampath, K. (2012). Dorsal activity of maternal squint is mediated by a non-coding function of the RNA. *Development*, *139*(16), 2903–2915. https://doi.org/10.1242/dev.077081.

Lin, C. M., Chen, H. J., Leung, C. L., Parry, D. A., & Liem, R. K. (2005). Microtubule actin crosslinking factor 1b: A novel plakin that localizes to the Golgi complex. *Journal of Cell Science*, *118*(Pt. 16), 3727–3738. https://doi.org/10.1242/jcs.02510.

Lippok, B., Song, S., & Driever, W. (2014). Pou5f1 protein expression and posttranslational modification during early zebrafish development. *Developmental Dynamics*, *243*(3), 468–477. https://doi.org/10.1002/dvdy.24079.

Little, S. C., & Mullins, M. C. (2009). Bone morphogenetic protein heterodimers assemble heteromeric type I receptor complexes to pattern the dorsoventral axis. *Nature Cell Biology*, *11*(5), 637–643. https://doi.org/10.1038/ncb1870.

Liu, J. X., Xu, Q. H., Yu, X., Zhang, T., Xie, X., & Ouyang, G. (2018). Eaf1 and Eaf2 mediate zebrafish dorsal-ventral axis patterning via suppressing Wnt/beta-catenin activity. *International Journal of Biological Sciences*, *14*(7), 705–716. https://doi.org/10.7150/ijbs.18997.

Li-Villarreal, N., Forbes, M. M., Loza, A. J., Chen, J., Ma, T., Helde, K., et al. (2015). Dachsous1b cadherin regulates actin and microtubule cytoskeleton during early zebrafish embryogenesis. *Development*, *142*(15), 2704–2718. https://doi.org/10.1242/dev.119800.

Long, S., Ahmad, N., & Rebagliati, M. (2003). The zebrafish nodal-related gene southpaw is required for visceral and diencephalic left-right asymmetry. *Development*, *130*(11), 2303–2316. https://doi.org/10.1242/dev.00436.

Lowe, L. A., Supp, D. M., Sampath, K., Yokoyama, T., Wright, C. V., Potter, S. S., et al. (1996). Conserved left-right asymmetry of nodal expression and alterations in murine situs inversus. *Nature*, *381*(6578), 158–161. https://doi.org/10.1038/381158a0.

Lowe, L. A., Yamada, S., & Kuehn, M. R. (2001). Genetic dissection of nodal function in patterning the mouse embryo. *Development*, *128*(10), 1831–1843.

Lu, F. I., Thisse, C., & Thisse, B. (2011). Identification and mechanism of regulation of the zebrafish dorsal determinant. *Proceedings of the National Academy of Sciences of the United States of America*, *108*(38), 15876–15880. https://doi.org/10.1073/pnas.1106801108.

Maegawa, S., Varga, M., & Weinberg, E. S. (2006). FGF signaling is required for {beta}-catenin-mediated induction of the zebrafish organizer. *Development*, *133*(16), 3265–3276. https://doi.org/10.1242/dev.02483.

Marlow, F. L., & Mullins, M. C. (2008). Bucky ball functions in Balbiani body assembly and animal-vegetal polarity in the oocyte and follicle cell layer in zebrafish. *Developmental Biology*, *321*(1), 40–50. https://doi.org/10.1016/j.ydbio.2008.05.557.

Marques, S., Borges, A. C., Silva, A. C., Freitas, S., Cordenonsi, M., & Belo, J. A. (2004). The activity of the nodal antagonist Cerl-2 in the mouse node is required for correct L/R body axis. *Genes & Development*, *18*(19), 2342–2347. https://doi.org/10.1101/gad.306504.

Mei, W., Lee, K. W., Marlow, F. L., Miller, A. L., & Mullins, M. C. (2009). hnRNP I is required to generate the Ca2+ signal that causes egg activation in zebrafish. *Development*, *136*(17), 3007–3017. https://doi.org/10.1242/dev.037879.

Melby, A. E., Beach, C., Mullins, M., & Kimelman, D. (2000). Patterning the early zebrafish by the opposing actions of bozozok and vox/vent. *Developmental Biology*, *224*(2), 275–285. https://doi.org/10.1006/dbio.2000.9780.

Melton, D. A. (1987). Translocation of a localized maternal mRNA to the vegetal pole of Xenopus oocytes. *Nature*, *328*(6125), 80–82. https://doi.org/10.1038/328080a0.

Milewski, R., Szpila, M., & Ajduk, A. (2018). Dynamics of cytoplasm and cleavage divisions correlates with preimplantation embryo development. *Reproduction*, *155*(1), 1–14. https://doi.org/10.1530/REP-17-0230.

Mintzer, K. A., Lee, M. A., Runke, G., Trout, J., Whitman, M., & Mullins, M. C. (2001). Lost-a-fin encodes a type I BMP receptor, Alk8, acting maternally and zygotically in dorsoventral pattern formation. *Development*, *128*(6), 859–869.

Mizuno, T., Yamaha, E., Kuroiwa, A., & Takeda, H. (1999). Removal of vegetal yolk causes dorsal deficencies and impairs dorsal-inducing ability of the yolk cell in zebrafish. *Mechanisms of Development*, *81*(1–2), 51–63. https://doi.org/10.1016/s0925-4773(98)00202-0.

Mo, S., Wang, L., Li, Q., Li, J., Li, Y., Thannickal, V. J., et al. (2010). Caveolin-1 regulates dorsoventral patterning through direct interaction with beta-catenin in zebrafish. *Developmental Biology*, *344*(1), 210–223. https://doi.org/10.1016/j.ydbio.2010.04.033.

Montague, T. G., Gagnon, J. A., & Schier, A. F. (2018). Conserved regulation of nodal-mediated left-right patterning in zebrafish and mouse. *Development*, *145*(24), dev171090. https://doi.org/10.1242/dev.171090.

Montague, T. G., & Schier, A. F. (2017). Vg1-nodal heterodimers are the endogenous inducers of mesendoderm. *eLife*, *6*, e28183. https://doi.org/10.7554/eLife.28183.

Monteiro, R., van Dinther, M., Bakkers, J., Wilkinson, R., Patient, R., ten Dijke, P., et al. (2008). Two novel type II receptors mediate BMP signalling and are required to establish left-right asymmetry in zebrafish. *Developmental Biology*, *315*(1), 55–71. https://doi.org/10.1016/j.ydbio.2007.11.038.

Mullins, M. C., Hammerschmidt, M., Kane, D. A., Odenthal, J., Brand, M., van Eeden, F. J., et al. (1996). Genes establishing dorsoventral pattern formation in the zebrafish embryo: The ventral specifying genes. *Development*, *123*, 81–93.

Muraoka, O., Shimizu, T., Yabe, T., Nojima, H., Bae, Y. K., Hashimoto, H., et al. (2006). Sizzled controls dorso-ventral polarity by repressing cleavage of the chordin protein. *Nature Cell Biology*, *8*(4), 329–338. https://doi.org/10.1038/ncb1379.

Nakamura, T., Saito, D., Kawasumi, A., Shinohara, K., Asai, Y., Takaoka, K., et al. (2012). Fluid flow and interlinked feedback loops establish left-right asymmetric decay of Cerl2 mRNA. *Nature Communications*, *3*, 1322. https://doi.org/10.1038/ncomms2319.

Nelson, A. C., Cutty, S. J., Niini, M., Stemple, D. L., Flicek, P., Houart, C., et al. (2014). Global identification of Smad2 and Eomesodermin targets in zebrafish identifies a conserved transcriptional network in mesendoderm and a novel role for Eomesodermin in repression of ectodermal gene expression. *BMC Biology*, *12*, 81. https://doi.org/10.1186/s12915-014-0081-5.

Nikaido, M., Tada, M., & Ueno, N. (1999). Restricted expression of the receptor serine/threonine kinase BMPR-IB in zebrafish. *Mechanisms of Development*, *82*(1–2), 219–222. https://doi.org/10.1016/s0925-4773(99)00023-4.

Nojima, H., Rothhamel, S., Shimizu, T., Kim, C. H., Yonemura, S., Marlow, F. L., et al. (2010). Syntabulin, a motor protein linker, controls dorsal determination. *Development*, *137*(6), 923–933. https://doi.org/10.1242/dev.046425.

Nojima, H., Shimizu, T., Kim, C. H., Yabe, T., Bae, Y. K., Muraoka, O., et al. (2004). Genetic evidence for involvement of maternally derived Wnt canonical signaling in dorsal determination in zebrafish. *Mechanisms of Development*, *121*(4), 371–386. https://doi.org/10.1016/j.mod.2004.02.003.

Norris, D. P., Brennan, J., Bikoff, E. K., & Robertson, E. J. (2002). The Foxh1-dependent autoregulatory enhancer controls the level of nodal signals in the mouse embryo. *Development*, *129*(14), 3455–3468.
Nusse, R., & Clevers, H. (2017). Wnt/beta-catenin signaling, disease, and emerging therapeutic modalities. *Cell*, *169*(6), 985–999. https://doi.org/10.1016/j.cell.2017.05.016.
Okabe, N., Xu, B., & Burdine, R. D. (2008). Fluid dynamics in zebrafish Kupffer's vesicle. *Developmental Dynamics*, *237*(12), 3602–3612. https://doi.org/10.1002/dvdy.21730.
Onichtchouk, D., Geier, F., Polok, B., Messerschmidt, D. M., Mossner, R., Wendik, B., et al. (2010). Zebrafish Pou5f1-dependent transcriptional networks in temporal control of early development. *Molecular Systems Biology*, *6*, 354. https://doi.org/10.1038/msb.2010.9.
Ornitz, D. M., & Itoh, N. (2015). The fibroblast growth factor signaling pathway. *Wiley Interdisciplinary Reviews: Developmental Biology*, *4*(3), 215–266. https://doi.org/10.1002/wdev.176.
Ota, S., Tonou-Fujimori, N., & Yamasu, K. (2009). The roles of the FGF signal in zebrafish embryos analyzed using constitutive activation and dominant-negative suppression of different FGF receptors. *Mechanisms of Development*, *126*(1–2), 1–17. https://doi.org/10.1016/j.mod.2008.10.008.
Oteiza, P., Koppen, M., Concha, M. L., & Heisenberg, C. P. (2008). Origin and shaping of the laterality organ in zebrafish. *Development*, *135*(16), 2807–2813. https://doi.org/10.1242/dev.022228.
Paridaen, J. T., Danesin, C., Elas, A. T., van de Water, S., Houart, C., & Zivkovic, D. (2009). Apc1 is required for maintenance of local brain organizers and dorsal midbrain survival. *Developmental Biology*, *331*(2), 101–112. https://doi.org/10.1016/j.ydbio.2009.04.022.
Pelliccia, J. L., Jindal, G. A., & Burdine, R. D. (2017). Gdf3 is required for robust nodal signaling during germ layer formation and left-right patterning. *eLife*, *6*, e28635. https://doi.org/10.7554/eLife.28635.
Peng, G., & Westerfield, M. (2006). Lhx5 promotes forebrain development and activates transcription of secreted Wnt antagonists. *Development*, *133*(16), 3191–3200. https://doi.org/10.1242/dev.02485.
Peterson, A. G., Wang, X., & Yost, H. J. (2013). Dvr1 transfers left-right asymmetric signals from Kupffer's vesicle to lateral plate mesoderm in zebrafish. *Developmental Biology*, *382*(1), 198–208. https://doi.org/10.1016/j.ydbio.2013.06.011.
Pezeron, G., Anselme, I., Laplante, M., Ellingsen, S., Becker, T. S., Rosa, F. M., et al. (2006). Duplicate sfrp1 genes in zebrafish: sfrp1a is dynamically expressed in the developing central nervous system, gut and lateral line. *Gene Expression Patterns*, *6*(8), 835–842. https://doi.org/10.1016/j.modgep.2006.02.002.
Pomreinke, A. P., Soh, G. H., Rogers, K. W., Bergmann, J. K., Blassle, A. J., & Muller, P. (2017). Dynamics of BMP signaling and distribution during zebrafish dorsal-ventral patterning. *eLife*, *6*, e25861. https://doi.org/10.7554/eLife.25861.
Ramel, M. C., & Hill, C. S. (2013). The ventral to dorsal BMP activity gradient in the early zebrafish embryo is determined by graded expression of BMP ligands. *Developmental Biology*, *378*(2), 170–182. https://doi.org/10.1016/j.ydbio.2013.03.003.
Ramel, M. C., & Lekven, A. C. (2004). Repression of the vertebrate organizer by Wnt8 is mediated by vent and Vox. *Development*, *131*(16), 3991–4000. https://doi.org/10.1242/dev.01277.
Rankin, C. T., Bunton, T., Lawler, A. M., & Lee, S. J. (2000). Regulation of left-right patterning in mice by growth/differentiation factor-1. *Nature Genetics*, *24*(3), 262–265. https://doi.org/10.1038/73472.

Reifers, F., Bohli, H., Walsh, E. C., Crossley, P. H., Stainier, D. Y., & Brand, M. (1998). Fgf8 is mutated in zebrafish acerebellar (ace) mutants and is required for maintenance of midbrain-hindbrain boundary development and somitogenesis. *Development, 125*(13), 2381–2395.

Reim, G., & Brand, M. (2006). Maternal control of vertebrate dorsoventral axis formation and epiboly by the POU domain protein Spg/Pou2/Oct4. *Development, 133*(14), 2757–2770. https://doi.org/10.1242/dev.02391.

Rohner, N., Bercsenyi, M., Orban, L., Kolanczyk, M. E., Linke, D., Brand, M., et al. (2009). Duplication of fgfr1 permits Fgf signaling to serve as a target for selection during domestication. *Current Biology, 19*(19), 1642–1647. https://doi.org/10.1016/j.cub.2009.07.065.

Rong, X., Chen, C., Zhou, P., Zhou, Y., Li, Y., Lu, L., et al. (2014). R-spondin 3 regulates dorsoventral and anteroposterior patterning by antagonizing Wnt/beta-catenin signaling in zebrafish embryos. *PLoS One, 9*(6), e99514. https://doi.org/10.1371/journal.pone.0099514.

Russ, A. P., Wattler, S., Colledge, W. H., Aparicio, S. A., Carlton, M. B., Pearce, J. J., et al. (2000). Eomesodermin is required for mouse trophoblast development and mesoderm formation. *Nature, 404*(6773), 95–99. https://doi.org/10.1038/35003601.

Ryan, K., Garrett, N., Mitchell, A., & Gurdon, J. B. (1996). Eomesodermin, a key early gene in Xenopus mesoderm differentiation. *Cell, 87*(6), 989–1000. https://doi.org/10.1016/s0092-8674(00)81794-8.

Sampaio, P., Ferreira, R. R., Guerrero, A., Pintado, P., Tavares, B., Amaro, J., et al. (2014). Left-right organizer flow dynamics: How much cilia activity reliably yields laterality? *Developmental Cell, 29*(6), 716–728. https://doi.org/10.1016/j.devcel.2014.04.030.

Sampath, K., Cheng, A. M., Frisch, A., & Wright, C. V. (1997). Functional differences among Xenopus nodal-related genes in left-right axis determination. *Development, 124*(17), 3293–3302.

Satija, R., Farrell, J. A., Gennert, D., Schier, A. F., & Regev, A. (2015). Spatial reconstruction of single-cell gene expression data. *Nature Biotechnology, 33*(5), 495–502. https://doi.org/10.1038/nbt.3192.

Schier, A. F., & Talbot, W. S. (2005). Molecular genetics of axis formation in zebrafish. *Annual Review of Genetics, 39*, 561–613. https://doi.org/10.1146/annurev.genet.37.110801.143752.

Schmidt, D., Reuter, H., Huttner, K., Ruhe, L., Rabert, F., Seebeck, F., et al. (2018). The integrator complex regulates differential snRNA processing and fate of adult stem cells in the highly regenerative planarian Schmidtea mediterranea. *PLoS Genetics, 14*(12), e1007828. https://doi.org/10.1371/journal.pgen.1007828.

Schneider, P. N., Slusarski, D. C., & Houston, D. W. (2012). Differential role of axin RGS domain function in Wnt signaling during anteroposterior patterning and maternal axis formation. *PLoS One, 7*(9), e44096. https://doi.org/10.1371/journal.pone.0044096.

Schneider, S., Steinbeisser, H., Warga, R. M., & Hausen, P. (1996). Beta-catenin translocation into nuclei demarcates the dorsalizing centers in frog and fish embryos. *Mechanisms of Development, 57*(2), 191–198. https://doi.org/10.1016/0925-4773(96)00546-1.

Scholpp, S., & Brand, M. (2004). Endocytosis controls spreading and effective signaling range of Fgf8 protein. *Current Biology, 14*(20), 1834–1841. https://doi.org/10.1016/j.cub.2004.09.084.

Schulte-Merker, S., Lee, K. J., McMahon, A. P., & Hammerschmidt, M. (1997). The zebrafish organizer requires chordino. *Nature, 387*(6636), 862–863. https://doi.org/10.1038/43092.

Schweickert, A., Vick, P., Getwan, M., Weber, T., Schneider, I., Eberhardt, M., et al. (2010). The nodal inhibitor coco is a critical target of leftward flow in Xenopus. *Current Biology, 20*(8), 738–743. https://doi.org/10.1016/j.cub.2010.02.061.

Seiliez, I., Thisse, B., & Thisse, C. (2006). FoxA3 and goosecoid promote anterior neural fate through inhibition of Wnt8a activity before the onset of gastrulation. *Developmental Biology*, *290*(1), 152–163. https://doi.org/10.1016/j.ydbio.2005.11.021.

Shamipour, S., Kardos, R., Xue, S. L., Hof, B., Hannezo, E., & Heisenberg, C. P. (2019). Bulk actin dynamics drive phase segregation in zebrafish oocytes. *Cell*, *177*(6), 1463–1479. e1418. https://doi.org/10.1016/j.cell.2019.04.030.

Shao, M., Lin, Y., Liu, Z., Zhang, Y., Wang, L., Liu, C., et al. (2012). GSK-3 activity is critical for the orientation of the cortical microtubules and the dorsoventral axis determination in zebrafish embryos. *PLoS One*, 7(5), e36655. https://doi.org/10.1371/journal.pone.0036655.

Sharma, D., & Kinsey, W. H. (2008). Regionalized calcium signaling in zebrafish fertilization. *The International Journal of Developmental Biology*, *52*(5–6), 561–570. https://doi.org/10.1387/ijdb.072523ds.

Shimizu, T., Bae, Y. K., Muraoka, O., & Hibi, M. (2005). Interaction of Wnt and caudal-related genes in zebrafish posterior body formation. *Developmental Biology*, *279*(1), 125–141. https://doi.org/10.1016/j.ydbio.2004.12.007.

Shimizu, T., Yamanaka, Y., Nojima, H., Yabe, T., Hibi, M., & Hirano, T. (2002). A novel repressor-type homeobox gene, ved, is involved in dharma/bozozok-mediated dorsal organizer formation in zebrafish. *Mechanisms of Development*, *118*(1–2), 125–138.

Shimizu, T., Yamanaka, Y., Ryu, S. L., Hashimoto, H., Yabe, T., Hirata, T., et al. (2000). Cooperative roles of Bozozok/dharma and nodal-related proteins in the formation of the dorsal organizer in zebrafish. *Mechanisms of Development*, *91*(1–2), 293–303. https://doi.org/10.1016/s0925-4773(99)00319-6.

Shinya, M., Eschbach, C., Clark, M., Lehrach, H., & Furutani-Seiki, M. (2000). Zebrafish Dkk1, induced by the pre-MBT Wnt signaling, is secreted from the prechordal plate and patterns the anterior neural plate. *Mechanisms of Development*, *98*(1–2), 3–17. https://doi.org/10.1016/s0925-4773(00)00433-0.

Shiratori, H., & Hamada, H. (2014). TGFbeta signaling in establishing left-right asymmetry. *Seminars in Cell & Developmental Biology*, *32*, 80–84. https://doi.org/10.1016/j.semcdb.2014.03.029.

Sidi, S., Goutel, C., Peyrieras, N., & Rosa, F. M. (2003). Maternal induction of ventral fate by zebrafish radar. *Proceedings of the National Academy of Sciences of the United States of America*, *100*(6), 3315–3320. https://doi.org/10.1073/pnas.0530115100.

Slagle, C. E., Aoki, T., & Burdine, R. D. (2011). Nodal-dependent mesendoderm specification requires the combinatorial activities of FoxH1 and eomesodermin. *PLoS Genetics*, 7(5), e1002072. https://doi.org/10.1371/journal.pgen.1002072.

Smith, K. A., Noel, E., Thurlings, I., Rehmann, H., Chocron, S., & Bakkers, J. (2011). Bmp and nodal independently regulate lefty1 expression to maintain unilateral nodal activity during left-right axis specification in zebrafish. *PLoS Genetics*, 7(9), e1002289. https://doi.org/10.1371/journal.pgen.1002289.

Song, J., Oh, S. P., Schrewe, H., Nomura, M., Lei, H., Okano, M., et al. (1999). The type II activin receptors are essential for egg cylinder growth, gastrulation, and rostral head development in mice. *Developmental Biology*, *213*(1), 157–169. https://doi.org/10.1006/dbio.1999.9370.

Songhet, P., Adzic, D., Reibe, S., & Rohr, K. B. (2007). fgf1 is required for normal differentiation of erythrocytes in zebrafish primitive hematopoiesis. *Developmental Dynamics*, *236*(3), 633–643. https://doi.org/10.1002/dvdy.21056.

Speer, K. F., Sommer, A., Tajer, B., Mullins, M. C., Klein, P. S., & Lemmon, M. A. (2019). Non-acylated Wnts can promote signaling. *Cell Reports*, *26*(4), 875–883. e875. https://doi.org/10.1016/j.celrep.2018.12.104.

Stachel, S. E., Grunwald, D. J., & Myers, P. Z. (1993). Lithium perturbation and goosecoid expression identify a dorsal specification pathway in the pregastrula zebrafish. *Development*, *117*(4), 1261–1274.

Sun, J., Yan, L., Shen, W., & Meng, A. (2018). Maternal Ybx1 safeguards zebrafish oocyte maturation and maternal-to-zygotic transition by repressing global translation. *Development*, *145*(19), dev166587. https://doi.org/10.1242/dev.166587.

Takahashi, K., Ishii, K., & Yamashita, M. (2018). Staufen1, Kinesin1 and microtubule function in cyclin B1 mRNA transport to the animal polar cytoplasm of zebrafish oocytes. *Biochemical and Biophysical Research Communications*, *503*(4), 2778–2783. https://doi.org/10.1016/j.bbrc.2018.08.039.

Tanaka, S., Hosokawa, H., Weinberg, E. S., & Maegawa, S. (2017). Chordin and dickkopf-1b are essential for the formation of head structures through activation of the FGF signaling pathway in zebrafish. *Developmental Biology*, *424*(2), 189–197. https://doi.org/10.1016/j.ydbio.2017.02.018.

Tatomer, D. C., Elrod, N. D., Liang, D., Xiao, M. S., Jiang, J. Z., Jonathan, M., et al. (2019). The integrator complex cleaves nascent mRNAs to attenuate transcription. *Genes & Development*, *33*(21 – 22), 1525–1538. https://doi.org/10.1101/gad.330167.119.

Tendeng, C., & Houart, C. (2006). Cloning and embryonic expression of five distinct sfrp genes in the zebrafish Danio rerio. *Gene Expression Patterns*, *6*(8), 761–771. https://doi.org/10.1016/j.modgep.2006.01.006.

Thisse, B., & Thisse, C. (2004). *Fast release clones: a high throughput expression analysis*. ZFIN Direct Data Submission.

Thisse, B., & Thisse, C. (2015). Formation of the vertebrate embryo: Moving beyond the Spemann organizer. *Seminars in Cell & Developmental Biology*, *42*, 94–102. https://doi.org/10.1016/j.semcdb.2015.05.007.

Thomsen, G. H., & Melton, D. A. (1993). Processed Vg1 protein is an axial mesoderm inducer in Xenopus. *Cell*, *74*(3), 433–441. https://doi.org/10.1016/0092-8674(93)80045-g.

Toretsky, J. A., & Wright, P. E. (2014). Assemblages: Functional units formed by cellular phase separation. *The Journal of Cell Biology*, *206*(5), 579–588. https://doi.org/10.1083/jcb.201404124.

Tran, L. D., Hino, H., Quach, H., Lim, S., Shindo, A., Mimori-Kiyosue, Y., et al. (2012). Dynamic microtubules at the vegetal cortex predict the embryonic axis in zebrafish. *Development*, *139*(19), 3644–3652. https://doi.org/10.1242/dev.082362.

Tsang, M., Friesel, R., Kudoh, T., & Dawid, I. B. (2002). Identification of Sef, a novel modulator of FGF signalling. *Nature Cell Biology*, *4*(2), 165–169. https://doi.org/10.1038/ncb749.

Tuazon, F. B., & Mullins, M. C. (2015). Temporally coordinated signals progressively pattern the anteroposterior and dorsoventral body axes. *Seminars in Cell & Developmental Biology*, *42*, 118–133. https://doi.org/10.1016/j.semcdb.2015.06.003.

Valenti, F., Ibetti, J., Komiya, Y., Baxter, M., Lucchese, A. M., Derstine, L., et al. (2015). The increase in maternal expression of axin1 and axin2 contribute to the zebrafish mutant ichabod ventralized phenotype. *Journal of Cellular Biochemistry*, *116*(3), 418–430. https://doi.org/10.1002/jcb.24993.

van de Water, S., van de Wetering, M., Joore, J., Esseling, J., Bink, R., Clevers, H., et al. (2001). Ectopic Wnt signal determines the eyeless phenotype of zebrafish masterblind mutant. *Development*, *128*(20), 3877–3888.

van Eeden, F. J., Granato, M., Schach, U., Brand, M., Furutani-Seiki, M., Haffter, P., et al. (1996). Genetic analysis of fin formation in the zebrafish, Danio rerio. *Development*, *123*, 255–262.

Varga, M., Maegawa, S., Bellipanni, G., & Weinberg, E. S. (2007). Chordin expression, mediated by nodal and FGF signaling, is restricted by redundant function of two beta-catenins in the zebrafish embryo. *Mechanisms of Development*, *124*(9–10), 775–791. https://doi.org/10.1016/j.mod.2007.05.005.

Veien, E. S., Grierson, M. J., Saund, R. S., & Dorsky, R. I. (2005). Expression pattern of zebrafish tcf7 suggests unexplored domains of Wnt/beta-catenin activity. *Developmental Dynamics*, *233*(1), 233–239. https://doi.org/10.1002/dvdy.20330.

Wagner, D. E., Weinreb, C., Collins, Z. M., Briggs, J. A., Megason, S. G., & Klein, A. M. (2018). Single-cell mapping of gene expression landscapes and lineage in the zebrafish embryo. *Science*, *360*(6392), 981–987. https://doi.org/10.1126/science.aar4362.

Wall, N. A., Craig, E. J., Labosky, P. A., & Kessler, D. S. (2000). Mesendoderm induction and reversal of left-right pattern by mouse Gdf1, a Vg1-related gene. *Developmental Biology*, *227*(2), 495–509. https://doi.org/10.1006/dbio.2000.9926.

Wang, G., Cadwallader, A. B., Jang, D. S., Tsang, M., Yost, H. J., & Amack, J. D. (2011). The rho kinase Rock2b establishes anteroposterior asymmetry of the ciliated Kupffer's vesicle in zebrafish. *Development*, *138*(1), 45–54. https://doi.org/10.1242/dev.052985.

Wang, G., Manning, M. L., & Amack, J. D. (2012). Regional cell shape changes control form and function of Kupffer's vesicle in the zebrafish embryo. *Developmental Biology*, *370*(1), 52–62. https://doi.org/10.1016/j.ydbio.2012.07.019.

Webb, S. E., Chan, C. M., & Miller, A. L. (2013). Introduction of aequorin into zebrafish embryos for recording Ca(2+) signaling during the first 48 h of development. *Cold Spring Harbor Protocols*, *2013*(5), 383–386. https://doi.org/10.1101/pdb.top066316.

Webb, S. E., Fluck, R. A., & Miller, A. L. (2011). Calcium signaling during the early development of medaka and zebrafish. *Biochimie*, *93*(12), 2112–2125. https://doi.org/10.1016/j.biochi.2011.06.011.

Webb, S. E., & Miller, A. L. (2006). Ca2+ signaling and early embryonic patterning during the blastula and gastrula periods of zebrafish and Xenopus development. *Biochimica et Biophysica Acta*, *1763*(11), 1192–1208. https://doi.org/10.1016/j.bbamcr.2006.08.004.

Weeks, D. L., & Melton, D. A. (1987). A maternal mRNA localized to the vegetal hemisphere in Xenopus eggs codes for a growth factor related to TGF-beta. *Cell*, *51*(5), 861–867. https://doi.org/10.1016/0092-8674(87)90109-7.

Wei, C. Y., Wang, H. P., Zhu, Z. Y., & Sun, Y. H. (2014). Transcriptional factors smad1 and smad9 act redundantly to mediate zebrafish ventral specification downstream of smad5. *The Journal of Biological Chemistry*, *289*(10), 6604–6618. https://doi.org/10.1074/jbc.M114.549758.

Welch, E., & Pelegri, F. (2014). Cortical depth and differential transport of vegetally localized dorsal and germ line determinants in the zebrafish embryo. *BioArchitecture*, *5*(1–2), 13–26. https://doi.org/10.1080/19490992.2015.1080891.

Whitaker, M. (2006). Calcium at fertilization and in early development. *Physiological Reviews*, *86*(1), 25–88. https://doi.org/10.1152/physrev.00023.2005.

White, R. J., Collins, J. E., Sealy, I. M., Wali, N., Dooley, C. M., Digby, Z., et al. (2017). A high-resolution mRNA expression time course of embryonic development in zebrafish. *eLife*, *6*, e30860. https://doi.org/10.7554/eLife.30860.

Wilk, K., Bilinski, S., Dougherty, M. T., & Kloc, M. (2005). Delivery of germinal granules and localized RNAs via the messenger transport organizer pathway to the vegetal cortex of Xenopus oocytes occurs through directional expansion of the mitochondrial cloud. *The International Journal of Developmental Biology*, *49*(1), 17–21. https://doi.org/10.1387/ijdb.041906kw.

Wilm, T. P., & Solnica-Krezel, L. (2003). Radar breaks the fog: Insights into dorsoventral patterning in zebrafish. *Proceedings of the National Academy of Sciences of the United States of America*, *100*(8), 4363–4365. https://doi.org/10.1073/pnas.0931010100.

Woods, I. G., Wilson, C., Friedlander, B., Chang, P., Reyes, D. K., Nix, R., et al. (2005). The zebrafish gene map defines ancestral vertebrate chromosomes. *Genome Research*, *15*(9), 1307–1314. https://doi.org/10.1101/gr.4134305.

Wu, S. Y., Shin, J., Sepich, D. S., & Solnica-Krezel, L. (2012). Chemokine GPCR signaling inhibits beta-catenin during zebrafish axis formation. *PLoS Biology*, *10*(10), e1001403. https://doi.org/10.1371/journal.pbio.1001403.

Xie, X. W., Liu, J. X., Hu, B., & Xiao, W. (2011). Zebrafish foxo3b negatively regulates canonical Wnt signaling to affect early embryogenesis. *PLoS One*, *6*(9), e24469. https://doi.org/10.1371/journal.pone.0024469.

Xing, Y. Y., Cheng, X. N., Li, Y. L., Zhang, C., Saquet, A., Liu, Y. Y., et al. (2018). Mutational analysis of dishevelled genes in zebrafish reveals distinct functions in embryonic patterning and gastrulation cell movements. *PLoS Genetics*, *14*(8), e1007551. https://doi.org/10.1371/journal.pgen.1007551.

Xiong, B., Rui, Y., Zhang, M., Shi, K., Jia, S., Tian, T., et al. (2006). Tob1 controls dorsal development of zebrafish embryos by antagonizing maternal beta-catenin transcriptional activity. *Developmental Cell*, *11*(2), 225–238. https://doi.org/10.1016/j.devcel.2006.06.012.

Xu, P., Zhu, G., Wang, Y., Sun, J., Liu, X., Chen, Y. G., et al. (2014). Maternal eomesodermin regulates zygotic nodal gene expression for mesendoderm induction in zebrafish embryos. *Journal of Molecular Cell Biology*, *6*(4), 272–285. https://doi.org/10.1093/jmcb/mju028.

Yamaguchi, T. P. (2001). Heads or tails: Wnts and anterior-posterior patterning. *Current Biology*, *11*(17), R713–R724. https://doi.org/10.1016/s0960-9822(01)00417-1.

Yamanaka, Y., Mizuno, T., Sasai, Y., Kishi, M., Takeda, H., Kim, C. H., et al. (1998). A novel homeobox gene, dharma, can induce the organizer in a non-cell-autonomous manner. *Genes & Development*, *12*(15), 2345–2353. https://doi.org/10.1101/gad.12.15.2345.

Yamauchi, H., Miyakawa, N., Miyake, A., & Itoh, N. (2009). Fgf4 is required for left-right patterning of visceral organs in zebrafish. *Developmental Biology*, *332*(1), 177–185. https://doi.org/10.1016/j.ydbio.2009.05.568.

Yan, L., Chen, J., Zhu, X., Sun, J., Wu, X., Shen, W., et al. (2018). Maternal Huluwa dictates the embryonic body axis through beta-catenin in vertebrates. *Science*, *362*(6417). eaat1045https://doi.org/10.1126/science.aat1045.

Yan, C., Zheng, W., & Gong, Z. (2015). Zebrafish fgf10b has a complementary function to fgf10a in liver and pancreas development. *Marine Biotechnology (New York, N.Y.)*, *17*(2), 162–167. https://doi.org/10.1007/s10126-014-9604-x.

Yao, S., Qian, M., Deng, S., Xie, L., Yang, H., Xiao, C., et al. (2010). Kzp controls canonical Wnt8 signaling to modulate dorsoventral patterning during zebrafish gastrulation. *The Journal of Biological Chemistry*, *285*(53), 42086–42096. https://doi.org/10.1074/jbc.M110.161554.

Yeo, C., & Whitman, M. (2001). Nodal signals to Smads through Cripto-dependent and Cripto-independent mechanisms. *Molecular Cell*, *7*(5), 949–957. https://doi.org/10.1016/s1097-2765(01)00249-0.

Yilmaz, O., Patinote, A., Nguyen, T. V., Com, E., Lavigne, R., Pineau, C., et al. (2017). Scrambled eggs: Proteomic portraits and novel biomarkers of egg quality in zebrafish (Danio rerio). *PLoS One*, *12*(11), e0188084. https://doi.org/10.1371/journal.pone.0188084.

Yoon, Y. J., & Mowry, K. L. (2004). Xenopus Staufen is a component of a ribonucleoprotein complex containing Vg1 RNA and kinesin. *Development*, *131*(13), 3035–3045. https://doi.org/10.1242/dev.01170.

Yu, S. R., Burkhardt, M., Nowak, M., Ries, J., Petrasek, Z., Scholpp, S., et al. (2009). Fgf8 morphogen gradient forms by a source-sink mechanism with freely diffusing molecules. *Nature*, *461*(7263), 533–536. https://doi.org/10.1038/nature08391.

Zayed, Y., Qi, X., & Peng, C. (2019). Identification of novel MicroRNAs and characterization of MicroRNA expression profiles in zebrafish ovarian follicular cells. *Frontiers in Endocrinology*, *10*, 518. https://doi.org/10.3389/fendo.2019.00518.

Zhou, Y., & King, M. L. (1996). Localization of Xcat-2 RNA, a putative germ plasm component, to the mitochondrial cloud in Xenopus stage I oocytes. *Development, 122*(9), 2947–2953.

Zhu, B., Pardeshi, L., Chen, Y., & Ge, W. (2018). Transcriptomic analysis for differentially expressed genes in ovarian follicle activation in the zebrafish. *Frontiers in Endocrinology, 9*, 593. https://doi.org/10.3389/fendo.2018.00593.

Zinski, J., Bu, Y., Wang, X., Dou, W., Umulis, D., & Mullins, M. C. (2017). Systems biology derived source-sink mechanism of BMP gradient formation. *eLife, 6*, e22199. https://doi.org/10.7554/eLife.22199.

Zinski, J., Tajer, B., & Mullins, M. C. (2018). TGF-beta family signaling in early vertebrate development. *Cold Spring Harbor Perspectives in Biology, 10*(6), a033274. https://doi.org/10.1101/cshperspect.a033274.

CHAPTER THIRTEEN

Maternal contributions to gastrulation in zebrafish

Lilianna Solnica-Krezel*
Department of Developmental Biology and Center of Regenerative Medicine, Washington University School of Medicine, St. Louis, MO, United States
*Corresponding author: e-mail address: solnical@wustl.edu

Contents

1. Introduction 392
   1.1 Crash course on gastrulation movements: Internalization 393
   1.2 Epiboly 394
   1.3 Convergence and extension 396
2. Maternal-to-zygotic transition and its timing relative to gastrulation: A genomic perspective 397
   2.1 Maternal regulation 398
   2.2 Zygotic genome activation (ZGA) 399
   2.3 Midblastula transition (MBT) 400
   2.4 Downregulation of maternal transcripts 400
3. Requirement for maternal and zygotic gene products for gastrulation: A genetic perspective 401
   3.1 Strict maternal (SM) genes 403
   3.2 Maternal zygotic (MZ) genes 403
   3.3 Strict maternal zygotic (SMZ) genes 409
   3.4 Zygotic (Z) genes 410
4. Instructive and permissive contributors to gastrulation from the maternal and zygotic genomes: An embryo perspective 411
   4.1 Cell proliferation 411
   4.2 Cell motility 413
   4.3 Axis specification 413
   4.4 Epiboly 414
   4.5 Mesendoderm internalization 415
   4.6 Convergence and extension 417
5. Perspective 418
Acknowledgments 419
References 419

Current Topics in Developmental Biology, Volume 140
ISSN 0070-2153
https://doi.org/10.1016/bs.ctdb.2020.05.001

Abstract

Gastrulation is a critical early morphogenetic process of animal development, during which the three germ layers; mesoderm, endoderm and ectoderm, are rearranged by internalization movements. Concurrent epiboly movements spread and thin the germ layers while convergence and extension movements shape them into an anteroposteriorly elongated body with head, trunk, tail and organ rudiments. In zebrafish, gastrulation follows the proliferative and inductive events that establish the embryonic and extraembryonic tissues and the embryonic axis. Specification of these tissues and embryonic axes are controlled by the maternal gene products deposited in the egg. These early maternally controlled processes need to generate sufficient cell numbers and establish the embryonic polarity to ensure normal gastrulation. Subsequently, after activation of the zygotic genome, the zygotic gene products govern mesoderm and endoderm induction and germ layer patterning. Gastrulation is initiated during the maternal-to-zygotic transition, a process that entails both activation of the zygotic genome and downregulation of the maternal transcripts. Genomic studies indicate that gastrulation is largely controlled by the zygotic genome. Nonetheless, genetic studies that investigate the relative contributions of maternal and zygotic gene function by comparing zygotic, maternal and maternal zygotic mutant phenotypes, reveal significant contribution of maternal gene products, transcripts and/or proteins, that persist through gastrulation, to the control of gastrulation movements. Therefore, in zebrafish, the maternally expressed gene products not only set the stage for, but they also actively participate in gastrulation morphogenesis.

Abbreviations

BMP bone morphogenetic protein
EVL enveloping layer
M maternal
MBT midblastula transition
MZ maternal zygotic
MZT maternal-to-zygotic transition
PCP planar cell polarity
YSL yolk syncytial layer
ZGA zygotic genome activation

1. Introduction

Animal embryogenesis initiates with a single-celled zygote, which unites one copy each of the maternal and paternal genomes and is also endowed with a large reservoir of maternally provided transcripts and proteins. Utilizing these resources, the zygote embarks on an intricate sequence of cell proliferation, fate specification and morphogenesis to establish a species-specific body plan. Early inductive and morphogenetic processes set aside the extraembryonic lineages and embryonic compartment of pluripotent stem cells, which can take diverse forms, including a mass of mesenchymal cells, in fish and frogs, or a single-cell thick epithelial sheet in *Drosophila* or amniotes. The subsequent signaling interactions between extraembryonic and embryonic tissues and/or within the embryonic tissues, specify the anteroposterior (AP) and dorsoventral (DV) axes, and the germ layers: mesoderm, endoderm, and ectoderm. Gastrulation is the next critical morphogenetic process during which the three germ layers are rearranged and shaped to form a body plan featuring head, trunk and tail with organ rudiments.

The focus of this chapter in the book "*Maternal effect genes in development*" is on the role of the maternally expressed genes in the process of gastrulation. This problem will be considered primarily from the zebrafish perspective, but comparisons will also be made to other species. In all the animals studied, gastrulation is initiated after the zygotic genome is activated. However, the relationship between the process of gastrulation and the maternal to zygotic transition, i.e., the process whereby the zygotic genome is activated, and the maternal gene products are cleared, is quite varied between species. In fact, across species, the maternal gene products variably contribute to the process of gastrulation, but as the accumulating genetic evidence reviewed here indicates, this contribution is quite significant in zebrafish.

1.1 Crash course on gastrulation movements: Internalization

At the onset of gastrulation, fate maps of animal embryos reveal prospective germ layers occupying the same tissue plane (Solnica-Krezel, 2005, 2020; Stern, 2004). Emboly or internalization of prospective mesoderm and endoderm rearranges them such that endoderm takes the most internal position in the nascent embryonic body, while mesoderm cells move between the endoderm and superficial ectoderm. In all animals, this process involves formation of an opening in the embryonic anlage, known as the blastopore in the well-studied amphibians, the ventral furrow in *Drosophila*, the blastoderm margin in fish and the primitive streak in amniotes. Endodermal and mesodermal precursors move through this opening, even though the underlying cellular behaviors differ (Solnica-Krezel & Sepich, 2012). In *Drosophila*, prospective mesoderm specified in the ventral aspect of the single-cell thick epithelial sphere invaginates into the embryo interior (Ko & Martin, 2020; Stathopoulos & Newcomb, 2020). There, the mesodermal cells undergo epithelial to mesodermal transition (EMT) to break away from the invaginated epithelium and migrate to form internal tissues (Ko & Martin, 2020; Leptin & Grunewald, 1990).

In fish mesenchymal embryonic tissues, or blastoderm, form a multilayered cup covering the animal hemisphere of the syncytial yolk cell and are covered by an epithelial enveloping layer (EVL) (Fig. 1) (Bruce & Heisenberg, 2020; Warga & Kimmel, 1990). The blastoderm margin in fish serves as the blastopore, through which endodermal and mesodermal cells ingress either as individuals or *en masse* (synchronized ingression) (Pinheiro & Heisenberg, 2020). In amniotes at the onset of gastrulation, embryonic tissues form a single-layered epithelium, which is flat in chick, and in many mammals including human, or cup shaped in mice. Mesodermal and endodermal cells internalize through the primitive streak, where they undergo EMT and ingress as individual mesenchymal cells (Nowotschin & Hadjantonakis, 2020; Voiculescu, 2020; Williams, Burdsal, Periasamy, Lewandoski, & Sutherland, 2012).

1.2 Epiboly

Depending on the shape and size of the embryonic tissues, the concurrent movements of epiboly spread and thin the germ layers to various degrees in different species. In fish, the three cellular compartments, EVL, deep

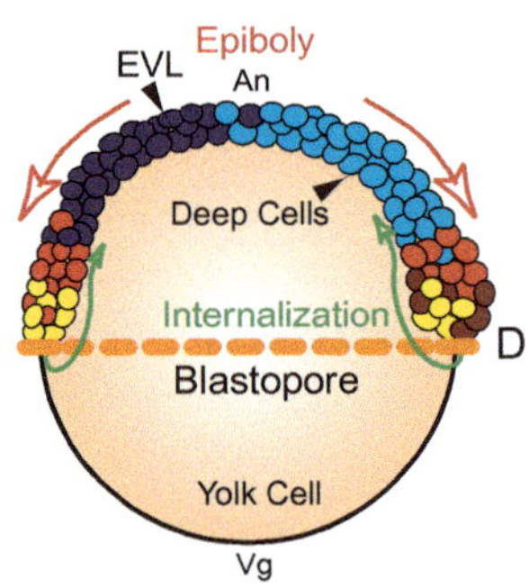

Fig. 1 Epiboly and internalization gastrulation movements in zebrafish. In early zebrafish gastrula, embryonic cells form a blastoderm cup covering the animal half of the large yolk cell. Mesenchymal deep cells are covered by a superficial enveloping layer (EVL). Prospective mesodermal (red) and endodermal (yellow) cells are positioned close to the blastoderm margin, blastopore. Epiboly movements thin the blastoderm and spread it around the yolk cell. Internalization movements entail ingression at the blastopore of individual or groups of mesodermal or endodermal cells, which subsequently move toward the animal pole; they are initiated on the dorsal side and spread laterally. An, animal; D, dorsal; Vg, vegetal.

blastoderm cells and yolk syncytial layer (YSL) engage in epiboly (Fig. 1) (Bruce & Heisenberg, 2020; Warga & Kimmel, 1990). The blastoderm comprised of a multilayer of mesenchymal deep cells thins and spreads around the yolk cell by radial intercalations that largely occur without directional bias (Bensch, Song, Ronneberger, & Driever, 2013). The EVL covers the blastoderm and is attached at its vegetal edge to the YSL via tight junctions (Koppen, Fernandez, Carvalho, Jacinto, & Heisenberg, 2006; Trinkaus, 1951). As the EVL, a simple squamous epithelium undergoes epiboly, its cells flatten and expand their apical surface while simultaneously decreasing their thickness, without significant volume change (Campinho et al., 2013; Marsal, Hernandez-Vega, & Martin-Blanco, 2017). This simultaneous thinning and spreading of EVL is thought to be passive and driven by cortical contractions of the actomyosin belt and removal of membrane in the YSL to which the EVL is attached (Marsal et al., 2017). The enormous YSL is involved in epiboly in several different ways. As a syncytium, it is comprised of a set of syncytial nuclei that are initially located in the animal portion of the YSL at the onset of epiboly. As epiboly ensues, the yolk syncytial nuclei spread vegetally, likely by transport of the nuclei along an intricate and dynamic network of microtubules (Fei et al., 2019; Solnica-Krezel & Driever, 1994; Strahle & Jesuthasan, 1993). In parallel, the fraction of YSL membrane that is not covered by EVL diminishes as the blastoderm front advances, in part by localized endocytosis (Betchaku & Trinkaus, 1986; Lepage, Tada, & Bruce, 2014; Marsal et al., 2017; Solnica-Krezel & Driever, 1994). As the EVL and blastoderm reach the embryo equator, a dynamic actomyosin ring forms within the YSL; progressive contraction of this belt is important for epiboly (Behrndt et al., 2012; Koppen et al., 2006). Actomyosin contraction is widely appreciated as a driving mechanism for morphogenesis, and consistent with this notion disruption of the actomyosin belt, also referred to as a cable, impairs epiboly progression (Behrndt et al., 2012); however, the cable mechanism alone could not account for the full force measured along the AV axis (Behrndt et al., 2012). In one model, the actomyosin ring generates an additional puling force through resistance to the retrograde actomyosin flow emanating from the vegetal to the animal pole (Behrndt et al., 2012). Alternatively, tissue surface tension differences between the stiffer yolk cortex and the softer EVL have been proposed to establish a cortical tension gradient that directs the actomyosin belt contractions in the YSL toward the vegetal pole. Here, recruitment of actin and myosin by cortical animally directed flow is thought to result in the circumferential contraction of the YSL in an animal-vegetal direction (Bruce & Heisenberg, 2020; Marsal et al., 2017). Although epiboly is completed nearly 7 h after ZGA and near the completion of the MZT, it relies extensively on cytoskeletal elements, including many maternally expressed factors discussed below.

1.3 Convergence and extension

Starting commonly at midgastrulation, the convergence and extension (C&E) gastrulation movements, narrow the germ layers along the mediolateral axis and elongate them from head to tail, thus establishing a blueprint of the body plan with organ rudiments (Fig. 2). This highly conserved morphogenetic movement is driven by a combination of epithelial or mesenchymal cell behaviors, including cell shape changes, directed cell migration, polarized radial and planar cell intercalation, oriented cell division, although the extent to which each of these different morphogenetic cell behaviors partake in C&E in different species from *Drosophila* (Pare & Zallen, 2020), through tunicates (Winkley, Kourakis, DeTomaso, Veeman, & Smith, 2020), zebrafish (Williams & Solnica-Krezel, 2020), and frog (Keller & Sutherland, 2020),

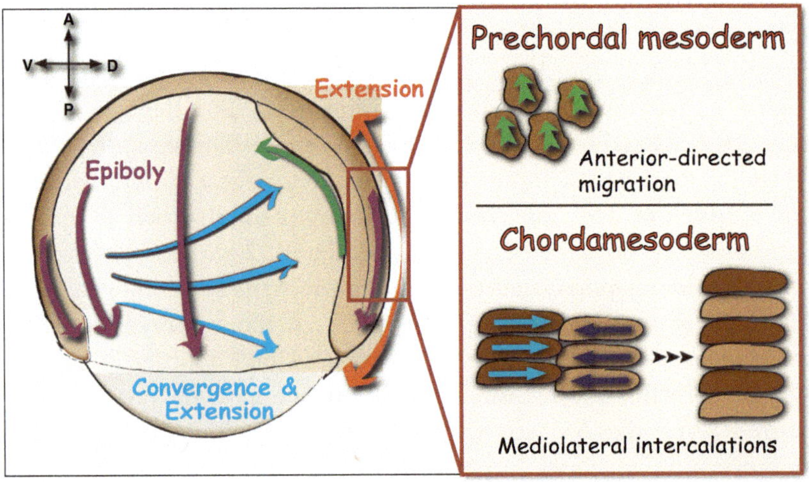

Fig. 2 Convergence and extension gastrulation movements in zebrafish. Convergence & Extension movements are initiated at midgastrulation, when epiboly movements continue to spread embryonic tissues around the yolk cell. C&E movements narrow the embryonic tissues toward the dorsal side and extend them anteroposteriorly. In the dorsal gastrula region, C&E movements are driven by a combination of anterior migration of prechordal mesoderm and mediolateral intercalations of mediolaterally elongated chordamesoderm cells. A, anterior; D, dorsal; P, posterior; V, ventral.

varies. In zebrafish, C&E is initially driven by anterior migration of internalized mesodermal cells toward the animal pole and random walk movements of endodermal cells in the lateral gastrula regions, followed by both cell types turning their trajectories toward dorsal at midgastrulation (7 hpf) (Pezeron et al., 2008; Sepich, Calmelet, Kiskowski, & Solnica-Krezel, 2005). C&E of dorsal mesoderm and prospective neuroectoderm is driven predominantly by polarized planar and radial intercalations of mediolaterally polarized mesenchymal cells that simultaneously lead to convergence and extension (Glickman, Kimmel, Jones, & Adams, 2003; Williams & Solnica-Krezel, 2020; Yin, Kiskowski, Pouille, Farge, & Solnica-Krezel, 2008). The essential and evolutionary regulator of these polarized cell migration and intercalation behaviors in chordates is the non-canonical Wnt/planar cell polarity pathway (Wnt/PCP) (Huebner & Wallingford, 2018; Keller & Sutherland, 2020). Wnt/PCP signaling is thought to work as a cellular compass that coordinates embryo patterning and morphogenesis, with some of its components becoming enriched at the anterior and others at the posterior cell edges (Gray, Roszko, & Solnica-Krezel, 2011; Yin et al., 2008).

2. Maternal-to-zygotic transition and its timing relative to gastrulation: A genomic perspective

As a fertilized egg initiates embryogenesis, its genome is transcriptionally inactive (Wu & Vastenhouw, 2020). Therefore, the earliest steps of embryogenesis are instructed by the gene products, RNAs and proteins, deposited in the developing egg by the female during oogenesis. Once the embryo achieves specific developmental milestones, zygotic genome activation (ZGA) occurs and gradually the embryonic genome takes control of development (Kane & Kimmel, 1993; Newport & Kirschner, 1982). However, the hand-off of genetic control of embryogenesis from the maternal to the zygotic genome is not as simple as transcriptional activation of the zygotic genome. This process known as the maternal-to-zygotic transition (MZT) also involves, indeed requires, downregulation of the maternal dowry. In different animals, the two processes take place at different times in terms of absolute developmental timing and also relative to the onset of gastrulation (reviewed in Vastenhouw, Cao, & Lipshitz, 2019; Wu & Vastenhouw, 2020; Yartseva & Giraldez, 2015).

2.1 Maternal regulation

During oogenesis in zebrafish, the majority of the genome is expressed and transcripts and/or proteins representing up to 70% of genes are deposited in the egg (Aanes et al., 2011; Harvey et al., 2013). Variable proportions of the genome are expressed in the eggs of different species based on global analyses, with *Drosophila* as well as frogs, similar to zebrafish, expressing up to three quarters of their genome maternally, while this proportion is significantly lower, about one-third in *C. elegans* and mammals (Vastenhouw et al., 2019).

Zebrafish eggs display an animal-vegetal polarity, with the animal pole featuring a micropyle, the sperm entry point (Kimmel, Ballard, Kimmel, Ullmann, & Schilling, 1995). Before activation of the egg, the cytoplasm and the yolk material appear to be thoroughly intermixed; however, contact with water or fertilization activates a cytoplasmic streaming process that results in segregation of nonyolky cytoplasm to form an animal cytoplasmic island positioned atop the vegetal yolk mass (Fig. 3). The zygote, undergoes the synchronous and meroblastic cleavages that maintain connections between the blastomeres (Kimmel et al., 1995). Many maternally deposited RNAs and proteins exhibit discrete asymmetric distribution, accumulating either at the vegetal pole, animal pole, or as proposed for *ndr1/squint* transcript encoding one of the Nodal ligands can be enriched in the prospective dorsal blastomeres (Escobar-Aguirre, Elkouby, & Mullins, 2017; Gore et al., 2005; Winata & Korzh, 2018). The expression of a large portion of the genome in the egg and asymmetric distribution of the maternal gene products, are hallmarks of so-called mosaic development, whereby individual blastomeres, or embryo regions do not possess equal developmental potential. Mammalian eggs, which are endowed with gene products representing a smaller proportion of their genomes, also feature an

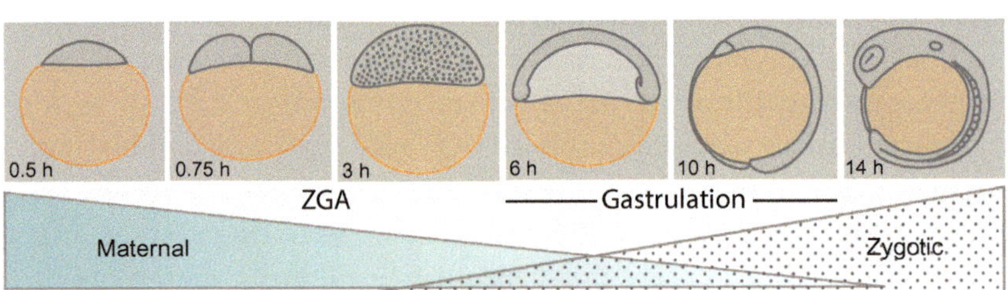

Fig. 3 Maternal and zygotic gene products during early zebrafish embryogenesis. The dynamic changes in the levels of maternal and zygotic gene products in the course of early zebrafish embryogenesis. h, hours post fertilization.

animal-vegetal axis. However, a fertilized mammalian zygote, during early cleavages, produces blastomeres that are largely pluripotent, a tell-tale sign of regulative development. Despite this highly regulative and pluripotent potential, accumulating evidence supports the existence of asymmetries in the murine zygote that could lead to molecular differences between blastomeres. Such epigenetic heterogeneity could in turn lead to differences in gene expression and eventually bias developmental potential of the blastomeres (Hupalowska et al., 2018; White et al., 2016).

Egg activation or fertilization triggers processes that stimulate expression of the maternal gene products. In the absence of transcriptional regulation before the ZGA, the zygote deploys many sophisticated mechanisms to regulate translation of the maternally deposited transcripts. The main levels of regulation include, the translation rate and the stability of the maternal RNAs, both of which are governed by the chemical RNA properties, including its sequence, the 7-methylguanylate cap at its 5′-end, and the length of the 5′-end poly(A) tail (Winata et al., 2018; Yartseva & Giraldez, 2015). The newly translated proteins contribute to the chores of the maternally contributed proteins to drive the early processes of embryogenesis, including cleavages, preparation of the zygotic genome for the ZGA, and dependent on when the latter occurs, specification of embryonic polarity, as exemplified by *D. melanogaster*, frog and fish (Escobar-Aguirre et al., 2017; Houston, 2017; Stathopoulos & Newcomb, 2020).

2.2 Zygotic genome activation (ZGA)

Regardless of when ZGA occurs relative to the process of axial symmetry breaking, it precedes and is absolutely required for gastrulation. ZGA was thought to occur in a step-wise fashion, with a minor wave of transcription followed by a major one, but current genomic studies indicate that the ZGA constitutes a period of gradual activation of the embryonic genome (Vastenhouw et al., 2019). In zebrafish, whole-sale activation of zygotic transcription was thought to initiate at embryonic cell cycle 10 (1000 cell stage; 3 hpf in zebrafish) (Fig. 3) (Kane & Kimmel, 1993). However, according to more recent reports zygotic transcription is initiated as early as 64 cell stage and is gradual and stochastic (Chan et al., 2019). In mouse, two waves of zygotic transcription have been detected, with some genes already transcribed in one-celled zygotes and transcription intensifying in 2-cell embryos (Hamatani, Carter, Sharov, & Ko, 2004). In humans, the zygotic genome becomes fully active at the 4–8 cell stage (Petropoulos et al., 2016; Rossant & Tam, 2018). ZGA mobilizes a significant fraction of the embryonic genome, with ca. 25% of all zebrafish genes being transcribed (Aanes et al., 2011; Harvey et al., 2013; Lee et al., 2013). Yet, extrapolating from the analysis of 5045 genes for which maternal and paternal alleles could be discerned, Harvey et al., inferred that in zebrafish only 5% (951 genes) are exclusively zygotically expressed during blastula and gastrula stages, 34% exclusively maternally and 61% showing both maternal and zygotic expression (Harvey et al., 2013). Distinguishing the newly transcribed transcripts based on exon/intron expression, Lee and colleagues estimated the number of exclusively zygotically expressed genes at the blastula/gastrula stage (4–6 hpf) to be higher, ca. 2400 genes and thus ca. 13% of zebrafish genes (Lee et al., 2013). A similar estimate, 14–19% of exclusively zygotically expressed genes during gastrulation, was obtained based on the high-resolution bulk RNA sequencing (Elizabeth Busch-Nentwich and Richard J. White, personal communication; White et al., 2017). These estimates suggest a significant contribution of the zebrafish maternal genome to embryogenesis, especially gastrulation.

2.3 Midblastula transition (MBT)

The initiation of zygotic transcription in fish and frogs is associated with additional significant changes in embryogenesis, including lengthening of the cell cycle, loss of cell division synchrony and initiation of cellular motility, collectively known as midblastula transition (MBT) (Kane & Kimmel, 1993; Newport & Kirschner, 1982). In zebrafish, the MBT is preceded or concurrent with another important process, formation of the extraembryonic yolk syncytial layer (YSL), when a set of the marginal blastomeres fuse with the underlying yolk cell. As discussed above, the YSL plays important roles in the process of epiboly as well as in inductive interactions with the embryonic blastomeres (Kimmel & Law, 1985; Mizuno, Yamaha, Wakahara, Kuroiwa, & Takeda, 1996).

2.4 Downregulation of maternal transcripts

The hand-off of the developmental controls from the maternal to the zygotic genome not only entails transcriptional activation of the zygotic genome, but also downregulation of a substantial fraction of the maternal transcripts deposited in the egg. This appears to be a highly regulated process, which is mediated by both maternally and zygotically encoded mechanisms (reviewed in Vastenhouw et al., 2019; Wu & Vastenhouw, 2020; Yartseva & Giraldez, 2015). In zebrafish a microRNA mediated mechanism of maternal RNA clearance that involves zygotically expressed miroRNA, *miR430*, plays an important role in eliminating maternally deposited RNAs (Giraldez et al., 2005). The level of maternal transcripts strongly declines before gastrulation: of the downregulated maternal transcripts, 60% are degraded between fertilization and 16 cell stage, and the remaining 40% are thought to decline soon after the onset of zygotic transcription (Fig. 3) (Aanes et al., 2011; Mathavan et al., 2005; Mishima & Tomari, 2016). That the clearance of maternal transcripts is required for normal gastrulation is evidenced by the analysis of MZ*dicer* zebrafish mutants, in which microRNA precursors are not processed into mature miRNAs. Mature miRNA deficiency interferes with maternal transcript clearance, which is associated with delayed epiboly and reduced C&E movements (Giraldez et al., 2005, 2006). The relevant targets of *mir430* that need to be downregulated to ensure normal gastrulation remain to be identified. These studies underscore that the majority of maternal transcripts decline before the onset of gastrulation in zebrafish and that this decline is required for normal gastrulation. However, the decline or absence of maternal transcripts is only an imperfect proxy for the extent of functional contribution of maternally expressed genes to gastrulation. Indeed, less is known about perdurance of maternally deposited proteins or translation products of the maternal transcripts that could extend the maternal contribution through gastrulation and beyond.

3. Requirement for maternal and zygotic gene products for gastrulation: A genetic perspective

Victor Hamburger's famously said, "*Our real teacher has been and still is the embryo, who is, incidentally, the only teacher who is always right.*" This continually relevant statement is interpreted by developmental geneticists in a unique way: let us do a genetic screen or gene disruption to ask the embryo about the relative contributions of the maternal and zygotic genomes to gastrulation. Unbiased genetic screens in animal research organisms, identified zygotic and maternal effect mutations affecting gastrulation (Nusslein-Volhard & Wieschaus, 1980). The proportion between these two classes differs between species with maternally acting gastrulation mutants more common in species characterized by mosaic development and late ZGA, and zygotic gastrulation mutants more frequent in species with regulative development and relatively early ZGA onset and MZT, as discussed above. The pioneering forward genetic screens for zygotically acting mutations affecting embryogenesis have been carried out in *Drosophila* (Nusslein-Volhard & Wieschaus, 1980),

C. elegans (Bucher & Greenwald, 1991), and later in vertebrates, including zebrafish (Driever et al., 1996; Haffter et al., 1996) and mouse (Kasarskis, Manova, & Anderson, 1998). Consistent with their mosaic development and heavy maternal gene expression, the zygotic screens in zebrafish identified relatively few mutations affecting gastrulation compared to thousands affecting post-gastrulation processes, such as segmentation and organogenesis (Hammerschmidt et al., 1996; Kane et al., 1996; Mullins et al., 1996; Solnica-Krezel et al., 1996). By contrast, the zygotic screens carried out in the mouse, especially those specifically analyzing embryonic phenotypes at midgestation and thus soon after gastrulation and neurulation, netted a higher proportion of genes affecting gastrulation (Kasarskis et al., 1998). Interestingly, the type of proteins discovered through analysis of zygotic gastrulation mutants identified in fish versus mouse also differs. The zygotic screens in zebrafish, identified components of signaling pathways such as BMP (Dpp in *Drosophila*) (Kramer et al., 2002; Tuazon & Mullins, 2015), planar cell polarity (PCP) (Heisenberg et al., 2000; Jessen et al., 2002), or Nodal signaling (Feldman et al., 1998; Gritsman et al., 1999; Sampath et al., 1998), cell adhesion molecules (Kane, McFarland, & Warga, 2005), or transcription factors such as Bozozok (Fekany et al., 1999), Notail/Brachyury (Halpern, Ho, Walker, & Kimmel, 1993; Schulte-Merker, van Eeden, Halpern, Kimmel, & Nusslein-Volhard, 1994), but hardly any genes encoding cytoskeletal components or chromatin factors (Williams et al., 2018).

By contrast, the forward and reverse zygotic screens for mutations affecting gastrulation and/or neurulation in the mouse not only uncovered signaling pathway components: Nodal (Conlon et al., 1994), Hedgehog (Chiang et al., 1996), and transcription factors including Brachyury (Beddington, Rashbass, & Wilson, 1992), but also genes encoding cytoskeletal regulators, such as Rac1 (Migeotte, Grego-Bessa, & Anderson, 2011), Cofilin (Grego-Bessa, Hildebrand, & Anderson, 2015), or P120 catenin (Hernandez-Martinez, Ramkumar, & Anderson, 2019). The maternal effect screens carried out in zebrafish (Abrams & Mullins, 2009; Dosch et al., 2004; Pelegri & Schulte-Merker, 1999; Wagner, Dosch, Mintzer, Wiemelt, & Mullins, 2004), have uncovered a broader spectrum of gene products, including cell cycle regulators, chromatin factors and cytoskeletal components (Eno, Solanki, & Pelegri, 2016; Gupta et al., 2010; Langdon et al., 2016; Yabe et al., 2009), in addition to the signaling pathway components that regulate embryonic axis formation such as Ichabod/β-catenin (Kelly, Chin, Leatherman, Kozlowski, & Weinberg, 2000) or Huluwa (Yan et al., 2018).

Mutations in genes expressed exclusively by the maternal genome can be only identified by maternal effect screens, as demonstrated in zebrafish by pioneering screens for maternal effect mutations in Mary Mullins laboratory (Fig. 4) (Dosch et al., 2004; Wagner et al., 2004). For some genes that are zygotically and maternally expressed, both maternally and zygotically produced transcripts can impact gastrulation in different ways, and can be identified by zygotic or/and maternal screens. These genes can be further subdivided into two sub-classes of maternal zygotic (MZ) genes, and strict MZ (SMZ) genes as described in the following sections (Figs. 5–7).

3.1 Strict maternal (SM) genes

SM genes are those for which the genotype of the female parent determines the phenotype of the progeny, i.e., a homozygous mutant female produces phenotypic progeny with the same penetrance and expressivity regardless of the genotype of the male parent (Fig. 5) (Abrams & Mullins, 2009). This category encompasses genes that are exclusively maternally expressed, and/or function in processes that occur before the ZGA. The current analyses of maternal and zygotic transcripts in early zebrafish embryos, indicate this class constitutes 35% of genes (Harvey et al., 2013). Functionally, maternally expressed genes contribute to gastrulation in many different ways. The first and most intuitive category includes those genes that affect processes preceding gastrulation, and mutations in these genes halt embryogenesis before the onset of gastrulation. Examples include mutations in genes specifying oocyte polarity, such as *bucky ball* that prevent normal fertilization (Bontems et al., 2009; Marlow & Mullins, 2008), or *cellular island* (Aurora B) that is essential for early cleavages (Yabe et al., 2009). An interesting group of SM-effect mutations are those that affect maternal expression and function of a gene that is also expressed zygotically. The presumed regulatory mutation *ichabod* that strongly reduces maternal expression of the *β-catenin2* gene resulting in deficiencies in specification of the embryonic axis is an example (Kelly et al., 2000).

3.2 Maternal zygotic (MZ) genes

Genes in this category are expressed maternally and zygotically and both maternal and zygotic expression are required to different degrees for normal gastrulation (Fig. 6). Many of the genes in this category were identified as zygotic mutations affecting gastrulation (Hammerschmidt et al., 1996b; Mullins et al., 1996; Solnica-Krezel et al., 1996). They are predominantly

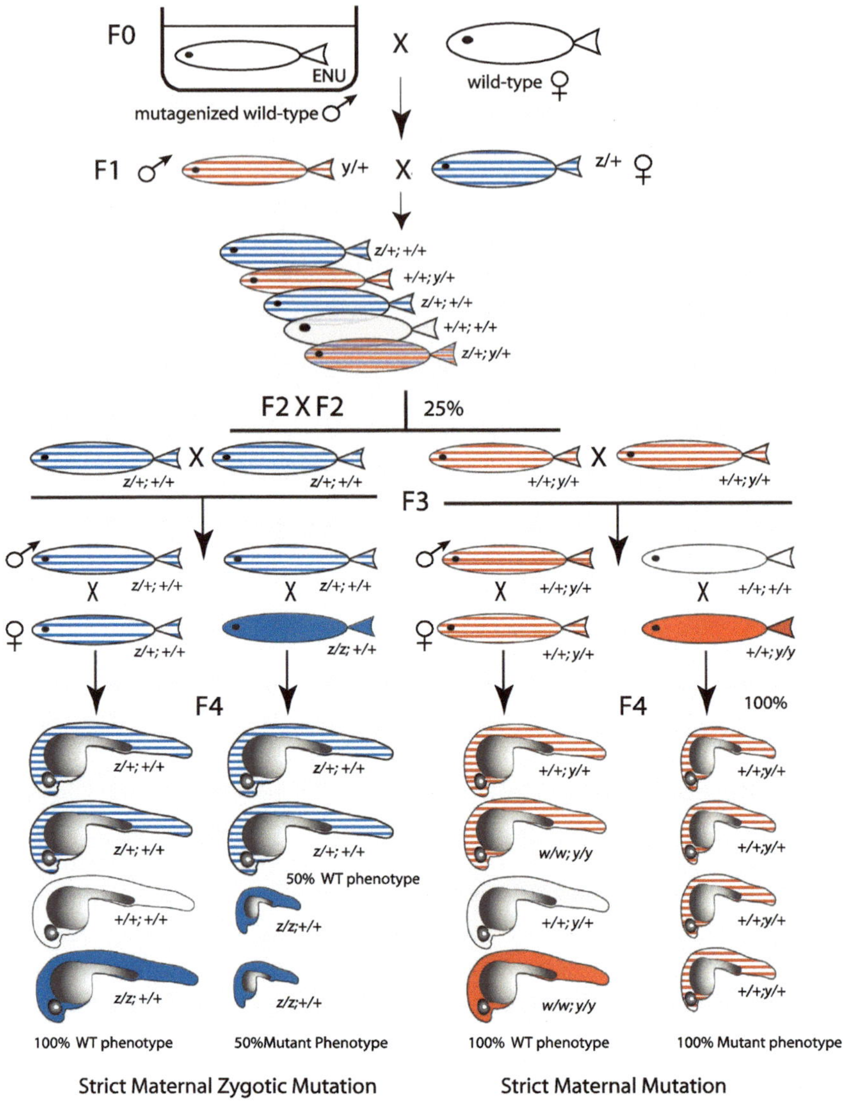

Fig. 4 Forward genetic screens for maternal effect mutations in zebrafish. A schematic illustrating how *N*-ethyl-*N*-nitrosourea (ENU)-induced mutations in F0 zebrafish males, are driven via breeding to homozygosity in F3 generation. Breeding of F3 animals uncovers strict maternal effect mutations, whereby homozygous *y/y* female produces 100% mutant embryos when crossed to WT male. Strict maternal zygotic mutations are manifested only by homozygous mutant embryos, *z/z*, generated by homozygous *z/z* females.

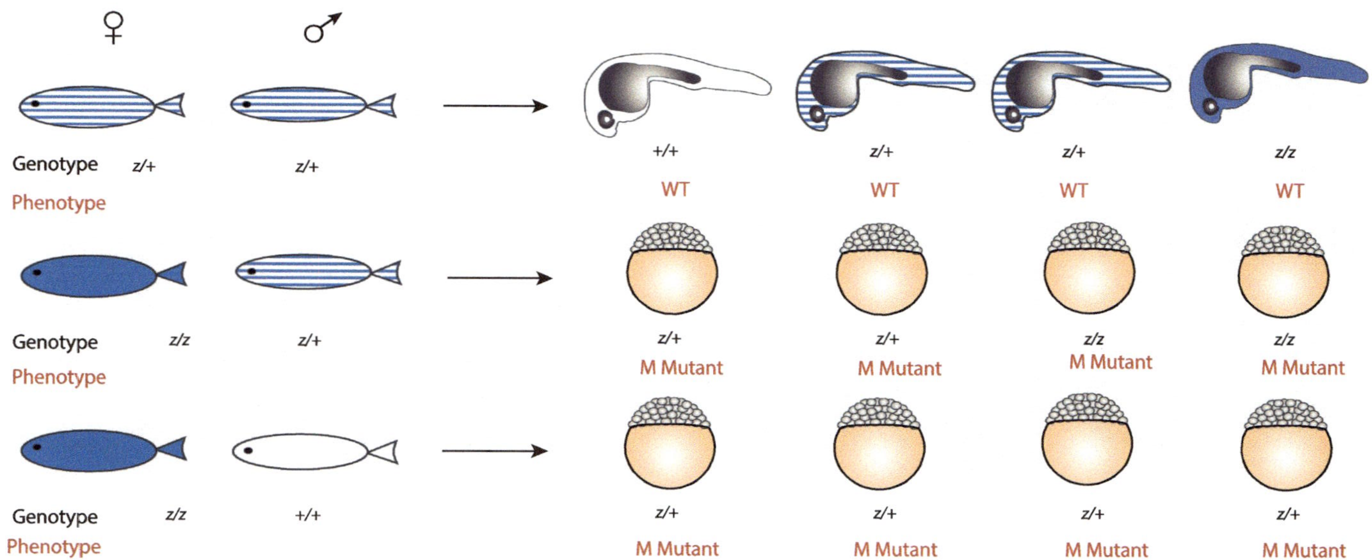

Fig. 5 Strict maternal mutations. Genotypes and phenotypes of embryos generated by females heterozygous or homozygous for a strict maternal effect mutation, *z*. Homozygous mutant females produce 100% phenotypic mutant embryos, regardless of the male and embryo genotype.

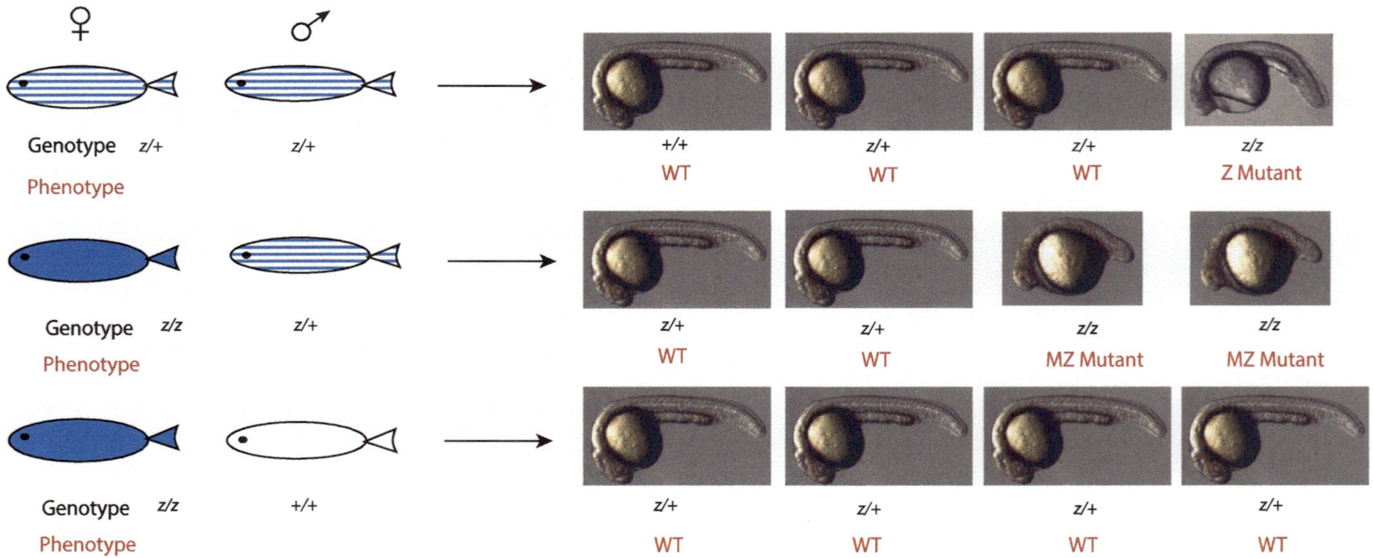

Fig. 6 Maternal zygotic mutations. Genotypes and phenotypes of embryos generated by females heterozygous or homozygous for a maternal zygotic mutation, in *udu/gon4l* gene (Williams et al., 2018). Crosses of heterozygous animals produce 25% homozygous mutant embryos with a shorter body and edema. Homozygous mutant females when mated with heterozygous males, produce 50% homozygous mutant embryos with a very short body. By contrast, *udu*/+ heterozygous embryos produced by homozygous mutant females do not manifest gastrulation phenotypes.

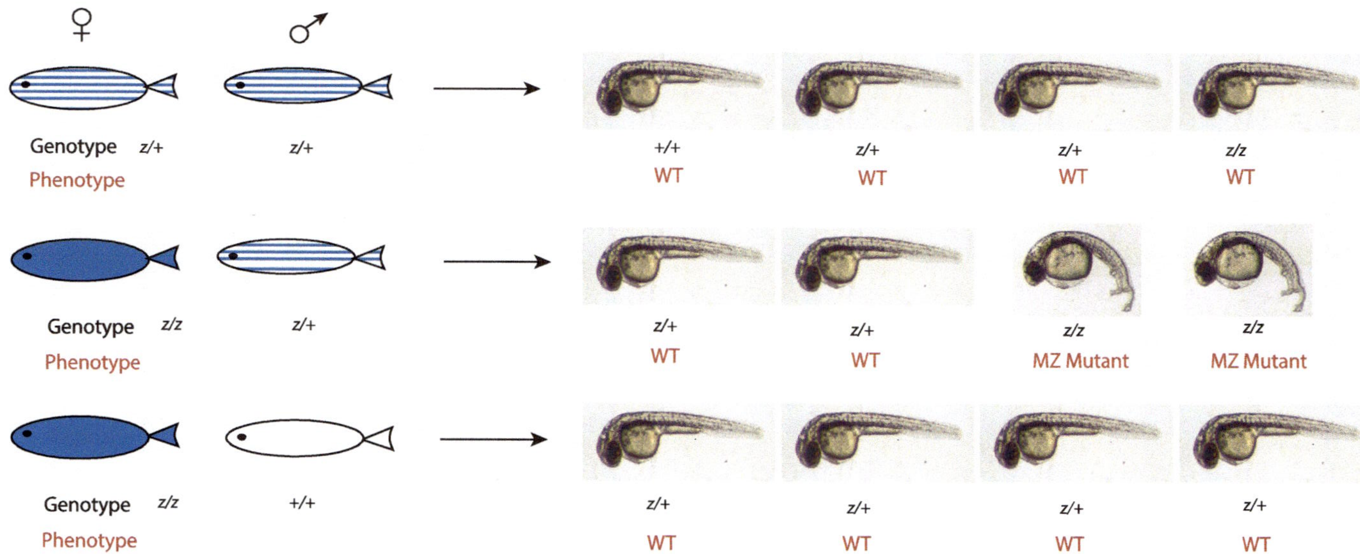

Fig. 7 Strict maternal zygotic mutations. Genotypes and phenotypes of embryos generated by females heterozygous or homozygous for a strict maternal effect mutation. Mutant phenotypes are observed only in homozygous mutant embryos produced by homozygous mutant females.

zygotically expressed but also exhibit low amounts of maternal RNA or protein expression, which persists during gastrulation, and in these cases, the maternal gene products are redundant with the zygotic ones. Effectively, whereas homozygous zygotic mutants produced by heterozygous females present with a gastrulation phenotype, this phenotype is exacerbated in homozygous mutant progeny of homozygous mutant females. Heterozygous progeny generated by crossing homozygous females with a wild-type male usually are aphenotypic or present with only mild phenotypes (Fig. 6). Uncovering the full function of these genes by examining their MZ phenotypes is often experimentally challenging if the zygotic phenotype is lethal, and thus requires strategies such as germline transplantation (Ciruna et al., 2002) or genome editing specifically in the germline.

The zebrafish *bozozok/dharma* gene is a representative of this class, with zygotic mutants showing variable defects in embryonic axis formation, and the penetrance and expressivity of the phenotype being slightly increased in mutant embryos generated by mutant females (Fekany et al., 1999). Other examples include genes encoding BMP antagonists, *chordin/chordino* (Schulte-Merker, Lee, McMahon, & Hammerschmidt, 1997) and *sizzled/ogon* (Miller-Bertoglio et al., 1999; Yabe et al., 2003), or the BMP effector *Smad5/Somitabun* (Kramer et al., 2002). Similarly, several mutations impairing C&E movements identified in the zygotic screens were later shown to inactivate Wnt/PCP pathway components, including Vangl2/Trilobite (Ciruna, Jenny, Lee, Mlodzik, & Schier, 2006; Jessen et al., 2002), Glypican4/Knypek (Topczewski et al., 2001), Wnt11/Silberblick (Heisenberg et al., 2000), and Wnt5/Pipetail (Rauch et al., 1997). These zygotic mutants manifest strong gastrulation phenotypes, which are exacerbated to various degrees in mutant embryos produced by homozygous mutant females (Ciruna et al., 2006; Topczewski et al., 2001). Therefore, although identified as zygotic mutants, these genes could be also classified as maternal zygotic with partially redundant maternal and zygotic functions. However, in these cases, the maternal component does not appear to be essential, as the progeny of homozygous mutant females and wild-type males undergo normal gastrulation.

Another example is the *gon4-like/ugly duckling* (*gon4l/udu*) gene encoding a chromatin factor, which has been identified in zygotic screens based on its tail morphogenesis phenotype (Hammerschmidt et al., 1996), red blood cell deficiency (Liu et al., 2007), and in synthetic zygotic screens as a recessive enhancer of *gpc4/kny* C&E mutant phenotype (Williams et al., 2018).

The mild *Zgon4l/udu* C&E mutant phenotypes become strongly exacerbated in MZ mutants (Fig. 6). However, females harboring a mutant germline produce overtly normal progeny when crossed with wild-type males, indicating that zygotic expression alone is sufficient for normal gastrulation (Fig. 6) (Williams et al., 2018). Yet a different example is *tdgf1/one-eyed pinhead*, which encodes an essential co-factor for Nodal ligands that was identified in zygotic screens based on the lack of endoderm, prechordal plate and floor plate (Schier, Neuhauss, Helde, Talbot, & Driever, 1997). However, homozygous mutant embryos produced by mutant females exhibit a dramatically stronger phenotype, including loss of most mesoderm, strongly impaired C&E movements and neural patterning defects, thus phenocopying a full loss of Nodal signaling (Gritsman et al., 1999). These phenotypes are as severe as those seen in *ndr1/squint;ndr2/cyclops* mutants, in which the two genes encoding the Nodal ligands expressed during gastrulation are inactivated (Feldman et al., 1998; Gritsman et al., 1999; Sampath et al., 1998).

The second subclass of this MZ category includes genes, of which loss of function produces a mild strict maternal gastrulation phenotype with a very mild or undetectable zygotic gastrulation mutant phenotype, but strong MZ phenotype. Here the expression and function of the maternal and zygotic products are overlapping and partially redundant. The *stat3* gene is in this category, with zygotic mutants undergoing gastrulation and embryogenesis without any overt defects, but later during larval stages inflammation and scoliosis are apparent in the mutants, which consequently fail to thrive (Liu, Sepich, & Solnica-Krezel, 2017). When homozygous females were generated by germline transplantations and crossed with mutant males, the resulting embryos showed reduced early cleavages, as well as reduced cell proliferation during blastula stages and early gastrulation, in turn resulting in reduced axial extension. Interestingly, females harboring a *stat3* mutant germline produce progeny that exhibit transient cell proliferation and axis extension defects, indicating that zygotic *stat3* transcripts can partially rescue these phenotypes (Liu et al., 2017).

3.3 Strict maternal zygotic (SMZ) genes

Genes in this group are expressed both maternally and zygotically, with maternal and zygotic expression having fully redundant function during gastrulation. Mutant phenotypes are only observed when a homozygous mutant female is mated with either a heterozygous or homozygous mutant male, producing 50% or 100% phenotypically mutant progeny, respectively. Neither crosses of two heterozygotes nor of homozygous mutant females to wild-type males result in phenotypic progeny (Fig. 7). Only a few examples, all from reverse genetic approaches, of such genes have been reported, including zebrafish *dchs2*, which encodes one of the three homologs of the atypical cadherin, Dachsous. Neither mutants lacking zygotic *dchs2*, $dchs2^{stl1/stl1}$ mutants, nor the progeny of $dchs2^{stl1/stl1}$ females and WT males, lacking maternal *dchs2* expression, exhibit overt morphologic gastrulation phenotypes. However, MZ$dchs2^{stl1/stl1}$ mutants manifest mild and transient epiboly and C&E defects (Li-Villareal et al., 2015).

3.4 Zygotic (Z) genes

The embryonic phenotype is solely determined by the embryonic genotype and thus encompasses genes expressed exclusively from the zygotic genome. However, as only ca. 5–15% of zebrafish genes are exclusively zygotically expressed (Harvey et al., 2013; Lee et al., 2013; White et al., 2017), this is a small proportion of the zebrafish genome. Nevertheless, most of the forward genetic screens in zebrafish focused on zygotic phenotypes and it is likely that a large proportion of such genes affecting gastrulation have been identified. This category includes one of the first gastrulation mutants identified through the pioneering screens in the Kimmel lab, *cyclops/nodal-related 2* (Hatta, Kimmel, Ho, & Walker, 1991), which encodes one of the three Nodal ligands in zebrafish (Rebagliati, Toyama, Haffter, & Dawid, 1998; Sampath et al., 1998) and is involved in mesoderm induction and pattering. Another example is *tbx16/spadetail*, the first zebrafish mutant shown to affect convergence movements of mesoderm cells in a cell-autonomous fashion (Ho & Kane, 1990), and later found to disrupt the Tbx16 transcription factor (Griffin, Amacher, Kimmel, & Kimelman, 1998). *notail/brachury* is another zygotically expressed gene encoding an evolutionarily conserved transcription factor that regulates trunk and tail fates and movements (Halpern et al., 1993; Schulte-Merker et al., 1994) and had been previously discovered in the mouse as Brachyury (Beddington et al., 1992; Kispert & Herrmann, 1993). Also among the zygotic genes discovered during the two large-scale screens for zygotic mutations in zebrafish (Driever et al., 1996; Haffter et al., 1996), is *sizzled/ogon*, which encodes a BMP antagonist that regulates DV patterning (Lee, Ambrosio, Reversade, & De Robertis, 2006; Solnica-Krezel et al., 1996; Yabe et al., 2003).

4. Instructive and permissive contributors to gastrulation from the maternal and zygotic genomes: An embryo perspective

Gastrulation movements are driven by morphogenetic behaviors of individual cells or cell collectives that include proliferation, cell shape changes, cell migration, and various types of cell intercalation (Solnica-Krezel, 2020). Therefore, cells are both building blocks and active participants in gastrulation morphogenesis. The cellular morphogenetic programs are specified both by cell-type or identity, positional information and/or other environmental signals. Therefore, the process of gastrulation must be preceded and/or accompanied by cell proliferation, specification of embryonic polarity, induction of germ layers and acquisition of cell motility by embryonic cells. In the following sections, I discuss the contribution of the maternal and zygotic genomes to these processes.

4.1 Cell proliferation

Cell proliferation and gastrulation movements must be coordinated. As mentioned above, embryonic cells are both building blocks and effectors of gastrulation movements. On the other hand, in animals such as fruit fly or zebrafish, rapid cell proliferation precedes gastrulation, during which cell divisions occur infrequently (Leise 3rd & Mueller, 2004). In fact, cell division and movement appear to be competing activities, likely because cell division and motility utilize common cytoskeletal machineries. As gastrulating zebrafish cells divide, they stall their movement, round up and abolish their planar polarized asymmetries (Ciruna et al., 2006). Accordingly, cell divisions must be limited for normal C&E of the paraxial mesoderm in *Xenopus* (Leise 3rd & Mueller, 2004) and posterior body elongation in zebrafish (Bouldin, Snelson, Farr 3rd, & Kimelman, 2014). Conversely, cell proliferation during gastrulation appears largely dispensable for axis elongation, as evidenced by predominantly normal body elongation in the zebrafish early mitotic regulator, *emi*, mutants in which mitosis ceases from early gastrulation, and in embryos where cell proliferation is chemically inhibited during gastrulation (Quesada-Hernandez et al., 2010; Zhang, Kendrick, Julich, & Holley, 2008). In addition, the relationship between cell division orientation and axis elongation in zebrafish remains unresolved. While some studies point out that oriented cell division under the regulation of Wnt/PCP signaling is a driving force for axis elongation (Gong, Mo, & Fraser, 2004), others argued against the importance of cell division orientation in axis extension (Quesada-Hernandez et al., 2010). Further support for the notion that cell division *during* gastrulation in zebrafish is not an essential contributor to morphogenesis comes from time-lapse analyses of C&E movements of paraxial mesoderm that reconstructed the observed cell morphogenetic behaviors, including cell division, dorsal migration and intercalations (Yin et al., 2008). As only a few cell divisions occurred in the course of the analysis, it was concluded that active cell proliferation during gastrulation does not significantly contribute to C&E (Yin et al., 2008).

Whereas cell division or its orientation *during* gastrulation are not significant contributors to morphogenesis, the totality of cell proliferation before and during gastrulation is. Zebrafish *stat3*, encoding a known promoter of cell division, is both maternally and zygotically expressed (Liu et al., 2017). In MZ*stat3* mutants both the early rapid and synchronous cell cleavages that are regulated by the maternal genome as well as the later asynchronous cell divisions that continue through the blastula period (Kane & Kimmel, 1993; Kane, Warga, & Kimmel, 1992), are reduced (Liu et al., 2017). The resulting ca. 22% reduction in cell number in MZ*stat3* gastrulae leads to AP shortened axial and paraxial mesoderm, and these morphogenetic defects can be partially suppressed by ectopic expression of Cdc25 cell cycle regulator (Liu et al., 2017). Both ca. 22% reduction of cell number and axis extension defects are phenocopied by chemical inhibition of cell proliferation starting at early gastrulation (Liu et al., 2017). The axis extension defect in the gastrulae associated with reduced cell number is due to an insufficient number of cells available to participate in ML cell intercalation underlying AP axis extension (Liu et al., 2017). Because both early synchronous cleavages and asynchronous cell divisions following MBT occur albeit with little growth (Dalle Nogare, Pauerstein, & Lane, 2009), cells in MZ*stat3* mutants or cells in embryos in which cell proliferation was chemically inhibited are bigger; however, having larger but fewer cells does not ensure normal axis extension. Interestingly, M*stat3* mutant embryos, which express one wild-type *stat3* allele after ZGA, exhibit cell proliferation deficiency comparable to that observed in MZ*stat3* mutants at early gastrulation (6 hpf) (Liu et al., 2017). However, at late gastrulation (10 hpf) M*stat3* mutants show reduced proliferation compared to WT but greater than the proliferation observed in MZ*stat3* mutants (Liu et al., 2017). Notably, this partial rescue of cell proliferation by zygotic expression of *stat3* is correlated with partial rescue of axial extension defects. Therefore, cell proliferation promoted by both maternal and zygotic *stat3* have partially redundant functions to ensure sufficient cell numbers are available to participate in ML cell intercalation underlying AP axis extension (Liu et al., 2017). In contrast to fish or frog embryos, mouse and human embryos initiate gastrulation with just a few hundred cells and undergo dramatic growth during the course of gastrulation (McDole et al., 2018). Murine zygotic *Stat3* mutants arrest by early gastrulation, but it is not clear whether and to what degree a cell proliferation defect is involved (Holland et al., 2007).

4.2 Cell motility

Cell motility is essential for any and all gastrulation movements. During zebrafish embryogenesis, motility emerges after the MBT and thus activation of the zygotic genome at 3 hpf (Kane & Kimmel, 1993). Early studies that deployed small molecule inhibitors of transcription, such as alpha-amanitin, demonstrated that in the presence of inhibitor, frog and zebrafish development halts soon after MBT and no gastrulation movements are initiated. Which genes are necessary and sufficient for embryonic cells to become motile, and whether maternally contributed gene products are also required remains to be determined.

4.3 Axis specification

Embryonic polarity, i.e., specification of the AP and DV embryonic axes, is not absolutely required for epiboly or internalization movements, but is essential for C&E movements. Clearly, embryonic polarity and C&E cell behavior movements have to be coordinated, to instruct dorsal convergence and anteroposterior extension of the germ layers (Gray et al., 2011). In zebrafish, the key process in breaking the radial animal-vegetal symmetry of the zygote and early blastula is a series of events that are controlled by the maternally expressed genes, which culminate in accumulation of the transcriptional effector of the canonical Wnt pathway, β-catenin, in nuclei on the dorsal side of the zebrafish blastula before ZGA (Langdon & Mullins, 2011; Marlow, 2020). These processes, starting with asymmetric transport of dorsal determinants by an asymmetric set of microtubules from the vegetal pole toward the future dorsal side, followed by interactions of many positive, like Huluwa (Yan et al., 2018) or Dvl (Xing et al., 2018) and negative regulators of β-catenin, culminate in dorsal enrichment of β-catenin, which will initiate a dorsal cascade of gene expression upon ZGA. These key patterning events that set the stage for gastrulation are extensively reviewed in (Marlow, 2020; Pinheiro & Heisenberg, 2020). Failure to specify the dorsal embryonic axis and subsequently the gastrula organizer, results in excess BMP signaling and severe inhibition of C&E movements, producing embryos with reduced axis extension and rostral body, and an excess ventroposterior tissues (Gonzalez et al., 2000; Myers, Sepich, & Solnica-Krezel, 2002a, 2002b).

4.4 Epiboly

Epiboly is the first morphogenetic movement initiated during zebrafish embryogenesis at 4 hpf, just 1 h after the ZGA/MBT takes place (Kimmel et al., 1995; Warga & Kimmel, 1990). Epiboly entails vegetal spreading of the three cellular compartments: blastoderm, enveloping layer (EVL), and yolk syncytial layer (YSL), and is associated with thinning of the blastoderm and EVL (Fig. 1). These compartments are established before MBT and thus develop under control of the maternal genome. Although epiboly initiation and its normal progression require zygotic transcription and MZT (Giraldez et al., 2006), current evidence indicates that the contribution of maternal gene products to this process is also significant. Interference with cytoplasmic polyadenylation of maternal transcripts using Cordycepin delays or blocks epiboly (Winata et al., 2018). The large-scale genetic screens for zygotic zebrafish mutants (Driever et al., 1996; Haffter et al., 1996), identified several mutations but only one gene affecting epiboly (Kane et al., 1996; Shimizu et al., 2005; Solnica-Krezel et al., 1996). Despite featuring many names, including *half-baked*, *avalanche* or *volcano*, to reflect the dramatic phenotype of embryos ceasing during mid to late epiboly and subsequently falling apart, these mutations all disrupt the *cdh1* gene encoding E-cadherin, one of the predominant cell adhesion molecules during zebrafish embryogenesis (McFarland, Warga, & Kane, 2005; Shimizu et al., 2005). However, some of these mutants exhibited dominant maternal effect phenotypes, consistent with both maternal and zygotic *cdh1* expression and function (Kane et al., 1996), which is required chiefly for adhesion between deep blastomeres and EVL for epiboly progression (Shimizu et al., 2005).

Likewise, maternal effect screens identified several mutations interfering with epiboly (Wagner et al., 2004). Maternal effect *betty boop* (*bbp*) mutants, have a dramatic phenotype characterized by precocious contraction of the actomyosin ring in the yolk cell that leads to yolk cell lysis. Further analyses identified *bbp* as the zebrafish homolog of the serine-threonine kinase mitogen activated protein kinase activated protein kinase 2, which regulates the activity of F-actin in the yolk cell to drive progressive movement of the blastoderm to enclose the yolk (Holloway et al., 2009). Given that actomyosin cytoskeleton and microtubules play such critical roles in the process of epiboly (Cheng, Miller, & Webb, 2004; Solnica-Krezel & Driever, 1994; Strahle & Jesuthasan, 1993), it is not surprising that many of the components and regulators of this process are maternally contributed, as exemplified by *bbp* (Holloway et al., 2009). Essential functions in epiboly for several such genes have recently been demonstrated by reverse genetic approaches. MZ*dchs1b* mutations in atypical cadherin result in delayed epiboly movements (in addition to the earlier strict maternal effect defects described already) that are associated with an abnormally bundled cytoskeleton (Li-Villareal et al., 2015). Morpholino and biochemical studies, identified maternally expressed *clip1a* (*clip-170*) as a target of Pregnenolone that promotes microtubule polymerization during epiboly (Hsu, Liang, Chen, & Chung, 2006; Weng et al., 2013). A different example of a maternal effect gene required for epiboly is *poky/ikk1* encoding a kinase required for proper differentiation and integrity of the EVL epithelium (Fukazawa et al., 2010). Consequently, M-effect mutant embryos that fail to initiate epiboly or initiate with delay, eventually lyse. Interestingly, mutant embryos in which zygotic transcription was blocked with α-amanitin exhibited EVL cells with normal morphology, supporting the notion that maternal *poky/ikk1* gene products are required not for early EVL differentiation, but also after the ZGA (Fukazawa et al., 2010).

4.5 Mesendoderm internalization

Mesendoderm internalization via a blastopore brings the mesendoderm beneath the surface ectoderm and is the defining and conserved movement of gastrulation (Fig. 1). However, cellular mechanisms of this process vary from the invagination of the ventral epithelium sheet in *Drosophila*, to involution of the mesenchymal layer in frogs, to ingression of individual mesenchymal cells following an epithelial to mesenchymal transition in sea urchin, chick and mammals (reviewed in Solnica-Krezel & Sepich, 2012). In zebrafish, mesendoderm internalization is initiated on the dorsal side of the gastrula at 5 hpf via ingression of individual cells (Montero et al., 2005; Shih & Fraser, 1995). Subsequently, the wave of internalization spreads laterally to reach the ventral side of the embryo, where it takes the form of synchronized ingression with many individual cells internalizing simultaneously (Kane & Adams, 2002; Keller, Schmidt, Wittbrodt, & Stelzer, 2008). As in frogs and fish, cells embarking on internalization are of mesenchymal character; however, this process is not preceded via a classic EMT, but likely by an EMT variation (Pinheiro & Heisenberg, 2020).

The process of EMT and internalization, as well as the ensuing migration away from the blastopore is dependent on mesoderm and endoderm induction (Carmany-Rampey & Schier, 2001; Keller et al., 2008). In vertebrate embryos this involves various combinations of the Nodal, Wnt/β-catenin, BMP and FGF signaling (Marlow, 2020; Nowotschin & Hadjantonakis, 2020; Pinheiro & Heisenberg, 2020; Schier, 2009). In zebrafish, Nodal signaling takes center stage. Nodal engages Ndr1/Squint and Ndr2/Cyclops ligands that form heterodimers with Vg1 (Feldman et al., 1998; Montague & Schier, 2017; Sampath et al., 1998), and signal via the receptor complex—comprised of two each of the type I and type II serine-threonine kinase Activin and the co-receptor Tdgf/Oep, which once activated phosphorylates the downstream transcriptional effectors Smad2/3 (Marlow, 2020; Schier, 2009). Nodal signaling acts as a morphogen, with its highest levels specifying endoderm, and prechordal mesoderm and lower levels chorda mesoderm, followed by paraxial mesoderm (Chen & Schier, 2001). To induce mesoderm and endoderm in the common progenitor pool, Nodal was proposed to induce long-range Fgf signaling but inhibit it locally (van Boxtel, Economou, Heliot, & Hill, 2018). Since *ndr1/squint* and *ndr2/cyclops* genes are primarily zygotically expressed and have been identified in zygotic mutant screens (Erter, Solnica-Krezel, & Wright, 1998; Feldman et al., 1998; Hatta et al., 1991; Sampath et al., 1998), the process of mesendoderm internalization is controlled by the zygotic genome. Yet, multiple Nodal and FGF signaling pathway components are both maternally and zygotically expressed, indicating the mother has some saying in this process as well. Indeed, zygotic *one-eyed pinhead* (*tdgf/oep*) mutants exhibit endoderm and prechordal mesoderm deficiencies (Schier et al., 1997), but MZ*oep/tdgf* mutants lack endoderm, most of the dorsolateral mesoderm, and consequently hardly any internalization ensues, and strong C&E defects arise, thus phenocopying *ndr1/squint;ndr2/cyclops* compound mutants (Feldman et al., 1998; Gritsman et al., 1999). Similarly overlapping roles of maternal and zygotic expression have been reported for *smad2* (Dubrulle et al., 2015). An important role of the maternal dowry for gastrulation has been shown for Gdf3/Vg1; maternally expressed Gdf3/Vg1 is thought to form heterodimers with zygotically expressed Ndr1/Squint and Ndr2/Cyclops to ensure rapid and robust Nodal signaling activation following the ZGA (Montague & Schier, 2017; Pelliccia, Jindal, & Burdine, 2017).

Nodal signaling is not only necessary but is even sufficient to induce cell-autonomous ectopic ingression of single cells (David & Rosa, 2001). Several downstream targets of Nodal signaling that mediate the dynamic cellular properties underlying the process of internalization and subsequent migration of mesendodermal cells have been identified, including Snail1, an evolutionarily conserved transcription factor that represses the *cdh1* gene encoding E-cadherin (Pinheiro & Heisenberg, 2020). Notably however, no zygotically required genes that encode cytoskeletal components or their regulators, which must be involved in mesendoderm migration and internalization, have been identified in zebrafish by forward genetic approaches. Such genes could have functionally redundant ohnologs or unrelated genes, and/or lack of their zygotic expression could be masked by maternal transcripts, or their protein products, which persist through gastrulation.

4.6 Convergence and extension

C&E movements mediolaterally narrow and anteroposteriorly elongate mesoderm, endoderm and the neural part of the ectoderm (Williams & Solnica-Krezel, 2020). Whereas the nascent axial mesoderm starts extending (through anterior migration) and converging to the dorsal midline soon after internalization, the massive convergence movements of lateral mesoderm, endoderm and neuroectoderm start only at midgastrulation, at ca. 70% epiboly (Fig. 2) (Pezeron et al., 2008; Sepich et al., 2005). At this stage internalization movements are largely complete (Keller et al., 2008). Consistent with this late onset, many zygotic mutants affecting C&E have been identified. Among these are the genes encoding components of the Wnt/PCP signaling pathway, which is essential for mediolateral cell elongation and their polarized migration and intercalation, including the genes *wnt11/silberblick* (Heisenberg et al., 2000), *wnt5/pipetail* (Rauch et al., 1997) encoding the ligands, as well as *vangl2/trilobite* (Roszko, Sepich, Jessen, Chandrasekhar, & Solnica-Krezel, 2015) and *glypican4/knypek* (Topczewski et al., 2001), encoding membrane associated proteins. Whereas these genes have clear and strong zygotic C&E defects, they are also maternally expressed. Indicating that this maternal expression has a partially overlapping role with the zygotic function, MZ*wnt11/silberblick*, MZ*vangl2/trilobite*, and MZ*glypican4/knypek* mutants exhibit exacerbated phenotypes (Ciruna et al., 2006; Heisenberg et al., 2000; Topczewski et al., 2001). For some Wnt/PCP components or other genes required for C&E, the maternal expression can completely suffice for normal C&E movements in zygotic mutants. This is exemplified by Scribble, a cytoplasmic protein that interacts with Wnt/PCP and apical/basal polarity regulators (Wada et al., 2005), Ptk7 receptor tyrosine kinase (Hayes, Naito, Daulat, Angers, & Ciruna, 2013), and the atypical cadherin, Dchs2 (Li-Villareal et al., 2015), or genes for which phenotypes were observed when translation blocking (thus interfering with translation of both maternal and zygotic transcripts) but not splice-blocking antisense morpholinos have been used. Within the latter group one finds genes encoding such proteins as Ryk (Lin, Baye, Westfall, & Slusarski, 2010) and Ror2 (Bai et al., 2014) receptor tyrosine kinases. In some cases, revealing a gene's function in C&E requires disrupting zygotic or maternal and zygotic expression and activity of two or more related genes. This group includes Disheveled, a core intracellular component of the Wnt/PCP pathway (Xing et al., 2018). Alternatively, rather than mutating multiple related genes or an entire gene family, dominant negative approaches have been effective in querying gene contribution to developmental processes, including gastrulation. Among examples of cytoskeletal regulators in this class is Rho kinase 2, whose function in ML cell elongation and polarity C&E, downstream of Wnt/PCP signaling, has been revealed by expression of its dominant negative form (Marlow, Topczewski, Sepich, & Solnica-Krezel, 2002).

5. Perspective

Gastrulation, through massive cell movements and rearrangements, places the three germ layers in their canonical positions and shapes them into a body plan with organ rudiments. In all animals, gastrulation starts after and absolutely requires activation of the zygotic genome. However, normal gastrulation depends on the maternal genome to establish the embryonic anteroposterior and dorsoventral embryonic polarities, which will instruct some gastrulation movements like C&E. To various degrees in different animals, the maternal genome needs to direct early cleavages to ensure sufficient cell numbers for normal progression of gastrulation. In some animals, like *Drosophila* and zebrafish, the process of gastrulation partially overlaps with the MZT, and many of the maternal transcripts must be degraded for normal gastrulation movements to occur. On the other hand, a subset of maternal transcripts or their translation products that persist during gastrulation can partially or even completely mask the requirement for zygotic expression and function of these genes.

The increasing availability of information on spatiotemporal gene expression from extensive bulk RNA-sequence analyses, single-cell RNA analyses facilitates identification of gene candidates that may function in gastrulation or its aspect and make it possible to assess the level of maternal contribution and expression of potentially redundant genes (Farrell et al., 2018; White et al., 2017). The ease of reverse genetics makes it easy to generate compound mutants to investigate functional redundancy. Recent advances also should facilitate generation of germline-only mutants by conducting genome engineering in the germline thus avoiding the technically challenging germline replacement approach. Finally, let us ask the embryo by continuing to conduct forward genetic screens to identify additional maternal and maternal zygotic genes that control gastrulation.

Acknowledgments

I would like to thank the volume editor, Florence Marlow, and Linwei Li for their constructive comments on the manuscript, as well as Linwei Li and Isa Roszko for contributing figures. I am also grateful to Elizabeth Busch-Nentwich and Richard White for discussions and sharing unpublished data. The work on gastrulation in my laboratory is supported in part by the grant R35 GM118179 from the National Institutes of Health/National Institute of General Medical Sciences.

References

Aanes, H., Winata, C. L., Lin, C. H., Chen, J. P., Srinivasan, K. G., Lee, S. G., et al. (2011). Zebrafish mRNA sequencing deciphers novelties in transcriptome dynamics during maternal to zygotic transition. *Genome Research*, *21*, 1328–1338.

Abrams, E. W., & Mullins, M. C. (2009). Early zebrafish development: It's in the maternal genes. *Current Opinion in Genetics & Development*, *19*, 396–403.

Bai, Y., Tan, X., Zhang, H., Liu, C., Zhao, B., Li, Y., et al. (2014). Ror2 receptor mediates Wnt11 signaling and affects convergence and extension movements in zebrafish. *The Journal of Biological Chemistry*, *289*, 20664–20676.

Beddington, R. S., Rashbass, P., & Wilson, V. (1992). Brachyury—A gene affecting mouse gastrulation and early organogenesis. *Development Supplement*, 157–165.

Behrndt, M., Salbreux, G., Campinho, P., Hauschild, R., Oswald, F., Roensch, J., et al. (2012). Forces driving epithelial spreading in zebrafish gastrulation. *Science*, *338*, 257–260.

Bensch, R., Song, S., Ronneberger, O., & Driever, W. (2013). Non-directional radial intercalation dominates deep cell behavior during zebrafish epiboly. *Biology Open*, *2*, 845–854.

Betchaku, T., & Trinkaus, J. P. (1986). Programmed endocytosis during epiboly of *Fundulus heteroclitus*. *American Zoologist*, *26*, 193–199.

Bontems, F., Stein, A., Marlow, F., Lyautey, J., Gupta, T., Mullins, M. C., et al. (2009). Bucky ball organizes germ plasm assembly in zebrafish. *Current Biology*, *19*, 414–422.

Bouldin, C. M., Snelson, C. D., Farr, G. H., 3rd, & Kimelman, D. (2014). Restricted expression of cdc25a in the tailbud is essential for formation of the zebrafish posterior body. *Genes & Development*, *28*, 384–395.

Bruce, A. E. E., & Heisenberg, C. P. (2020). Mechanisms of zebrafish epiboly: A current view. *Current Topics in Developmental Biology*, *136*, 319–341.

Bucher, E. A., & Greenwald, I. (1991). A genetic mosaic screen of essential zygotic genes in Caenorhabditis elegans. *Genetics*, *128*, 281–292.

Campinho, P., Behrndt, M., Ranft, J., Risler, T., Minc, N., & Heisenberg, C. P. (2013). Tension-oriented cell divisions limit anisotropic tissue tension in epithelial spreading during zebrafish epiboly. *Nature Cell Biology, 15*, 1405–1414.

Carmany-Rampey, A., & Schier, A. F. (2001). Single-cell internalization during zebrafish gastrulation. *Current Biology, 11*, 1261–1265.

Chan, S. H., Tang, Y., Miao, L., Darwich-Codore, H., Vejnar, C. E., Beaudoin, J. D., et al. (2019). Brd4 and P300 confer transcriptional competency during zygotic genome activation. *Developmental Cell, 49*, 867–881, e868.

Chen, Y., & Schier, A. F. (2001). The zebrafish nodal signal squint functions as a morphogen. *Nature, 411*, 607–610.

Cheng, J. C., Miller, A. L., & Webb, S. E. (2004). Organization and function of microfilaments during late epiboly in zebrafish embryos. *Developmental Dynamics, 231*, 313–323.

Chiang, C., Litingtung, Y., Lee, E., Young, K. E., Corden, J. L., Westphal, H., et al. (1996). Cyclopia and defective axial patterning in mice lacking Sonic hedgehog gene function. *Nature, 383*, 407–413.

Ciruna, B., Jenny, A., Lee, D., Mlodzik, M., & Schier, A. F. (2006). Planar cell polarity signalling couples cell division and morphogenesis during neurulation. *Nature, 439*, 220–224.

Ciruna, B., Weidinger, G., Knaut, H., Thisse, B., Thisse, C., Raz, E., et al. (2002). Production of maternal-zygotic mutant zebrafish by germ-line replacement. *Proceedings of the National Academy of Sciences of the United States of America, 99*, 14919–14924.

Conlon, F. L., Lyons, K. M., Takaesu, N., Barth, K. S., Kispert, A., Herrmann, B., et al. (1994). A primary requirement for nodal in the formation and maintenance of the primitive streak in the mouse. *Development, 120*, 1919–1928.

Dalle Nogare, D. E., Pauerstein, P. T., & Lane, M. E. (2009). G2 acquisition by transcription-independent mechanism at the zebrafish midblastula transition. *Developmental Biology, 326*, 131–142.

David, N. B., & Rosa, F. M. (2001). Cell autonomous commitment to an endodermal fate and behaviour by activation of Nodal signalling. *Development, 128*, 3937–3947.

Dosch, R., Wagner, D. S., Mintzer, K. A., Runke, G., Wiemelt, A. P., & Mullins, M. C. (2004). Maternal control of vertebrate development before the midblastula transition: Mutants from the zebrafish I. *Developmental Cell, 6*, 771–780.

Driever, W., Solnica-Krezel, L., Schier, A. F., Neuhauss, S. C. F., Malicki, J., Stemple, D. L., et al. (1996). A genetic screen for mutations affecting embryogenesis in zebrafish. *Development, 123*, 37–46.

Dubrulle, J., Jordan, B. M., Akhmetova, L., Farrell, J. A., Kim, S. H., Solnica-Krezel, L., et al. (2015). Response to Nodal morphogen gradient is determined by the kinetics of target gene induction. *eLife, 4*, e05042.

Eno, C., Solanki, B., & Pelegri, F. (2016). aura (mid1ip1l) regulates the cytoskeleton at the zebrafish egg-to-embryo transition. *Development, 143*, 1585–1599.

Erter, C. E., Solnica-Krezel, L., & Wright, C. V. (1998). Zebrafish nodal-related 2 encodes an early mesendodermal inducer signaling from the extraembryonic yolk syncytial layer. *Developmental Biology, 204*, 361–372.

Escobar-Aguirre, M., Elkouby, Y. M., & Mullins, M. C. (2017). Localization in oogenesis of maternal regulators of embryonic development. *Advances in Experimental Medicine and Biology, 953*, 173–207.

Farrell, J. A., Wang, Y., Riesenfeld, S. J., Shekhar, K., Regev, A., & Schier, A. F. (2018). Single-cell reconstruction of developmental trajectories during zebrafish embryogenesis. *Science, 360*, eaar3131.

Fei, Z., Bae, K., Parent, S. E., Wan, H., Goodwin, K., Theisen, U., et al. (2019). A cargo model of yolk syncytial nuclear migration during zebrafish epiboly. *Development, 146*, dev169664.

Fekany, K., Yamanaka, Y., Leung, T., Sirotkin, H. I., Topczewski, J., Gates, M. A., et al. (1999). The zebrafish *bozozok* locus encodes Dharma, a homeodomain protein essential for induction of gastrula organizer and dorsoanterior embryonic structures. *Development, 126*, 1427–1438.

Feldman, B., Gates, M. A., Egan, E. S., Dougan, S. T., Rennebeck, G., Sirotkin, H. I., et al. (1998). Zebrafish organizer development and germ-layer formation require nodal-related signals. *Nature, 395*, 181–185.

Fukazawa, C., Santiago, C., Park, K. M., Deery, W. J., Gomez de la Torre Canny, S., Holterhoff, C. K., et al. (2010). poky/chuk/ikk1 is required for differentiation of the zebrafish embryonic epidermis. *Developmental Biology, 346*, 272–283.

Giraldez, A. J., Cinalli, R. M., Glasner, M. E., Enright, A. J., Thomson, J. M., Baskerville, S., et al. (2005). MicroRNAs regulate brain morphogenesis in zebrafish. *Science, 308*, 833–838.

Giraldez, A. J., Mishima, Y., Rihel, J., Grocock, R. J., Van Dongen, S., Inoue, K., et al. (2006). Zebrafish MiR-430 promotes deadenylation and clearance of maternal mRNAs. *Science, 312*, 75–79.

Glickman, N. S., Kimmel, C. B., Jones, M. A., & Adams, R. J. (2003). Shaping the zebrafish notochord. *Development, 130*, 873–887.

Gong, Y., Mo, C., & Fraser, S. E. (2004). Planar cell polarity signalling controls cell division orientation during zebrafish gastrulation. *Nature, 430*, 689–693.

Gonzalez, E. M., Fekany-Lee, K., Carmany-Rampey, A., Erter, C., Topczewski, J., Wright, C. V., et al. (2000). Head and trunk in zebrafish arise via coinhibition of BMP signaling by bozozok and chordino [in process citation]. *Genes & Development, 14*, 3087–3092.

Gore, A. V., Maegawa, S., Cheong, A., Gilligan, P. C., Weinberg, E. S., & Sampath, K. (2005). The zebrafish dorsal axis is apparent at the four-cell stage. *Nature, 438*, 1030–1035.

Gray, R. S., Roszko, I., & Solnica-Krezel, L. (2011). Planar cell polarity: Coordinating morphogenetic cell behaviors with embryonic polarity. *Developmental Cell, 21*, 120–133.

Grego-Bessa, J., Hildebrand, J., & Anderson, K. V. (2015). Morphogenesis of the mouse neural plate depends on distinct roles of cofilin 1 in apical and basal epithelial domains. *Development, 142*, 1305–1314.

Griffin, K. J., Amacher, S. L., Kimmel, C. B., & Kimelman, D. (1998). Molecular identification of spadetail: Regulation of zebrafish trunk and tail mesoderm formation by T-box genes. *Development, 125*, 3379–3388.

Gritsman, K., Zhang, J., Cheng, S., Heckscher, E., Talbot, W. S., & Schier, A. F. (1999). The EGF-CFC protein one-eyed pinhead is essential for Nodal signaling. *Cell, 97*, 121–132.

Gupta, T., Marlow, F. L., Ferriola, D., Mackiewicz, K., Dapprich, J., Monos, D., et al. (2010). Microtubule actin crosslinking factor 1 regulates the Balbiani body and animal-vegetal polarity of the zebrafish oocyte. *PLoS Genetics, 6*, e1001073.

Haffter, P., Granato, M., Brand, M., Mullins, M. C., Hammerschmidt, M., Kane, D. A., et al. (1996). The identification of genes with unique and essential functions in the development of the zebrafish, Danio rerio. *Development, 123*, 1–36.

Halpern, M. E., Ho, R. K., Walker, C., & Kimmel, C. B. (1993). Induction of muscle pioneers and floor plate is distinguished by the zebrafish no tail mutation. *Cell, 75*, 99–111.

Hamatani, T., Carter, M. G., Sharov, A. A., & Ko, M. S. (2004). Dynamics of global gene expression changes during mouse preimplantation development. *Developmental Cell, 6*, 117–131.

Hammerschmidt, M., Pelegri, F., Mullins, M. C., Kane, D. A., Brand, M., van Eeden, F. J., et al. (1996a). Mutations affecting morphogenesis during gastrulation and tail formation in the zebrafish, Danio rerio. *Development, 123*, 143–151.

Hammerschmidt, M., Pelegri, F., Mullins, M. C., Kane, D. A., van Eeden, F. J., Granato, M., et al. (1996b). dino and mercedes, two genes regulating dorsal development in the zebrafish embryo. *Development*, *123*, 95–102.

Harvey, S. A., Sealy, I., Kettleborough, R., Fenyes, F., White, R., Stemple, D., et al. (2013). Identification of the zebrafish maternal and paternal transcriptomes. *Development*, *140*, 2703–2710.

Hatta, K., Kimmel, C. B., Ho, R. K., & Walker, C. (1991). The cyclops mutation blocks specification of the floor plate of the zebrafish central nervous system. *Nature*, *350*, 339–341.

Hayes, M., Naito, M., Daulat, A., Angers, S., & Ciruna, B. (2013). Ptk7 promotes non-canonical Wnt/PCP-mediated morphogenesis and inhibits Wnt/beta-catenin-dependent cell fate decisions during vertebrate development. *Development*, *140*, 1807–1818.

Heisenberg, C. P., Tada, M., Rauch, G. J., Saude, L., Concha, M. L., Geisler, R., et al. (2000). Silberblick/Wnt11 mediates convergent extension movements during zebrafish gastrulation. *Nature*, *405*, 76–81.

Hernandez-Martinez, R., Ramkumar, N., & Anderson, K. V. (2019). p120-catenin regulates WNT signaling and EMT in the mouse embryo. *Proceedings of the National Academy of Sciences of the United States of America*, *116*, 16872–16881.

Ho, R. K., & Kane, D. A. (1990). Cell-autonomous action of zebrafish spt-1 mutation in specific mesodermal precursors. *Nature*, *348*, 728–730.

Holland, S. M., DeLeo, F. R., Elloumi, H. Z., Hsu, A. P., Uzel, G., Brodsky, N., et al. (2007). STAT3 mutations in the hyper-IgE syndrome. *The New England Journal of Medicine*, *357*, 1608–1619.

Holloway, B. A., Gomez de la Torre Canny, S., Ye, Y., Slusarski, D. C., Freisinger, C. M., Dosch, R., et al. (2009). A novel role for MAPKAPK2 in morphogenesis during zebrafish development. *PLoS Genetics*, *5*, e1000413.

Houston, D. W. (2017). Vertebrate axial patterning: From egg to asymmetry. *Advances in Experimental Medicine and Biology*, *953*, 209–306.

Hsu, H. J., Liang, M. R., Chen, C. T., & Chung, B. C. (2006). Pregnenolone stabilizes microtubules and promotes zebrafish embryonic cell movement. *Nature*, *439*, 480–483.

Huebner, R. J., & Wallingford, J. B. (2018). Coming to consensus: A unifying model emerges for convergent extension. *Developmental Cell*, *46*, 389–396.

Hupalowska, A., Jedrusik, A., Zhu, M., Bedford, M. T., Glover, D. M., & Zernicka-Goetz, M. (2018). CARM1 and paraspeckles regulate pre-implantation mouse embryo development. *Cell*, *175*, 1902–1916, e1913.

Jessen, J. R., Topczewski, J., Bingham, S., Sepich, D. S., Marlow, F., Chandrasekhar, A., et al. (2002). Zebrafish trilobite identifies new roles for Strabismus in gastrulation and neuronal movements. *Nature Cell Biology*, *4*, 610–615.

Kane, D., & Adams, R. (2002). Life at the edge: Epiboly and involution in the zebrafish. *Results and Problems in Cell Differentiation*, *40*, 117–135.

Kane, D. A., Hammerschmidt, M., Mullins, M. C., Maischein, H. M., Brand, M., van Eeden, F. J., et al. (1996). The zebrafish epiboly mutants. *Development*, *123*, 47–55.

Kane, D. A., & Kimmel, C. B. (1993). The zebrafish midblastula transition. *Development*, *119*, 447–456.

Kane, D. A., McFarland, K. N., & Warga, R. M. (2005). Mutations in E-cadherin block cell behaviors that are necessary for teleost epiboly. *Development*, *132*, 1105–1116.

Kane, D. A., Warga, R. M., & Kimmel, C. B. (1992). Mitotic domains in the early embryo of the zebrafish. *Nature*, *360*, 735–737.

Kasarskis, A., Manova, K., & Anderson, K. V. (1998). A phenotype-based screen for embryonic lethal mutations in the mouse. *Proceedings of the National Academy of Sciences of the United States of America*, *95*, 7485–7490.

Keller, P. J., Schmidt, A. D., Wittbrodt, J., & Stelzer, E. H. (2008). Reconstruction of zebrafish early embryonic development by scanned light sheet microscopy. *Science*, *322*, 1065–1069.

Keller, R., & Sutherland, A. (2020). Convergent extension in the amphibian, Xenopus laevis. *Current Topics in Developmental Biology*, *136*, 271–317.

Kelly, C., Chin, A. J., Leatherman, J. L., Kozlowski, D. J., & Weinberg, E. S. (2000). Maternally controlled (beta)-catenin-mediated signaling is required for organizer formation in the zebrafish. *Development*, *127*, 3899–3911.

Kimmel, C. B., Ballard, W. W., Kimmel, S. R., Ullmann, B., & Schilling, T. F. (1995). Stages of embryonic development of the zebrafish. *Developmental Dynamics*, *203*, 253–310.

Kimmel, C. B., & Law, R. D. (1985). Cell lineage of zebrafish blastomeres. II. Formation of the yolk syncytial layer. *Developmental Biology*, *108*, 86–93.

Kispert, A., & Herrmann, B. G. (1993). The Brachyury gene encodes a novel DNA binding protein. *EMBO Journal*, *12*, 3211–3220.

Ko, C. S., & Martin, A. C. (2020). The cellular and molecular mechanisms that establish the mechanics of Drosophila gastrulation. *Current Topics in Developmental Biology*, *136*, 141–165.

Koppen, M., Fernandez, B. G., Carvalho, L., Jacinto, A., & Heisenberg, C. P. (2006). Coordinated cell-shape changes control epithelial movement in zebrafish and Drosophila. *Development*, *133*, 2671–2681.

Kramer, C., Mayr, T., Nowak, M., Schumacher, J., Runke, G., Bauer, H., et al. (2002). Maternally supplied smad5 is required for ventral specification in zebrafish embryos prior to zygotic bmp signaling. *Developmental Biology*, *250*, 263.

Langdon, Y. G., Fuentes, R., Zhang, H., Abrams, E. W., Marlow, F. L., & Mullins, M. C. (2016). Split top: A maternal cathepsin B that regulates dorsoventral patterning and morphogenesis. *Development*, *143*, 1016–1028.

Langdon, Y. G., & Mullins, M. C. (2011). Maternal and zygotic control of zebrafish dorsoventral axial patterning. *Annual Review of Genetics*, *45*, 357–377.

Lee, H. X., Ambrosio, A. L., Reversade, B., & De Robertis, E. M. (2006). Embryonic dorsal-ventral signaling: Secreted frizzled-related proteins as inhibitors of tolloid proteinases. *Cell*, *124*, 147–159.

Lee, M. T., Bonneau, A. R., Takacs, C. M., Bazzini, A. A., DiVito, K. R., Fleming, E. S., et al. (2013). Nanog, Pou5f1 and SoxB1 activate zygotic gene expression during the maternal-to-zygotic transition. *Nature*, *503*, 360–364.

Leise, W. F., 3rd, & Mueller, P. R. (2004). Inhibition of the cell cycle is required for convergent extension of the paraxial mesoderm during Xenopus neurulation. *Development*, *131*, 1703–1715.

Lepage, S. E., Tada, M., & Bruce, A. E. (2014). Zebrafish dynamin is required for maintenance of enveloping layer integrity and the progression of epiboly. *Developmental Biology*, *385*, 52–66.

Leptin, M., & Grunewald, B. (1990). Cell shape changes during gastrulation in Drosophila. *Development*, *110*, 73–84.

Lin, S., Baye, L. M., Westfall, T. A., & Slusarski, D. C. (2010). Wnt5b-Ryk pathway provides directional signals to regulate gastrulation movement. *The Journal of Cell Biology*, *190*, 263–278.

Liu, Y., Du, L., Osato, M., Teo, E. H., Qian, F., Jin, H., et al. (2007). The zebrafish udu gene encodes a novel nuclear factor and is essential for primitive erythroid cell development. *Blood*, *110*, 99–106.

Liu, Y., Sepich, D. S., & Solnica-Krezel, L. (2017). Stat3/Cdc25a-dependent cell proliferation promotes embryonic axis extension during zebrafish gastrulation. *PLoS Genetics*, *13*, e1006564.

Li-Villareal, N., Forbes, M. M., Loza, A. J., Chen, J., Ma, T., Helde, K., et al. (2015). Dachsous1b cadherin regulates actin and microtubule cytoskeleton during early zebrafish embryogenesis. *Development, 142*, 2704–2718.

Marlow, F. L. (2020). Setting up for gastrulation in zebrafish. *Current Topics in Developmental Biology, 136*, 33–83.

Marlow, F. L., & Mullins, M. C. (2008). Bucky ball functions in Balbiani body assembly and animal-vegetal polarity in the oocyte and follicle cell layer in zebrafish. *Developmental Biology, 321*, 40–50.

Marlow, F., Topczewski, J., Sepich, D., & Solnica-Krezel, L. (2002). Zebrafish Rho kinase 2 acts downstream of Wnt11 to mediate cell polarity and effective convergence and extension movements. *Current Biology, 12*, 876–884.

Marsal, M., Hernandez-Vega, A., & Martin-Blanco, E. (2017). Contractility, differential tension and membrane removal lead zebrafish epiboly biomechanics. *Cell Cycle, 16*, 1328–1335.

Mathavan, S., Lee, S. G., Mak, A., Miller, L. D., Murthy, K. R., Govindarajan, K. R., et al. (2005). Transcriptome analysis of zebrafish embryogenesis using microarrays. *PLoS Genetics, 1*, 260–276.

McDole, K., Guignard, L., Amat, F., Berger, A., Malandain, G., Royer, L. A., et al. (2018). In toto imaging and reconstruction of post-implantation mouse development at the single-cell level. *Cell, 175*, 859–876.e833.

McFarland, K. N., Warga, R. M., & Kane, D. A. (2005). Genetic locus half baked is necessary for morphogenesis of the ectoderm. *Developmental Dynamics, 233*, 390–406.

Migeotte, I., Grego-Bessa, J., & Anderson, K. V. (2011). Rac1 mediates morphogenetic responses to intercellular signals in the gastrulating mouse embryo. *Development, 138*, 3011–3020.

Miller-Bertoglio, V., Carmany-Rampey, A., Furthauer, M., Gonzalez, E. M., Thisse, C., Thisse, B., et al. (1999). Maternal and zygotic activity of the zebrafish *ogon* locus antagonizes BMP signaling. *Developmental Biology, 214*, 72–86.

Mishima, Y., & Tomari, Y. (2016). Codon usage and 3' UTR length determine maternal mRNA stability in zebrafish. *Molecular Cell, 61*, 874–885.

Mizuno, T., Yamaha, E., Wakahara, M., Kuroiwa, A., & Takeda, H. (1996). Mesoderm induction in zebrafish. *Nature, 383*, 131–132.

Montague, T. G., & Schier, A. F. (2017). Vg1-Nodal heterodimers are the endogenous inducers of mesendoderm. *eLife, 6*, e28183.

Montero, J. A., Carvalho, L., Wilsch-Brauninger, M., Kilian, B., Mustafa, C., & Heisenberg, C. P. (2005). Shield formation at the onset of zebrafish gastrulation. *Development, 132*, 1187–1198.

Mullins, M. C., Hammerschmidt, M., Kane, D. A., Odenthal, J., Brand, M., van Eeden, F. J. M., et al. (1996). Genes establishing dorsoventral pattern formation in the zebrafish embryo: The ventral specifying genes. *Development, 123*, 81–93.

Myers, D., Sepich, D. S., & Solnica-Krezel, L. (2002a). BMP activity gradient regulates convergent extension during zebrafish gastrulation. *Developmental Biology, 243*, 81–98.

Myers, D. C., Sepich, D. S., & Solnica-Krezel, L. (2002b). Convergence and extension in vertebrate gastrulae: Cell movements according to or in search of identity? *Trends in Genetics, 18*, 447–455.

Newport, J., & Kirschner, M. (1982). A major developmental transition in early Xenopus embryos: II. Control of the onset of transcription. *Cell, 30*, 687–696.

Nowotschin, S., & Hadjantonakis, A. K. (2020). Guts and gastrulation: Emergence and convergence of endoderm in the mouse embryo. *Current Topics in Developmental Biology, 136*, 429–454.

Nusslein-Volhard, C., & Wieschaus, E. (1980). Mutations affecting segment number and polarity in Drosophila. *Nature, 287*, 795–801.

Pare, A. C., & Zallen, J. A. (2020). Cellular, molecular, and biophysical control of epithelial cell intercalation. *Current Topics in Developmental Biology*, *136*, 167–193.
Pelegri, F., & Schulte-Merker, S. (1999). A gynogenesis-based screen for maternal-effect genes in the zebrafish, Danio rerio. *Methods in Cell Biology*, *60*, 1–20.
Pelliccia, J. L., Jindal, G. A., & Burdine, R. D. (2017). Gdf3 is required for robust Nodal signaling during germ layer formation and left-right patterning. *Elife*, *6*, e28635.
Petropoulos, S., Edsgard, D., Reinius, B., Deng, Q., Panula, S. P., Codeluppi, S., et al. (2016). Single-cell RNA-Seq reveals lineage and X chromosome dynamics in human preimplantation embryos. *Cell*, *167*, 285.
Pezeron, G., Mourrain, P., Courty, S., Ghislain, J., Becker, T. S., Rosa, F. M., et al. (2008). Live analysis of endodermal layer formation identifies random walk as a novel gastrulation movement. *Current Biology*, *18*, 276–281.
Pinheiro, D., & Heisenberg, C. P. (2020). Zebrafish gastrulation: Putting fate in motion. *Current Topics in Developmental Biology*, *136*, 343–375.
Quesada-Hernandez, E., Caneparo, L., Schneider, S., Winkler, S., Liebling, M., Fraser, S. E., et al. (2010). Stereotypical cell division orientation controls neural rod midline formation in zebrafish. *Current Biology*, *20*, 1966–1972.
Rauch, G. J., Hammerschmidt, M., Blader, P., Schauerte, H. E., Strahle, U., Ingham, P. W., et al. (1997). Wnt5 is required for tail formation in the zebrafish embryo. *Cold Spring Harbor Symposia on Quantitative Biology*, *62*, 227–234.
Rebagliati, M. R., Toyama, R., Haffter, P., & Dawid, I. B. (1998). cyclops encodes a nodal-related factor involved in midline signaling. *Proceedings of the National Academy of Sciences of the United States of America*, *95*, 9932–9937.
Rossant, J., & Tam, P. P. L. (2018). Exploring early human embryo development. *Science*, *360*, 1075–1076.
Roszko, I., Sepich, D. S., Jessen, J. R., Chandrasekhar, A., & Solnica-Krezel, L. (2015). A dynamic intracellular distribution of Vangl2 accompanies cell polarization during zebrafish gastrulation. *Development*, *142*, 2508–2520.
Sampath, K., Rubinstein, A. L., Cheng, A. M., Liang, J. O., Fekany, K., Solnica-Krezel, L., et al. (1998). Induction of the zebrafish ventral brain and floorplate requires cyclops/nodal signalling. *Nature*, *395*, 185–189.
Schier, A. F. (2009). Nodal morphogens. *Cold Spring Harbor Perspectives in Biology*, *1*, a003459.
Schier, A. F., Neuhauss, S. C., Helde, K. A., Talbot, W. S., & Driever, W. (1997). The one-eyed pinhead gene functions in mesoderm and endoderm formation in zebrafish and interacts with no tail. *Development*, *124*, 327–342.
Schulte-Merker, S., Lee, K. J., McMahon, A. P., & Hammerschmidt, M. (1997). The zebrafish organizer requires chordino. *Nature*, *387*, 862–863.
Schulte-Merker, S., van Eeden, F. J., Halpern, M. E., Kimmel, C. B., & Nusslein-Volhard, C. (1994). no tail (ntl) is the zebrafish homologue of the mouse T (Brachyury) gene. *Development*, *120*, 1009–1015.
Sepich, D. S., Calmelet, C., Kiskowski, M., & Solnica-Krezel, L. (2005). Initiation of convergence and extension movements of lateral mesoderm during zebrafish gastrulation. *Developmental Dynamics*, *234*, 279–292.
Shih, J., & Fraser, S. E. (1995). Distribution of tissue progenitors within the shield region of the zebrafish gastrula. *Development*, *121*, 2755–2765.
Shimizu, T., Yabe, T., Muraoka, O., Yonemura, S., Aramaki, S., Hatta, K., et al. (2005). E-cadherin is required for gastrulation cell movements in zebrafish. *Mechanisms of Development*, *122*, 747–763.
Solnica-Krezel, L. (2005). Conserved patterns of cell movements during vertebrate gastrulation. *Current Biology*, *15*, R213–R228.
Solnica-Krezel, L. (2020). Gastrulation: From embryonic pattern to form. In P. M. Wassarman (Ed.), *Current topics in developmental biology*. Academic Press.

Solnica-Krezel, L., & Driever, W. (1994). Microtubule arrays of the zebrafish yolk cell: Organization and function during epiboly. *Development, 120*, 2443–2455.

Solnica-Krezel, L., & Sepich, D. S. (2012). Gastrulation: Making and shaping germ layers. *Annual Review of Cell and Developmental Biology, 28*, 687–717.

Solnica-Krezel, L., Stemple, D. L., Mountcastle-Shah, E., Rangini, Z., Neuhauss, S. C. F., Malicki, J., et al. (1996). Mutations affecting cell fates and cellular rearrangements during gastrulation in zebrafish. *Development, 123*, 117–128.

Stathopoulos, A., & Newcomb, S. (2020). Setting up for gastrulation: D. melanogaster. *Current Topics in Developmental Biology, 136*, 3–32.

Stern, C. D. (2004). Gastrulation. In *From cells to embryo*. Cold Spring Harbor Laboratory Press.

Strahle, U., & Jesuthasan, S. (1993). Ultraviolet irradiation impairs epiboly in zebrafish embryos: Evidence for a microtubule-dependent mechanism of epiboly. *Development, 119*, 909–919.

Topczewski, J., Sepich, D. S., Myers, D. C., Walker, C., Amores, A., Lele, Z., et al. (2001). The zebrafish glypican knypek controls cell polarity during gastrulation movements of convergent extension. *Developmental Cell, 1*, 251–264.

Trinkaus, J. P. (1951). A study of the mechanism of epiboly in the egg of fundulus heteroclitus. *The Journal of Experimental Zoology, 118*, 269–320.

Tuazon, F. B., & Mullins, M. C. (2015). Temporally coordinated signals progressively pattern the anteroposterior and dorsoventral body axes. *Seminars in Cell & Developmental Biology, 42*, 118–133.

van Boxtel, A. L., Economou, A. D., Heliot, C., & Hill, C. S. (2018). Long-range signaling activation and local inhibition separate the mesoderm and endoderm lineages. *Developmental Cell, 44*, 179–191, e175.

Vastenhouw, N. L., Cao, W. X., & Lipshitz, H. D. (2019). The maternal-to-zygotic transition revisited. *Development, 146*, dev161471.

Voiculescu, O. (2020). Movements of chick gastrulation. *Current Topics in Developmental Biology, 136*, 409–428.

Wada, H., Iwasaki, M., Sato, T., Masai, I., Nishiwaki, Y., Tanaka, H., et al. (2005). Dual roles of zygotic and maternal Scribble1 in neural migration and convergent extension movements in zebrafish embryos. *Development, 132*, 2273–2285.

Wagner, D. S., Dosch, R., Mintzer, K. A., Wiemelt, A. P., & Mullins, M. C. (2004). Maternal control of development at the midblastula transition and beyond: Mutants from the zebrafish II. *Developmental Cell, 6*, 781–790.

Warga, R. M., & Kimmel, C. B. (1990). Cell movements during epiboly and gastrulation in zebrafish. *Development, 108*, 569–580.

Weng, J. H., Liang, M. R., Chen, C. H., Tong, S. K., Huang, T. C., Lee, S. P., et al. (2013). Pregnenolone activates CLIP-170 to promote microtubule growth and cell migration. *Nature Chemical Biology, 9*, 636–642.

White, M. D., Angiolini, J. F., Alvarez, Y. D., Kaur, G., Zhao, Z. W., Mocskos, E., et al. (2016). Long-lived binding of Sox2 to DNA predicts cell fate in the four-cell mouse embryo. *Cell, 165*, 75–87.

White, R. J., Collins, J. E., Sealy, I. M., Wali, N., Dooley, C. M., Digby, Z., et al. (2017). A high-resolution mRNA expression time course of embryonic development in zebrafish. *eLife, 6*, e30860.

Williams, M., Burdsal, C., Periasamy, A., Lewandoski, M., & Sutherland, A. (2012). Mouse primitive streak forms in situ by initiation of epithelial to mesenchymal transition without migration of a cell population. *Developmental Dynamics, 241*, 270–283.

Williams, M. L. K., Sawada, A., Budine, T., Yin, C., Gontarz, P., & Solnica-Krezel, L. (2018). Gon4l regulates notochord boundary formation and cell polarity underlying axis extension by repressing adhesion genes. *Nature Communications, 9*, 1319.

Williams, M. L. K., & Solnica-Krezel, L. (2020). Cellular and molecular mechanisms of convergence and extension in zebrafish. *Current Topics in Developmental Biology*, *136*, 377–407.
Winata, C. L., & Korzh, V. (2018). The translational regulation of maternal mRNAs in time and space. *FEBS Letters*, *592*, 3007–3023.
Winata, C. L., Lapinski, M., Pryszcz, L., Vaz, C., Bin Ismail, M. H., Nama, S., et al. (2018). Cytoplasmic polyadenylation-mediated translational control of maternal mRNAs directs maternal-to-zygotic transition. *Development*, *145*, dev159566.
Winkley, K. M., Kourakis, M. J., DeTomaso, A. W., Veeman, M. T., & Smith, W. C. (2020). Tunicate gastrulation. *Current Topics in Developmental Biology*, *136*, 219–242.
Wu, E., & Vastenhouw, N. (2020). From Mother to Embryo: A molecular perspective on zygotic genome activation. *Current Topics in Developmental Biology*, *140*.
Xing, Y. Y., Cheng, X. N., Li, Y. L., Zhang, C., Saquet, A., Liu, Y. Y., et al. (2018). Mutational analysis of dishevelled genes in zebrafish reveals distinct functions in embryonic patterning and gastrulation cell movements. *PLoS Genetics*, *14*, e1007551.
Yabe, T., Ge, X., Lindeman, R., Nair, S., Runke, G., Mullins, M. C., et al. (2009). The maternal-effect gene cellular island encodes aurora B kinase and is essential for furrow formation in the early zebrafish embryo. *PLoS Genetics*, *5*, e1000518.
Yabe, T., Shimizu, T., Muraoka, O., Bae, Y. K., Hirata, T., Nojima, H., et al. (2003). Ogon/secreted frizzled functions as a negative feedback regulator of Bmp signaling. *Development*, *130*, 2705–2716.
Yan, L., Chen, J., Zhu, X., Sun, J., Wu, X., Shen, W., et al. (2018). Maternal Huluwa dictates the embryonic body axis through beta-catenin in vertebrates. *Science*, *362*, eaat1045.
Yartseva, V., & Giraldez, A. J. (2015). The maternal-to-zygotic transition during vertebrate development: A model for reprogramming. *Current Topics in Developmental Biology*, *113*, 191–232.
Yin, C., Kiskowski, M., Pouille, P. A., Farge, E., & Solnica-Krezel, L. (2008). Cooperation of polarized cell intercalations drives convergence and extension of presomitic mesoderm during zebrafish gastrulation. *The Journal of Cell Biology*, *180*, 221–232.
Zhang, L., Kendrick, C., Julich, D., & Holley, S. A. (2008). Cell cycle progression is required for zebrafish somite morphogenesis but not segmentation clock function. *Development*, *135*, 2065–2070.

FSC
www.fsc.org
MIX
Papier | Fördert gute Waldnutzung
FSC® C083411

Kolibri 360 GmbH
Ferdinand-Jühlke-Straße 7
99095 Erfurt, Deutschland
produktsicherheit@kolibri360.de